Rotifer Symposium IV

Developments in Hydrobiology 42

Series editor
H.J. Dumont

Rotifer Symposium IV

Proceedings of the Fourth Rotifer Symposium, held in Edinburgh, Scotland, August 18–25, 1985

Edited by
L. May, R. Wallace and A. Herzig

Reprinted from Hydrobiologia, vol. 147

1987 **DR W. JUNK PUBLISHERS**
a member of the KLUWER ACADEMIC PUBLISHERS GROUP
DORDRECHT / BOSTON / LANCASTER

Distributors

for the United States and Canada: Kluwer Academic Publishers, P.O. Box 358, Accord Station, Hingham, MA 02018-0358, USA
for the UK and Ireland: Kluwer Academic Publishers, MTP Press Limited, Falcon House, Queen Square, Lancaster LA1 1RN, UK
for all other countries: Kluwer Academic Publishers Group, Distribution Center, P.O. Box 322, 3300 AH Dordrecht, The Netherlands

Library of Congress Cataloging in Publication Data

International Rotifer Symposium (4th : 1985 : Edinburgh, Scotland)
Rotiger Symposium IV.

(Developments in hydrobiology ; 42)
"Reprinted from Hydrobiologia."
1. Rotifera--Congresses. I. May, L. II. Wallace, R. III. Herzig, A. IV. Hydrobiologia. V. Title. VI. Series.
QL391.R8I57 1985 595.1'81 87-3307
ISBN 90-6193-645-4

ISBN 90-6193-645-4 (this volume)
ISBN 90-6193-751-5 (series)

Copyright

PRINTED IN THE NETHERLANDS

Preface

The fourth international rotifer symposium was held in Edinburgh, Scotland, August 18–25, 1985, hosted by the Institute of Terrestrial Ecology. This meeting continued the tradition of holding rotifer symposia at three-year intervals. The first announcement of the fourth meeting was circulated at the end of 1983 to almost 300 people whose names appeared on the mailing list of the international newsletter, Rotifer News. In total, 68 people from 23 countries attended the meeting. It is interesting to note that, of these 68 participants, 21 had attended the first meeting, held in Linz, Austria, 1976, and 13 had attended all three previous meetings.

As in previous symposia, some research topics were identified in advance of the meeting as being in need of special attention. This resulted in the presentation of 7 invited review papers. In addition, 44 short communications and 6 poster displays on a wide variety of topics were presented. A film on the use of rotifers as larval food in an Israeli fish farm was also shown. Most evenings were set aside for general discussions of the papers presented during the day.

Participants were accommodated at the Pollock Halls of Residence, Edinburgh, which also provided facilities for lecture and poster sessions. On Wednesday afternoon a tour of Edinburgh, including a visit to the Palace of Holyrood, was arranged. This was followed by an evening banquet with traditional Scottish entertainment. On Thursday evening most participants attended a fireworks display which was part of the Edinburgh Festival. Finally, an excursion to Loch Lomond and the Trossach hills was arranged for the Saturday after the meeting.

The organisers would like to thank Mr. C. J. Place and colleagues at the Institute of Terrestrial Ecology for their invaluable help in organising the meeting and preparing the symposium volume for publication. We are also grateful for financial support from the Royal Society, the British Council and British Petroleum (Scotland).

Dr W. Junk BV Publishers and Dr H. J. Dumont kindly arranged for the conference proceedings to be published in *Hydrobiologia* and as a volume in the series 'Developments in Hydrobiology'. Manuscripts were revised by the editors and several anonymous reviewers to whom the editors offer their grateful thanks. Language and style were changed where necessary to improve clarity of meaning and to reduce length. The editors hope that this volume will make a useful contribution to rotifer research.

Plate 1. Upper panel: Joyce Mehling, Richard Stemberger, Linda May, Russell Shiel, Stephen Threlkeld, Anne Lyche, Kennedy Roche; Upper middle panel: Margaretha de Ridder, James Litton, Pierre Clément, Robert Wallace, Radovan Erben, Jean-Claude Rostan, Minoru Sudzuki, Liliana Ramella, Lois Bateman, Rosalind Pontin, Reidar Grundström; James Litton, Robert Wallace, Esther Lubzens, Wolfgang Hofmann, Ramesh Gulati, B. K. Sharma, Pierre Clément;
Lower middle panel: Lois Bateman, Walter Koste, Anthony Saunders-Davies; Heidi Berner-Fankhauser; Richard Stemberger, Joyce Mehling, Yvette Edmondson, Tommy Edmondson, Ylva Pejler, Birger Pejler, Alois Herzig, Rama Chengelath;
Lower panel: Martin Scott; Nan Duncan; Eric Holloway, Jens Petter Nilssen, T. A. Rezeq, Claudia Ricci, Alois Herzig.

Plate 2. Upper panel: John Gilbert; Reidar Grundström, Diane Hewitt, Luis Cruz-Pizarro, Lesley Stewart, Jean-Claude Rostan; H. Schrimpf, Eric Holloway, Ruth Laxhuber, Norbert Walz;
Upper middle panel: Anne Lyche, Terry Snell, Liliana Ramella, Tommy Edmondson, Birger Pejler, Thomas Nogrady; back row: Carlo Robotti, Manuel Yufera; centre row: Hachiro Hirata, Claudia Ricci, Luis Cruz-Pizarro, Davi Ehlert, Anne Lyche, Charles King, Maria Miracle, Sif Johansson, Robert Wallace, Jim Litton, H. Schrimpf, Anthony Saunders-Davies, Agnes Ruttner-Kolisko; front row: Mrs. Hirata, Kazutsugu Hirayama, Alois Herzig, Jens Petter Nilssen, Henri Dumont, Elizabeth Wurdak, Terry Snell;
Lower middle panel: Charles Hussey, Minoru Sudzuki, Charles James, Thomas Nogrady; Lesley Stewart, Sif Johansson, Alisdair Scott; Peter Starkweather, Sif Johansson, James Green and wife;
Lower panel: Manuel Serra, Manuel Yufera, Marjanca Markic, Radovan Erben, Galina Galkovskaja, Rosalind Pontin, Anthony Saunder-Davies, Jean-Claude Rostan; the conference centre.

Contents

PART THREE: COLONIAL ROTIFERS

PART FOUR: POPULATION DYNAMICS AND SPATIAL DISTRIBUTION

PART FIVE: AQUACULTURE, FEEDING AND NUTRITION

Participants of the Fourth International Rotifer Symposium

Bailey-Watts, A. E., Institute of Terrestrial Ecology, Bush Estate, Penicuik, Midlothian EH26 0QB, SCOTLAND

Bateman, L., Sir Wilfred Grenfell College, Memorial University, Department of Biology, Corner Brook, Newfoundland, CANADA A2H 6P9

Berner-Fankhauser, H., Zoologisches Institut, Universität Bern, Baltzerstrasse 3, CH-3012 Bern, SWITZERLAND

Brain, C. K., Transvaal Museum, P.O. Box 413, Pretoria 0001, S. AFRICA

Chengalath, R., National Museum of Natural Sciences, National Museums of Canada, Ottawa, CANADA K1A 08

Clément, P., Laboratoire d'Histologie et Biologie Tissulaire, Université Claude Bernard Lyon 1, 43 Boulevard du 11 Nov., F-69622 Villeurbanne Cedex, FRANCE

Cruz-Pizarro, L., Departamento de Ecologia, Facultad de Ciencias, Universidad de Granada, Granada, SPAIN

Denayer, B., Laboratory for Biological Research in Aquatic Pollution, J. Plateaustraat 22, 9000 Gent, BELGIUM

Dumont, H. J., Rijksuniversiteit Gent, Instituut v. Dierkunde, Lab. Ekologie, Ledeganckstraat 35, B-9000 Gent, BELGIUM

Duncan, A., Department of Zoology, Royal Holloway and Bedford New College, Englefield Green, Surrey TW20 9TY, ENGLAND

Edmondson, W. T., Department of Zoology NJ-15, University of Washington, Seattle, Washington 98195, U.S.A.

Erben, R., Zoologijski zavod, Rooseveltor trg 6, 41001 Zagreb, YUGOSLAVIA

Galkovskaya, G., Institute of Zoology of Academy of Sciences of BSSR, Academytcheskaya 27, Minsk 220072, USSR

George, D. G., Freshwater Biological Association, The Ferry House, Ambleside, Cumbria, ENGLAND

Gilbert, J. J., Department of Biological Sciences, Dartmouth College, Hanover, New Hampshire 03755, U.S.A.

Green, J., School of Biological Sciences, Queen Mary College, Mile End Road, London E1 4NS, ENGLAND

Grundström, R., Institute of Limnology, Box 557, S-751 22 Uppsala, SWEDEN

Gulati, R., Limnologisch Instituut van de Koninklijke Nederlandse Akademie van Wetenschappen, Vijverhof, Nieuwersluis, THE NETHERLANDS

Herzig, A., Limnologisches Institut, Gaisberg 116, A-5310 Mondsee, AUSTRIA

Hewitt, D., Freshwater Biological Association, The Ferry House, Ambleside, Cumbria, ENGLAND

Hirata, H., Faculty of Fisheries, Kagoshima University, Kagoshima 890, JAPAN

Hirayama, K., Faculty of Fisheries, Nagasaki University, 1-14 Bunkyo Machi, Nagasaki 852, JAPAN

Hofmann, W., Max-Planck-Institut für Limnologie, Postfach 165, 2320 Plön, GERMANY

Hollowday, E. D., 45 Manor Road, Aylesbury, Bucks HP20 1JB, ENGLAND

Hussey, C. G., Department of Zoology, The British Museum (Nat. Hist.), Cromwell Road, South Kensington, London SW7 5BD, ENGLAND

James, C. M., Kuwait Institute for Scientific Research, Mariculture and Fisheries Dept., P.O. Box 1638, Salmiya, KUWAIT

Johansson, S., Askö Laboratory, University of Stockholm, S-10691 Stockholm, SWEDEN

King, C. E., Department of Zoology, Oregon State University, Corvallis, Oregon 97331, U.S.A.
Kirika, A., Institute of Terrestrial Ecology, Bush Estate, Penicuik, Midlothian EH26 0QB, SCOTLAND
Koste, W., Ludwig-Brill-Strasse 5, D 4570 Quakenbrück, GERMANY
Laxhuber, R., Zoologisches Institut der Universität München. Seidlstrasse 25, 8000 München 2, GERMANY
Litton, J. R. Jr., Biology Department, Saint Mary's College, Notre Dame, Indiana 46556, U.S.A.
Lubzens, E., Israel Oceanographic & Limnological Research, Tel-Shikmona, P.O. Box 8030, Haifa, ISRAEL
Lyche, A., Department of Limnology, Box 1027, Blindern, 0315 Oslo 3, NORWAY
Markič, M., 620000 Maribor, Truborjéva 25a, YUGOSLAVIA
May, L., Institute of Terrestrial Ecology, Bush Estate, Penicuik, Midlothian EH26 0QB, SCOTLAND
Minkoff, G., Dunstaffnage Marine Research Station, P.O. Box 3, Oban, Argyll PA34 4AD, SCOTLAND
Miracle, M. R., Dpto. Ecologia, Facultad de Ciencias Biologicas, Universidad de Valencia, Burjasot (Valencia), SPAIN
Morales-Baquero, R., Catedra de Ecologia, Facultad de Ciencias, Universidad de Granada, SPAIN
Nilssen, J. P., Norsk Biotech A/S, P.O. Box 788, Krossen, N-4301 Sandnes, NORWAY
Nogrady, T., Department of Biology, Queen's University, Kingston, CANADA K7L 3N6
Pejler, B., Institute of Limnology, Box 557, S – 751 22 Uppsala 1, SWEDEN
Pontin, R. M., 26 Hermitage Woods Crescent, St. John's Woking, Surrey, ENGLAND
Radwan, S., Academy of Agriculture, Department of Zoology and Hydrobiology, Akademicka 13, 20-934 Lublin, POLAND
Rezeq, T. A., Kuwait Institute for Scientific Research, Mariculture and Fisheries Dept., P.O. Box 1638, Salmiya, KUWAIT
Ricci, C., Dipartimento di Biologia, Sezione di Ecologia, Universita Degli Studie, Via Celoria 26, 20133 Milano, ITALY
Ridder, M. de, V. van Malderlaan 37/3, B-1710 Dilbeek, BELGIUM
Robotti, C., Instituto di Zoologia – D.B.A., Universita di Torino, Via Accademia Albertina 17, 10123 Torina, ITALY
Roche, K., Laboratoria voor Ekologie, K.L. Ledeganckstraat 35, B-9000 Gent, BELGIUM
Rostan, J. C., Département de Biologie Animale et Ecologie, Université Claude Bernard Lyon 1, 43 Boulevard du 11 Nov., F-69622 Villeurbanne, FRANCE
Ruttner-Kolisko, A., Biologische Station, A-3293 Lunz am See, AUSTRIA
Saunders-Davies, A., 33 Park Road, Esher, Surrey, ENGLAND
Schrimpf, A., Wasserwirtschaftsamt Rosenheim, Königstrasse 19, D-8200 Rosenheim, GERMANY
Scott, A., Freshwater Biological Association, River Laboratory, East Stoke, Wareham, Dorset BH20 6BB, ENGLAND
Scott, J. M., Dunstaffnage Marine Research Station, P.O. Box 3, Oban, Argyll PA34 4AD, SCOTLAND
Serra, M., Dto. Ecologia, Fac. Ciencias Biologicas, Univ. Valencia, Burjasot (Valencia) SPAIN
Sharma, B. K., Department of Zoology, North-Eastern Hill University, Shillong – 793014, Meghalaya, INDIA
Shiel, R. J., The University of Adelaide, Department of Botany, G.P.O. Box 498, Adelaide, SOUTH AUSTRALIA 5001
Snell, T. W., Division of Science and Mathematics, University of Tampa, Tampa, Florida 33606 U.S.A.
Starkweather, P. L., Department of Biological Sciences, University of Nevada, Las Vegas, Nevada 89109, U.S.A.
Stemberger, R., Department of Biology, Dartmouth College, Hanover, New Hampshire, U.S.A.
Stewart, L., Freshwater Biological Association, The Ferry House, Ambleside, Cumbria, ENGLAND
Stich, H. B., Institut für Seenforschung und Fischereiwesen, Untere Seestr. 81, D-7995 Langenargen, GERMANY

Sudzuki, M., Nihon Daigaku, Higashi-Arai 557, Omiya-shi, Saitama-ken, JAPAN 330
Threlkeld, S., University of Oklahoma, Biological Station, Kingston, OK 73439, U.S.A.
Vicente, E., Dto. de Microbiologia, Fac. Ciencias, Biologicas Universidad de Valencia, Burjasot (Valencia), SPAIN
Wallace, R. L., Department of Biology, Ripon College, Ripon, Wisconsin 54971, U.S.A.
Walz, N., Zoologisches Institut der Universität München, Seidlstrasse 25, 8 München 2, GERMANY
Wurdak, E. S., Department of Biology, Saint John's University, Collegeville, MN 56321, U.S.A.
Yúfera, M., Instituto de Investigaciones, Pesqueras de Cádiz, Puorto Pesquero s/n, Cádiz, SPAIN

Hydrobiologia 147: 1–2 (1987)
May, L., Wallace, R. & Herzig, A. (eds), Rotifer Symposium IV

Bruno Bērziņš in memoriam, 1909–1985

On September 5*th*, 1985, Bruno Bērziņš passed away after a remarkable life. He was born in Latvia in 1909, but World War I brought him to Nizjnij-Novgorod in central Russia. After returning to Latvia he studied at the University of Riga, and later joined the staff of its hydrobiological institue. During this time he was engaged in various research projects including, migrations of Baltic fishes, biology of perch, and hydrography of Latvian rivers. However, his most comprehensive study concerned the chemistry and biology of 34 lakes in southeast Latvia.

During World War II, Bērziņš immigrated to Sweden but was able to bring with him only fragments of his scientific manuscripts that later formed the basis of several publications. In Sweden he identified zooplankton for the Fishery Board, and after some years became director of a station for fishery research. Later Bērziņš joined the staff of the Limnological Institute in Lund, where he worked on several large projects as an expert of plankton.

Bērziņš produced many publications dealing with fishery biology. However, he is most well known for his studies on microfauna, especially rotifers. His efforts led to the description of about 150 new species or subspecies and increased the number of species known from south and central Sweden by several times. Besides his taxonomic work he had a wide interest in rotifers. He published a paper on the rotifers that live on sponges and one where he concurrently analyzed the horizontal and vertical distributions of rotifers in Lake Skärshultsjön in southern Sweden. However, the largest body of material collected by Berzins still waits publication. During his long career, he collected a vast amount of biological and chemical data from 350 lakes, 50 ponds, 20 pools, 150 running-water locations, and 15 mires. Analysis of these data has now been completed and the work of publication has begun [see

Bērziņš & Pejler, in this volume]. With these future contributions, Bērziņš research will provide a much needed body of base-line ecological data from which important theoretical extrapolations may be made.

Although Bruno Bērziņš was a refugee, and during his later years suffered from a severe disease, he was never heard to complain. Always contented, good-tempered, helpful, and filled with good humour, he enriched the lives of those who knew him well.

Birger Pejler
Limnology Institute
Box 557
751 22 Uppsala
Sweden

Hydrobiologia 147: 3–8 (1987)
May, L., Wallace, R. & Herzig, A. (eds), Rotifer Symposium IV

Keratella cochlearis (Gosse) in Africa

J. Green
School of Biological Sciences, Queen Mary College, Mile End Road, London E1 4NS

Keywords: *Keratella*, distribution, biogeography, ecology, Africa, morphometry.

Abstract

Keratella cochlearis occurs over the complete latitudinal range of Africa, and over a considerable altitudinal range, but not in the highest mountain tarns. Physical, chemical, and biological factors may affect the distribution of this species. Lack of suitable food, and competition or predation by larger zooplankters are probably important limiting factors. Morphometric variation in samples from Zimbabwe is much less than in samples from the Auvergne. In particular, posterior spine length shows a much lower relative growth in relation to lorica size.

Introduction

Keratella cochlearis is one of the most widespread rotifers in the world, particularly in the western Palaearctic. One might expect that, in the tropics, this species would occur most frequently in upland areas where temperature is similar to that of warm temperate regions, but this is not the case. This paper examines the distribution of *K. cochlearis* in Africa, and discusses why this species is not common in African lakes. It also reports on morphological variation in populations in Zimbabwe.

The distribution of *Keratella cochlearis* in Africa

Keratella cochlearis is found over the complete latitudinal range in Africa (Fig. 1) and over a considerable altitudinal range (Table 1). However, it was not found in high mountain lakes at altitudes between 3800 and 4782 m [Löffler, 1968].

Distribution maps may be misleading as they give no information about the relative abundance of species. For instance, the record of *K. cochlearis* from the northern end of Lake Albert, Uganda, results from a one-year survey of the whole lake, based on monthly samples from 8 stations, but *K. cochlearis* was found in small numbers in only two samples from one station.

For such a cosmopolitan species, it is surprising that *K. cochlearis* is absent from many African lakes (Table 2). These lakes are probably unsuitable for the species, since there are few barriers to its dispersal and the species generally has a widespread distribution. Physical, chemical and biological characteristics of a lake probably account for the absence of *K. cochlearis*.

Physical and chemical factors

Permanent low temperature may render a water body unsuitable for *K. cochlearis*. This may be the situation in the high tarns on Mount Kenya [Löffler, 1968], many of which ice over at night. Bosselmann [1979] has shown that, in the presence of abundant food, *K. cochlearis* in Lake Esrom, Denmark, reproduces only when the temperature is above 4 °C. The upper limit for reproduction is not known, but thermal races may have evolved, and the population in Lake Asejire, Nigeria, might be expected to show different responses to high tem-

Table 1. Altitudinal range of some localities for *Keratella cochlearis* in Africa.

Locality	Latitude	Altitude (m)	Surface temperature °C
Lake Asejire, Nigeria	7°30 N	137	25 – 32
Lake Kainji, Nigeria	9 – 10° N	141	23 – 31
Lake Kariba, Zimbabwe	17°S	483	22 – 30
Lake Victoria	0 – 3°S	1134	25 – 27
Lake Moodie, Zimbabwe	18°S	1768	13.2
Purdon Dam, Zimbabwe	18°S	1920	15
Gulliver Dam, Zimbabwe	18°S	1951	12.8
Connemara Lake, Zimbabwe	18°S	2362	11.5

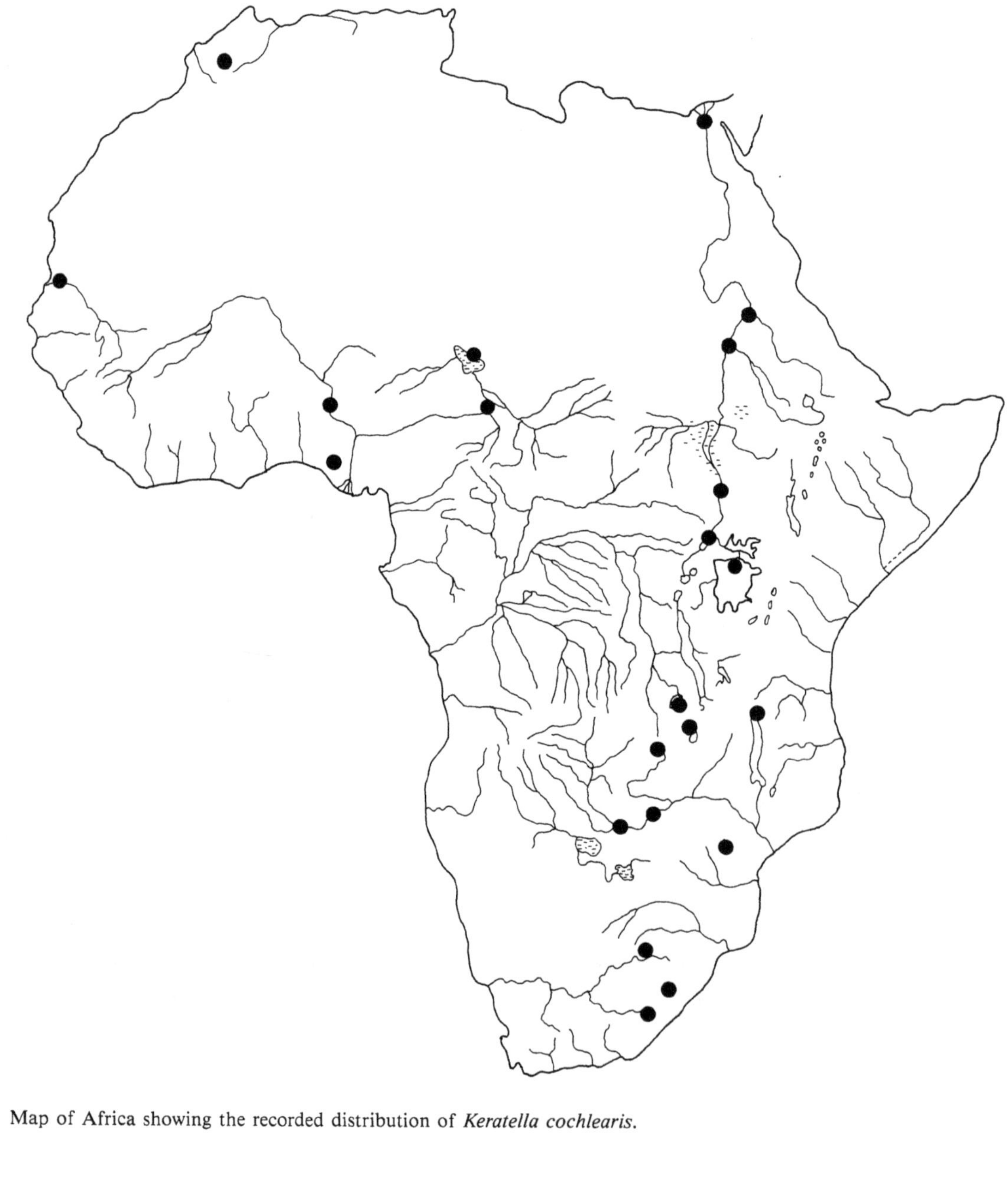

Fig. 1. Map of Africa showing the recorded distribution of *Keratella cochlearis*.

Table 2. A selection of African lakes where *Keratella cochlearis* was found to be absent.

Lake	Latitude	Altitude (m)	Remarks	Authority
Large Dariba, Sudan	13°N	2200	Saline: K_{20} = 27000 μS *Spirulina* dominant	1
Small Dariba, Sudan	13°N	2230	Saline: K_{20} = 6000 μS	1
Pawlo, Ethiopia	9°N	1870	pH 9.2; *Microcystis* dominant	2
Bishoftu, Ethiopia	9°N	1870	pH 9.2; *Microcystis* dominant	2
Aranguadi, Ethiopia	9°N	1900	pH 10.3; *Spirulina* dominant	2
Kilotes, Ethiopia	9°N	2000	pH 9.6; *Chroococcus* dominant	2
Awassa, Ethiopia	7°N	1675	*K. tropica* present	2
Shala, Ethiopia	7°30N	1540	Saline: K_{20} = 29000 μS	2
Kyoga, Uganda	1–2°N	1036	*K. tropica* present	3
George, Uganda	0°	912	*K. tropica* present	3
Edward, Uganda	0°30 S	912	*K. tropica* present	3
Albert, Uganda (midlake)	1–2°N	619	*K. tropica* present; midlake dominated by *Daphnia lumholtzi.*	3
Bunyonyi, Uganda	1°S	1950	*K. tropica* present	4
Mulehe, Uganda	1°S	1803	*K. tropica* present	4
Mutanda, Uganda	1°S	1791	*K. tropica* present	4
Naivasha, Kenya	0°45 S	1890	*K. tropica* present	5, 10, 11
Chilwa, Malawi	15°S	620	level fluctuates widely	9
Barombi Mbo, Cameroon	4°30 N	300	K_{20} = 39 μS	6
Soden, Cameroon	4°30 N	450	K_{20} = 27 μS	6
Kotto, Cameroon	4°30 N	110	K_{20} = 150 μS; *Microcystis* dominant	6
Debundsha, Cameroon	4°N	80	K_{20} = 12μS; *Botryococcus* dominant	8
Sonfon, Sierra Leone	9°N	549	K_{20} = 30 μS	7
d'Ayame, Ivory Coast	5°30 N	cal00	*K. tropica* present	12

Authorities: 1 – Green, Moghraby & Ali [1979]. 2 – Green, unpublished observations. 3 – Green [1967]. 4 – Green [1965 & 1976]. 5 – Mavuti [1984]. 6 – Green [1972]. 7 – Green [1979]. 8 – Green, Corbet & Betney [1974]. 9 – Kalk *et al.* [1979]. 10 – Pejler [1974]. 11 – Nogrady [1983]. 12 – Yte *et al.* [1983].

perature from those shown by a population from temperate regions.

Highly saline lakes are chemically unsuitable for *K. cochlearis*. This limitation probably excludes the species from the Large Dariba Lake in the Sudan, and Lake Shala in Ethiopia (Table 2).

Biological factors

There are other African lakes (Table 2) that do not have low temperature or high salinities, and yet *K. cochlearis* is not present in them. Some are characterised by dense blooms of cyanophytes, particularly *Microcystis,* but this in itself does not render a lake unsuitable for *Keratella* species. For instance, *K. tropica* occurs in Lake George, Uganda, which has a permanent dense bloom of *Microcystis.* But the presence of *Microcystis* may inhibit small flagellates, such as *Chrysochromulina* and *Rhodomonas* upon which *K. cochlearis* feeds [Edmondson, 1965; Pourriott, 1964, 1965, 1977; Lindstrom, 1983]. In Table 2 there are ten lakes in which *K. tropica* is found which may indicate that it has different food requirement from *K. cochlearis.* The two species are similar in size and might be expected to be competitors. However, these two species co-occur in many lakes e.g. Lakes Asejire, Kainji, and Kariba, and do not show population trends characteristic of competing species. In 1972, *K. tropica* was more abundant than *K. cochlearis* in Lake Kariba [Begg, 1976]. In 1983 the numbers of

K. tropica were reduced to about 2% of the 1972 densities, while the numbers of *K. cochlearis* remained about the same [Green, 1985]. If the two were competitors, *K. cochlearis* would be expected to increase when *K. tropica* declined since more food would be available for *K. cochlearis*. Clearly, in Lake Kariba, this did not happen. In Lake Asejire, *K. tropica* and *K. cochlearis* often show simultaneous peak abundances [Egborge, 1981] which further suggests that they do not compete directly.

Competition with other larger zooplankters may influence rotifer populations. Experimental evidence that populations of *K. cochlearis* can be controlled by interference competition by *Daphnia* has recently been produced by Gilbert and Stemberger [1985]. Not only does *Daphnia* compete for food, but it can sweep rotifers into its feeding chamber where they are injured during rejection, sometimes mortally. Large calanoid copepods may act in a similar way. Anderson & Raasveldt [1974] found rotifers to be sparse in the presence of large populations of *Hesperodiaptomus* and *Aglaodiaptomus*. Green [1986] found that in some Ethiopian lakes rotifer populations were diminished in the presence of large populations of *Lovenula africana*. Some calanoid copepods actually ingest rotifers [cf. Williamson, 1983 and references therein].

Keratella cochlearis seems to have a high distribution potential in Africa which is only partly realised. Physical, chemical and biological factors limit its distribution either separately or in combination. For instance Lake Aranguadi in Ehthiopia is highly saline, has a dense bloom of *Spirulina*, and a large population of *Lovenula africana* [Green, 1986]. Thus this lake is triply unsuitable for *K. cochlearis*.

The morphometry of *Keratella cochlearis* in Zimbabwe

An important characteristic of *K. cochlearis* in Europe is the variable length of the posterior spine [Lauterborn, 1900, 1903]. Several well marked forms may co-exist in the same lake and could be regarded as seperate species [Hofman, 1980; Dumont, 1983]. Therefore it is of interest of examine the range of morphological variation of the posterior spine in African populations. Recently I examined specimens from four impoundments in the Nyanga Hills, and from Lake Kariba, Zimbabwe.

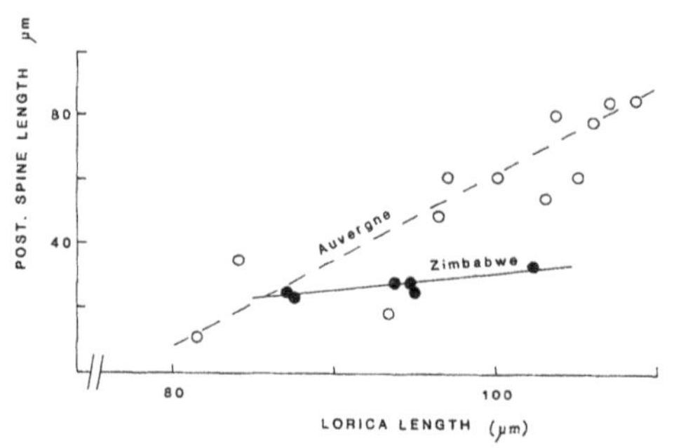

Fig. 2. Relationship between lorica length and posterior spine length of *Keratella cochlearis* in Zimbabwe and the Auvergne. The lines have been fitted by eye.

Figure 2 shows the relationship between mean lorica length and mean posterior spine length for the Zimbabwe samples compared with similar samples from the Auvergne [Green, 1981]. In the Zimbabwe samples the posterior spine increased by only 7 µm for an increase of 10 µm in lorica length. In contrast, the posterior spines of specimens from the Auvergne increased by about 27 µm for each 10 µm increase in lorica length. The standard errors of the mean lorica length vary from 0.5 to 1.47 µm

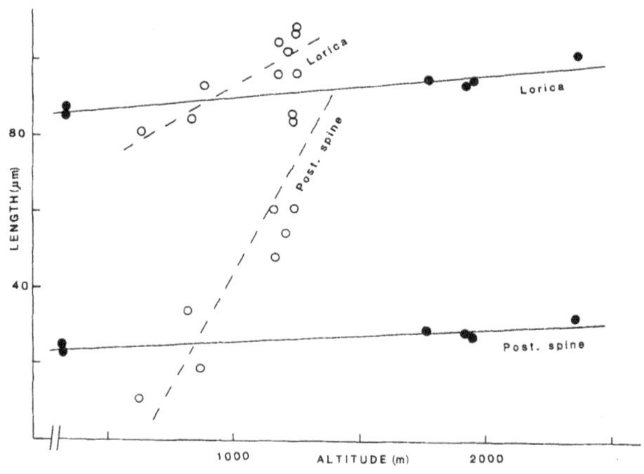

Fig. 3. Relationship between altitude of lake and lorica and posterior spine lengths of *Keratella cochlearis* from Zimbabwe (———) and the Auvergne (– – –). The lines have been fitted by eye.

in the Auvergne samples, and from 0.4 to 0.7 μm in the Zimbabwe samples.
Posterior spine lengths were more variable, with standard errors ranging from 0.8 to 4.3 μm in the Auvergne samples, and from 0.4 to 1.2 μm in the Zimbabwe samples.

In Fig. 3 mean lorica and posterior spine lengths are plotted as a function of altitude. There is a marked contrast between the two localities. In the Auvergne, posterior spine length increased more rapidly than lorica length with increasing altitude. In contrast, the increase in dimensions with altitude in Zimbabwe was less marked, although specimens from the highest altitude (2362 m) were significantly larger than those from the lowest altitude (483 m) ($p < 0.001$).

Relative growth of the posterior spine in relation to lorica length was clearly much lower in the Zimbabwe populations. This small response may have developed because of the limited range of temperature in any one locality in Zimbabwe, assuming that body size and temperature are major factors in regulating spine length. Biological agents also influence spine growth in *K. cochlearis* [Stemberger & Gilbert, 1984] and it would be interesting to see if the populations from Zimbabwe also respond to such agents.

Acknowledgements

The work in Zimbabwe was supported by a grant from the Central Research Fund of the University of London.

References

Anderson, R. S. & L. G. Raarsveldt, 1974. Gammarus predation and the possible effects of Gammarus and Chaoborus feeding on the zooplankton composition in some small lakes and ponds in western Canada. Canad. Wildlife Serv. Occ. Pap. 18: 1–23.

Begg, G. W., 1976. The relationship between the diurnal movements of some of the zooplankton and the sardine Limnothrissa miodon in Lake Kariba, Rhodesia. Limnol. Oceanogr. 21: 529–539.

Bosselmann, S., 1979. Population dynamics of Keratella cochlearis in Lake Esrom. Arch. Hydrobiol. 87: 152–165.

Dumont, H. J., 1983. Biogeography of rotifers. Hydrobiologia 104: 19–30.

Edmondson, W. T., 1965. Reproductive rate of planktonic rotifers as related to food and temperature in Nature. Ecol. Monogr. 35: 61–111.

Egborge, A. B. M., 1981. The composition, seasonal variation and distribution of zooplankton in Lake Asejire, Nigeria. Rev. Zool. Afr. 95: 136–180.

Gilbert, J. J. & R. S. Stemberger, 1985. Control of Keratella populations by interference competition from Daphnia. Limnol. Oceanogr. 30: 180–188.

Green, J., 1965. Zooplankton of Lakes Mutanda, Bunyonyi and Mulehe. Proc. zool. Soc. Lond. 144: 383–402.

Green, J., 1967. Associations of Rotifera in the zooplankton of the lake sources of the White Nile. J. Zool., Lond. 151: 343–378.

Green, J., 1972. Ecological studies on crater lakes in West Cameroon. Zooplankton of Barombi Mbo, Mboandong, Lake Kotto and Lake Soden. J. Zool. Lond. 166: 283–301.

Green, J., 1976. Changes in the zooplankton of Lakes Mutanda, Bunyonyi and Mulehe (Uganda). Freshwat. Biol. 6: 433–436.

Green, J., 1979. The fauna of Lake Sonfon, Sierra Leone. J. Zool., Lond. 187: 113–133.

Green, J., 1981. Altitude and seasonal polymorphism of Keratella cochlearis (Rotifera) in lakes of the Auvergne, Central France. Biol. J. Linn. Soc. Lond. 16: 55–61.

Green, J., 1985. Horizontal variations in associations of zooplankton in Lake Kariba. J. Zool., Lond. 206: 225–239.

Green, J., 1986. Zooplankton associations in some Ethiopian crater lakes. Freshwat. Biol. 16.

Green, J., S. A. Corbet & E. Betney, 1974. Ecological studies on crater lakes. Freshwat. Biol. 16: 495–499.
Lond. 173: 199–223.

Green, J., A. I. Moghraby & O. M. M. Ali, 1979. Biological observations on the crater lakes of Jebel Marra, Sudan. J. Zool., Lond. 189: 493–502.

Hofmann, W., 1980. On morphological variation in Keratella cochlearis population from Holstein lakes (Northern Germany). Hydrobiologia 73: 255–258.

Kalk, M., A. J. McLachlan & C. Howard Williams, 1979. Lake Chilwa, studies of change in a tropical ecosystem. Monographiae Biologicae 35: 1–462.

Lauterborn, R., 1900. Der Formenkreis von Anuraea cochlearis. Ein Beitrag zur Kenntnis der Variabilität bei Rotatorien I Teil: Morphologische Gliederung des Formenkreises. Verh. Naturh. Med. Ver. Heidelb. N.F. 6: 412–448.

Lauterborn, R., 1903. Der Formenkreis von Anuraea cochlearis. Ein Beitrag zur Kenntnis der Variabilität bei Rotatorien. II Teil: Die cyclische oder temporale Variation von Anuraea cochlearis. Verh. Naturh. Med. Ver. Heidelb. N.F. 7: 529–621.

Lindstrom, K., 1983. Changes in growth and size of Keratella cochlearis (Gosse) in relation to some environmental factors in cultures. Hydrobiologia 104: 325–328.

Löffler, H., 1968. Die Hochgebirgseen Ostafrikas. Hochgebirgsforschung 1: 1–65.

Mavuti, K., 1984. Studies on the zooplankton of Lake Naivasha, Kenya. PhD thesis, University of Nairobi.

Nogrady, T., 1983. Succession of planktonic rotifer populations in some lakes of the Eastern Rift Valley, Kenya. Hydrobiologia 98: 45–54.

Pejler, B., 1974. On the rotifer plankton of some East African lakes. Hydrobiologia 44: 389–396.

Pourriot, R., 1964. Étude experimentale de variations morphologiques chez certaines espèces de Rotifères. Bull. Soc. zool. Fr. 89: 555–561.

Pourriot, R., 1965. Recherches sur l'écologie des Rotifères. Vie Milieu (Suppl.) 21: 1–224.

Pourriot, R., 1977. Food and feeding habits of Rotifera. Arch. Hydrobiol. Beih. Ergebn. Limnol. 8: 243–260.

Stemberger, R. S. & J. J. Gilbert, 1984. Spine development in the rotifer Keratella cochlearis: induction by cyclopoid copepods and Asplanchna. Freshwat. Biol. 14: 639–647.

Williamson, C. E., 1983. Invertebrate predation on planktonic rotifers. Hydrobiologia 104: 385–396.

Yte, W. A., J. Rey & R. Pourriot, 1983. Peuplement zooplanktonique d'un lac de barrage de Côte d'Ivoire. Annls Limnol. 19 (1): 3–8.

Hydrobiologia 147: 9–14 (1987)
May, L., Wallace, R. & Herzig, A. (eds), Rotifer Symposium IV

Distribution of rotifers in African fresh and inland saline waters

M. De Ridder
Institute of Ecology, The State University of Gent, Belgium

Keywords: Rotifera, historical review, distribution, Africa

Abstract

The distribution of rotifers in African fresh and inland saline waters is discussed in the light of historical expeditions and modern limnological research. Number of papers published and number of rotifer species described for each country are given. Some peculiarities of African rotifer fauna are described.

Historical review

Our knowledge of African rotifers began with the adventurous journey of C.G. Ehrenberg in 1820, through Egypt and the Sudan. Most studies of African rotifers have been associated with more comprehensive expeditions to the continent and the most iomportant of these are given in chronological order in Table 1. A full bibliography of scientific papers describing the rotifer fauna of Africa can be found in de Ridder (in prep.). The more recent studies have tended to take a more ecological or environmental approach to limnology than the earlier studies. Man-made impoundments have been studied for their industrial and/or agricultural potential. Lakes have been developed for their fisheries resources. The impact of pollutants and insecticides on aquatic food chains has been studied. Increasingly, African limnologists have been playing a more important part in pure and applied aspects of the biology of their continent.

Number of papers

Some 220 papers dealing with African rotifers have been published since 1829, although their distribution over the continent as a whole is uneven (Fig. 1). Many areas are still *terra incognita* as far as rotifers are concerned, particularly Western and Central Africa.

Number of taxa per country

More than 510 taxa, identified to species level, are now known from Africa. These belong to the following genera:

Class Digononta
Order Bdelloidea

Adineta	Mniobia
Didymodactylus	Otostephanos
Dissotrocha	Philodina
Habrotrocha	Pleuretra
Henoceros	Rotaria
Macrotrachela	Scephanotrocha

Class Monogononta
Orders Ploïma, Monimotrocha, Paedotrocha

Anuraeopsis	Lindia
Ascomorpha	Lophocharis
Ascomorphella	Macrochaetus
Aspelta	Manfredium
Asplanchna	Microcodon
Asplanchnella	Monommata
Asplanchnopus	Mytilina
Beauchampia	Notholca

Brachionus	Notommata
Cephalodella	Paracolurella
Collotheca	Platyias
Colurella	Pleurotrocha
Conochilus	Ploesoma
Cupelopagis	Polyarthra
Cyrtonia	Pompholyx
Dicranophorus	Proales
Dipleuchlanis	Proalides
Diplois	Proalinopsis
Encentrum	Pseudoploesoma
Eosphora	Ptygura
Eothinia	Repauliana
Epiphanes	Resticula
Euchlanis	Scaridium
Filinia	Sinantherina
Floscularia	Sphyrias
Gastropus	Squatinella
Hexarthra	Synchaeta
Horaella	Taphrocampa
Itura	Testudinella
Kellicottia	Tetrasiphon
Keratella	Trichocerca
Lacinularia	Trichotria
Lecane	Tripleuchlanis
Lepadella	Trochosphaera
Limnias	Wolga

Generally, there is a direct relationship between the number of papers devoted to a given country and the number of rotifer species recorded from that country (Fig. 1).

Peculiarities of the African rotifer fauna

The most striking features of the African rotifer fauna are described below:

(a) high percentage of cosmopolitan species (70%); small number of thermophiles (16%); small number of tropical and sub-tropical species (14%),

(b) among the cosmopolitan species, several cold-stenothermal species (Fig. 2) usually associated with an arctic or subarctic distribution [see Dumont, 1983],

(c) some species with a very limited range and strict ecological requirements e.g. taxa of soda lakes:

- *Brachionus dimidiatus*, found in Ethiopia, Kenya, Chad and Sudan (Jebel Marra, Dariba Lakes),
- *Brachionus novaezelandiae*, found in the RSA,
- *Cephalodella elmenteita*, recorded from Kenya and Chad,
- *Hexarthra jenkinae*, known from Kenya, Chad, Sudan (Jebel Marra, Dariba Lakes) and Senegal,

(d) a very low proportion of endemic species of the class Monogononta (0.5%). Seventy new species have been found only in Africa; 23 of these have been confirmed as valid taxa, mostly of the genera *Lecane* and *Testudinella*. There are also 3 African sub-species of the genera *Macrochaetus* and *Testudinella*,

(e) of the 49 species of the order Bdelloidea described from Africa, 31 are apparently limited to this continent, and 23 are confined to South Africa. The scarcity of data concerning the world-wide distribution of Bdelloids precludes any statement on endemism among these species. Further investigations are clearly needed.

References

Beauchamp, P. de, 1932a. Scientific results of the Cambridge expedition to the East African Lakes, 1930–1931. 6. Rotifères et gastrotriches. J. Linn. Soc. 38: 231–248.

Beauchamp, P. de, 1932b. Rotifères des Lacs de la Vallées du Rift (Kenya). Ann. Mag. nat. Hist. Sér. 10, 9: 158–165.

Beauchamp, P. de, 1939. Rotifères. In: Exploration du Parc national Albert (mission H. Damas, 1935–1936), Fasc. 5: 1–11. Inst. Parcs nat. Congo belge, Bruxelles.

Berzins, B., 1957. Beitrag zur Rotatorienfauna Gambiens. Acta trop. 14: 230–235.

Berzins, B., 1959. Rotatorien aus Französisch Westafrika. Bull. I.F.A.N. 21(A): 921–933.

Berzins, B., 1973. Rotatorien aus den Azoren. Bol. Museu mun. Funchal 27 (Art. 122): 84–89.

Brunelli, G., & G. Cannicci, 1938. Le caratteristische biologiche del Lago Tana. Rend. Acad. nat. Lincei - Cl. Scienze fis., mat. nat. Estr. vol. 27, ser. 6a, fasc. 7: 319–323.

Brunelli, G. & G. Cannicci, 1940. Le caratteristiche biologiche del Lago Tana. Miss. Stud. Lago Tana. Ric. limnol. – B. Chimica e biol.: 71–116.

Brunelli, G. & G. Cannicci, 1941. Ricerche sul plancton e sulle caratteristiche biolimnoliche del Lago Margherita. Esplor. Laghi della Fossa Galla, I: 3–26.

Bryce, D., 1931: Report on the Rotifera. Mr. Omer Cooper's investigation of the Abessynian Freshwaters. Proc. zool. Soc. Lond. 1931: 865–878.

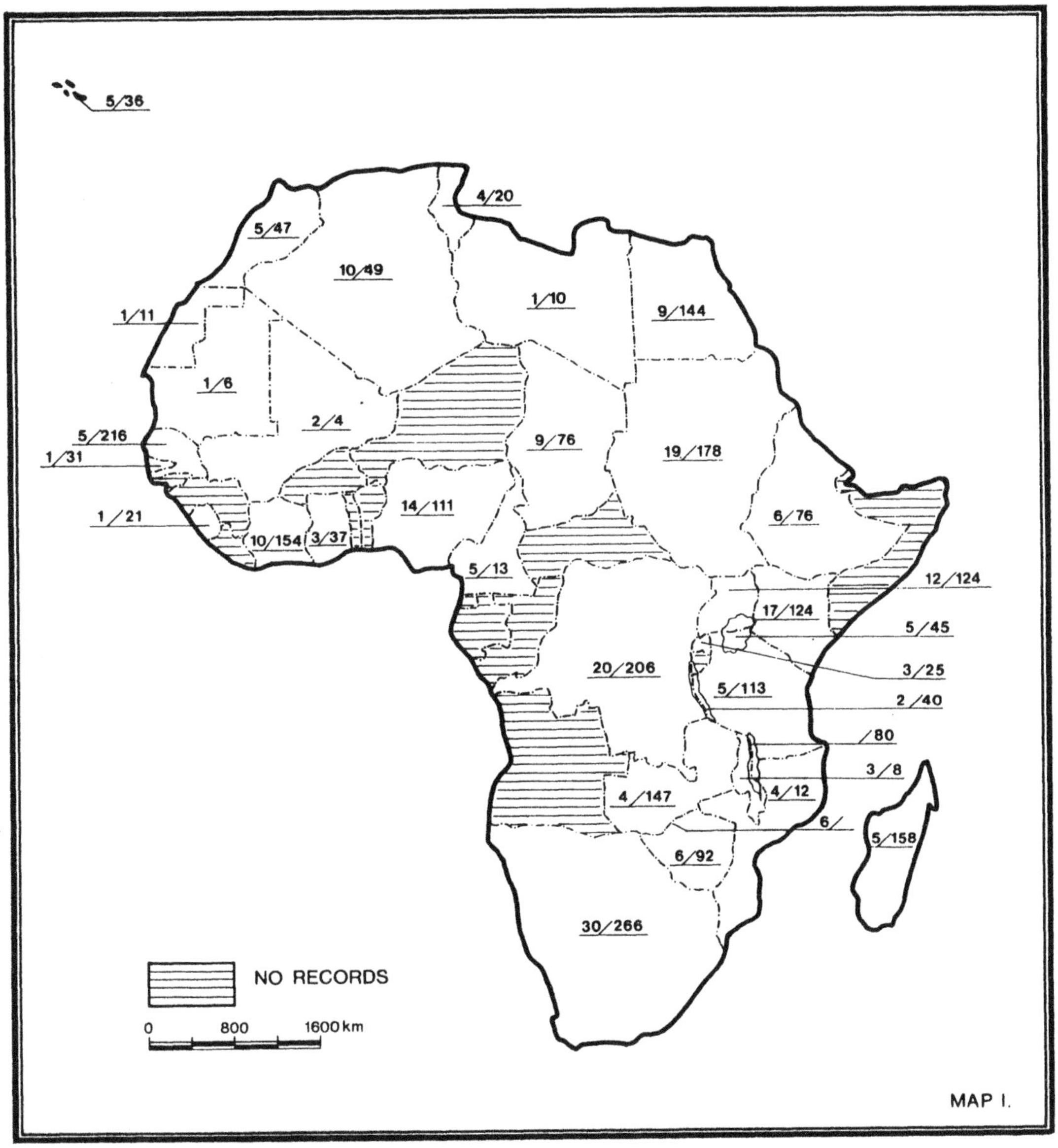

Fig. 1. Map of Africa showing the number of papers published describing the rotifer fauna of each geographical area (first figure) and the total number of rotifer species recorded (second figure). The shaded areas have not been studied.

Cannicci, G. & F. Almagia, 1947. Notizie sulla 'facies' planctonica di alcuni laghi della fossa galla. Boll. Pesca, Piscic. idrobiol. 23: 5–28.

Collin, A., 1897. Rotatorien, Gastrotichen und Entozoen Ostafrikas. Tierwelt Ostafrikas, IV, 3, Teil 3: 13 pp.

Coussement, M., 1975. Rotatoria van Marokko. Report Ghent State University. Unpublished.

Coussement, M. & H. J. Dumont, 1980. Some peculiar elements in the rotifer fauna of the Atlantic Sahara and of the Atlas Mountains. Dev. Hydrobiol. 1: 249:254.

Daday, E. von, 1907. Planktontiere aus dem Victoria-Nyanza. Sammelausbeute von A. Borgert 1904–1905. Zool. Jb. Syst. 25: 245–254.

Daday, E. von, 1910a. Beiträge zur Kenntnis der Mikrofauna des Nils. Sitzungsber. K. Akad. Wiss. Wien, 119, Abt. 1: 537–589.

Daday, E. von, 1910b. Untersuchungen über die Süszwasser-mikrofauna Deutsch-Ostafrikas. Zool. Stuttgart. Bd. 23, H. 59: 59–106.

De Ridder, M., 1981. Rotifères. In: J.J. Symoens (ed), Explora-

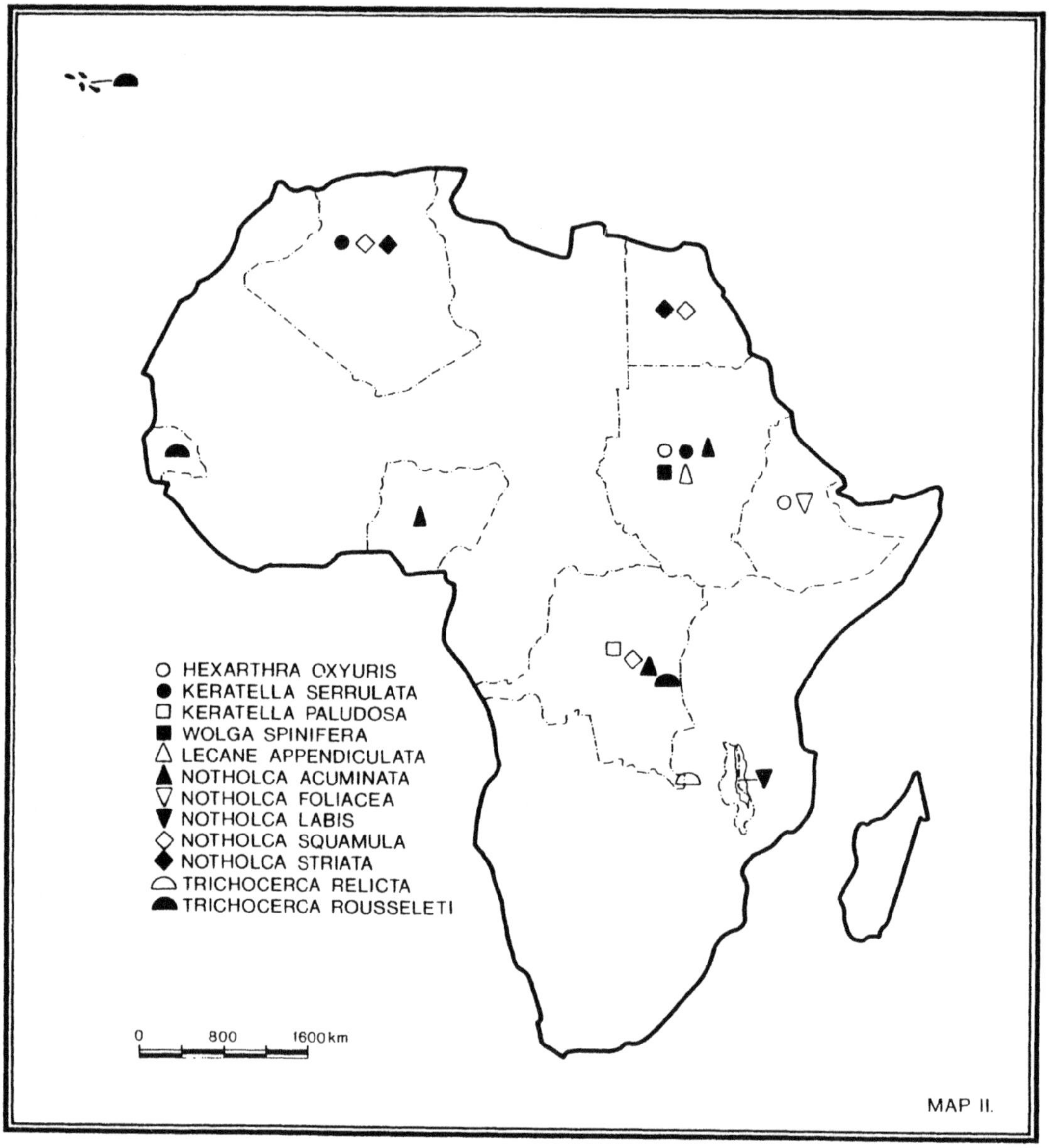

Fig. 2. Map of Africa showing the distribution of cold stenothermal rotifer species.

tion hydrobiologique du bassin du Lac Bangwéolo et du Luapula. Vol. 11, fasc. 4: 166 pp. Cercle hydrobiologique de Bruxelles.

De Ridder, M., 1984a. A review of the rotifer fauna of the Sudan. In: Dumont, H. J., A. I. El Moghrabi & L. A. Desougi (eds), Limnology and marine biology of the Sudan: 113–130. Dev. Hydrobiol. 21: 363 pp.

De Ridder, M., 1984b. Onderzoekingen over de verspreiding der raderdieren in Afrika. 2. Raderdieren uit Zuid-Tunesië en uit Zuidelijk Afrika. Natuurwet. Tijdschr. 65 (1983): 155–163.

De Ridder, M., 1984c. Taxonomical and zoogeographical remarks on Rotifers from the Ivory Coast (W.-Africa). Revue Hydrobiol. trop. 17: 295–299.

De Ridder, M., 1985. Contributions to the knowledge of African Rotifers. 1. Rotifers from Senegal. Hydrobiologia 120: 47–51.

De Ridder, M., in prep. Annotated checklist of non-marine rotifers (Rotifera) from African inland waters. Koninklijk Museum voor Midden-Africa, Tervuren, België. Zoologische Documentatie.

Table 1. The most important expeditions to collect African rotifers, in chronological order.

Sampling area	Expedition leader	Resulting publications
Egypt, Sudan (1820)	Ehrenberg	Ehrenberg [1829, 1838]
Egypt (1853)	Von Fridau & Schmarda	Schmarda [1854]
East Africa (1887 – 1890)	Stuhlman	Collin [1897]
Açores (1887)	Prince of Monaco	de Guerne [1888]
German East Africa (1898 – 1900)	Fülleborn	von Daday [1910b]
Lake Victoria (1904 – 1905)	Borgert	von Daday [1907]
Egypt & Sudan (1904)	Werner	von Daday [1910a]
Tanganyika (1904 – 1905)	Cunnington	Rousselet [1910]
South Africa	Shackleton	Murray [1911a, b, c]
SE Belgian Congo (1911 – 1912)	Stappers	Evens [1947, 1949]
Lake Victoria (1911 – 1912)	Alluaud & Jeannel	Virieux [1913]
Algeria, Tunisia (1924 – 1927)	Gauthier	Gauthier [1928]
The Hoggar (1928)	Seurat	Gauthier [1931]
Abyssinia (1926 – 1927)	Scott	Bryce [1931]
Rift Valley lakes, Kenya (1929)	Sladen	de Beauchamp [1932b]
East African lakes (1930 – 1931)	–	de Beauchamp [1932a]
Belgian Congo (1935 – 1936)	Damas	de Beauchamp [1939]
South Africa (1928)	Hutchinson, Pickford & Schuurman	Hutchinson, Pickford & Schuurman [1932]
Ethiopia (1937)	–	Brunelli & Cannici [1938, 1940, 1941]; Cannicci & Almagia [1947]
Lake Tanganyika (1946 – 1947)	Capart	Gillard [1957]
Gambia (1950)	–	Berzins [1957]
French West-Africa	Monod	Berzins [1959]
Açores & Madeira (1957)	University of Lund, Sweden	Berzins [1973]
North Sudan (1962)	Löffler	Löffler [1963]
Lake Chad, Lake Léré and adjacent waters (1965 – 1977)	–	Pourriot [1968, 1971, 1980]; Pourriot, Iltis & Lévêque-Duwat [1967]; Pourriot & Rougier [1975]
River Sokoto, Nigeria	Holden	Green [1960]; Wulfert [1965]
Bangweulu & Luapula Basins, Zaïre, Zambia	Symoens	Thomasson [1966]; de Ridder [1981]
Morocco (1971), Sahara and adjacent ares (1975, continuing)	University of Gent, Belgium; mostly led by Dumont	Coussement [1975]; Coussement & Dumont [1980]; Dumont, Miron, dall 'Asta, Decraemer & Claus [1973]; Dumont & Coussement [1976]; de Ridder [1984a, b, c, 1985]

Dumont, H. J., 1983. Biogeography of Rotifers. Hydrobiologia 113: 19 – 30.

Dumont, H. J. & M. Coussement, 1976. Rotifers from Rio de Oro (North-western Sahara). Hydrobiologia 51: 109 – 112.

Dumont, H. J., I. Miron, U. dall'Asta, W. Decraemer & Chr. Claus, 1973. Limnological aspects of some Moroccon Atlas Lakes, with reference to some physical and chemical variables, the nature and distribution of phyto- and zooplancton, including a note on possibilities for the development of an inland fishery. Int. Revue ges. Hydrobiol. 58: 33 – 60.

Evens, F., 1947. Quelques Rotateurs nouveaux observés au Congo Belge. Rev. Zool. Bot. afr. 40: 175 – 184.

Evens, F., 1949. Le Plancton du lac Moéro at de la région d'Elisabethville. Rev. Zool. Bot. afr. 41: 233 – 277; id. 42: 1 – 64.

Gauthier, H., 1928. Recherches sur la faune des eaux continentales de l'Algérie et de la Tunisie. Imprim. Minerva, Alger: 419 pp.

Gauthier, H., 1931. Faune aquatique du Sahara central. Récoltes de M.L.G. Seurat au Hoggar en 1928. Bull. Soc. Hist. nat. Afr. N. 22: 350 – 400.

Gillard, A., 1957. Exploration hydrobiologique du lac Tanganika. Rotifères. Mém. Inst. roy. Sc. nat. Belg. vol. III, fasc. 6: 3 – 26.

Green, J., 1960. Zooplankton of the River Sokoto. The Rotifers. Proc. zool. Soc. Lond. 135: 491–523.

Guerne, J. de, 1888. Excursions zoologiques dans les îles de Fayal et de San Miguel (Açores). Camp. scient. yacht monégasque l'Hirondelle, 3e année, 1887. Ed. Gauthier-Villars, Paris, 1888: 113 pp.

Hutchinson, G. E., G. E. Pickford & J.F.M. Schuurman, 1932. A. contribution to the hydrobiology of pans and other inland waters of South Africa. Arch. Hydrobiol. 24: 1–154.

Löffler, H., 1963. Ergebnisse der zoologischen Nubien-Expedition 1962. Teil XVIII: Zur Binnenwasserfauna einiger Kleingewässer und Brunnen im nördlichen Sudan. Ann. naturhist. Mus. Wien 66. 489–494.

Murray, J., 1911a. Some African Rotifers: Bdelliodea of Tropical Africa. J. r. miscrosc. Soc. 1911: 1–18.

Murray, J., 1911b. South African Rotifera, collected by the Shackleton Antarctic Expedition, 1907. J. r. microsc. Soc. 1911: 584–587.

Murray, J., 1911c. Bdelloid Rotifera of South Africa. Ann. Transvaal Mus. 3: 1–19.

Pourriot, R., 1968. Rotifères du lac Tchad. Bull. I.F.A.N., 30, sér. A, 2: 471–496.

Pourriot, R., 1971. Prospection hydrobiologique du lac de Léré et des mares avoisinantes. II. Rotifères. Cah. O.R.S.T.O.M., sér. hydrobiol. 5, 2: 171–174.

Pourriot, R., 1980. Rotifères. In: Durand, J.-R. & C. Lévêque (eds.), Flore et fauna aquatiques de l'Afrique sahélo-soudanaise. T.1. Ed. O.R.S.T.O.M. Collection Iniatiations-Documentations techniques 44: 389 pp.: 219–244.

Pourriot, R., A.Iltis & S. Lévêque-Duwat, 1967. Le plancton des mares natronées du Tchad. Int. Revue ges. Hydrobiol. 52: 535–543.

Pourriot, R. & C. Rougier, 1975. Dynamique d'une population expérimentale de Brachionus dimidiatus (Bryce) (Rotifère) en fonction de la nourriture et de la température. Ann. Limnol. 11: 125–143.

Rousselet, C. F., 1910. Zoological Results of the Third Tanganyika Expedition, conducted by Dr. W. A. Cunnington, F.R.S., 1904-1905. Report on the Rotifera. Proc. zool. Soc. Lond. 1910: 792–799.

Schmarda, L. K., 1854. Zur Naturgeschichte Aegyptens. Denkschr. Akad. Wiss. Wien, Math. Naturw. Kl. 7, 2: 1–28.

Thomasson, K., 1966. Le phytoplancton du Lac Shiwa Ngandu. In: J. J. Symoens (ed), Exploration hydrobiologique du bassin du Lac Bangweólo et du Luapula, Vol. 4, fasc. 2: 92 pp. Cercle hydrobiologique, Bruxelles.

Virieux, J., 1913. Plancton du lac Victoria-Nyanza, In Voyage de C.Alluaud et R. Jeannel en Afrique orientale, 1911-1912: 20 pp.

Wulfert, K., 1965. Rädertiere aus einigen afrikanischen Gewässern. Limnologica 3: 347–366.

Hydrobiologia 147: 15–22 (1987)
May, L., Wallace, R. & Herzig, A. (eds), Rotifer Symposium IV

On species of genus *Lepadella* (Eurotatoria:Monogononta:Colurellidae) from North-Eastern India, with remarks on Indian taxa

B. K. Sharma & Sumita Sharma[1]
Department of Zoology, North-Eastern Hill University, Shillong-793014, INDIA
[1]*Zoological Survey of India, Risa Colony, Shillong-793003, INDIA*

Keywords: *Lepadella,* North-Eastern India, new taxa, Indian species, distribution.

Abstract

Ten species of *Lepadella* Bory de St. Vincent, 1826, including one new species and one new form, are documented from North-Eastern India. Two of these species are new records from this country and six are new reports from N.E. region. Comments are also made on the status and distribution of various Indian taxa.

Introduction

Very little is known about the distribution of Indian species of the genus *Lepadella.* The only significant previous contributions to this field are those of Edmondson & Hutchinson [1934], Wulfert [1966], Vasisht & Battish [1971] and Sharma [1978].

This communication is part of a wider investigation of the rotifer fauna of North-Eastern India. It deals with ten species of the genus *Lepadella* (Syn. *Eulepadella* Hauer 1926), including descriptions of a new species and one new form. Two species are new records from this country and six are new to North-Eastern region. In all, eighteen species are so far known from India. The various taxa examined are illustrated and commented upon. Additional comments are made about other Indian taxa and on the distribution of each species.

Material and methods

The material for this study was collected from various localities in North-Eastern region (Fig. 1). The specimens were preserved in 5% formalin and mounted in Polyvinyl-lactophenol for detailed examination. All drawings, except cross-sections, were made with a camera lucida. Measurements are given in micrometers (μm).

Systematic account

The genus *Lepadella* is divided into three subgenera – *Lepadella* (Syn. *Eulepadella* Hauer 1926), *Heterolepadella* Bartos 1955, and *Xenolepadella* Hauer 1926 which have equally long and free toes, unequal and free toes, and partly fused toes respectively [Kutikova, 1970; Koste, 1978]. All of the taxa examined from North-Eastern region, and those previously reported from India [Edmondson & Hutchinson, 1934; Naidu, 1967; Nayar, 1968; Wulfert, 1966; Nayar & Nair, 1969; Vasisht & Gupta, 1967; Vasisht & Battish, 1971; Sharma, 1976, 1978, 1980a, 1980b; Sharma & Sharma, 1984; Sharma & Sharma, in press], belong to the first two subgenera.

Taxonomic notes on taxa from the North-Eastern region

Subgenus: *Lepadella* (Syn. *Eulepadella* Hauer 1926)

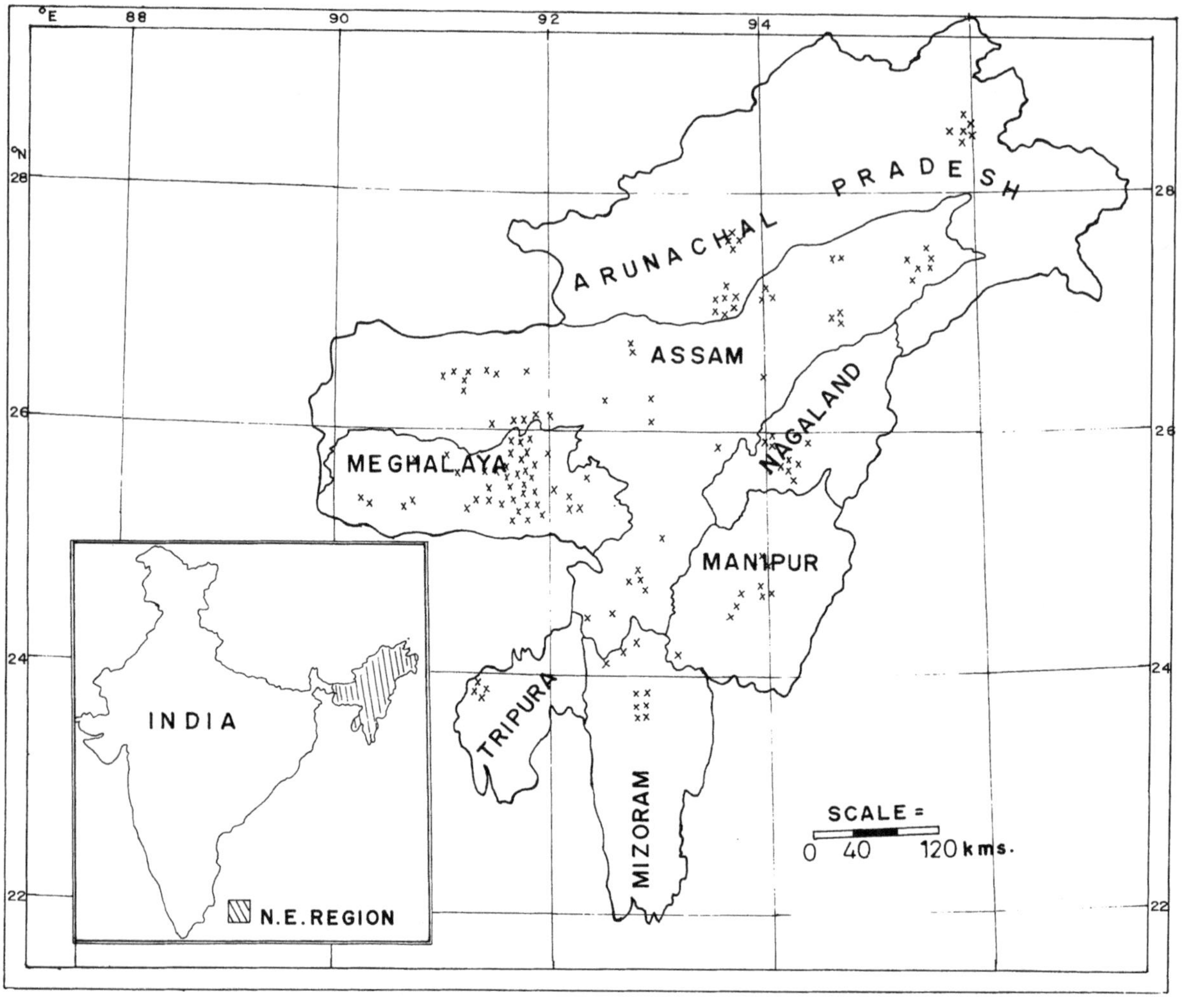

Fig. 1. Map of North-Eastern India showing sites of collection. (inset map of India, indicating North-Eastern region).

Lepadella acuminata (Ehrenberg 1834) (Figs. 2, 3)

Material: Meghalaya, Assam, Arunachal Pradesh

Characteristics: Posterior pointed projection of variable length.

Measurements: Lorica length 94–104; maximum width 58–66; anterior width 24–32; toes 22–26.

Distribution: Nilgiri Hills, West Bengal and Assam.

Lepadella cristata (Rousselet 1893) (Figs. 4–6)

Material: Meghalaya

Characteristics: Lorica with a dorsal median keel and an anterior backwardly directed crest.

Measurements: Lorica length 92–108; maximum width 68–70; dorsal crest 66–78; toes 28–35.

Distribution: New record for India.

Lepadella dactyliseta (Stenroos 1898) (Figs. 7–9)

Material: Meghalaya

Characteristics: Dorsal plate semicircular in cross-section. Lorica produced posteriorly into distinct widely separated corners. Foot-groove almost parallel. Foot small; toes elongated and pointed.

Measurements: Lorica length 78–100; maximum width 54–68; anterior width 30–40; toes 24–36.

Distribution: New record for India.

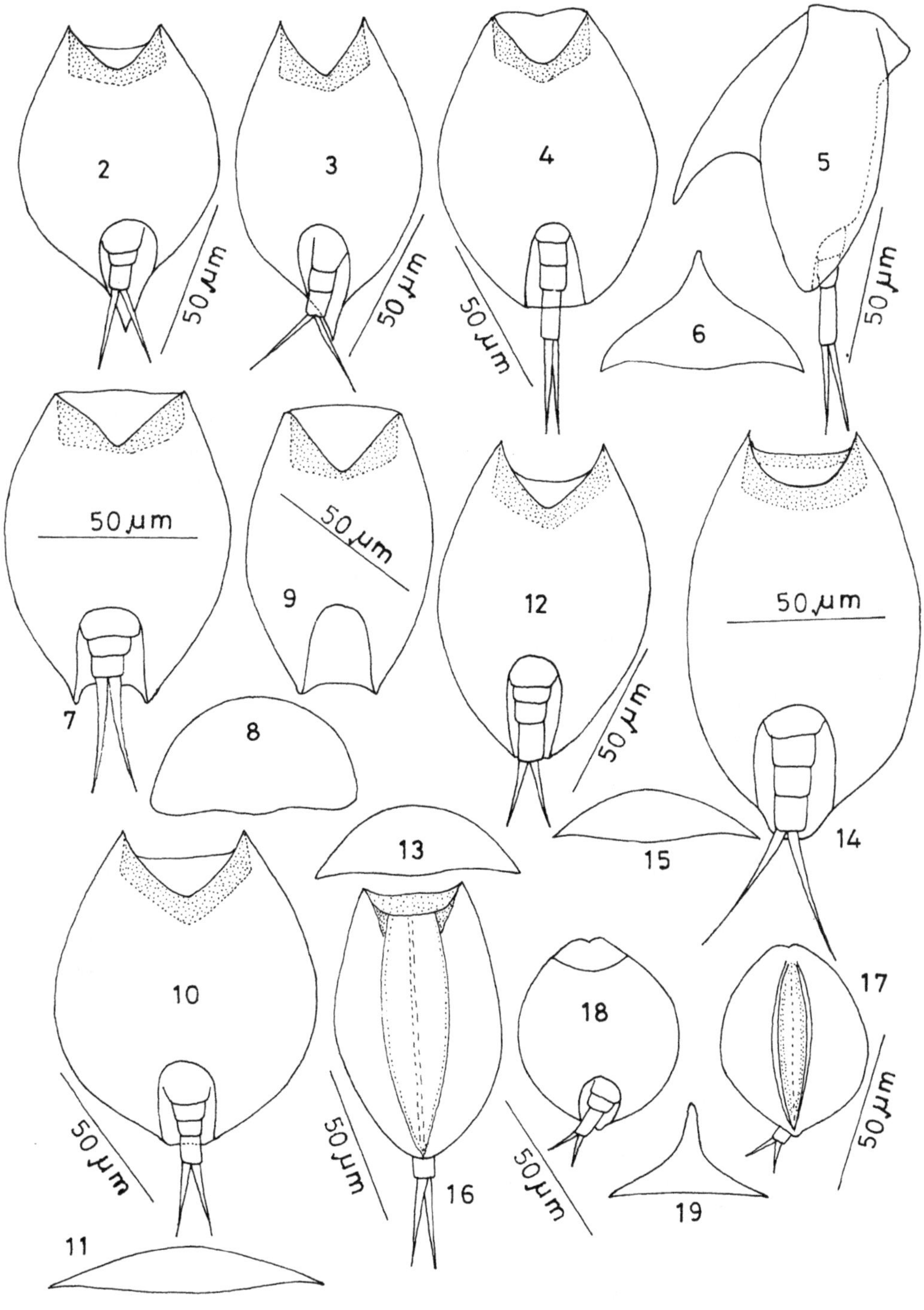

Figs. 2–19; 2, 3: *Lepadella acuminata* (Ehrb.), ventral views; 4–6: *Lepadella cristata* (Rousselet), ventral view, lateral view and cross-section; 7–9: *Lepadella dactyliseta* (Stenroos), ventral view, cross-section, lorica (ventral view); 10, 11: *Lepadella ovalis* (Müller), ventral view and cross-section; 12, 13: *Lepadella patella* (Müller), ventral view and cross-section; 14, 15: *Lepadella patella* f. *elongata* **new form,** ventral view and cross-section; 16: *Lepadella rhomboides* (Gosse), dorsal view; 17–19: *Lepadella triptera* Ehrb., dorsal view, ventral view and cross-section.

Lepadella ovalis (Müller 1786) (Figs. 10, 11)

Material: Meghalaya, Assam, Nagaland, Mizoram, Manipur, Arunachal Pradesh, Tripura

Characteristics: Lorica oval to circular in outline: compressed dorso-ventrally. Examined specimens smaller than those from West Bengal and Orissa.

Measurements: Lorica length 90–112; maximum width 68–84; anterior width 24–36; toes 22–26.

Distribution: Widely distributed, reported previously from Kashmir, Ladak, Panjab, Haryana, Kerala, West Bengal, Orissa and Assam.

Lepadella patella (Müller 1773) (Figs. 12, 13)

syn.: *L. similis* Lucks: Naidu 1967, p. 386, Plate II, Fig. 7

Material: Meghalaya, Assam, Nagaland, Mizoram, Manipur, Arunachal Pradesh, Tripura

Characteristics: Differs from *L. ovalis* by the small size of lorica and strongly arched dorsal plate. Present specimens smaller than those reported elsewhere from this country. *L. similis* reported from Andhra Pradesh [Naidu, 1967] is apparently synonymous with *L. patella.*

Measurements: Lorica length 75–108; maximum width 58–74; anterior width 22–30; toes 20–26.

Distribution: Kashmir, Ladak, Panjab, Gujarat, Rajasthan, West Bengal, Orissa and Assam.

Lepadella patella f. *elongata* **new form** (Figs. 14, 15)

Material examined: 3 parthenogenetic females (on slides), Regd. No. R. NE 101/3, Department of Zoology, North-Eastern Hill University, Shillong. Loc.: Khanduli, Jaintia Hills, Meghalaya State. Date: 18.3.1984. Coll.: B. K. Sharma

Characteristics: Differs from *L. patella* f. *elongata* (Ehrb. 1834) in (i) lorica relatively elongated; (ii) anterior sinus broader; and (iii) toes elongated.

Measurements: Lorica length 128; maximum width 75; anterior width 36; foot 38; toes 42.

Lepadella rhomboides (Gosse 1884) (Fig. 16)

Material: Meghalaya, Assam, Nagaland, Manipur, Mizoram

Characteristics: Lorica broadly ovate; dorsal median keel broad, forming a faint median ridge.

Measurements: Lorica length 84–90; maximum width 52–56; toes 24–26.

Distribution: Panjab, Gujarat, Nilgiri Hills, West Bengal, Orissa and Assam.

Lepadella triptera Ehrenberg 1830 (Figs. 17–19)

syn.: *L. crestata* **n.sp.** Vasisht & Battish 1971, p. 191, Figs. 7–9 *L. crestata* Vasisht & Battish: Sharma 1976, p. 257.

Material: Meghalaya, Assam

Characteristics: Lorica almost circular and with slightly produced posterior end. Present specimens longer than those from West Bengal.

Measurements: Lorica length 58–62; maximum width 48–50; toes 12–14.

Distribution: Ladak, Nilgiri Hills, Panjab and West Bengal.

Lepadella nartiangensis **new species** (Figs. 20, 21)

Holotype (on slide): Regd. No. R. NE. 102/1, Department of Zoology, North-Eastern Hill University, Shillong. Loc.: Nartiang, Jaintia Hills, Meghalaya State. Date: 21.9.1984. Coll.: B. K. Sharma

Paratype (on slide): Regd. No. R. NE. 103/1, other details as per Holotype.

Description: Lorica elongate-oval in outline, moderately compressed dorso-ventrally. Anterior dorsal sinus broad, slightly concave and with a broad collar. Anterior ventral sinus deep and V-shaped, with a stippled collar and flanked with two triangular spines at its external angles. Posterior end of lorica almost straight and produced into two widely separated and posteriorly projecting blunt ends. Foot-groove characteristic, almost semicircular for about first half of its length, then broadening out and diverging to posterior projections of lorica. Foot three-jointed, not projecting beyond lorica; last foot-joint longest. Toes moderately long and acutely pointed.

Measurements: Lorica length 118; maximum width 76; anterior width (ventral sinus) 34; anterior width (dorsal sinus) 45; length foot-groove 35;

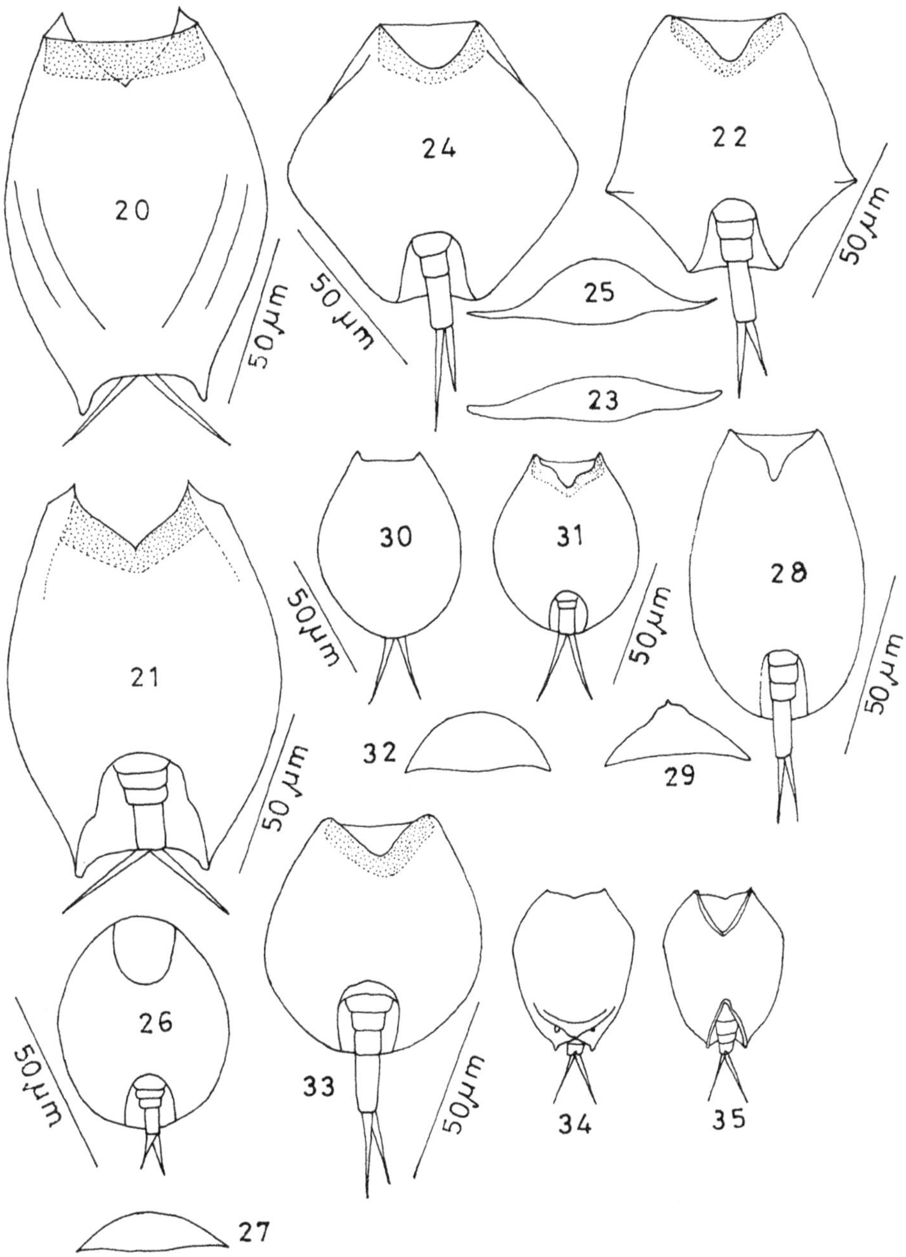

Figs. 20–35; 20, 21: *Lepadella nartiangensis* **new species,** dorsal view and ventral view; 22, 23: *Lepadella (H.) ehrenbergi* (Perty), ventral view, and cross-section; 24, 25: *Lepadella (H.) heterostyla* (Murray), ventral view and cross-section; 26, 27: *Lepadella aspida* Harring, ventral view and cross-section [after Sharma & Sharma, 1984]; 28, 29: *Lepadella imbricata* Harring, ventral view and cross-section [after Sharma, 1978]; 30–32: *Lepadella longiseta* Myers, dorsal view, ventral view and cross-section [after Sharma, 1980b]; 33: *Lepadella (H.) aspicora* Myers, ventral view [after Sharma, 1978]; 34, 35: *Lepadella kostei* Wulfert, dorsal view and ventral view [after Wulfert, 1966].

maximum width foot-groove 35; foot 28; toes 30.

Remarks: Broadly resembles with *L. latusinus* (Hilgendorf 1899) but differs from this species as: (i) lorica elongated; (ii) characteristic shape of the anterior end; (iii) distinct shape of the foot-groove.

Subgenus: *Heterolepadella* Bartos 1955

Lepadella (H.) ehrenbergi (Perty 1850) (Figs. 22, 23)

Material: Meghalaya, Assam, Nagaland

Characteristics: Lorica produced laterally into two broad triangular spines. Posterior end of lorica of the examined specimens broader than those reported from West Bengal.

Measurements: Lorica length 70–84; maximum width 72–84; anterior width 28–38; right toe 22–26, left toe 18–22.

Distribution: West Bengal.

Lepadella (H.) heterostyla (Murray 1913) (Figs. 24, 25)

Material: Meghalaya, Assam, Mizoram

Characteristics: Lorica rhomboidal, with its edges curving upwards from the blunt lateral angles. Anterior width variable in the studied material.

Measurements: Lorica length 78–82; maximum width 70–82; anterior width 26–38; right toe 24–26; left toe 18–22.

Distribution: West Bengal.

Notes on other Indian taxa

Of the other Indian taxa of this genus, *Lepadella aspida* (Figs. 26, 27), *L. imbricata* (Figs. 28, 29), *L. longiseta* (Figs. 30–32) and *L. (H.) aspicora* (Fig. 33) can be clearly distinguished. However, some other interesting taxa are commented on below.

Lepadella kostei Wulfert 1966 (Figs. 34, 35)

Material: Locality – Nimeta water works, Baroda (Gujarat).

Type specimens: not designated

Characteristics: Distinguishable by the general outline and the distinct shape of the posterior end.

Measurements: Lorica length 75; maximum width 56; toes 23.

Distribution: Gujarat.

Lepadella bicornis Vasisht & Battish 1971 (Figs. 36, 37)

Material: Locality – Govt. Fish Farm Tanks, Chandigarh (U.T.)

Type specimens: not designated.

Characteristics: Closely resembles with *L. (H.) heterostyla* but has toes of equal length. Specimens were not presently available for examination of the toes. Hence, the status of this species needs further confirmation.

Measurements: Lorica length 76; maximum width 78; toes 24–26.

Distribution: Chandigarh (U.T.).

Lepadella ovalis f. *larga* Sharma 1978 (Figs. 38–40)

L. ovalis f. *larga* **new form:** Sharma 1978, p. 85, Figs. 19–21.

Material: 2 specimens (on slides), Regd. Nos. P. 3021/1 & P. 3022/1, Zoological Survey of India, Calcutta. Loc.: Maheshtala, 24-Parganas (West Bengal). Date: 13.12.1976. Coll.: B. K. Sharma

Characteristics: Differs from *L. ovalis* in (i) lorica broader than its length; (ii) dorsal plate more arched; (iii) shape of posterior margin of lorica; and (iv) shape of the foot-groove.

Measurements: Lorica length 94; maximum width 100; anterior width 32; length of foot-groove 26; foot 32; toes 25.

Distribution: West Bengal.

Lepadella triprojectus Sharma 1978 (Figs. 41, 42)

L. triprojectus Sharma 1978, p. 87, Figs. 34–35.

Holotype (on slide): Regd. No. P. 3020/1, Zoological Survey of India, Calcutta. Loc.: Indian Museum Tank, Calcutta (West Bengal). Date: 14.8.1976. Coll.: B. K. Sharma.

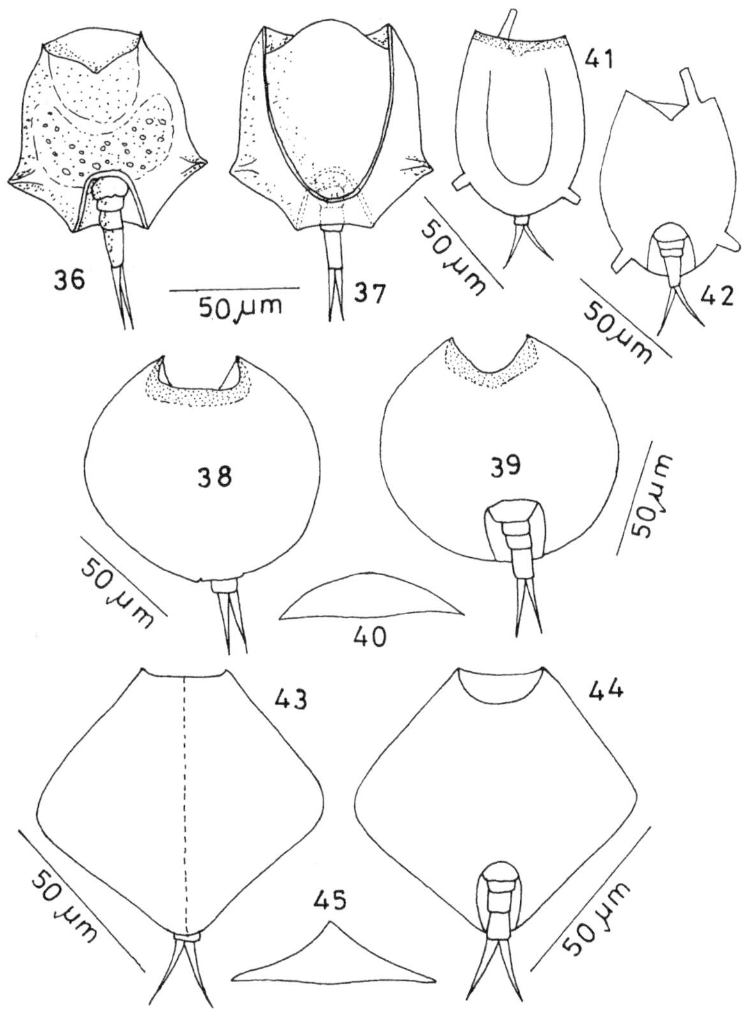

Figs. 36–45; 36, 37: ? *Lepadella bicornis* Vasisht & Battish, ventral view and dorsal view [after Vasisht & Battish, 1971]; 38–40: *Lepadella ovalis* f. *larga* Sharma, dorsal view, ventral view and cross-section [after Sharma, 1978]; 41,42: *Lepadella triprojectus* Sharma, dorsal view and ventral view [after Sharma, 1978]; 43–45: *Lepadella rhomboidula* (Bryce), dorsal view, ventral view and cross-section [after Sharma, 1978].

Characteristics: With three distinct tube-like projections. A vary rare species of this genus.

Measurements: Lorica length 70; maximum width 55; anterior width 38; anterior projection 15; postero-lateral projections 9–9; foot-groove 20; foot 26; toes 20.

Distribution: West Bengal.

Lepadella rhomboidula (Bryce 1890) (Figs. 43–45)

syn.: *L. triptera* f. *rhomboidula* (Bryce): Koste 1978, p. 192, T. 62, 5a–d.

Characteristics: Lorica broadly rhomboid, with a median dorsal keel. Koste (1978) treats this species as a form of *L. triptera.*

Measurements: Lorica length 72; maximum width 78; toes 21.

Distribution: West Bengal.

Discussion

The material examined from North-Eastern revealed ten species (11 taxa) of *Lepadella,* including

L. nartiangensis **new species** and *L. patella* f. *elongata* **new form.** In all, eighteen species (20 taxa) are documented from this country; this comprises only about 0.7% of the Indian Rotifera. *Lepadella* spp. from North-Eastern region show 58.3% similarity (vide Sorenson Index) to those reported from West Bengal. The taxonomic status of *L. bicornis* Vasisht & Battish needs to be ascertained.

A majority of the species studied show cosmopolitan distribution and comprise an important component of the periphytic rotifers. *L. ovalis* and *L. patella* are also found in plankton samples and tolerate a wide range of pH. *L. cristata, L. patella* f. *elongata, L. triptera* and *L. nartiangensis* were collected from acidic waters (pH 5.2–6.5). *L. acuminata, L. ehrenbergi, L. dactyliseta* and *L. heterostyla* were found in acidic to nearly alkaline waters (pH 5.5–7.2). *L. rhomboides* was found in habitats with pH ranging between 6.8–7.6.

L. ovalis and *L. patella* show a wide distribution in this region and in India as a whole. *L. rhomboides, L. triptera, L. heterostyla, L. nartiangensis, L. patella* f. *elongata* are very rare in the studied samples. *L. acuminata, L. dactyliseta* and *L. cristata* are fairly common in the material from paddy-fields of Meghalaya State. Of the other Indian taxa, *L. aspicora, L. aspida, L. imbricata, L. longiseta, L. rhomboidula, L. kostei, L. ovalis* f. *larga* and *L. triprojectus* are notable for their rare and restricted occurrence i.e., between 20°N–24°N. However, *L. aspida* was also collected from localities around 30°N. *L. kostei, L. triprojectus, L. ovalis* f. *larga, L. nartiangensis, L. patella* f. *elongata* represent endemic elements in the Indian fauna.

The distribution of *Lepadella* spp. in different States/Union Territories of India clearly indicates insufficient attention to the periphytic forms. West Bengal tops the list with the report of 12 species (13 taxa), followed by Meghalaya (10 species) and Assam (6 species) respectively. The samples from the paddy-fields of this region and particularly those from Meghalaya State indicated highest rotifer diversity/sample of 32–35 species which also included 6–8 species of genus *Lepadella.* Further, extensive collections from the paddy-fields of North-Eastern India especially those from hilly areas are likely to yield some more rare and interesting (even endemic) species of *Lepadella.*

Acknowledgements

One of the authors (BKS) is grateful to the Indian National Science Academy and Committee for Science and Technology in Developing Countries for awarding him an INSA-COSTED Travel Grant Fellowship for young scientists to enable him to participate in this symposium. The senior author also thanks the British Council who generously contributed towards his expenses in Edinburgh.

References

Edmondson, W. T. & G. E. Hutchinson, 1934. Report on Rotatoria. 9. Yale North India Expedition. Mem. Conn. Acad. Arts Sci. 10: 153–186.

Koste, W., 1978. Rotatoria, Monogonota. Die Rädertiere Mitteleuropas. Begründent von Max Voigt. Gebrüder Borntraeger, Stuttgart, Vols. 1 & 2, 673 pp. 234 pls.

Kutikova, L. A., 1970. Rotifer Fauna USSR. Fauna USSR. 104. Acad. Nauk. SSSR, Leningrad, 744 pp. (in Russian).

Naidu, K. V., 1967. A contribution to the Rotatorian fauna of South India. J. Bombay. nat. Hist. Soc. 64: 384–388.

Nayar, C. K. G., 1968. Rotifer fauna of Rajasthan. Hydrobiologia 31: 168–185.

Nayar, C. K. G. & K. K. N. Nair, 1969. A collection of brachionid rotifers from Kerala: Proc. Indian Acad. Sci. 69: 223–233.

Sharma, B. K., 1976. Rotifers collected from North-West India. Newsl. zool. Surv. India 2: 255–259.

Sharma, B. K., 1978. Contributions to the rotifer fauna of West Bengal. II. Genus Lepadella Bory de St. Vincent, 1826. Hydrobiologia 58: 83–88.

Sharma, B. K., 1980a. New records of Eurotatoria from Assam State, India. Bangladesh J. Zool. 8: 73–74.

Sharma, B. K., 1980b. Contributions to the rotifer fauna of Orissa, India. Hydrobiologia 70: 225–233.

Sharma, B. K. & Sumita Sharma (nee Das), 1984. A note on some Eurotatoria from Panjab State, India. Hydrobiologia 109: 279–282.

Sharma, B. K. & Sumita Sharma (in press). On some collections of monogonont rotifers (Rotifera: Eurotatoria) from Haryana State, India. J. Bombay nat. Hist. Soc.

Vasisht, H. S. & S. K. Battish, 1971. The rotifer fauna of North India. Lepadella and Colurella. Res. Bull. (N.S.) Panjab Univ. 22: 189–192.

Vasisht, H. S. & C. L. Gupta, 1967. The rotifer fauna of Chandigarh. Res. Bull. (N.S.) Panjab Univ. 18: 485–496.

Wulfert, K., 1966. Rotatorien aus dem Stausse Azwa und der Trinkwasser Aufbereitung der Stadt Baroda (Indien). Limnologica 4: 53–93.

Hydrobiologia 147: 23–29 (1987)
May, L., Wallace, R. & Herzig, A. (eds), Rotifer Symposium IV

The Rotifera of impoundments in Southeastern Australia

R. J. Shiel, C. J. Merrick & G. G. Ganf
Botany Department, University of Adelaide, Box 498 G.P.O. Adelaide, South Australia 5001

Keywords: Rotifera, species diversity, population dynamics, impoundments, storage retention-time

Abstract

Seasonal changes in species composition of rotifers in seven southeastern Australian impoundments are documented. Reservoirs had different dominant species and low overall community similarity. Rotifers were dominant only in unstable, short retention-time reservoirs. Reservoirs with longer retention-times had microcrustacean communities with calanoid copepods most abundant. Intensive sampling of Mt Bold Reservoir demonstrated extremely rapid changes in the rotifer community.

Introduction

Much of continental Australia is characterized by extreme periodicity of standing freshwaters and absence of natural freshwater lakes. Typical lacustrine zooplankton communities as described from the northern hemisphere [cf. Hutchinson, 1967] probably were rare prior to impoundment of many Australian rivers this century. The most common plankton assemblages accross southern Australia may have been those of billabongs (shallow lentic floodplain waters, cf. Shiel & Koste [1983]) or halophilic assemblages dominated by the cosmopolitan taxa *Brachionus plicatilis* and *Hexarthra fennica/H. jenkinae* [cf. Walker, 1973]. Damming major rivers provided lacustrine conditions where previously they did not occur, opening new habitats to plankton communities previously adapted to ephemeral conditions.

Replacement of opportunist rotifer species by microcrustacean zooplankton in the more stable conditions of River Murray reservoirs was noted by Shiel & Walker [1984], although species composition and seasonality of reservoir rotifer species were not described. The only study of the impact of rotifers in an Australian reservoir is that of Ganf *et al.* [1983], who described the apparent decimation of a *Volvox* population by *Ascomorphella volvocicola* in Mt Bold Reservoir (South Australia).

To determine the extent to which the rotifer fauna of relatively recent impoundments resembles the communities of floodplain waters which the reservoirs submerged, this paper analyzes the density, diversity and seasonality of the rotifer communities of seven reservoirs in southeastern Australia. To do this we have drawn on an earlier study of River Murray impoundments [Shiel, 1981] and our continuing study of plankton dynamics in Mt Bold Reservoir. Although detailed analyses of the Mt Bold plankton await completion of a more extensive project [Ganf, Merrick, unpubl. obs.], we include summary data here to contrast events in a pumped storage reservoir with those of R. Murray impoundments.

In the Australian context, an impoundment is formed behind an in-stream impediment to river flows, i.e. a dam or retaining wall. For political expediency (but not limnological precision) these impoundments are called 'Lakes', although they exhibit riverine features and may be subject to extreme seasonal drawdown, to the detriment of the biotic community. Reservoirs may be off-stream, not necessarily behind a dam, e.g. a natural or artificial depression, and may be filled by natural

Table 1. Reservoirs in southeastern Australia from which rotifer data are summarized. Most abundant rotifer species are listed. Mean species diversity is measured by the Shannon-Wiener index (H′).

Reservoir	Co-ordinates	Capacity (Ml)	Depth (m)	Retention time (yr)	n collections	n rotifer spp.	% littoral	Species comprising >20% of the plankton (seasonally)	H′
Burrinjuck	34°58′S/ 148°42′E	1.03×10^6	60	1	5	16	25	*Ascomorphella volvocicola* *Polyarthra dolichoptera* *Hexarthra mira*	1.24
Dartmouth	36°35′S/ 147°31′E	4.0×10^6	170	4	28	38	42	*P. dolichoptera* *Synchaeta* spp. *Lacinularia ismaeloviensis*	1.80
Hume	36°01′S/ 147°07′E	3.0×10^6	15	1	26	26	4	*Brachionus urceolaris* *Asplanchna priodonta* *Conochilus dossuarius*	1.99
Mulwala	36°01′S/ 146°05′E	117×10^3	14	<0.04	23	67	55	*Keratella cochlearis* *Synchaeta pectinata* *Polyarthra vulgaris* *C. dossuarius*	2.61
Eildon	37°11′S/ 144°54′E	3.4×10^6	76	2	11	22	9	*K. cochlearis* *P. dolichoptera* *P. vulgaris*	2.10
Menindee	32°20′S/ 142°20′E	1.8×10^6	5	>0.5	4	21	10	*Brachionus angularis* *Keratella australis* *Filinia pejleri*	2.76
Mt Bold	35°07′S/ 138°43′E	47×10^3	47	<0.5	1500	37	14	*A. volvocicola* *Synchaeta* spp. *A. priodonta* *H. mira*	1.95

flows or pumping, i.e. impoundments are reservoirs, but not necessarily vice versa. 'Storage' is a general term to encompass any stored water volume from tanks to large multipurpose impoundments.

Sampling sites

Seven reservoirs in southeastern Australia were sampled for this study (Fig. 1, Table 1). Five of these: Burrinjuck (Murrumbidgee River), Dartmouth (Mitta Mitta River) Hume and Mulwala (River Murray) and Eildon (Goulburn River) are in-stream impoundments on River Murray tributaries. Moveable barrages divert Darling River flows into shallow off-river storages at Menindee. Mt Bold (which in some years receives >90% of its

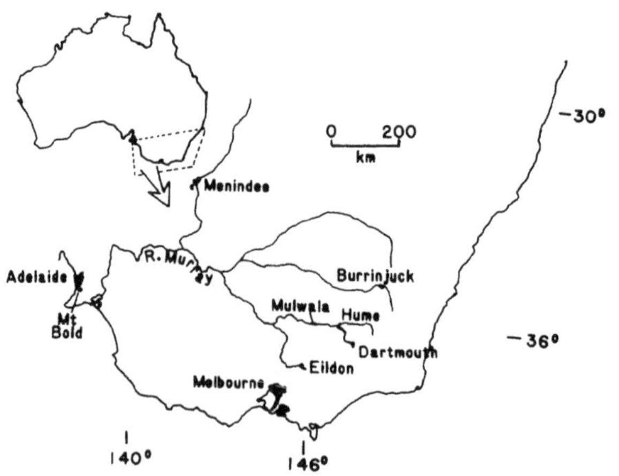

Fig. 1. Impoundments from which samples were collected for rotifers. All the impoundments, except Mt Bold Reservoir, are on Murray-Darling tributaries.

water by pumping ca. 80 km from the River Murray) is a small in-stream bulk-storage reservoir on the Onkaparinga River, South Australia [cf. Ganf, 1982].

Methods

The sampling methods used in River Murray reservoirs are described by Shiel *et al.* [1982]. Net tows and volume samples (53 μm-mesh) were collected monthly from Hume, Mulwala and Dartmouth, and seasonally from Burrinjuck, Eildon and Menindee. Mt Bold was sampled twice weekly September 1981–February 1982 and weekly 1982–1983 at 2 m intervals in the euphotic zone and 5 m intervals thereafter with a 13 litre trap fitted with a 40 μm-mesh stainless steel net. After collection, concentrated samples were preserved with 4% (v/v) formalin. Zooplankton numbers were estimated by subsampling.

Results and discussion

Table 1 reports total rotifer species recorded, dominant taxa, proportion of non-planktonic incursions and mean community diversity (H′) from each reservoir. Three of the storages had rotifer-dominated communities; Mulwala, with a 3–14 day retention time and strong influence of shallow morphometry and fringing weedbeds (hence a high proportion of littoral incursions in the 67 species recorded); Dartmouth, which was filling for the first time during the study and was inoculated from diverse submerged floodplain habitats, and Menindee, which was subject to the influence of Darling River flows with a riverine rotifer plankton [Shiel, 1985]. The other three Murray-Darling reservoirs were more stable; microcrustacean communities with calanoid copepods (*Boeckella* and/or *Calamoecia*) were numerically dominant. Rotifers rarely constituted >20% of zooplankton in these storages; when population pulses occurred they generally were short-lived (days). Extremely rapid fluctuations in rotifer species composition also were recorded in Mt Bold, and a comparable number of species to those of the Murray reservoirs, despite the greater intensity of sampling (Table 1).

Table 2. Relative similarity (CC values) of overall rotifer composition in seven reservoirs in southeastern Australia.

Dartmouth	Hume	Mulwala	Eildon	Menindee	Mt Bold	
0.14	0.10	0.15	0.15	0.13	0.23	Burrinjuck
	0.24	0.22	0.20	0.15	0.34	Dartmouth
		0.16	0.25	0.15	0.20	Hume
			0.19	0.16	0.30	Mulwala
				0.17	0.37	Eildon
					0.27	Menindee

Notably, species dominants and seasonality differed for each storage. Table 2 provides a measure of community similarity (Jaccard community coefficient $CC = \frac{c}{a + b - c}$, where a = no. of species in the first sample, b = no. of species in the second sample, and c = no. of species common to both). Of a total 107 rotifer species identified, only one, *Conochilus dossuarius*, occurred in all seven reservoirs, despite their relatively close (latitudinal) proximity. Three species (*Polyarthra vulgaris, Pompholyx complanata* and *Filinia longiseta*) occurred in six reservoirs, five species (*Keratella australis, K. cochlearis, K. procurva, Synchaeta pectinata* and *Asplanchna brightwelli*) in five reservoirs and eight species (*K. slacki, Euchlanis dilatata, Lecane luna, S. oblonga, S. stylata, Hexarthra intermedia* and *H. mira*) occurred in four reservoirs. With the exception of the endemics *K. australis* and *K. slacki*, these rotifers apparently are cosmopolitan, widely-tolerant taxa. Of the remaining species, 63 were recorded from only one habitat and 27 from two or three of the storages, i.e. >80% of the Rotifera were of relatively restricted distribution, accounting for the generally low community similarity evident in Table 2. Somewhat higher values for Mt Bold reflect the intensive sampling effort there.

Seasonality

Figure 2 shows the sequential appearance of domi-

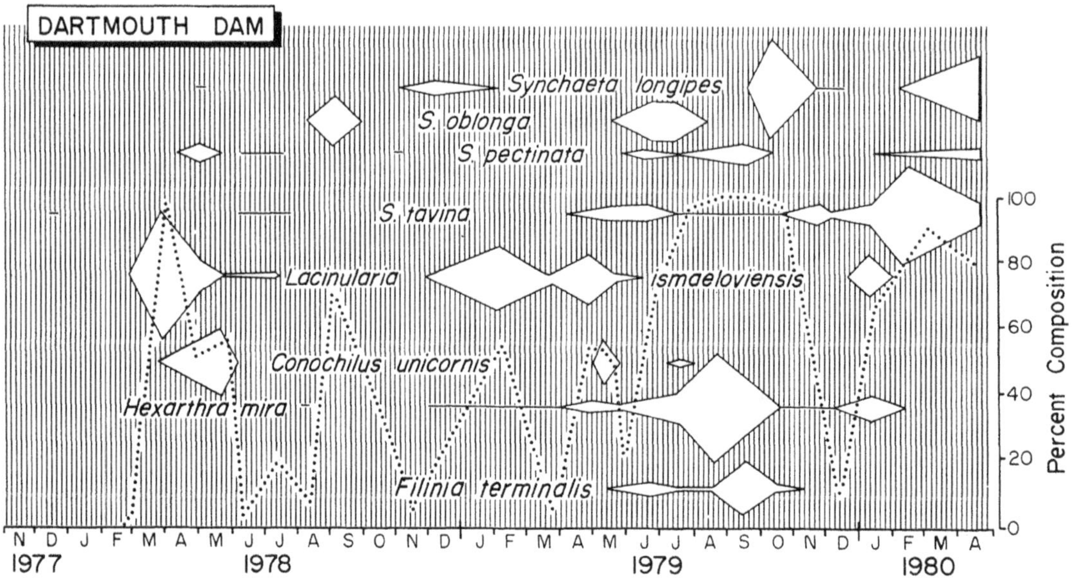

Fig. 2. Seasonality of dominant rotifer species in Dartmouth Reservoir 1977–1980. Rotifers as % of zooplankton.

nant rotifers in Dartmouth following closure of the dam in late 1977. Two colonial species (*Lacinularia ismaeloviensis* and *Conochilus dossuarius*) were the first limnetic rotifers to appear, and both occurred subsequently in summer-autumn. Four species of *Synchaeta* occurred sequentially: *Synchaeta pectinata* (autumn), *S. oblonga* (winter), *S. longipes* (spring) and *S. tavina* (perennial with a spring-summer peak). *Hexarthra mira* was perennial in the reservoir with a winter maximum, while the cold stenotherm *Filinia terminalis* appeared in winter. The dominant rotifers of Hume and Mulwala are compared for 1976–1978 in Fig. 3. Rotifers were a minor component of the Hume limnoplankton, only three species were important (*Brachionus urceolaris, Asplanchna priodonta, C. dossuarius*). In Mulwala the dominant rotifers (11 species) either were perennial or had asynchronous population peaks. The planktonic community apparently was flushed out by higher through-flows in winter 1977.

Sampling of the other Murray-Darling reservoirs was seasonal and insufficient for analysis of community changes. In one of these, Burrinjuck, the highest density of rotifers recorded from the Murray-Darling storages occurred in summer 1977, when *A. volvocicola* reached >2000 ind. l^{-1} in a *Volvox* bloom [Shiel, 1981]. Otherwise, rotifer densities rarely exceeded 200 ind. l^{-1}.

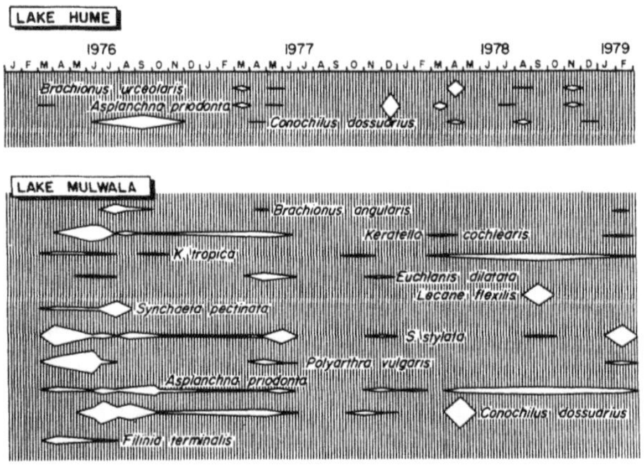

Fig. 3. Comparison of the rotifers (1976–1979) of Hume (retention time ≃1 yr), and Mulwala (retention time ≃2 weeks).

The Mt Bold rotifer community was diverse in the early part of the sampling programme, with up to 16 spp. occurring at all depths in the water column (40 m). With the exception of *Hexarthra mira*, which tended to concentrate at or about the thermocline, the dominant rotifers were most abundant in the top 2 m of the euphotic zone. *H. mira* also was the only species recorded perennially in Mt Bold; a typical depth profile for the species, e.g.

23.10.81: none at the surface, 0.3 ind. l^{-1} (2 m), 0.8 ind. l^{-1} (4 m), 2.5 ind. l^{-1} (6 m), 7 ind. l^{-1} (8 m), 10 ind. l^{-1} (10 m), decreasing thereafter to 3.6 ind. l^{-1} (35 m). Figure 4c compares the number of species in the water column (solid bars) with the number recorded at 2 m (open bars). In general, a single sample contained most of the taxa present in the whole column; only rarely were more than 3 species absent.

The relatively short sampling interval in Mt Bold highlighted the rapidity with which the rotifer community changed, and suggests that the details of community fluctuations were missed by monthly sampling in the other Murray-Darling reservoirs. For example, the spring 1981 bloom of *Synchaeta* in Mt Bold was not apparent on November 23, when two individuals of *S. stylata* occurred at 10 m. On November 27 *S. stylata/S. pectinata* reached ca. 24000 ind. l^{-1} at the surface, with >2000 ind. l^{-1} (10 m) and >1000 ind. l^{-1} (25 m). However, on November 30 only a few *Synchaeta* occurred in the top 2 m (2 ind. l^{-1}) and were not present below that depth. During this time the reservoir was almost full, throughflow at this time was minimal, and there were no abrupt changes in any of the parameters being monitored which would account for the population decrease (Fig. 4).

During this 'bloom' (Fig. 4b) overall community

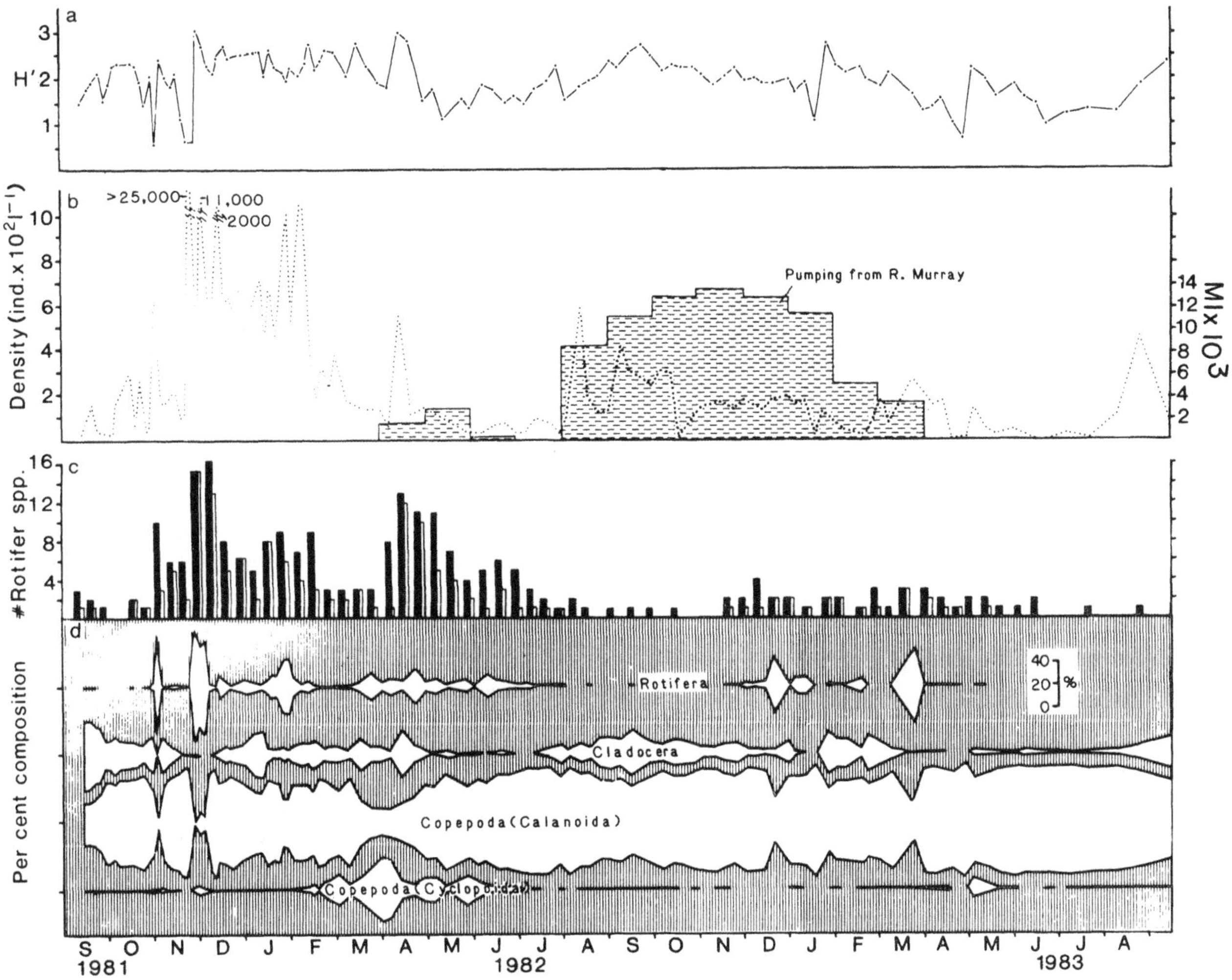

Fig. 4. Plankton dynamics in Mt Bold Reservoir. a, Community diversity, Shannon-Wiener H′. b, Total plankton density (at 2 m depth) and period of pumping into the reservoir from the River Murray. c, Total number of rotifer species in water column (solid bars) and at 2 m depth. d, Relative contribution of microcrustacea and rotifers to the plankton community.

diversity (Fig. 4a) decreased due to dominance by a single species. The rotifer population peaks and decline in diversity evident in Figs. 4a – d were due to *Synchaeta stylata* (99%, November 1981), *Keratella cochlearis/K. tropica* (83%, December 1981), *Conochilus dossuarius* (47%, January 1982), *Ascomorphella volvocicola* (58%, December 1982), and *Asplanchna priodonta* (65%, March 1983). The devastating influence of high rotifer density on the phytoplankton community is demonstrated by the interaction of *Ascomorphella* and *Volvox* [Ganf *et al.*, 1983].

In the second year of the Mt Bold study community events were complicated by commencement of pumping River Murray water into the Onkaparinga River upstream of the reservoir. Rotifers appeared most susceptible to changes in water quality incurred by pumping; both density and diversity of rotifers declined, leaving microcrustacean dominants. Further, the spring rotifer bloom was not seen the following year (1982). Only *H. mira* persisted through the period, with sporadic appearances of *Keratella* spp., a brief pulse of *A. volvocicola* (December 1982) and *Asplanchna priodonta* (March 1983). Increases in zooplankton populations after the onset of pumping (April/August 1982, Fig. 4b) were due to microcrustaceans (*Bosmina meridionalis* and copepods). The decline in rotifer populations in Mt Bold is the reverse of the rotifer response in a pumped storage [Potter & Meyer, 1982]. It may be significant that pumped River Murray water is chlorinated in the pipelines prior to release into domestic supply reservoirs [Garman, 1983]. Extremely high levels of trihalomethanes (200 – 300 $\mu g\ l^{-1}$) may be produced, and although these may dissipate rapidly, the levels and fate of less volatile chlorinated hydrocarbons reaching the reservoir are unknown. Rotifers may be more susceptible than the microcrustacean community to these substances. Further study clearly is warranted.

Conclusions

Apart from seasonal incursions of littoral taxa, Rotifera of the seven impoundments studied are not the same taxa abundant in the shallow billabongs of nearby floodplains [Shiel & Koste, 1983]. Although the billabong communities undoubtedly provided colonizing populations to the filling reservoirs, these did not persist in the altered environment. With the exception of *Keratella australis*, a eurytopic pancontinental endemic, dominant species of impoundments are cosmopolitan eurytopic or warm-stenothermal taxa. Unpublished data from a study of a tropical reservoir [Dr P. Hawkins, pers. comm.] indicates that the same taxa are perennial in northern Australia (e.g. *K. cochlearis, K. tropica, P. vulgaris, C. dossuarius*). Synchrony of seasonal occurrence is less distinct than reported from studies in the northern hemisphere, perhaps because environmental cues also are less distinct. Perhaps impoundments with short retention times have species whose short life cycles are adaptive in habitats with greater variability. Events in Mt Bold suggest that the rotifer community is highly sensitive to manipulation of flows and/or water quality.

Acknowledgements

Sampling in the Murray-Darling system comprised part of a Ph.D. programme at the University of Adelaide (RJS). Thanks are due to K. F. Walker, W. D. Williams for supervision, T. J. Hillman (Albury) and I. J. Powling (Melbourne) for collections and Ruth Evans for redrawing Figs. 2 & 3. The Mt Bold data were derived in part from a continuing Ph.D. programme at the university of Adelaide (CJM), supervised by G. G. Ganf, and in part from research funded by the Australian Research Grants Scheme. The Engineering and Water Supply Department, S. Aust. provided access to Mt Bold Reservoir and data on pumping. We thank W. Koste (W. Germany) for his help with rotifer identifications.

References

Ganf, G. G., 1982. Influence of added nutrients on the seasonal variation of algal growth potential in Mt Bold Reservoir, South Australia. Aust. J. Mar. Freshwat. Res. 33: 475 – 490.

Ganf, G. G., R. J. Shiel & C. J. Merrick, 1983. Parasitism: the possible cause of the collapse of a Volvox population in Mt Bold Reservoir, South Australia. Aust. J. Mar. Freshwat. Res. 34: 489–494.

Garman, D. E., 1983. Water quality issues in Australia. In: Water 2000: Consultants Rept. 7. Aust. Govt Publ. Service, Canberra: 1–265.

Hutchinson, G. E., 1967. A Treatise on Limnology. Vol. 2. Wiley, N.Y. 1115 pp.

Potter, D. U. & J. L. Meyer, 1982. Zooplankton communities of a new pumped storage reservoir. Wat. Res. Bull. 18: 635–642.

Shiel, R. J., 1981. Plankton of the Murray-Darling river system. Ph.D. thesis, University of Adelaide. 287 pp.

Shiel, R. J., 1985. Zooplankton of the Darling River system, Australia. Verh. Internat. Verein Limnol. 22: 2136–2140.

Shiel, R. J. & W. Koste, 1983. Rotifer communities of billabongs in northern and southeastern Australia. Hydrobiologia 104: 41–47.

Shiel, R. J. & K. F. Walker, 1984. Zooplankton of regulated and unregulated rivers: the Murray-Darling River system, Australia. In: A. Lillehammer & S. J. Saltveit (eds), Regulated Rivers. University of Oslo Press, pp. 263–270.

Shiel, R. J., K. F. Walker & W. D. Williams, 1982. Plankton of the lower River Murray, South Australia. Aust. J. Mar. Freshwat. Res. 33: 301–327.

Walker, K. F., 1973. Studies on a saline lake ecosystem. Aust. J. Mar. Freshwat. Res. 24: 21–71.

Hydrobiologia 147: 31–43 (1987)
May, L., Wallace, R. & Herzig, A. (eds), Rotifer Symposium IV

Tasmanian rotifera: Affinities with the Australian fauna

Walter Koste[1] & Russell J. Shiel[2]
[1]*Ludwig-Brill-Strasse 5, Quakenbrück D-4570, Federal Republic of Germany*
[2]*Botany Dept, University of Adelaide, Box 498, G.P.O. Adelaide, S. Aust. 5001*

Keywords: Rotifera, Tasmania, new species, biogeography

Abstract

Plankton was collected from more than 100 lentic and lotic habitats in Tasmania. Of the 131 rotifer taxa identified, 126 are first records for the island, and 9 are first records for Australia. Three new species, *Aspelta tilba,* sp. nov., *Lepadella tyleri* sp. nov., and *Testudinella unicornuta* sp. nov. are described and figured. *Horaëlla brehmi*, which has a disjunct distribution (equatorial regions and southeastern Australia), also occurs in Tasmanian waters. The morphology and development of this rare species are discussed. Geographical affinities of the Tasmanian assemblage are considered. There appears to be a significant tropical component in the Tasmanian Rotifera.

Introduction

In recent years, there has been a substantial increase in information concerning the rotifer fauna of Australia [Ganf *et al.,* 1983; Shiel & Koste, 1986; Koste & Shiel 1986a; Shiel *et al.,* in press]. Six hundred and twenty taxa are now known from Australia [Shiel & Koste, 1985; Koste & Shiel, 1986b]. However, compared to the northern hemisphere, our knowledge of the Australian rotifer fauna is poor. Only a small proportion of mainland Australia has been survyved, and it seems likely that only one-third to one-half of the rotifers which occur here have been recorded. Nevertheless, marked latitudinal (and to a lesser extent longitudinal) differences in species composition are evident, and a substantial number of endemic taxa have been identified.

Tasmania, the southernmost Australian state, deserves particular attention [Dumont, 1983]. This island has abundant, perennial, standing freshwaters, relatively high rainfall and an equable climate, and yet only three reports of rotifers from this area have been published [Sudzuki, 1967, 1985; De Deckker & Williams, 1982]. The present study represents the first extensive survey of Tasmanian Rotifera. Three new species are described and figured, and biogeographical affinities of the Tasmanian fauna are considered.

Study sites and methods

Tasmania (Fig. 1) is an island of approximately 55 000 km^2 separated from the mainland by a 240 km wide strait. The island has a temperate maritime climate [Williams, 1974] with considerably higher rainfall than the mainland (50–250 cm yr^{-1}). Standing fresh waters are abundant. On the Central Plateau there are several thousand lakes of up to 150 km^2, and larger impoundments have been created in the southwest by damming. These waterbodies generally are moderate in temperature (0–25 °C) and mostly very dilute (< 50 ppm total dissolved solids) [Tyler, 1974].

Samples were collected on two occasions (8–13 May 1980, 21–30 April 1984) from as wide a range of habitats as possible. These included cloud-forest

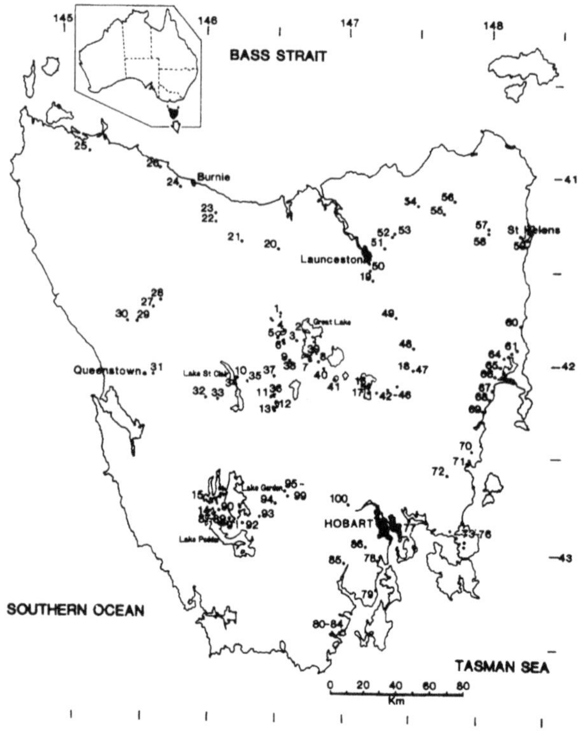

Fig. 1. Sampling site in Tasmania. Numbers refer to sampling stations.

moss-squeezings, roadside ditches, stock dams, natural lakes and large impoundments. The sampling sites are shown in Fig. 1. Plankton samples usually were taken at a single point by towing a 50 μm net near the shore (water depth). Two or three sites were sampled in larger lakes. Quantitative samples are were collected with a 30 ℓ plankton trap from some waterbodies. Small pools were sampled by pouring 20 – 30 ℓ of water through a 35 μm mesh stainless-steel net. All samples were concentrated to 120 mℓ, preserved in 4% (v/v) formalin, and returned to the laboratory in Whirl-Pak sample bags.

Some physical and chemical data were collected during this study. Temperature was measured at 30 cm depth with a hand-held mercury-in-glass thermometer. A portable pH meter (Radiometer 29) was used to measure pH. Water samples (200 mℓ) were returned to the laboratory, where conductivity and transparency were determined using a Radiometer CDM2e meter and a Hach 2100A turbidimeter, respectively.

On return to the laboratory the rotifers were identified using LP (60×) and HP (500×) (Olympus dissecting microscope and Zeiss Research microscope respectively). At least 200 individuals were counted from random subsamples and the Shannon-Wiener information theory index (H′) [Shannon, 1948] was calculated for each site.

Table 1. Rotifera recorded from Tasmania.

BDELLOIDEA	
1.	*Dissotrocha macrostyla* (Ehrbg)
2.	*Mniobia scarlatina* (Zelinka)
3.	Unidentified spp.
MONOGONONTA	
Fam. Epiphanidae	
4.	*Proalides (L.) subtilis* (Rodewald)
5.	*Epiphanes brachionus* (Ehrbg)
6.	*E.* cf. *clavulata* (Ehrbg)
Fam. Brachionidae	
7.	*Brachionus angularis* Gosse
8.	*B. calyciflorus* Pallas
9.	*B. lyratus* Shephard
10.	*B. quadridentatus melheni* Barrois & Daday
11.	*Keratella australis* (Berzins)
12.	*K. cochlearis* (Gosse)
13.	*K. hispida* (Lauterborn)
14.	*K. javana* Hauer
15.	*K. procurva* (Thorpe)
16.	*K. proc. robusta* Koste & Shiel
17.	*K. quadrata* (Muller)
18.	*K. slacki* (Berzins)
19.	*K. tropica* (Apstein)
20.	*K. valga* (Ehrbg)
21.	*Anuraeopsis fissa* (Gosse)
22.	*A. navicula* Rousselet
23.	*Notholca* cf. *squamula* (Muller)
Fam. Euchlanidae	
24.	*Manfredium eudactylotum* (Gosse)
25.	*Euchlanis deflexa* (Gosse)
26.	*E. deflexa larga* Kutikova
27.	*E. dilatata* Ehrbg
28.	*E. dilatata lucksiana* Hauer
29.	*E. meneta* Myers
Fam. Mytilinidae	
30.	*Mytilina ventralis* (Ehrbg)
31.	*Lophocharis salpina* (Ehrbg)
Fam. Trichotridae	
32.	*Wolga spinifera* (Western)
33.	*Trichotria tetractis* (Ehrbg)
Fam. Colurellidae	
34.	*Colurella adriatica* Ehrbg
35.	*C. colurus* (Ehrbg)
36.	*C. uncinata* (Muller)
37.	*C. uncinata bicuspidata* (Ehrbg)
38.	*C. uncinata deflexa* (Ehrbg)

Table 1 (cont.).

39. *Lepadella amphitropis* Harring
40. *L. patella* (Müller)
41. *L. rhomboides* (Gosse)
42. **L. tyleri* n. sp.

Fam. Lecanidae

43. *Lecane bulla* (Gosse)
44. *L. closterocerca* (Schmarda)
45. *L. flexilis* (Gosse)
46. **Lecane glypta* Harring & Myers
47. **L. haliclysta* Harring & Myers
48. *L. hamata* (Stokes)
49. *L. inopinata sympoda* (Hauer)
50. *L. lamellata thalera* (Harring & Myers)
51. *L. luna* (Müller)
52. *L. lunaris* (Ehrbg)
53. *L. ohioensis* (Herrick)
54. *L. perplexa* (Ahlstrom)
55. *L. quadridentata* (Ehrbg)
56. *L. scutata* (Harring & Myers)
57. *L. stichaea* Harring
58. **L. tasmaniensis* Koste & Shiel

Fam. Proalidae

59. *Proales* cf. *decipiens* (Ehrbg)

Fam. Lindiidae

60. *Lindia torulosa* Dujardin

Fam. Notommatidae

61. *Tetrasiphon hydrocora* Ehrbg (eggs only)
62. *Scaridium longicaudum* (Muller)
63. **Monommata* cf. *dentata* Wulfert
64. *M. grandis* Tessin
65. *Monommata* sp.
66. *Taphrocampa selenura* Gosse
67. *Eosphora* cf. *anthadis* Harring & Myers
68. *Notommata aurita* (Muller)
69. *Cephalodella catellina* (Muller)
70. *C. exigua* (Gosse)
71. *C. forficula* (Ehrbg)
72. *C. gibba* (Ehrbg)
73. *C.* cf. *nana* Myers
74. *C. ventripes* Dixon-Nuttall

Fam. Trichocercidae

75. *Trichocerca brachyura* (Gosse)
76. *T. cavia* (Gosse)
77. *T. capucina* Wierz. & Zach.
78. *T.* cf. *collaris* (Rousselet)
79. *T. intermedia* (Stenroos)
80. *T. longiseta* (Schrank)
81. *T. myersi* (Hauer)
82. *T. porcellus* (Gosse)
83. *T. pusilla* (Jennings)
84. *T. rattus* (Muller)
85. *T. similis* (Wierzejski)
86. *T. tenuior* (Gosse)
87. *T. tigris* (Muller)

Fam. Gastropodidae

88. *Ascomorpha ecaudis* (Perty)
89. *A. saltans* Bartsch

Fam. Synchaetidae

90. *Synchaeta longipes* (Gosse)
91. *S. oblonga* Ehrbg
92. *S. pectinata* Ehrbg
93. *S.* cf. *tremula* (Muller)
94. *Synchaeta* sp.
95. *Polyarthra dolichoptera* (Idelson)
96. *P. remata* (Skorikow)
97. *P. vulgaris* Carlin
98. *Ploesoma lenticulare* (Herrick)

Fam. Asplanchnidae

99. *Asplanchna sieboldi* (Leydig)

Fam. Dicranophoridae

100. *Dicranophorus hauerianus* Wiszniewski
101. *D. hercules* var. *intermedia* Wiszniewski
102. *D. robustus* Harring & Myers
103. **Aspelta tilba* n. sp.
104. **Encentrum* cf. *cruentum* Harring & Myers

Fam. Microcodinidae

105. *Microdon clavus* Ehrbg

Fam. Testudinellidae

106. *Testudinella amphora* Hauer
107. *T. husseyi* Koste & Shiel
108. **T. unicornuta* n. sp.
109. *Pompholyx complanata* Gosse
110. *P. sulcata* (Hudson)

Fam. Flosculariidae

111. *Limnias ceratophylli* Schrank
112. *L. melicerta* Weisse
113. *Lacinularia flosculosa* (Muller)

Fam. Conochilidae

114. *Conochilus dossuarius* (Hudson)
115. *C. natans* (Seligo)
116. *C. unicornis* Rousselet

Fam. Hexarthridae

117. *H. oxyuris* (Sernov)
118. *Hexarthra mira* (Hudson)

Fam. Filiniidae

119. *Filinia australiensis* Koste
120. *F. longiseta* (Ehrbg)
121. *F. opoliensis* (Zacharias)
122. *F. pejleri* Hutchinson
123. *F. pejleri grandis* Koste
124. *F. terminalis* (Plate)

Fam. Trochosphaeridae

125. *Horaëlla brehmi* Donner

Fam. Collothecidae

126. *Collotheca campanulata* (Dobie)
127. **C. campanulata longicaudata* (Hudson)
128. *C. ornata* (Ehrbg)
129. *C. pelagica* (Rousselet)
130. *C. trilobata* (Collins)
131. *Stephanoceros fimbriatus* (Goldfuss)

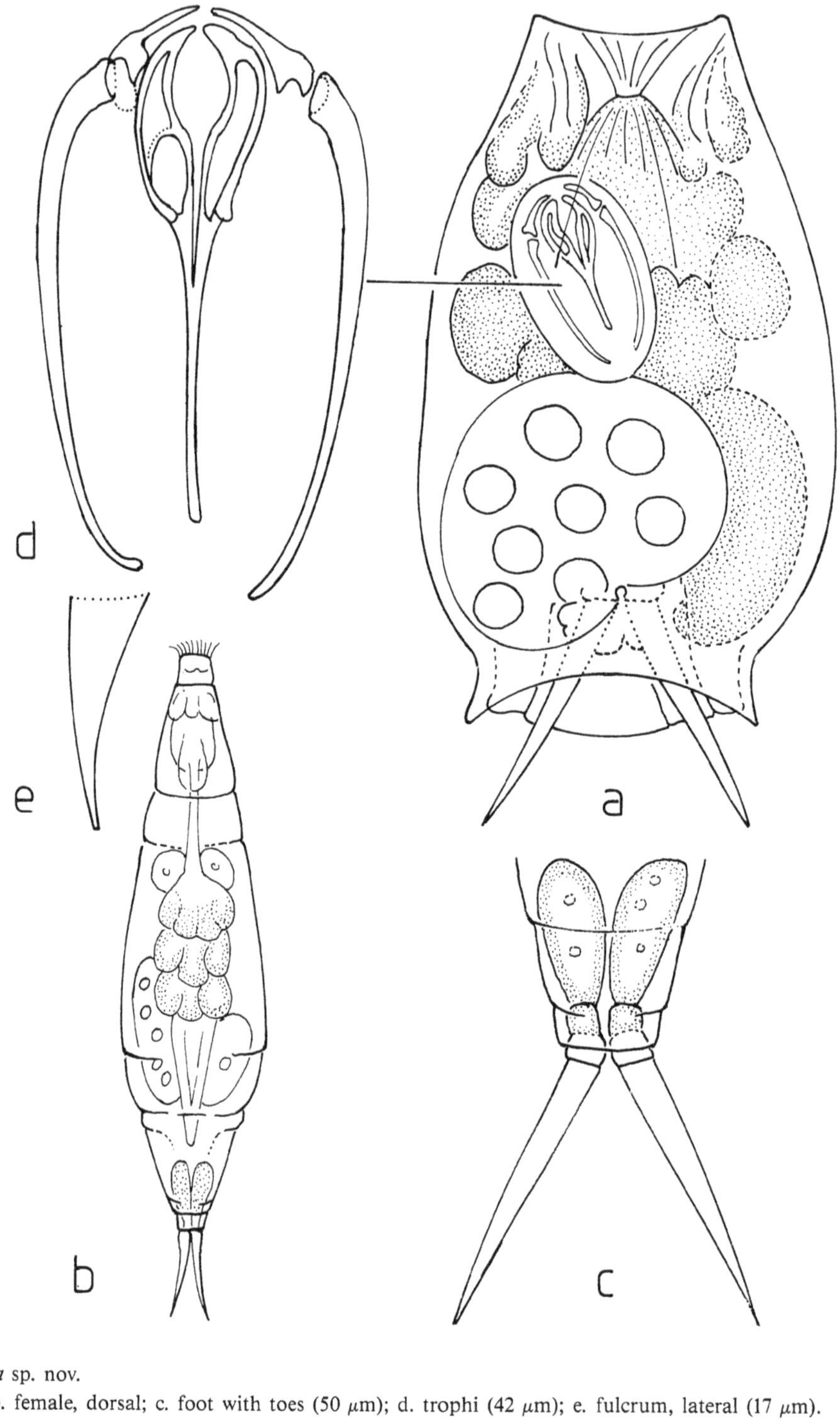

Fig. 2. Aspelta tilba sp. nov.
a. female, ventral; b. female, dorsal; c. foot with toes (50 μm); d. trophi (42 μm); e. fulcrum, lateral (17 μm).

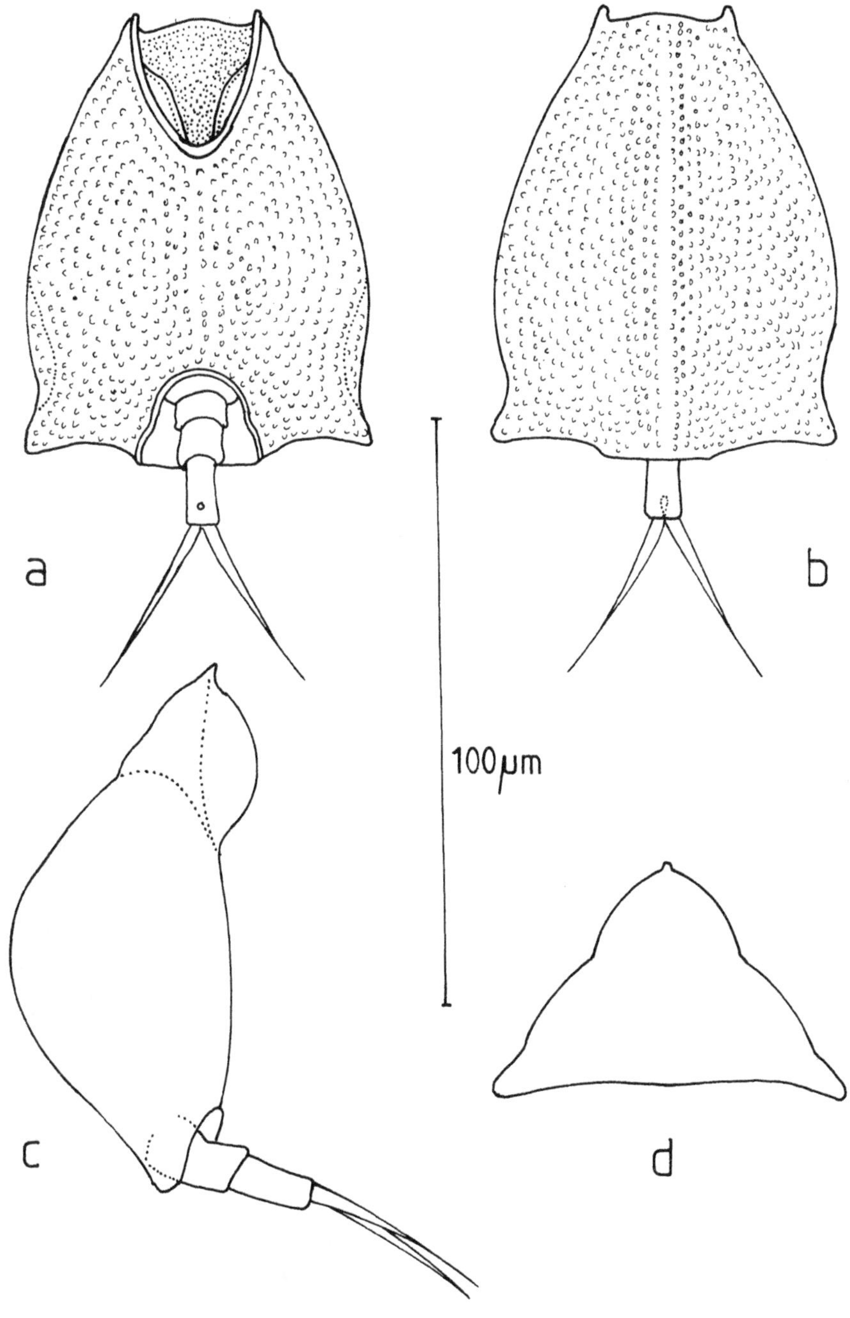

Fig. 3. Lepadella tyleri sp. nov.
a. ventral; b. dorsal; c. lateral; d. cross-section.

Systematic list

All taxa, apart from some specimens of Bdelloidea which were contracted in the preservative and unidentifiable, are recorded in Table 1. *Keratella tropica, Keratella valga, Polyathra dolichoptera, Polyarthra vulgaris* and *Pompholyx complanata* were recorded from Tasmania by Sudzuki [1967]. The remaining species are first records from the island. At least 132 taxa are now known from Tasmania, including *Brachionus plicatilis,* which was not recorded in our survey [De Deckker & Williams, 1982]. First records for Australia are indicated by an asterisk (*) in Table 1.

Descriptions of new species. Three new species are described here: *Aspelta tilba* sp. nov., *Lepadella tyleri* sp. nov., and *Testudinella unicornuta* sp. nov. *Lecane tasmaniensis* Koste & Shiel and *Testudinella husseyi* Koste & Shiel are described by Shiel & Koste, 1985.

Aspelta tilba Koste & Shiel sp. nov. (Figs. 2a–3)

Type material: Four contracted females in formalin (sample no. 1056), collected 8.0v.80 by R. J. Shiel, Botany Department, University of Adelaide. Holotype and paratypes are not designated because the available material was prepared for trophi analysis. The slide is lodged with the South Australian Museum, Adelaide (SAM) V 4024.

Type locality: Cameron's Lagoon (42°00′S/146°44′E), south of Great Lake, Tasmania (locality 7, Fig. 1).

Description: Soft, thin integument; head long; well-marked neck fold (Fig. 2a); body moderately elongated; rostrum prominent; long cilia visible dorsally (Fig. 2b); foot conical (Fig. 2c); toes end in long tips. Forcipate mastax; uncus asymmetric, rami narrow basally (Fig. 2d). The left ramus is stout, with an elongated inner opening; the right has a smaller oval opening. Outside borders of the rami thickened and knobbed at the base, the left larger than the right. Fulcrum long, with a slightly curved triangular lamellar plate (Fig. 2e). Internal organization without peculiarities.

Dimensions: Total length 350 μm, toes 50 μm, trophi 43 μm, manubria 37/41 μm, rami 19/22 μm, fulcrum 17 μm, unci 12/10 μm.

Discussion: This species resembles *Aspelta aper* [Harring, 1913] and *Aspelta egregia* Myers 1936 in its habit. Both also have long toes and similar trophi length. The moderately elongated straight toes of *A. tilba* are distinctive (ventrally curved in *A. aper,* parallel-sided ending in acute tips in *A. egregia*). Size and shape of trophi elements also are distinctive. Both *A. aper* and *A. egregia* have a lamellar plate under the rami which is absent in the new species; their unci are highly asymmetric and shorter (25–38 μm), whereas the unci of *A. tilba* are nearly symmetrical (43 μm). The outside borders of the rami also are thickened and knobbed at the base.

Etymology: The name *Tilba* is derived from an aboriginal phrase tilba-tilba meaning 'place of many waters'.

Lepadella tyleri Koste & Shiel sp. nov. (Figs. 3a–d, 4)

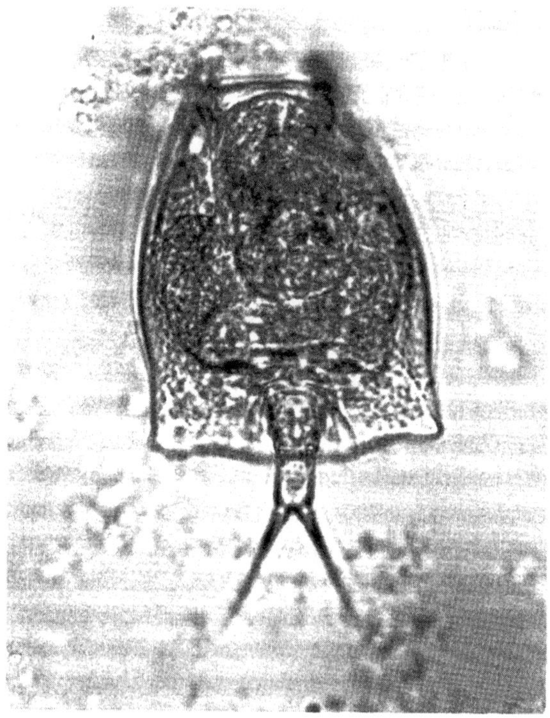

Fig. 4. Photomicrograph of *Lepadella tyleri,* in ventral aspect.

Type material: Two contracted females in formalin (sample No. 1155).

Holotype: Loricate female, sample No. 1155, collected 25.iv.84 by R. J. Shiel, Botany Department University of Adelaide. SAM V 4022.

Paratype: SAM V 4023. Date and place as for holotype.

Type locality: Stock dam, near turnoff at Mayfield Jetty Rd, 19 km south of Swansea, Tas. 42°19′S/148°01E (locality 69 in Fig. 1).

Description: Rigid lorica of trapezoidal outline (Fig. 3a, b, 4). Cross-section nearly triangular (Fig. 3d). Surface of both plates finely pustulated. Dorsal plate with median keel. Foot-opening with relatively broad base (Fig. 3a, c). Head aperture deep with lateral lamella. Caudal lorica with slight constriction above posterior, blunt lateral corners. Three foot segments, last with caudal antenna. Toes very thin, elongated and pointed.

Measurements: Total length 123–125 μm; lorica length 78–80 μm; greatest width 58 μm; head aperture width 21 μm, depth 22 μm; lorica height 36 μm; terminal foot segment 12 μm; toes 32 μm; foot-opening 18 × 18 μm.

Discussion: In cross-section this species appears to belong to the *Lepadella triptera* group. Different forms of this group were discussed by Koste [1981: Figs. 4–7]; *L. tyleri* somewhat resembles *Lepadella minorui.* The sp. nov. has the following morphological characteristics which are new to the taxonomic group: only a weak prolongation of the dorsal caudal lorica at the broad foot-opening base; trapezoidal outline of the lorica with caudal blunt corners; deep head aperture with lateral lamella (Fig. 3c: la) and absence of a notch at the dorsal anterior border.

Etymology: Lepadella tyleri sp. nov. is dedicated to Dr P. A. Tyler, University of Tasmania, with gratitude for his interest and support during fieldwork.

Ecology: L. tyleri was collected from open water in a small turbid farm stock dam (no emergent macrophytes) at 13° C, pH 6.0, transparency 82 NTU, conductivity 390 μs cm^{-1}.

Testudinella unicornuta Koste & Shiel sp. nov. (Fig. 5a–c, 6a, b).

Type material: Twenty eight contracted females in sample No. 1176, two in sample No. 1178, both in formalin.

Holotype: Loricate female, sample No. 1176 collected 28.iv.84 by R. J. Shiel, Botany Department, University of Adelaide. SAM V 4025.

Paratypes: 3 specimens in Koste collection, Gen. *Testudinella.* Date and place as for holotype.

Type locality: Roadside pool, Strathgordon road 0.5 km east of McPartlan Pass, Tas. 42°48′S/146°24′E (locality 90, Fig. 1).

Description: Lorica pear-shaped in lateral and dorsal view (Fig. 5a, 6a); dorsum convex, arched; ventral plate concave. Transverse section of midlorica dorsally concave, greatly depressed, ventrally convex (Fig. 5c). Foot-opening nearly triangular, situated at posterior margin of lorica within a short tube, continuing by a semi-circular structure at the end of ventral plate. Dorsal margin of head aperture with flat tongue, ventral margin with two symmetrical curves which connect behind closure folds of head-opening. At this point is a single sharply-pointed median horn (Fig. 5a, b, 6b: t) unknown in the genus. Lateral antenna posterior to middle, nearly 72 μm from anterior point of lorica.

Measurements: Total lorica length 126–130 μm; lorica width 80–86 μm; anterior margin width 36–39 μm; foot-opening 18 × 18 μm.

Ecology: Testudinella unicornuta was collected from open water in a peaty shallow pool; pH 6.2; transparency 1.5 NTU; conductivity 80 μs cm^{-1}.

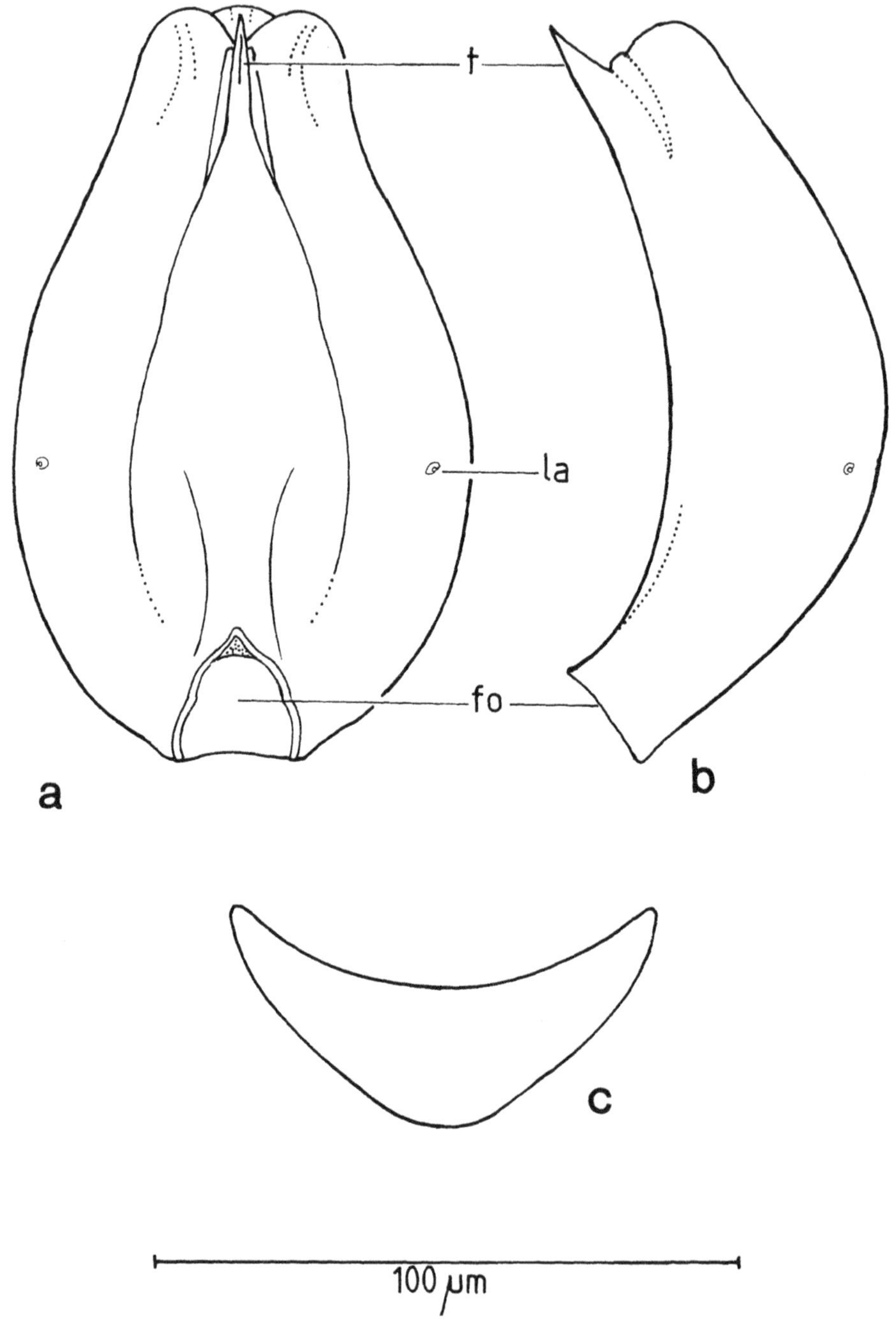

Fig. 5. Testudinella unicornuta
a. sp. nov.
a. dorsal; b. lateral; c. cross-section

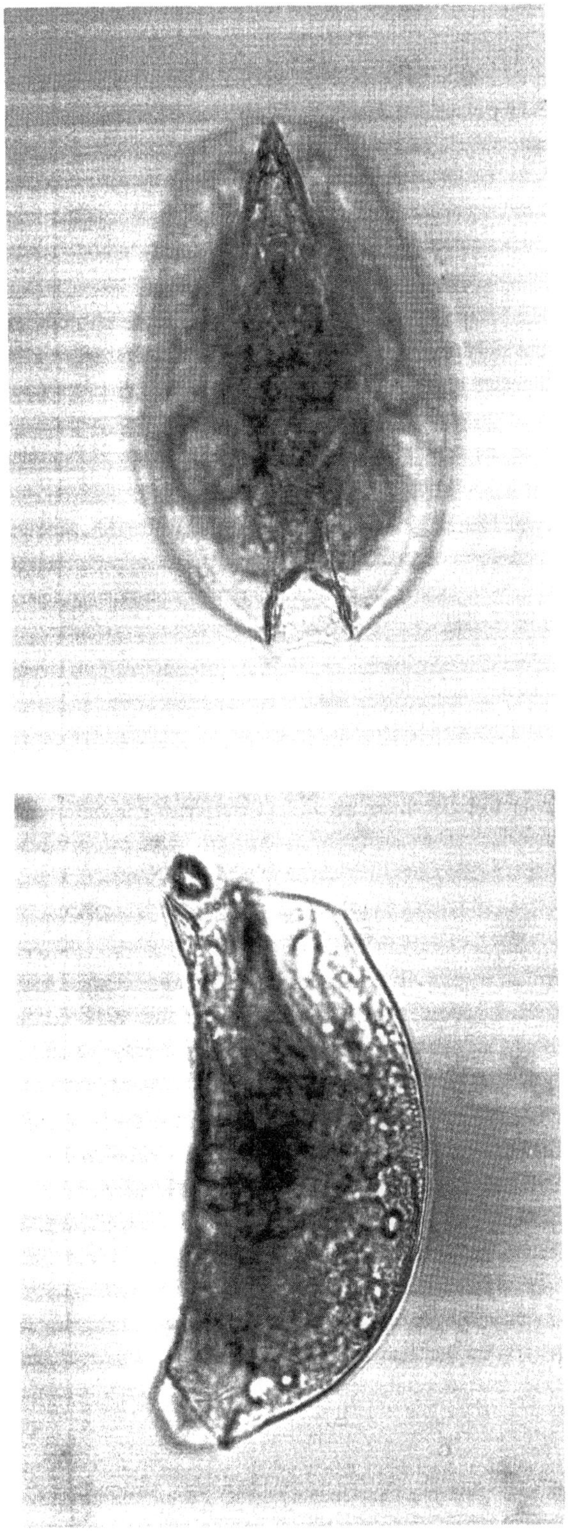

Fig. 6. Photomicrographs of *Testudinella unicornuta* a. ventral; b. lateral

Horaëlla brehmi Donner, a rare species

The distribution of this phylogenetically interesting species of the Family Trochosphaeridae is shown in Fig. 7. There are records from India, Sri Lanka, Equatorial Africa, Malaysia and southeastern Australia. The Tasmanian populations, at 42°S on the east coast (Fig. 1: Sites 66, 67 and 70) considerably extend the known distribution of *H. brehmi.*

The morphology and anatomy of *H. brehmi* was described by Donner [1949]. Koste [1973] compared it with the congeneric *Horaëlla thomassoni* Koste, a neotropical endemic. Little was known of the biology of *H. brehmi,* although both Ruttner-Kolisko [1974] and Koste [1973] suggested that the species was viviparous. Our material from Tasmania shows that this is not true. Mictic and amictic eggs develop in the genital tract of the female, but the embryos and resting eggs are surrounded by a shell. The young emerges as the egg is laid (Figs. 8a, 8g: em_5). In truly viviparous genera like *Asplanchna* no egg shells are formed; *H. brehmi* is better described as ovoviviparous.

The resting egg (Fig. 8c–F, i, j), unknown to date, resembles those of *Filinia.* There are two larger vacuolar wreaths (Fig. 8d–f) and smaller vesicular structures. Between the two rows of great vacuoles is a ring (Fig. 8f) which opens at the moment of hatching. The egg has a diameter of 110 μm, and appears to remain in the body cavity of the parent after death. We did not see males in the preserved samples, but conical shaped embryos were evident in some females (Fig. 8h: em_4, em_5). These may have been male embryos, although their coronas resembled those of females. These eggs were 54 μm in diameter. Other anatomical details are given in Fig. 8a–j. The adult females of the Tasmanian populations had larger body lengths than previously recorded. We observed a range of 320–340μm cf. up to 280 μm reported by Donner [1949] and Koste [1973].

Community composition

Of 46 spp. identified from the 1980 survey (20 sites), only one species, *Keratella cochlearis,* oc-

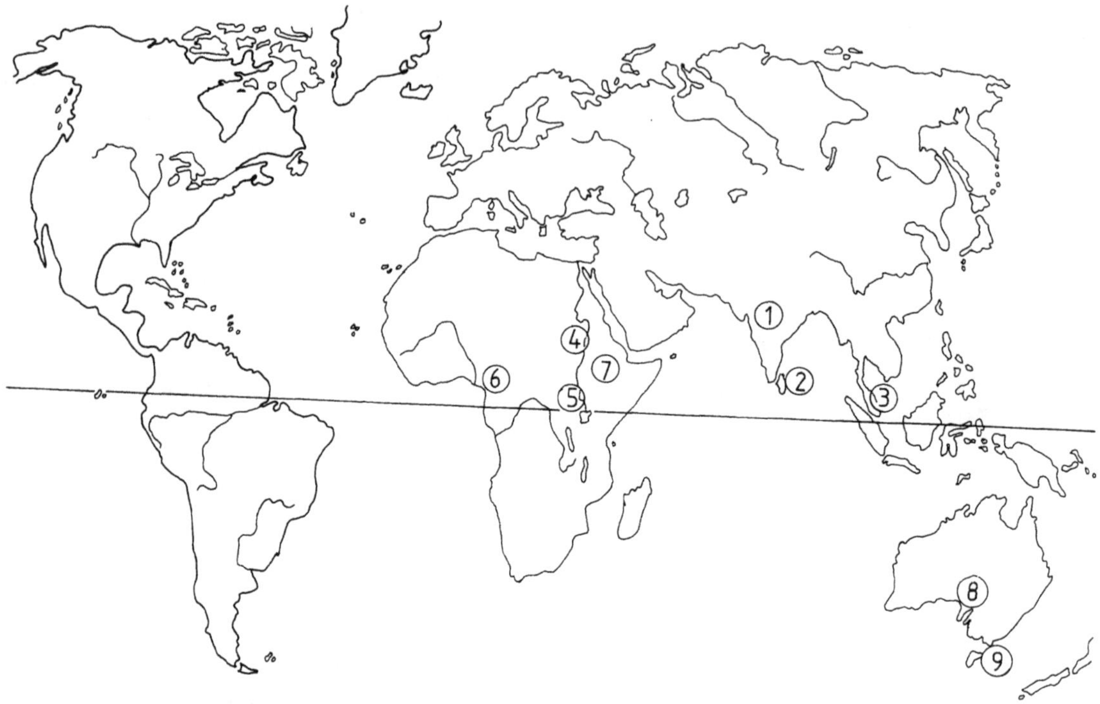

Fig. 7. Distribution records of *Horaëlla brehmi* Donner 1. N. India (Donner, 1949; George, 1966; Sharma, 1980); 2. Sri Lanka (Chengalath *et al.*, 1974); 3. Uganda (Beadle, 1963); 4. Sudan (Green, 1984); 5. Ethiopia (Wodajo & Belay, 1984); 6. Nigeria (Clarke, 1974); 7. Malaysia (Karunakaran & Johnson, 1978); 8. SE Australia (Shiel & Koste, 1979); 9. Tasmania (this study).

curred in more than 20% of samples. Most taxa (36) occurred in only one or two localities. In the 1984 survey (82 sites), 118 species were identified, including 32 also recorded in 1980. Species occurring in more than 20% of localities (in order of abundance) were: *Keratella slacki, K. cochlearis, Trichocera similis, K. australis, Polyarthra dolichoptera, Filinia longiseta, P. vulgaris, K. procurva, F. pejleri, Lecane lunaris.* These were commonly the dominants (> 10% of rotifers present) at sites where they occurred. More than half the taxa (N = 71, 60%) were collected in only one or two habitats.

Of the species listed in Table 1, 11 seem to be restricted to Australia, i.e. about 12%. This agrees with the level of endemism for the mainland reported by Dumont [1983]. Nine species are first records for the continent, while four seem to be endemic to Tasmania. Notably absent are species of the genus *Brachionus.* On the mainland, 3–6 taxa commonly occur in autumn collections, from predominantly alkaline waters. In Tasmania, only one member of this genus (*Brachionus angularis*) was found. It occurred in about 10% survey sites, all of which were acid waters.

From 1–25 taxa occurred at any site (mean 6.9

Fig. 8. Horaëlla brehmi Donner →

a. female, ventral, amictic with parthenogenetic embryo. Total length 324 μm; b. lateral antenna, enlarged; c. female, lateral, with developing resting egg (72 × 72 μm); d. free resting egg, diameter 108 μm; e. female with developed resting egg; f. resting egg, lateral; g. female with resting egg in development, ventral. Total length 342 μm; h. mictic female with five haploid male eggs. em_3 and em_5 males nearly ready to leave genital tract of the parent; i. photomicrograph of specimen figured in g; j. resting egg (HP).

(Abbreviations: an = anus, ap = apical field, dt = dorsal antenna, em = embryo, ey = eye, ex = excretory system, ga = ganglion, gg = gastric gland, h = head contracted, la = lateral antenna, m = mouth, ml = muscle, mx = mastax, ovd = oviduct, n = nerv, ph = pharynx, regg = resting egg, st = stomach).

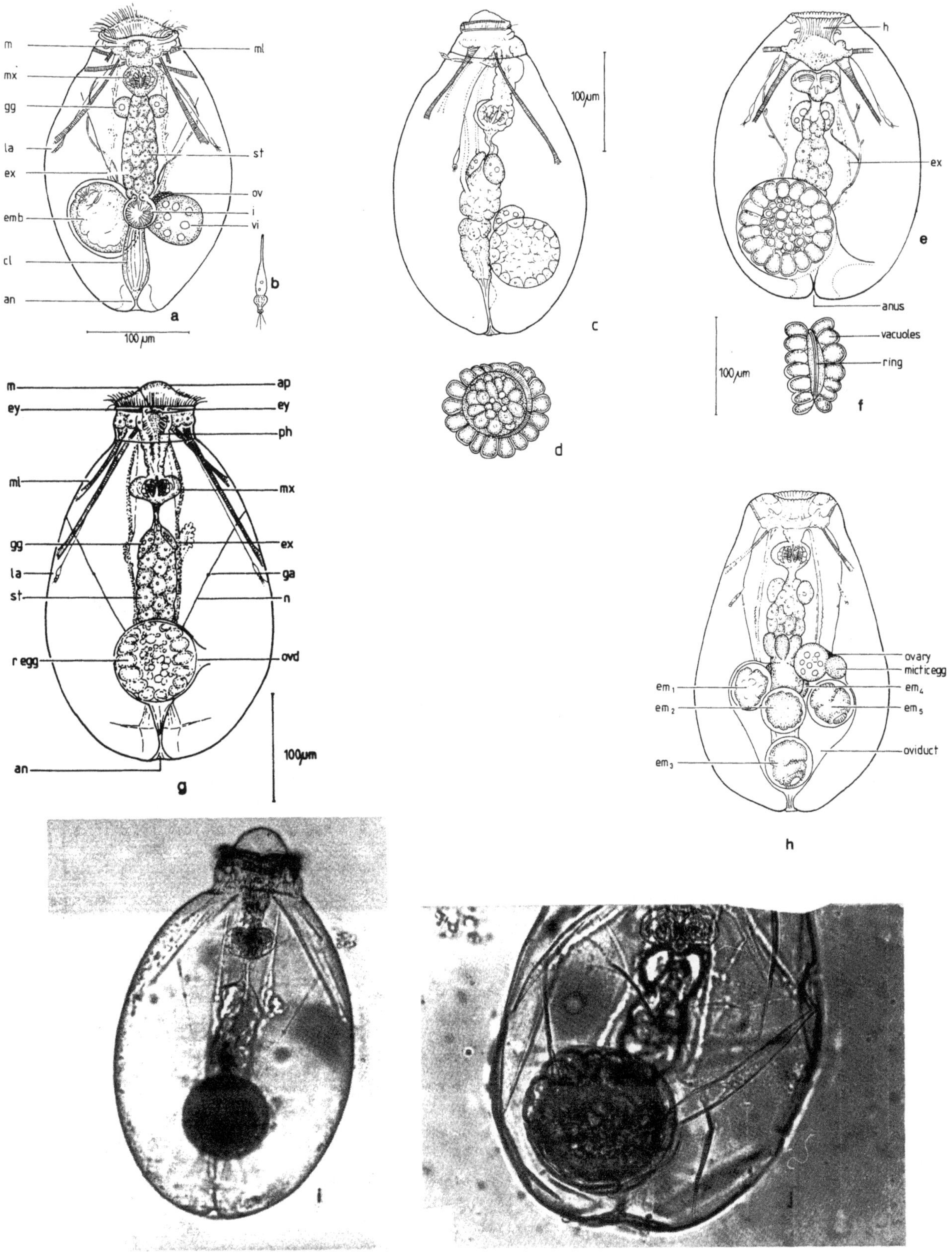

m
ml
mx
gg
la
ex
st
ov
emb
i
vi
cl
an
a
b
100 μm
c
100μm
d
h
ex
e
anus
100μm
vacuoles
ring
f
m
ap
ey
ey
ph
ml
mx
gg
ex
la
ga
st
n
r egg
ovd
an
g
100μm
ovary
micticegg
em1
em4
em2
em5
oviduct
em3
h
i
j

taxa/site), however within each site-category (i.e lake, stock-dam) there was considerable variation. Lake Pedder (Fig. 1: No. 87), for example, had eight rotifer species in the plankton on 28.iv.84; *F. pejleri/K. cochlearis* comprised 73% of the community (< 50 individuals ℓ^{-1}; H′ = 1.87). Nearby Lake Gordon (site No. 89) had a similar plankton (9 spp., same dominants, H′ = 1.85) whereas L. Dobson (site No. 96) on the same day, had only four species in the plankton (*T. similis* dominant; < 10 individuals l^{-1}; H′ = 1.7). In contrast, two stock dams sampled on 27.iv.84 had the following characteristics. Site No. 85 (no macrophytes) had 18 rotifer taxa with *K. australis/K. slacki* comprising 58% of the assemblage (> 200 individuals ℓ^{-1}; H′ = 2.73) and site No. 78 (*Myriophyllum*) had 20 rotifer taxa with *L. bulla/K. quadrata/F. longiseta* comprising 64% (> 200 individuals ℓ^{-1}; H′ = 3.16). Of 38 recorded species, only four taxa (11%) were shared! In general, rotifer density and diversity was lower in lakes than in smaller stock dams. In this study stock dams were found to have a rich rotifer plankton (in many cases with an equally diverse microcrustacean fauna). Comparable community differences between closely situated habitats also were seen in our earlier studies of Australian billabongs [Shiel & Koste, 1983]. Mainland billabongs tended to show greater rotifer diversity, perhaps because of greater habitat partitioning.

One surprising feature of Tasmanian rotifer assemblages was the relative abundance of warm stenothermal, cosmotropical species which are unexpected at 42–43° S. On the mainland, tropical species such as *Filinia opoliensis, F. pejleri* and *Horaëlla brehmi* are known from Darling River waters flowing from tropical Queensland into the River Murray system. River transport cannot be invoked in Tasmania and yet populations of cosmotropical rotifers apparently are widely-dispersed over the island and well-established. Cosmotropical microcrustacea (e.g. *Ceriodaphnia cornuta*) also were widespread on Tasmania. Species indicative of cold waters on the mainland, e.g. *Keratella quadrata, K. valga, Filinia australiensis* and *F. pejleri grandis,* which occur only at high altitude or in winter, co-occurred with eurytopic endemics (*K. australis, K. slacki*) and the tropical forms noted above. On present evidence we can only attribute these disjunct distributions to Tasmania's unusually mild climate [cf. Tyler, 1974].

Our comments here are only brief summaries of our autumn survey work. Nevertheless, it is evident that a diverse rotifer fauna occurs in Tasmania and that there are regional differences between the rotifer plankton of the island and that of the mainland. The full extent of regional differences has yet to be determined.

Acknowledgements

The support of the Australian Biological Resources Study for fieldwork in Tasmania is acknowledged. We also thank Dr P. A. Tyler, University of Tasmania for his hospitality and offer of access to this plankton collections, and to Ross Kaires and Judy Pointon, University of Adelaide, for assistance in making the collections under difficult conditions. Autor W. K. is grateful for the microscope facilities provided by the Deutsche Forschungsgemeinschaft, Bonn Bad Godesberg, FRG.

References

Beadle, I. C., 1963. Anaerobic life in a tropical crater lake Nature 200: 1223–1224.

Chengalath, R., C. H. Fernando & W. Koste, 1974. Rotifera from Sri Lanka (Ceylon). Bull. Fish. Res. Stn., Sri Lanka (Ceylon). 25: 83–96.

Clarke, N. V., 1978. A comparison of the zooplankton of Lake Kainji and of the River Niger and Swashi. Hydrobiologia 58: 17–23.

De Deckker, P. & W. D. Williams, 1982. Chemical and biological features of Tasmanian salt lakes. Aust. J. mar. Freshwat. Res. 33, 1127–32.

Donner, J., 1949. Horaëlla brehmi nov. gen. nov. spec., ein neues Rädertier aus Indien. Hydrobiologia 2: 134–140.

Dumont, H. J., 1983. Biogeography of rotifers. Hydrobiologia 104: 19–30.

Ganf, G. G., R. J. Shiel & C. J. Merrick, 1983. Parasitism: the possible cause of the collapse of a Volvox population in Mt Bold Reservoir, South Australia. Aust. J. mar. Freshwater Res. 34: 489–494.

George, M. G., 1966. Comparative plankton ecology of five fish tanks in Delhi, India. Hydrobiologia 27: 81–108

Green, J., 1984. Zooplankton associations in the swamps of southern Sudan. Hydrobiologia 113: 93–98.

Karunakaran, L. & A. Johnson, 1978. A contribution to the rotifer fauna of Singapore and Malaysia. Malayan natur. J., Kuala Lumpur 32: 173–208.

Koste, W., 1973. Horaëlla thomassoni n. sp., ein neues Rädertier aus Gewässern der Guiana-Brasilianischen Region der Neotropis. Arch. Hydrobiol. 72: 375–383.

Koste, W., 1981. Zur Morphologie, Systematik und Ökologie von neuen monogononten Rädertieren (Rotatoria) aus dem überschwemmungsgebiet des Magela Ck in der Alligator River Region, Australiens. I. Osnabrücker naturwiss. Mitt., 8: 97–126.

Koste, W. & R. J. Shiel 1986a: Rotifera from Australian inland waters. I. Bdelloidea (Rotifera: Digononta), II. Epiphanidae & Brachionidae (Rotifera: Monogononta). Aust. J. mar Freshwater Res. 37: 765–792 (I.). in press.

Koste, W. & R. J. Shiel, 1986b. New Rotifera (Aschelminthes) from Tasmania. Trans. R. Soc. S. Aust. 110(3): 93–109.

Ruttner-Kolisko, A., 1974. Plankton Rotifers; biology and taxonomy. Die Binnengewässer 26, Suppl., 126 pp.

Shannon, C. E., 1948. A mathematical theory of communication. Bell Systems Tech. J. 27: 379–423.

Sharma, B. K., 1980. Contributions to the rotifer fauna of Orissa, India. Hydrobiologia 70: 225–233.

Shiel, R. J. & W. Koste, 1979. Rotifera recorded from Australia. Trans. R. Soc. S. Aust. 103: 57–68.

Shiel, R. J. & W. Koste, 1983. Rotifer communities of billabongs in northern and southeastern Australia. Hydrobiologia 104: 41–47.

Shiel, R. J. & W. Koste, 1985. New species and new records of Rotifera (Aschelminthes) from Australian waters. Trans. R. Soc. S. Aust. 109: 1–15.

Shiel, R. J. & W. Koste, 1986. Australian Rotifera: ecology and biogeography. pp. 000–000 in Limnology in Australia (Ed. P. De Deckker & W. D. Williams). CSIRO/Junk BV.

Shiel, R. J., C. J. Merrick & G. G. Ganf, in press. The Rotifera of impoundments in southeastern Australia. Hydrobiologia.

Sudzuki, M., 1985. Tasmanian rotifers in mid-August (winter). Zool. Sci. 2: 1005.

Tyler, P. A., 1974. III. Limnological studies. Monogr. Biol. 25: 29–61.

Williams, W. D., 1974. (Ed.) Biogeography and ecology in Tasmania. Monogr. Biol. 25: 1–498.

Wodajo, K. & A. Belay, [1984]. Species composition and seasonal abundance of zooplankton in two Ethiopian Rift Valley lakes – Lakes Abiata and Langano. Hydrobiologia 113: 129–136.

Hydrobiologia 147: 45–47 (1987)
May, L., Wallace, R. & Herzig, A. (eds), Rotifer Symposium IV

Intraspecific variability of *Brachionus plicatilis*

Minoru Sudzuki
Biological Laboratory, Nihon Daigaku-University, Omiya-shi, Saitama-Ken, Japan

Keywords: Rotifera, *Brachionus plicatilis*, variability, taxonomy, morphology

Abstract

An extensive study of frontal margins of the lorica of *Brachionus plicatilis* was undertaken in an attempt to define its variability within this species. Specimens from mass cultures and from the natural environment were examined.

Introduction

According to Fukusho [1983] two distinct forms of *Brachionus plicatilis* [Müller 1786] are recognizable in mass-cultured populations. One population is 130 to 340 μm in length ($\bar{X}$ = 238.9 μm), possesses a saccate, soft lorica and is inclined to appear at temperatures below 20 °C. The other population is 100 to 210 μm in length ($\bar{X}$ = 163.3 μm), possesses a loricated, ovoid body, and tends to occur at temperatures above 20 °C.

Even based on size alone, the variability of this species is quite large. The purpose of this paper is to present a synopsis of my analysis of variability in the frontal margin of the lorica of *Brachionus plicatilis*.

Materials

I examined a large number of specimens (ca 6000) of *B. plicatilis* that I had collected from a wide variety of sources or that were kindly provided by colleagues. The study material included specimens from Bangkok, Israel, Japan, Singapore, and USSR (Estonia). A detailed account of the sources of the study material is given by Sudzuki [1986].

Results and discussions

From an analysis of a large number of specimens of *B. plicatilis*, and of the literature, I propose six suggestions that require further inquiry. 1. The specimens of Murray [1913: Fig. 47a], Koste [1980: Fig. 3], and Nogrady [1983: Fig. 3] may warrant detailed investigation, as they did not occur among the 6000 specimens which I examined. 2. *Brachionus orientalis* and *Brachionus baylyi* may be safely regarded as good species. 3. Fadeew's *decemcornis* and *murrayi* could be treated as *B. rotundiformis decemcornis* Fadeew 1925 and *B. rotundiformis murrayi* Fadeew 1925, respectively. 4. Fadeew's *longicornis* may be included within the extent of variation of *rotundiformis murrayi* Fadeew 1925. 5. Further work on both Müller's *plicatilis* and Ehrenberg's *mulleri* is necessary before we may be certain of their status. 6. The specimen with triangular projections at the lateral side of the pectoral margin of the lorica may be regarded as a new species. I propose the name *Brachionus rotundiformis estoniana* for this specimen.

Key of *Brachionus plicatilis* and related species:

I. Species and subspecies distinguishable based on variation of the pectoral margin of the lorica.

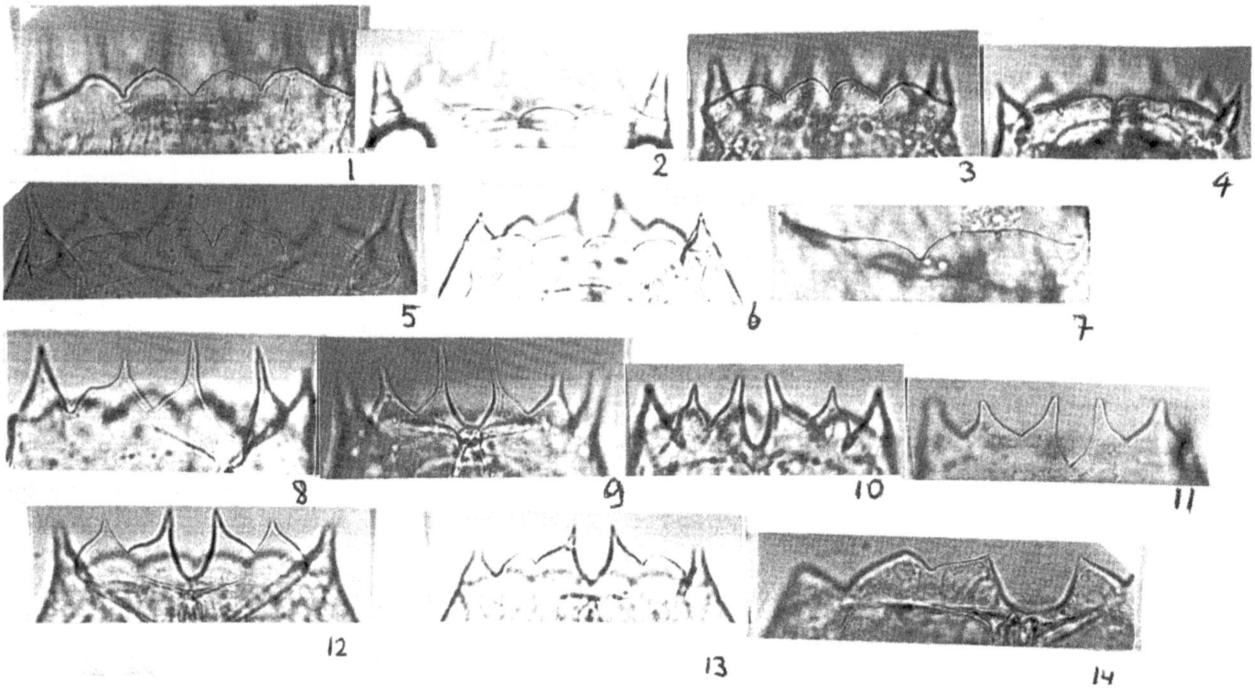

Fig. 1. Some representative examples of frontal margins of loricas discussed in the test. 1.1–*decemcornis;* 1.2–*estoniana;* 1.3–*rotundiformis murrayi;* 1.4–Ehrenberg's type; 1.5–Althaus' type; 1.6–*hepatotomus;* 1.7–Ehrenberg's type(?); 1.8–*rotundiformis;* 1.9–*rotundiformis;* 1.10–*rotundiformis;* 1.11–*plicatilis* (?); 1.12–*hepatotomus;* 1.13–*hepatotomus;* 1.14–Yamamoto's type.

1. Pectoral projections triangular in shape 2
–. Pectoral projections otherwise 6
2. All projections triangular, medians equilateral 3
–. Projections not as above 4
3. Medians with papillae at the tip ...... *baylyi*
–. Medians without papillae at the tip .... *decemcornis* (Fig. 1.1) [Fadeew, 1925: Fig. 7]
4. Medians triangular 5
–. Laterals triangular............... *estoniana* (Fig. 1.2)
5. Pectoral projections with papillae at tip...... .... see Ahlstrom [1940: Pl. 2, Figs. 1, 4]
–. Pectoral projections not as above .... *murrayi* (Fig. 1.3) [Fadeew, 1925: Fig. 4]
6. Most projections more or less circular in shape 7
–. Most projections reduced............ *mülleri* (Figs. 1.4, 1.7) [Ehrenberg, 1838: Fig. 5]
7. Laterals or medians undulated or toothed .... see Althaus [1956: Fig. 8] (Fig. 1.5)
–. Both laterals and medians not as above 8
8. Projections deeply scalloped or semi-circular . ... *rotundiformis* (Figs. 1.8–1.10) [Tschugunoff, 1921: Fig. 1/Ĥig. 1.2]
–. Projections not as above .... *hepatotomus* (Figs. 1.2, 1.3, 1.6) [Hudson & Gosse, 1886: Figs. 7, 8]

II Species and sub-species distinguishable based on variation of the occipital margin of the lorica.

1. Most spines equilateral triangular 2
–. Most spines not as above, rather scalene

triangular 4
2. Median projections pointed to form spines . 3
–. Median projections not as above .... *plicatilis* (Fig. 1.11) [Müller, 1786: Figs. 2, 4, 5]
3. Median spines curved outwards .... *orientalis* [Rodewald, 1937: Fig. 7]
–. Median spines not as above.... *rotundiformis* (Fig. 1.8) [Murray, 1913: Fig. 47]
4. Spines with swellings 5
–. Spines without swellings ...... see Yamamoto (Fig. 1.14) [1961: Fig. 5]
5. Median spines with external swellings........ .... *hepatotomus* (Figs. 1.2, 1.3, 1.6) [Hudson & Gosse, 1886: Fig. 7]
6. Spines mostly with internal swellings . *mülleri* [Ehrenberg, 1838: Fig. 5]

Clearly further study on these and other questions regarding the variability of the lorica of *B. plicatilis* is required. Current biochemical techniques allow isozyme analysis of different populations of rotifers [e.g. King, 1980]. Such work is necessary to resolve the systematic status of the different populations presented here.

Acknowledgements

I wish to express my deepest thanks to my colleagues who supplied the study materials: Dr. Fukusho, Mr. Okauchi, Mr. Hagiwara, Dr. Hino, Professor Egami, Dr. Haberman, and Mr. Tanaka. Thanks also are due to Professor Pejler, Dr. Tzschaschel, and Mr. Hollowday for providing copies of certain research papers unavailable to me. I appreciate the help of three anonymous reviewers who read and improved the manuscript.

References

Ahlstrom, E. H., 1940. A revision of the rotatorian genera Brachionus and Platyias with descriptions of one new species and two new varieties. Bull. am. Mus. nat. Hist. 77: 143–184.

Althaus, B., 1956. Faunistische-ökologische Studien an Rotatorien salzhaltiger Gewässer Mitteldeutschlands Wiss. Z. Martin-Luther Univ. Halle-Wittenberg. 6: 117–157.

Ehrenberg, C. G., 1838. Die Infusionstierchen als vollkommene Organismen. Ein Blick in das tiefere organische Leben der Natur. Leipzig, 547 pp.

Fadeew, N. N., 1925. Sur la connaissance de la faune de lacs du Transcaucasie (In Russian, French summary). Raboty Sev. Kavkazskoj Gidrobiol. Stancii. 1: 17–25.

Fukusho, K., 1983. Morphology and variation (in Japanese). In: Brachionus plicatilis, its biology and mass culture. Koseikaku, Tokyo pp: 35–51.

Hudson, C. T. & P. H. Gosse, 1886. The Rotifera or Wheel-animalcules. Longmans, Green & Co., London. 2 vols.

King, C. E., 1980. The genetic structure of zooplankton populations. In: W. C. Kerfoot (ed.), Evolution and Ecology of Zooplankton Communities. The University Press of New England, Hanover (N.H.); Lond.: 315–328.

Koste, W., 1980. Das Rädertier-Porträt. B. plicatilis ein Salzwasserrädertier. Mikrokosmos 151: 148–155.

Müller, O. F., 1786. Animalcula Infusoria fluviatilia et marine. Hauniae, 367 pp.

Murray, J., 1913. South American rotifera. III Conclusion. J.r. microsc. Soc. 18: 449–453.

Nogrady, T., 1983. Succession of planktonic rotifer populations in some lakes of the Eastern Rift Valley. Hydrobiologia 98: 45–54.

Rodewald, L., 1937. Rädertierfauna Rumaniens. II. Neue und bemerkenswerte Rädertiere aus Rumanien. Zool. Anz. 118: 235–243.

Sudzuki, M., 1986. Some problems of taxonomic treatment in the so-called Brachionus plicatilis (In Japanese). Iden (Heredity) 39: 52–57.

Tschugenov, H. L., 1921. A study of the plankton of the arctic regions of the Caspian Sea (In Russian). Report of the Volga Biological Station 6: 109–162.

Yamamoto, K., 1961. Plankton rotatoria in Japanese inland waters. Hydrobiologia 16: 364–411.

Hydrobiologia 147: 49–56 (1987)
May, L., Wallace, R. & Herzig, A. (eds), Rotifer Symposium IV

Rotifera from Northwestern Canada

R. Chengalath[1] & W. Koste[2]
[1]*Invertebrate Zoology Division, National Museum of Natural Sciences, Ottawa, Canada K1A OM8*
[2]*Ludwig-Brill-Strasse 5, D 4570 Quakenbrück, Federal Republic of Germany*

Keywords: Rotifera, Canada, Yukon Territory, British Columbia, additions to known fauna, species composition

Abstract

Ninety seven species of rotifers belonging to 32 genera are reported from British Columbia and the Yukon Territory of Canada. Three species are new records for Canada, and 55 and 51 species, respectively, are new records for British Columbia and Yukon Territory. Several interesting species and some that exhibited morphological differences from the original descriptions are illustrated and critical taxonomic features of these species are discussed. The species composition is compared with that from other regions of Canada and points of interest in the distribution of species are noted.

Introduction

The Rotifera of Canada have been insufficiently studied. This is especially true for the many bodies of water which are located in sparsely populated northern areas. Exploration of these vast aquatic environments is both costly and difficult because of long distances and poor accessibility. There are only a few studies that deal with rotifers from northwestern Canada. Murray [1911] reports 22 species from southern British Columbia, of which 21 are bdelloids, in material collected by the Shackleton Antarctic Expedition. Fifty-one species are recorded by Harring [1921] from Yukon Territory based on the plankton collections made by F. Johansen during the Canadian Arctic Expedition of 1913–1918. Hooper [1947] lists 19 species of rotifers found in plankton collections from the Yukon and Mackenzie river systems. More recently, other rotifer faunal lists have been compiled sporadically as part of general zooplankton or fishery studies, mainly from southern British Columbia. In most of these studies, rotifers were identified only to the generic level, and no morphological or ecological data were provided [Chengalath, 1984]. This paper documents the rotifer fauna of northwestern Canada, providing some points of interest in the distribution of species as well as illustrating the morphological differences found in some species.

Materials and methods

The sampling localities are shown in Fig. 1. Qualitative sampling of the littoral zone of ponds and lakes was carried out in the summer of 1984. A total of 90 samples were obtained with a 64 μm mesh conical plankton net 15 cm in diameter attached to a metal rod 3 m long. The waterbodies in the region under study varied from small, highly coloured, ponds that had a low content of electrolytes and profuse vegetation to alpine lakes surrounded by snow with clear blue waters and sandy bottom with little or no vegetation. The pH value of the waters ranged from 6.0 to 9.5. The collecting sites were located between 457 m and 2438 m above sea level, although the majority of them were in the range of 600–800 m.

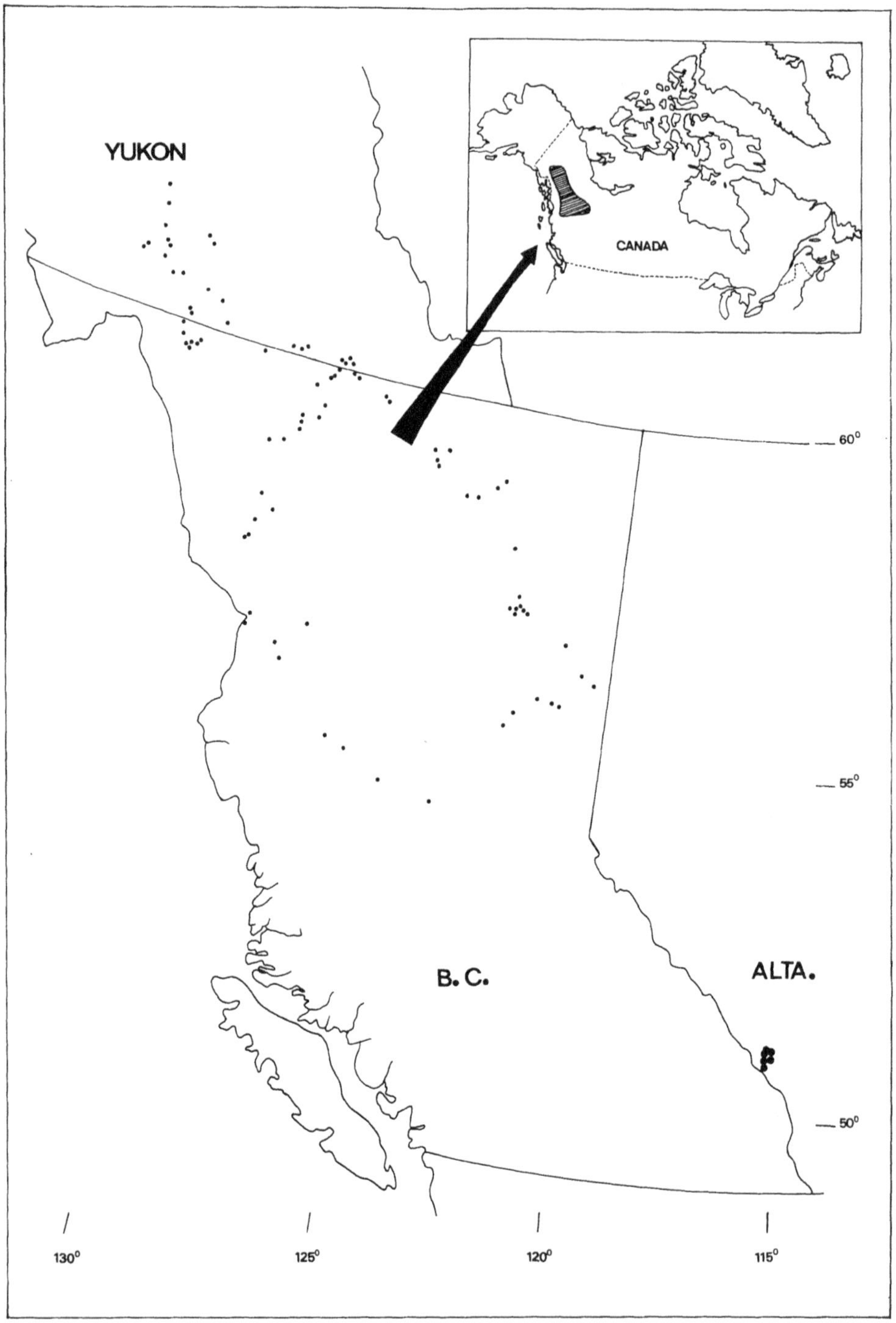

Fig. 1. Localities sampled in northwestern Canada.

Results and discussion

A list of the 97 species found during this study is given in Table 1, with percentages of samples in which each occurred as well as the new records for Canada, for the province of British Columbia, and for the Yukon Territory. Three species of rotifers were recorded for the first time in Canda and 55

Table 1. Occurrence and frequency of rotifer species in Yukon Territory and British Columbia during this study (B = New record for British Columbia, C = New record for Canada, Y = New record for Yukon).

Species		% Samples containing species	Species		% Samples containing species
Asplanchna priodonta	Y	2.7	*Conochilus unicornis*	–	20.7
Asplanchnopus multiceps	B	0.9	*Conochiloides dossuarius*	Y, B	0.9
Brachionus pterodinoides	Y	1.8	*Euchlanis dilatata*	–	27.0
Brachionus quadridentatus	Y	1.8	*Euchlanis incisa*	Y, B	6.3
Cephalodella catellina	B	1.8	*Euchlanis lyra*	Y, B	14.4
Cephalodella gibba	B	11.7	*Euchlanis meneta*	Y, B	19.8
Cephalodella nana	Y, B	0.9	*Euchlanis oropha*	B	8.1
Collotheca mutabilis	Y	1.8	*Euchlanis triquetra*	B	8.1
Collotheca ornata	Y, B	2.7	*Filinia longiseta*	Y	3.6
Collotheca pelagica	Y, B	1.8	*Floscularia janus*	Y, B	4.5
Colurella adriatica	Y, B	0.9	*Floscularia melicerta*	Y, B	2.7
Colurella obtusa	Y, B	0.9	*Gastropus hyptopus*	Y, B	0.9
Colurella uncinata	Y	3.6	*Hexarthra polyodonta*	Y, B	1.8
Conochilus hippocrepis	–	4.5	*Kellicottia longispina*	–	36.0
Keratella c. cochlearis	–	9.0	*Lecane quadridentata*	Y, B	0.9
Keratella c. faluta	–	10.8	*Lecane scutata*	Y, B	1.8
Keratella c. hispida	Y, B	6.3	*Lecane stichaea*	Y, B	4.5
Keratella c. recurvispina	Y	2.7	*Lecane ungulata*	B	1.8
Keratella crassa	*	0.9	*Lepadella acuminata*	B	2.7
Keratella hiemalis	B*	0.9	*Lepadella ovalis*	–	7.2
Keratella quadrata	–	5.4	*Lepadella patella*	–	10.8
Keratella testudo	Y	1.8	*Limnias melicerta*	Y, B	2.7
Lecane bulla	–	8.1	*Lophocharis salpina*	Y, B	3.6
Lecane closterocerca	B	5.4	*Microcodon clavus*	B*	0.9
Lecane cornuta	B	2.7	*Monommata dentata*	B*	1.8
Lecane flexilis	Y, B	4.5	*Mytilina mucronata*	B	3.6
Lecane hamata	–	2.7	*Mytilina ventralis*	B	9.9
Lecane hornemanni	B	1.8	*Notholca acuminata*	–	27.0
Lecane levistyla	Y, B	4.5	*Notholca caudata*	C*	0.9
Lecane ludwigii	–	3.6	*Notholca foliacea*	Y, B	14.4
Lecane luna	–	13.5	*Notholca labis*	Y, B	5.4
Lecane lunaris	–	34.2	*Notholca lapponica*	C*	1.8
Notholca squamula	Y, B	15.3	*Testudinella incisa*	Y, B	2.7
Notommata copeus	B	5.4	*Testudinella parva*	B	5.4
Notommata glyphura	Y, B	4.5	*Testudinella patina*	–	5.4
Notommata pachyura	Y, B	2.7	*Trichocerca bidens*	Y, B	2.7
Ploesoma hudsoni	Y	1.8	*Trichocerca dixonnutalli*	Y, B	0.9
Ploesoma lenticulare	B	6.3	*Trichocerca elongata*	Y, B	1.8
Ploesoma truncatum	–	3.6	*Trichocerca insignis*	Y, B	0.9
Polyarthra dolichoptera	Y	6.3	*Trichocerca intermedia*	Y, B	0.9
Polyarthra major	Y, B	0.9	*Trichocerca lata*	Y	2.7
Polyarthra vulgaris	Y	10.8	*Trichocerca longiseta*	–	20.7
Scaridium longicaudum	Y, B	3.6	*Trichocerca porcellus*	–	1.8
Synchaeta grandis	Y, B	2.7	*Trichocerca rattus*	B	7.2
Synchaeta longipes	Y, B	1.8	*Trichocerca tenuoir*	B	1.8
Synchaeta oblonga	Y	2.7	*Trichocerca weberi*	Y, B	4.5
Synchaeta pectinata	Y	1.8	*Trichotria pocillum*	–	10.8
Taphrocampa selenura	B*	0.9	*Trichotria tetractis*	–	20.7
			Wolga spinifera	C	4.5

* not found in Yukon during this study.

and 51 species, respectively, are new records for British Columbia and Yukon Territory. Bdelloid rotifers were present in some samples but most could not be positively identified to species level and are, therefore, not included in this study. The number of species found in any one sample varied from 0–26, with the majority of samples containing 6–12 species. Six samples contained no rotifers at all. Of the total 97 species found, only 7 species, *Euchlanis dilatata, Kellicottia longispina, Lecane lunaris, Notholca acuminata, Trichocerca longiseta, Trichotria tetractis*, and *Conochilus unicornis*, were present in more than 20% of the samples. Among these, *K. longispina* and *L. lunaris* were predominant, occurring in 36% and 34%, of the localities respectively, followed by *N. acuminata* which occurred in 27% of the samples. The largest numbers of species, 14 and 11, were recorded for the typical littoral genera *Lecane* and *Trichocerca*, respectively.

The family Brachionidae was represented by four genera, *Brachionus, Kellicottia, Keratella*, and *Notholca*. Only two species of *Brachionus, Brachionus pterodinoides* and *Brachionus quadridentatus*, were recorded and these were found in very few samples (Table 1). Harring [1921] did not find any *Brachionus* species in the Yukon and they are not a significant element of the plankton of northern localities. The interesting rotifer *B. pterodinoides* was last reported by Ahlstrom [1940] from highly alkaline waters of the Americas. During the present study, it was found in two small, muddy, enriched farm ponds devoid of any vegetation, with pH values of 9.5. The specimens we examined differed slightly from the description given by Ahlstrom [1940] in having a distinct median sinus in the anterior ventral margin and no stippling on the lorica (Figs. 2 & 3).

The genus *Keratella* was well represented with 8 taxa. *Keratella quadrata* and the 4 subspecies of *Keratella cochlearis* were the most common. *Keratella* is found throughout the world in freshwaters and is successful in brackish and marine habitats as well. However, not all species of *Keratella* are cosmopolitan in distribution. *K. cochlearis*, which was once considered cosmopolitan, is now known to have several subspecies each of which may have a limited distribution. Although considerable attention has been given to the various forms of *K. cochlearis*, the status of many of these forms still remains unclear. Several populations of *K. c. cochlearis, K. c. faluta, K. c. recurvispina* and *K. c. hispida* were found during this study (Figs. 4–9). *K. c. faluta* was the dominant form occurring in nearly 11% of the samples examined. This large *Keratella* is characterised by a relatively wide lorica which terminates in a moderately long, stout, median posterior spine that tapers gradually to a blunt point and can be easily recognised (Fig. 9). As with *K. c. faluta, K. c. recurvispina* has been reported rarely in the literature [Ahlstrom, 1943; Koste, 1978]. All the known records of *K. c. recurvispina* from Europe [Koste, 1978] and North America [Ahlstrom, 1943] are from marine and brackish waters. However, in the present study it was found in interior freshwater lakes and ponds. The most diagnostic feature of this *Keratella* is that the anterior intermediate spines are long, stout and extend clearly to the dorsal side and are divergent laterally (Figs. 7 & 8). It is possible that some previous workers may have identified this species as *Keratella taurocephala*, however, these two species are quite distinct and easily recognisable. Furthermore, from the distribution records that are available, it is clear that *K. taurocephala* is confined to acidic waters of southern and eastern Canada [Chengalath & Fernando, 1973; Chengalath & Koste, 1983]. *K. c. hispida* has a wider intercontinental distribution and several forms of *hispida*-like taxa have been described from North America and Europe [Ahlstrom, 1943; Pejler, 1962a; Kutikova, 1970; Koste, 1978]. As the name implies, *K. c. hispida* is characterised by a densely hispid lorica (Fig. 5). Ahlstrom [1943] in his revision of the genus states that this species is apparently uncommon in the United States, but mentions that he has seen good material from Karluk Lake, Kodiak Island, Alaska. As it was seen fairly commonly, and at times in large numbers, during this study from adjacent Yukon Territory and British Columbia, it perhaps enjoys wider distribution in northwestern North America.

The pattern of dorsal sculpture on the lorica is of great importance in differentiating species of *Keratella*. However, on some specimens the dorsal pat-

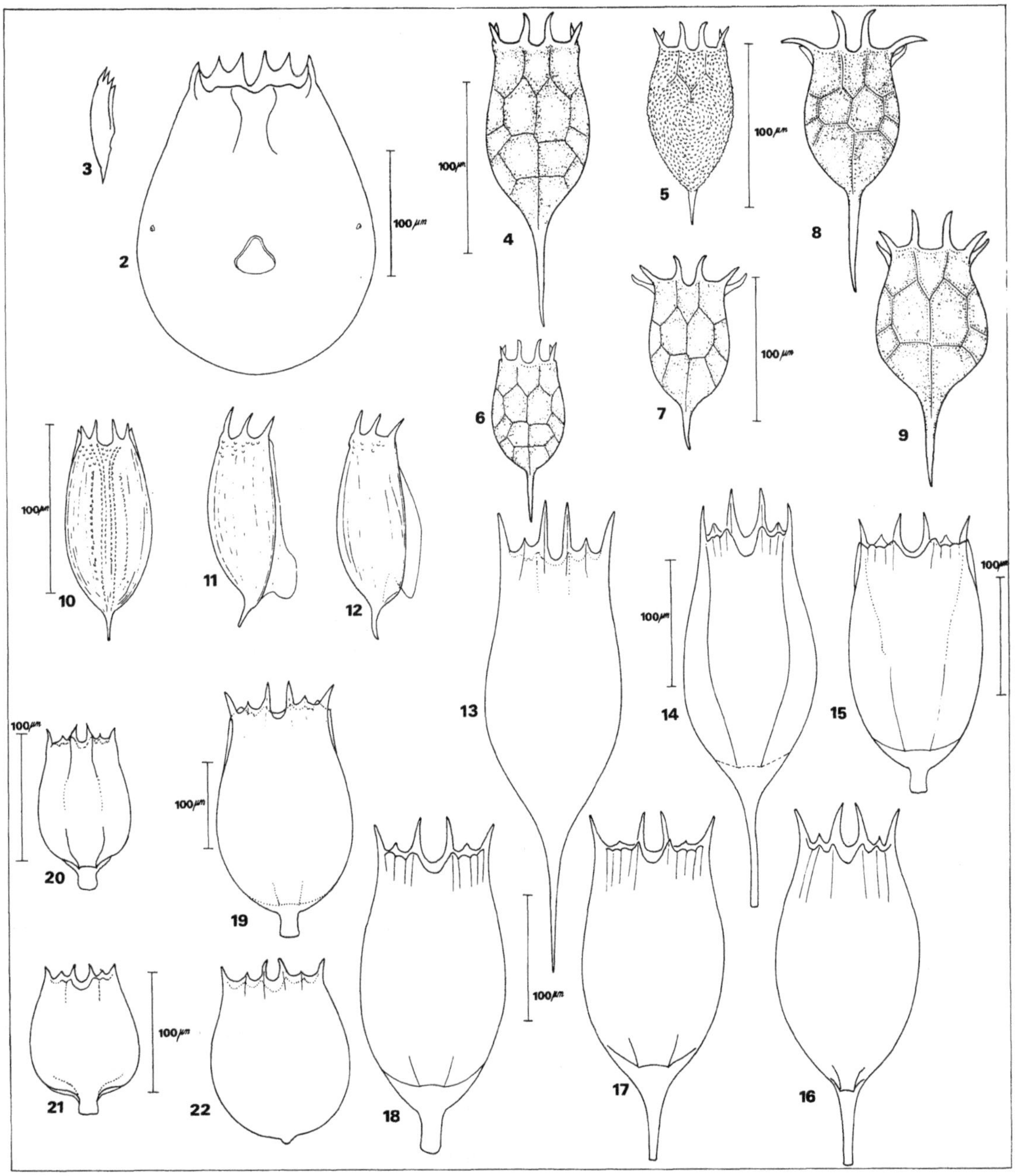

Fig. 2. *Brachionus pterodinoides*. Fig. 3. *B. pterodinoides*, lateral view. Figs. 4 & 6. *Keratella cochlearis*. Fig. 5. *K. cochlearis hispida*. Figs. 7 & 8. *K. cochlearis recurvispina*. Fig. 9. *K. cochlearis faluta*. Fig. 10. *Notholca foliacea*. Figs. 11 & 12. *N. foliacea*, lateral view. Fig. 13. *N. caudata*. Figs. 14–18. *N. acuminata*. Fig. 19. *N. labis*. Figs. 20 & 21. *N. lapponica*. Fig. 22. *N. squamula*.

tern is indistinct, and there are some species that cannot be separated from each other on this character alone. The two pairs of species, *K. quadrata* and *K. valga*, and *K. taurocephala* and *K. cochlearis*, provide good examples of this. *K. c. faluta, K. c. recurvispina*, and *K. c. hispida* all have the same foundation pattern on the dorsal plate as that of *K. c. cochlearis*, although it may be difficult to discern the dorsal pattern of *K. c. hispida*. However, other taxonomically important characters (i.e., size, ornamentation, lorica shape, and length of anterior and posterior spines) of the populations of *K. c. faluta, K. c. recurvispina* and *K. c. hispida* examined during this study, clearly show that these taxa are distinct from *K. c. cochlearis*. As has been done for *K. taurocephala*, these subspecies also may have to be elevated to the level of species. Uncertainty regarding the status of these taxa arose mainly because there were only a few records of their occurrence and also because of lack of good material for study. From information that is available in the literature concerning distribution of these subspecies, it appears that *K. c. faluta* has a limited distribution in northern North America, while *K. c. recurvispina* and *K. c. hispida* show a disjunct distribution in North America and Europe, mainly in northern latitudes. It is possible that these taxa have diverged morphologically from what was originally the same species and also may well have differentiated ecologically, physiologically and genetically.

Six species of *Notholca* were encountered during this survey and form an important component of the rotifer plankton of this region. *Notholca acuminata* was one of the dominant rotifers, found in 27% of the samples, sometimes in great numbers. Variability in shape and size of the lorica of *N. acuminata* is well known [Ruttner-Kolisko, 1974]. Some forms found in the present study are illustrated (Figs. 14–18). *Notholca squamula* was also common and a form with a small posterior segment on the more-or-less round lorica was observed (Fig. 22). Anterior spines of this form were finger shaped, rather than having the more normal pointed spines. Another common *Notholca, Notholca foliacea*, was found in more than 14% of the localities sampled. The loricas of specimens collected during this study were not as compressed laterally as in specimens from southern latitudes. They also had strong wave-like striations and granulations in the anterior dorsal end as well as along the mid-dorsal line (Fig. 10). The caudal appendage was stout and prominent and the median ventral ridge varied in size from about one-quarter to two-thirds of the body size (Figs. 11 & 12). Figures of this species in the literature show the median ventral ridge ranging in size from a small projection to less than one-half of the body size. *Notholca labis*, when found, usually occurred abundantly. It resembles *N. acuminata*, but is generally smaller in size and is much more rounded posteriorly with a short spatulate caudal appendage, narrower at the base (Fig. 19). Two species, *N. caudata* and *Notholca lapponica*, were found rarely during this investigation and are the first authentic records for North America. These species have been reported rarely in the literature and their known distribution was limited to Scandinavia and Finland [Ruttner-Kolisko, 1974; Koste, 1978]. A single deformed specimen of what could be *Notholca caudata* was found by Stemberger [1979] from the Great Lakes; he speculated that its occurrence in the Great Lakes might be accidental, as a result of being carried to North America by ships from European waters. *N. caudata* is one of the largest members of the genus and is considered as a possible glacial relict species [Pejler, 1962b]. Although it was found in only one sample from northern British Columbia, several specimens were available for study. It is easily distinguished by its large size and the pointed posterior spine (Fig. 13). *N. lapponica* on the other hand is one of the smallest *Notholca* species and has so far been reported only from Scandinavia [Koste, 1978] (Figs. 20–21). It is regarded as benthic, psammophilic, and cold stenothermic. The two localities where it was found were both relatively cold (5 °C & 10 °C), and had fine sandy bottom with no macrophytic vegetation.

Six species of the genus *Euchlanis* were found fairly commonly in both the Yukon and British Columbia. Among these, *Euchlanis dilatata* and *Euchlanis meneta* were the dominant species, being present in 27% and 20% of the samples examined. The genera *Lecane* and *Trichocerca* were represented by 14 and 11 species respectively. Although most species belonging to these two genera occurred sporadically, *L. lunaris* and *T. longiseta* were very common (Table 1).

The rotifer fauna of the study area is typically arctic-temperate, as described by Green [1972]. It is dominated by *Notholca* and *Keratella* species but has a near absence of *Brachionus* species. However, when the species composition of northwestern

Canada is compared with northeastern Canda, considerable differences are noted. All the samples were collected in the same manner by one of the present authors (R.C.) and therefore such a comparison is reasonable. A survey of the rotifers of Labrador and Newfoundland in northeastern Canada yielded 131 species belonging to 40 genera [Chengalath & Koste, 1983]. In northwestern Canada only 97 species belonging to 32 genera, were found, although more than twice the number of samples were examined. *Brachionus* spp. were extremely rare in both regions. *Notholca* species were predominant in the northwest whereas they were almost absent in northeastern Canada. The genus *Keratella* was represented by at least 8 recognisable taxa in the present study, while only 5 species were recorded from the northeast. Furthermore, of the total 13 taxa of *Keratella* only *K. cochlearis* was present in both regions. Species of the genus *Ascomorpha* were entirely absent from the northwest, but were common in the northeast, whereas species of *Euchlanis* were common in both areas and occurred abundantly. However, apart from simple presence or absence, species abundance should also be taken into account when discussing distributions. Thus *Gastropus*, while represented in northwestern Canada, was not nearly as abundant as in the northeast.

Rotifers like *Euchlanis dilatata* and *Keratella cochlearis* are generally thought to be cosmopolitan in their distribution and these species were found during the present study. However, the widely held concept of cosmopolitanism in rotifers is based mainly on the idea that the resting eggs can be transported from one waterbody to another. Too little is known about the production, morphology, and dispersal of these eggs for any conclusions to be drawn. The same is true for those rotifers that do not produce resting eggs, but undergo anabiosis. Passive dispersal of rotifers does occur, as evidenced by the colonization of farm ponds, reservoirs, and other artificial waterbodies, but it seems that colonization does not occur indiscriminately. It is becoming increasingly clear that the species composition of rotifers in different parts of the world, or in the same geographic region, can be quite different, and that certain species have very limited distribution [Green, 1972; Pejler, 1977; Dumont, 1983] which is influenced by local environmental conditions such as altitude, latitude, water chemistry, food availability, predation, and competition [Green, this volume].

Acknowledgements

We thank Diana Laubitz, Chief Curator, Invertebrate Zoology Division, National Museum of Natural Sciences for reading the manuscript and for her support and encouragement. The help of Bonnie Smith, Elaine Simon, and C. C. Chinnappa of Department of Biology, University of Calgary in collecting the samples is gratefully acknowledged. We also thank Rolande Gaulin for typing the manuscript.

References

Ahlstrom, E. H., 1940. A revision of the rotatorian genera Brachionus and Platyias with descriptions of one new species and two new varieties. Bull. amer. Mus. nat. Hist. 77: 148–184.

Ahlstrom, E. H., 1943. A revision of the rotatorian genus Keratella with descriptions of three new species and five new varieties. Bull. amer. Mus. nat. Hist. 80: 411–457.

Chengalath, R., 1984. Synopsis Speciorum: Rotifera. Bibliographia Invertebratorum Aquaticorum Canadensium 3. National Museum of Natural Sciences, National Museums of Canada, Ottawa, 102 pp.

Chengalath, R. & C. H. Fernando, 1973. The planktonic Rotifera of Ontario with records of distribution and notes on some morphological variation. Can. Field-Nat. 87: 267–277.

Chengalath, R. & W. Koste, 1983. Rotifera from northeastern Quebec, Newfoundland and Labrador, Canada. Hydrobiologia 104: 49–56.

Dumont, H. J., 1983. Biogeography of rotifers. Hydrobiologia 104: 19–30.

Green, J., 1972. Latitudinal variation in associations of planktonic Rotifera. J. Zool., Lond. 167: 31–39.

Green, J., 1986. Keratella cochlearis (Gosse) in Africa. Hydrobiologia (This volume).

Harring, H. K., 1921. The Rotifera of the Canadian Arctic Expedition, 1913–1918. Rep. Can. Arct. Exped. 1913–18, 8E: 1–23.

Hooper, F. F., 1947. Plankton collections from the Yukon and Mackenzie river systems. Trans. am. microsc. Soc. 66: 74–84.

Koste, W., 1978. Rotatoria. Borntraeger, Stuttgart-Berlin, 2

vols., 673 pp., 234 plates.
Kutikova, L. A., 1970. (Kolavratki) Rotifera fauna SSSR. Subclass Eurotatoria. Zool. Inst. Akad. Nauk, Leningrad 104: 742 pp. (In Russian).
Murray, J., 1911. Canadian Rotifera: collected by the Shackleton Antarctic Expedition, 1909. J.r. microsc. Soc. (1911): 284–297.
Pejler, B., 1962a. On the variation of the rotifer Keratella cochlearis (Gosse). Zool. Bidr. Upps. 35: 1–17.
Pejler, B., 1962b. Notholca caudata Carlin, a new presumed glacial relict. Zool. Bid. Upps. 33: 453–457.
Pejler, B., 1977. General problems on rotifer taxonomy and global distribution. Arch. Hydrobiol. Beih. 8: 212–220.
Ruttner-Kolisko, A., 1974. Plankton rotifers, biology and taxonomy. Suppl. Die Binnengewasser, XXVI/i: 1–164.
Stemberger, R. S., 1979. A guide to rotifers of the Laurentian Great Lakes. United States Environmental Protection Agency, Environmental Monitoring and Support Laboratory. EPA-600/4-79-021: 1–186.

Hydrobiologia 147: 57–64 (1987)
May, L., Wallace, R. & Herzig, A. (eds), Rotifer Symposium IV

Coexistence of rotifer (*Brachionus plicatilis*) clones in Soda Lake, Nevada

Charles E. King & Yuqi Zhao
Department of Zoology, Oregon State University, Corvallis, OR 97331, U.S.A.

Keywords: genetic variation, population structure, rotifera, isozymes

Abstract

Patterns of genetic variation among *Brachionus plicatilis* clones collected at different times from Soda Lake, Nevada (U.S.A.) were determined using polyacrylamide gel electrophoresis of isozymes. Six of the eleven enzymes assayed were variable. Three of these – malic dehydrogenase (MDH), cholinesterase (CHO), and phosphoglucomutase (PMG) – were selected to screen the 138 clones established from the field collections. Each enzyme revealed variation both within and between field samples. Thirty-eight of the 100 potential composite electromorphs formed by combining isozyme variants were found in the 138 clones. A substantial amount of genetic variation is therefore present. Patterns of electromorph frequency change resembled those reported for cladocerans. No evidence was found for a seasonal succession of distinct electromorphs. We suggest that the *B. plicatilis* in Soda Lake constitute a single population.

Introduction

Two models of rotifer population structure were proposed by King [1972, 1977] to account for adaptation to temporal environmental variation. One model, that of 'complete genetic discontinuity', depicts a rotifer species as being subdivided into a number of spatially overlapping but temporally discrete populations that are both genetically and ecologically isolated. Successional replacement of populations takes place as the lake environment undergoes its seasonal changes during the year. Evidence has been presented by King [1977], Snell and King [1977], and Snell [1979] that this model is the most appropriate description of the population structure of *Asplanchna*. Under the model of complete genetic discontinuity, clones from different populations co-occur only during periods of clonal replacement. Critical elements in this model are the prevention of gene flow via sexual recombination between clones from different populations and the prevention of coexistence by either differential hatching times of the sexual resting eggs or strong selection against individuals that hatch into the 'wrong' environment.

A second model, that of 'incomplete genetic discontinuity,' predicts coexistence of clones that undergo great changes in their relative frequencies during the year. For example, clones adapted to cool-water conditions maintain high frequency during the early part of the year, while their warm-water counterparts are at low frequency. As the season progresses and lake temperature increases, warm-water genotypes increase in frequency while cold-water genotypes decrease. Rotifers appropriately considered by this model are not structured into a succession of isolated populations. The model therefore predicts relative frequency changes among the clones in a single population, and not absolute changes in presence or absence of the different clones as in the model of complete genetic discontinuity.

A major goal of the present research has been to investigate the genetic structure of the rotifer *Brachionus plicatilis* in Soda Lake, Nevada, U.S.A. We have made four collections in the period from

March to September 1984. We will present in this paper the results of an analysis of the isozyme variation among clones established from these collections, and from very small samples from five earlier collections. Additional experiments performed to analyze the life table characteristics of these clones will be presented elsewhere.

Materials and methods

Soda Lake is a permanently stratified alkaline crater lake in northern Nevada (39°31′N, 118°52′W). In 1933, when G. E. Hutchinson [1937] visited the lake and reported *Brachionus plicatilis* to occur there, the lake had a maximum depth of 64.5 m and an average depth of 26.2 m. More recently the lake has become deeper and less alkaline due to irrigation of the surrounding land. Additional details on the lake are available in Morrison [1969], Koenig *et al.* [1971], and Kimmel *et al.* [1978].

The samples of *B. plicatilis* that form the major part of this report were established from collections made in March, May, July, and September 1984. Data from a few clones from earlier collections (August 1980, October 1981, April 1982, and May and June 1983) have also been included, but we must emphasize that the extremely small sizes of these samples renders them of little significance in indicating temporal patterns.

Individuals used to start the clones were isolated at the field site and were then brought back to the laboratory for culture and later analysis. Culture work was done in a defined artificial sea water at 20 ppt salinity and pH 8.0. Both *Dunaliella salina* and *Chlorella* sp. were used to feed stock cultures.

Isozyme variation was detected by using polyacrylamide gel electrophoresis in a 1.5 mm vertical slab gel. Eleven enzymes have been surveyed: malate dehydrogenase (MDH), malate enzyme (ME), 6-phosphogluconate dehydrogenase (6-PGDH), lactate dehydrogenase (LDH), glutamate dehydrogenase (GDH), cholinesterase (CHO), fumarase (FUM), phosphoglucomutase (PGM) and tetrazolium oxidase (TO). For details of the techniques employed, refer to King [1977], Snell [1979], and Snell and Winkler [1983]. The consistency of detected isozyme patterns was established and confirmed by repeated analysis of some clonal lines. An Austrian clone of *B. plicatilis,* kindly provided by Dr. A. Ruttner-Kolisko, was used as a marker in all runs. The resulting bands were scored and relative mobilities were calculated by measuring the migration distance of the brom phenol blue dye band, the enzyme bands, and the length of the gel.

Results

A. Electromorphs

138 clones were examined by electrophoresis. No isozyme variation among clones was found for five of the eleven enzymes assayed. These enzymes were: FUM, 6PGDH, ME, LDH and TO. In addition, only one clone (from May 1983) had a variant pattern for PGI. The remaining 5 enzymes were variable among clones. Variations in EST and GDH patterns were obvious, but difficult to interpret because of variable staining intensities and numerous faint bands. From two to eight bands of EST activity were present and two to six bands were observed in GDH. These two enzymes were omitted from analysis. The invariant enzymes and EST and GDH were examined in the 54 clones collected through March, 1984. The remaining three variable enzymes (MDH, CHO and PGM) were assayed on all clones.

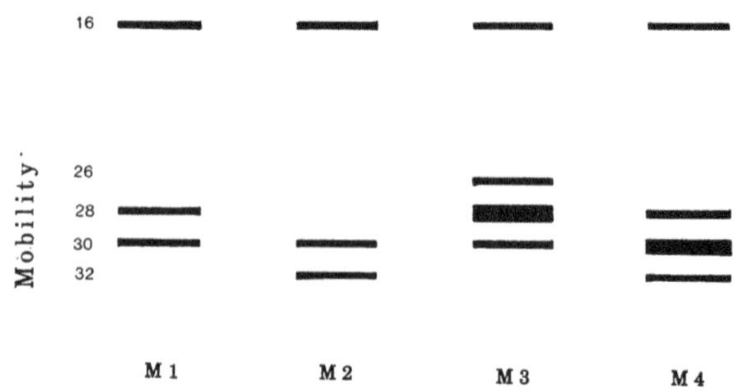

Fig. 1. Electromorphs of malic dehydrogenase (MDH). M1-M4 indicate the four observed patterns. Mobilities are expressed as percentages and indicate migration distances of each band relative to the distance migrated by the brom-phenol blue tracking dye.

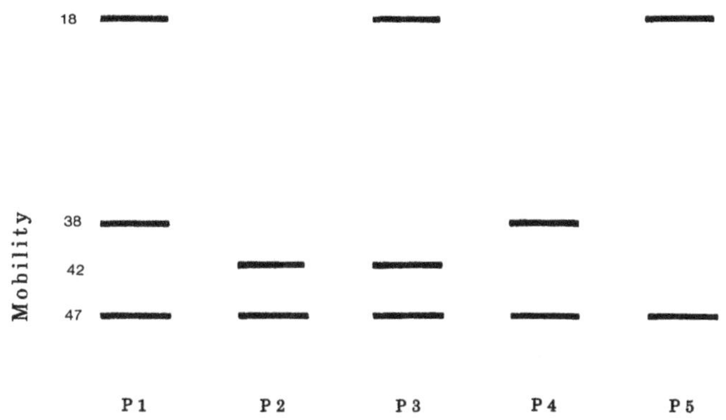

Fig. 2. Electromorphs of phosphoglucomutase (PGM). Pl-P5 indicate the five observed patterns. Mobilities are expressed as percentages and indicate migration distances of each band relative to the distance migrated by the brom-phenol blue tracking dye.

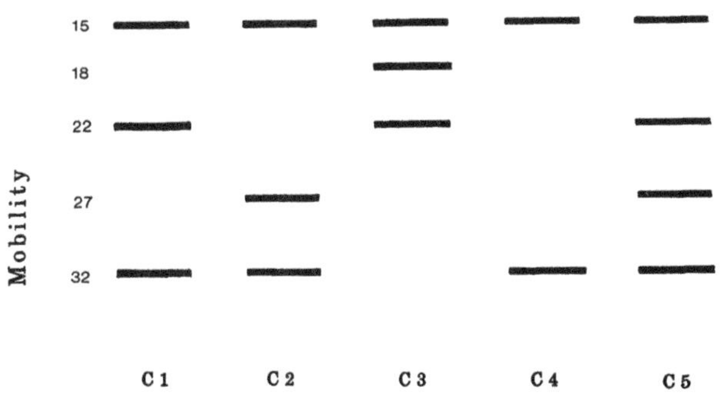

Fig. 3. Electromorphs of cholinesterase (CHO). Cl-C5 indicate the five observed patterns. Mobilities are expressed as percentages and indicate migration distances of each band relative to the distance migrated by the brom-phenol blue tracking dye.

Electrophoresis revealed four electromorph patterns for the enzyme malic dehydrogenase (Fig. 1). MDH showed two banding zones, MDH-1 and MDH-2. MDH-1 always had a single band with an invariant mobility of 17% relative to the distance migrated by the brom-phenol blue tracking dye. The MDH-2 zone was variable and had four electromorphs that have been designated Ml-M4. In the absence of genetic crosses we cannot provide a concrete explanation for the genetic basis of these electromorphs. It is, however, reasonable to assume that clones having different isozyme patterns are of different genotypes.

Phophoglucomutase (PGM) had two banding zones and five different patterns (Fig. 2) which have been designated Pl-P5. Cholinesterase (CHO) had five patterns (Fig. 3) which have been designated Cl-C5.

B. Clonal Electromorphic Frequencies

The extent of electrophoretically-detectable temporal variation among the clones is remarkable. Electromorphs at MDH, PGM, and CHO loci show striking changes in frequency. Since only a few clones were obtained in 1980–1983, these collections can only indicate the existence of certain electromorphs and cannot provide information about electromorph frequencies. More extensive information is available from collections made in March of 1984 and thereafter. A substantial measure of caution is still needed in discussing frequency data from these later collections for sample sizes are not large and it is not possible to measure variation due to sampling effects.

Electromorph frequencies of MDH are presented in Fig. 4. M4 was the most common pattern in May 1983, and was at the same approximate frequency of fifty percent one month later in June. In the following spring collection (March 1984), M4 was at a very low frequency and another MDH electromorph, Ml, was dominant. In the last three collections of 1984, M4 appeared to have regained its dominance and was at a frequency of approximately 50% from May through September.

Five electromorphs of PGM were found (Fig. 5). One of the five, Pl, was present at high frequency in all samples. Another, P3, was present in the four samples taken before May, 1984, but appeared to be absent thereafter. P2 occurred only once, in March, 1984.

Frequencies of CHO electromorphs are presented in Fig. 6. Substantial oscillations occurred for each of the five electromorphs. Three of the five electromorphs were present in most of the collections; the other two had more restrictive distributions.

In summary, most electromorphs were present in most collections. There is no evidence for seasonal succession of a single or a few dominant clones in Soda Lake. Given adequency of samples, replace-

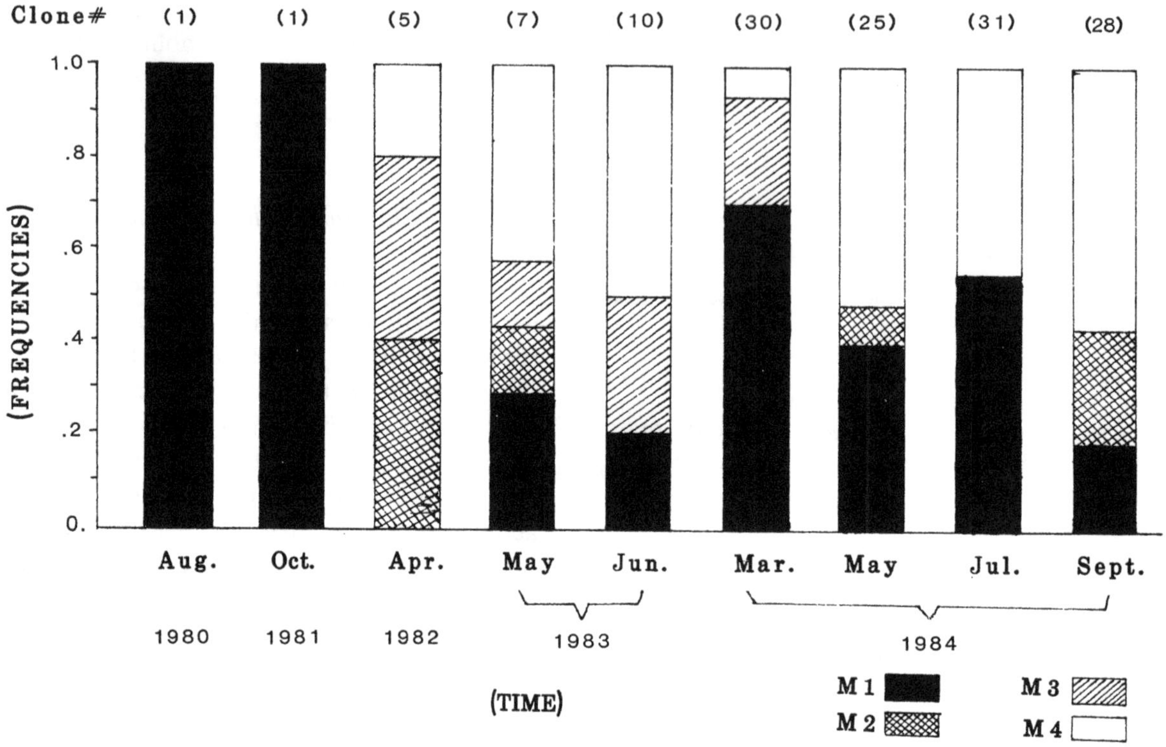

Fig. 4. Temporal pattern of MDH frequencies in Soda Lake. M1-M4 are the four MDH electromorphs presented in Fig. 1. The number of clones scored from each collection is indicated at the top of each bar.

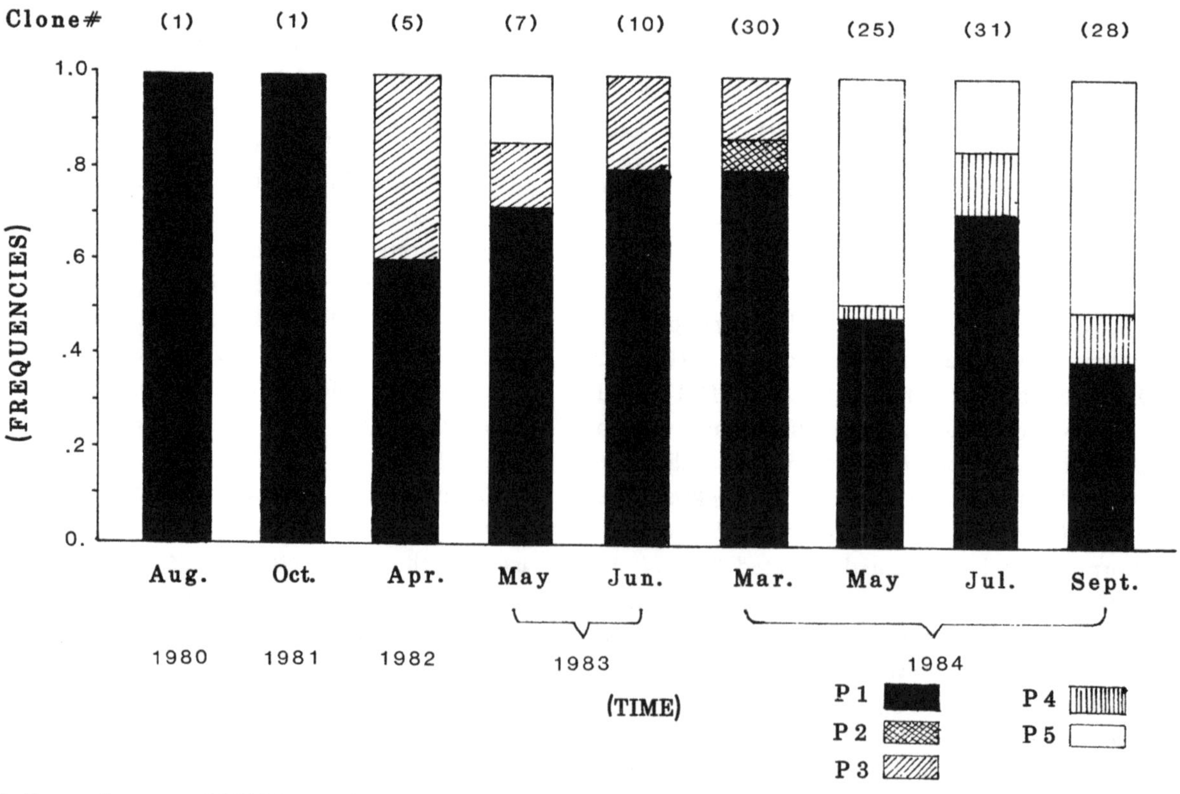

Fig. 5. Temporal pattern of PGM frequencies in Soda Lake. P1-P5 are the five PGM electromorphs presented in Fig. 2. The number of clones scored from each collection is indicated at the top of each bar.

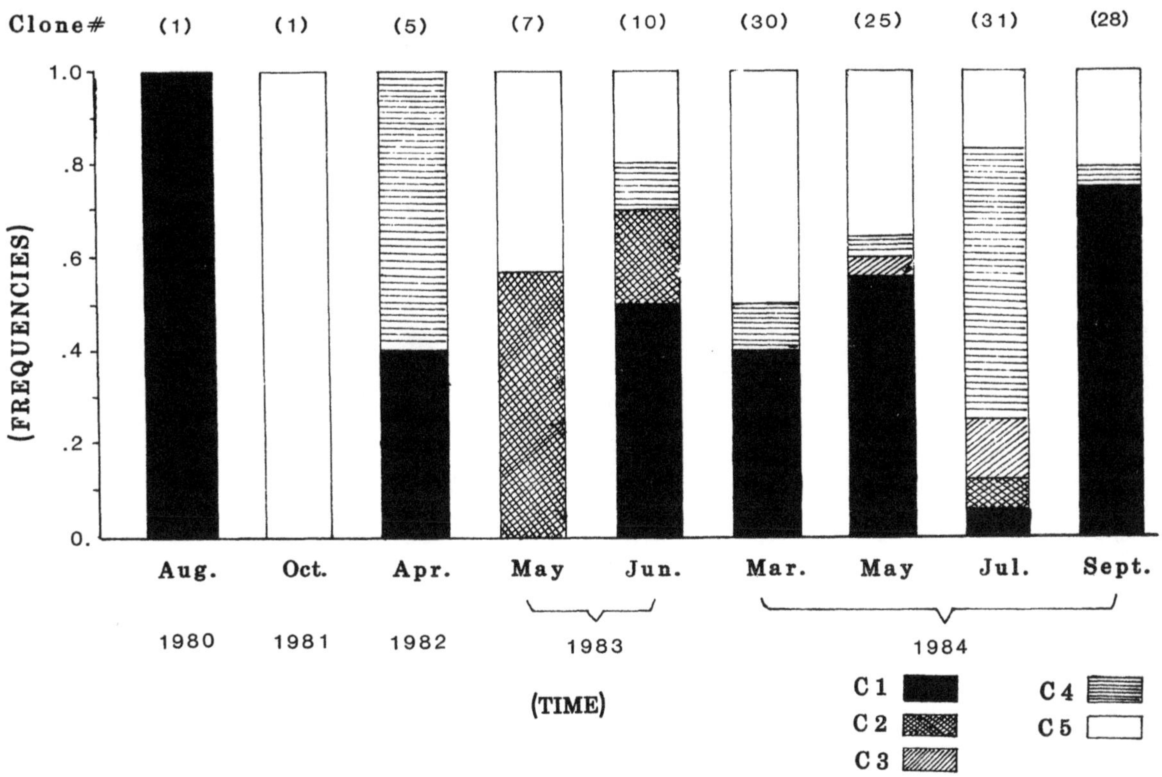

Fig. 6. Temporal pattern of CHO frequencies in Soda Lake. C1-C5 are the five CHO electromorphs presented in Fig. 3. The number of clones scored from each collection is indicated at the top of each bar.

ment of dominant clones also exists as evidenced by major changes in electromorph frequencies.

C. Composite electromorphs

Composite electromorphs (i.e., combinations of electrophoretic phenotypes) may also be used to identify clonal associations that occur repeatedly through time. Frequencies of the composite electromorphs are presented in Table 1. At least 11 and as many as 13 composite electromorphs were found in each of the 1984 samples. Since four electromorph patterns were identified for MDH and five each were found for CHO and PGM, there are potentially 100 different composite electromorphs for the three enzymes. The 38 composite electromorphs collected therefore represent 38% of the possible combinations. Clearly, this is a substantial portion of the potential variation.

Temporal changes in the frequencies of composite electromorphs were also observed among *B. plicatilis* clones in Soda Lake. In early March of 1984, M1-P1-C5 constituted 37 percent of the population sample; it had declined to 4% in the May sample and in mid-July it was 10%. This composite electromorph was absent from the collection taken at the beginning of September. A new composite electromorph M1-P1-C4 appeared in July at a frequency of 35 percent, and another, M4-P5-C1, was the dominant clone in the population in early September at a frequency of 29%. No single composite electromorph constituted over 37 percent of the population in any collection, nor did any composite electromorph maintain a frequency of greater than 16% in any two successive collections.

The picture obtained from this analysis is that there is a great deal of variation in single collections and even more variation through time. Clearly, the clones of *B. plicatilis* inhabiting Soda Lake constitute a very diverse assemblage.

Table 1. Percentage distribution of composite electromorphs in Soda Lake.

No. composite	1980	1981	1982	1983		1984			
	Aug	Oct	Apr	May	Jun	Mar	May	Jul	Sep
M_1-P_1-C_1	100				10	20	16	3	7
M_1-P_1-C_2				14					
M_1-P_1-C_3								3	
M_1-P_1-C_4					10			35	
M_1-P_1-C_5		100				37	4	10	
M_1-P_2-C_1						3			
M_1-P_2-C_4						3			
M_1-P_3-C_5						3			
M_1-P_4-C_1									4
M_1-P_5-C_1							16		4
M_1-P_5-C_4								3	4
M_1-P_5-C_5							4		
M_2-P_1-C_1			40						14
M_2-P_2-C_5				14			4		4
M_2-P_5-C_1									7
M_2-P_5-C_4							4		
M_3-P_1-C_1					10	13			
M_3-P_1-C_2				14	10				
M_3-P_1-C_4			20						
M_3-P_1-C_5						7			
M_3-P_3-C_4			20			3			
M_3-P_3-C_5					10				
M_4-P_1-C_1					20	3	12		4
M_4-P_1-C_2				14	10			6	
M_4-P_1-C_3							4	6	
M_4-P_1-C_4								6	
M_4-P_1-C_5				14	10		4		11
M_4-P_3-C_1					10				
M_4-P_3-C_2				14					
M_4-P_3-C_4			20			3		6	
M_4-P_3-C_5						3			
M_4-P_4-C_1							4	3	7
M_4-P_4-C_3								3	
M_4-P_4-C_4								6	
M_4-P_4-C_5				14					
M_4-P_5-C_1							8		29
M_4-P_5-C_4								3	
M_4-P_5-C_5							20	6	7
No. composite electromorphs	1	1	4	7	9	11	12	13	12
No. clones	1	1	5	7	10	30	25	31	28

Discussion

Isozyme analyses of *B. plicatilis* from Soda Lake indicate that a number of clones were collected multiple times from different seasons and may coexist over an extensive period. In this respect, the Soda Lake study produces similar results to a number of studies of Cladocera [Hebert, 1980; Hebert & Crease, 1983; Lynch, 1983, 1984; Young, 1979, 1984]. The coexistence of clones in Cladocera can, at least in part, be ascribed to their relatively long life spans. However, since *B. plicatilis* has a much shorter life cycle, extended coexistence of different clones raises interesting questions about both their genetical and ecological structures.

Two questions must be answered before we can achieve even a rudimentary understanding of the structure of this system. First, are the clones of *B. plicatilis* in Soda Lake genetically isolated from one another? Neither males of *B. plicatilis* nor females carrying resting eggs were observed in the Soda Lake collections so isolation is possible if there is no sexual reproduction. We regard this possibility as remote, however, since clones returned to the laboratory produced both males and resting eggs that underwent normal development and hatching. King [1977] suggested the existence of seasonally isolated populations in *Asplanchna girodi* in Golf Course Pond near Tampa, Florida. In that study sexual reproduction occurred frequently and populations were observed to undergo succession. In contrast, there is an apparent coexistence of clones in Soda Lake. It is therefore reasonable to conclude that the *B. plicatilis* in Soda Lake constitute a single diverse population.

The second question is what are the forces shaping the pattern of genetic diversity observed in this study? A definitive answer would require more extensive sampling and larger sample sizes. Significant ecological differences among clones and genotypes have been found in our life table experiments therefore it is clear that not all clones are selectively equivalent. These results will be presented in a separate paper.

Co-occurence does not necessarily demonstrate stable coexistence. In fact, over half of the identified composite electromorphs were found only once. Thirty-one composite electromorphs were identified in the four samples of 1984. Of these, 19 occurred once, 7 occurred twice, 4 occurred three times, and only one was present in all four samples. Although part of this variation is likely to be due to sampling effects, some of the clones that occurred only once were represented multiple times in their samples. It therefore seems likely that both clonal replacements and shifts in relative frequencies are taking place.

Clearly, the model of complete genetic discontinuity proposed by King [1972, 1977] does not apply to many of the clones in this system. An alternative model, that of incomplete genetic discontinuity may be a more appropriate statement of *B. plicatilis* population structure in Soda Lake. That is, clones adapted to a certain environment maintain high frequency when that environment is present, and are at low frequency in other environmental states.

Acknowledgements

This research was supported by grants from the National Science Foundation and National Institutes of Health.

References

Hebert, P. D. N., 1980. Clonal coexistence in Daphnia pulex (Leydig): another plankton paradox. Science 207: 1363–1365.

Hebert, P. D. N. & T. Crease, 1983. Clonal diversity in populations of Daphnia pulex reproducing by obligate parthenogenesis. Heredity 51: 353–369.

Hutchinson, G. E., 1937. A contribution to the limnology of arid regions. Transactions of the Connecticut Academy of Arts and Sciences 33: 50–132.

Kimmel, B. L., R. M. Gersberg, L. J. Paulson, R. P. Axler & C. R. Goldman, 1978. Recent changes in the meromictic status of Big Soda Lake, Nevada. Limnol. Oceanogr. 23: 1021–1025.

King, C. E., 1972. Adaption of rotifers to seasonal variation. Ecology 53: 408–418.

King, C. E., 1977. Genetics of reproduction, variation and adaptation in rotifers. Arch. Hydrobiol. Beih. 8: 187–201.

Koenig, E. R., J. R. Baker, L. J. Paulson & R. W. Tew, 1971. Limnological status of Big Soda Lake, Nevada, October 1970. Great Basin naturalist 31: 107–111.

Lynch, M., 1983. Ecological genetics of Daphnia pulex. Evolution 37: 358–374.

Lynch, M., 1984. The genetic structure of a cyclic parthenogen. Evolution 38: 186–203.

Morrison, R. B., 1969. Lake Lahontan: geology of southern Carson desert, Nevada. U.S. Geological Survey, Professional paper 401: 1–156.

Snell, T. W., 1979. Intraspecific competition and population structure in rotifers. Ecology 60: 494–502.

Snell, T. W. & C. E. King, 1977. Lifespan and fecundity patterns in rotifers: the cost of reproduction. Evolution 31: 882–890.

Snell, T. W. & B. C. Winkler, 1983. Isozyme analysis of rotifer protein. Biochemical systematics and ecology 12: 199–202.
Young, J. P. W., 1979. Enzyme polymorphism and cyclic parthenogenesis in Daphnia magna. I. Selection and clonal diversity. Genetics 92: 199–202.
Young, J. P. W., 1984. The population structure of cyclic parthenogens. In: G. S. Oxford & D. Rollinson (eds.) Systematics Association Special Volume No. 24, Protein polymorphism: Adaptive and Taxonomic Significance. Academic Press, Lond.; N.Y.: 361–378.

Hydrobiologia 147: 65–73 (1987)
May, L., Wallace, R. & Herzig, A. (eds), Rotifer Symposium IV

Rotifers from Turkey

Henri J. Dumont & M. De Ridder
Zoological Institute, State University of Gent, Belgium

Keywords: Rotifers, taxonomy, biogeography, anatolia, Turkey

Abstract

Seventy-eight rotifer species are reported from 19 localities in Turkey. About half are first records for that country. Some interesting species are illustrated and the composition of the regional fauna is discussed. Wide-ranging taxa are dominant, with notbale additions of tropical, subtropical, afrotropical and circum-boreal species. Many are restricted to saline environments, where some show an unusual vertical distribution. Endemism is suspected in a few taxa.

Introduction

The proportion of rotifer species endemic to old and remote continents and major islands like Australia, South America, Madagascar, and New Zealand is about 5–8% of their total fauna, by present-day taxonomic standards. In contrast, the vast continental masses of Eurasia, Africa, and North American seem to be relatively free of barriers to rotifer dispersal, and have a lower proportion of endemism. However, this statement should be applied with caution [Dumont, 1983], because most rotifer species have been described from this part of the world, and about 30% of all species have only been recorded once. There is a scarcity of data from many parts of the world and this certainly applieds to Anatolia (Turkey-in-Asia). This major continental block projecting westward on the Asian landmass, separating the Black Sea from the Mediterranean. It is situated at the crossroads of the Palaearctic, Oriental, and Afrotropical regions and, because of its complicated relief, presents environments ranging from Mediterranean shrublands to mountain forests and arid steppes.

Anatolia has many large lakes. Some are fresh (lakes Beysehir and Egridir), while other are saline (lakes Burdur, Aksehir, Van) or brines which may dry out in summer (lakes Aci, Tuz). In spite of this variety in aquatic biotopes, little is known about their rotifer fauna. Daday [1903] was the first to publish on Anatolian rotifers (lakes Apolyont and Iznik in West Anatolia), followed by Vavra [1903] and Zederbauer & Brehm [1907] (the east Anatolian mountain lake Sary Göl on Erdschias mountain). Much later, Geldiay [1949], Hauer [1957], and Tokat [1976] identified rotifers from Emir lake and Tchoubouk dam near Ankara, lake Van (Van Gölü), and Hazar Lake (Göl çük), respectively. Zederbauer & Brehm [1907] did not find rotifers in the saline crater lake Kratergöl, but Dumont [1981] reported one species from this site in central Anatolia (= loc 11 of this paper). Short lists of rotifer species from various lakes are also found in Mann [1940], Margaritora & Cottarelli [1970], and Margaritora *et al.* [1977]. Figure 1 shows the location of the most important lakes previously studied.

These studies resulted in a short checklist of 47 species and two subspecies (see appendix). Clearly, a taxonomic account of the rotifer material collected by the Gent University expedition to Turkey in July 1973 would be useful. Samples collected in

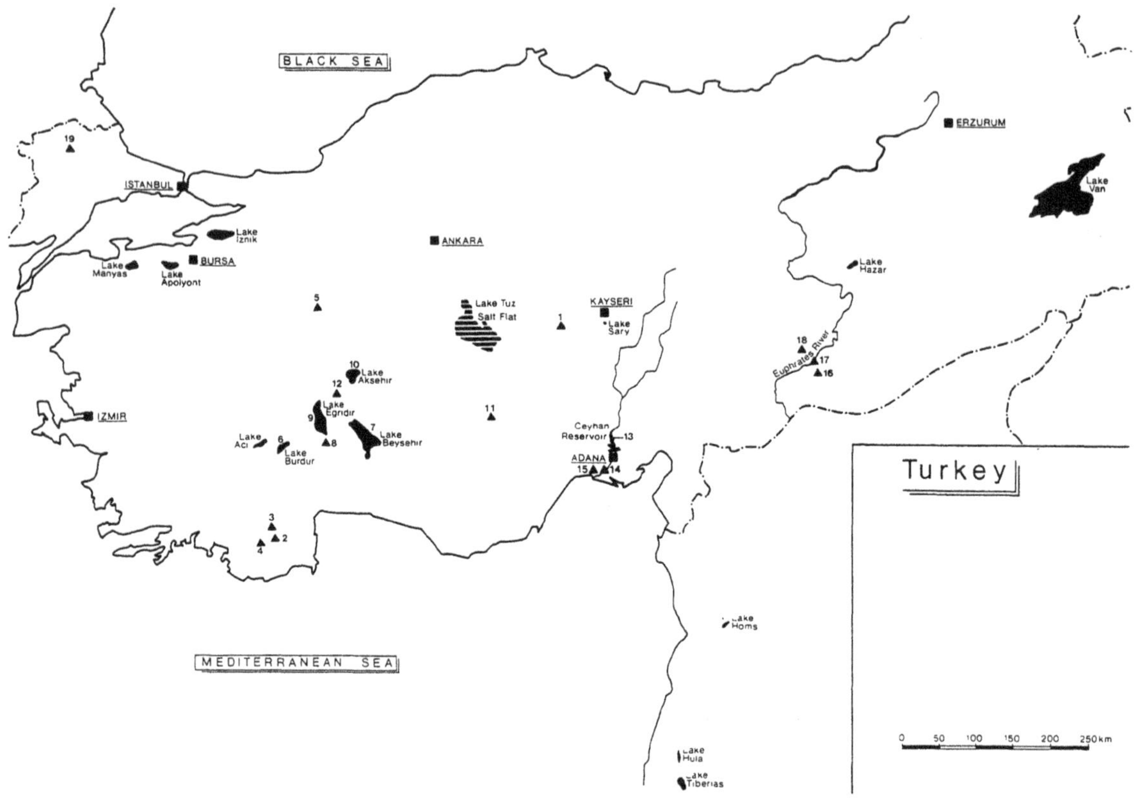

Fig. 1. Sampling localities in Turkey. Sites of important previous records also indicated.

May 1972, and in July 1975 were also included. Most samples were collected in the major lakes of Central Anatolia, but others from pools, marshes, and ditches.

List of Localities (see Fig. 1).

Samples from localities not containing rotifers are not mentioned.

1. Göreme, Central Anatolia, pool, 12.5.1972.
2. Kanle Çesme, West Anatolia, spring-fed pool, 20.7.1973.
3. Golova, West Anatolia, ditch overgrown with reeds, 21.7.1973.
4. Buda Çay near Elmali, West Anatolia, ditch, 22.7.1973.
5. Çifteler, West Anatolia, river, 17.7.1973.
6. Lake Burdur, 21.7.1973. Collection from the littoral, and a vertical series collected with a closing net: 0–5 m, 5–10 m, 10–20 m, 20–30 m, 30–40 m, 40–50 m, 50–60 m. pH 9.0–9.6, conductivity 25000 $\mu S\ cm^{-1}$, and chloride content c. 5g l^{-1}.
7. Lake Beysehir, 25.7.1973, freshwater lake with many swampy bays rich in submerged macrophytes. Samples collected in the open water, but a particular effort was made to study the littoral fauna. Conductivity 350 $\mu S\ cm^{-1}$; c. 35 mg l^{-1} Cl^-; pH 7.5–7.7.
8. Pond 6 km W of Egridir town, Central Anatolia, 23.7.1973.
9. Lake Egridir, 23.7.1973. The second-largest freshwater lake of Anatolia, with open water and littoral on its eastern side. Conductivity 310 $\mu S\ cm^{-1}$, 24 mg l^{-1} Cl^-, pH 7.1.
10. Lake Aksehir, 26.7.1973. Brackish water lake. Horizontal and vertical hauls (0–4 m) in open water and littoral with *Phragmites*. Conductivity 1900 $\mu S\ cm^{-1}$, 3.3 g $l^{-1}Cl^-$, pH 7.6.

11. Krater Göl, 27.7.1973. For details, see Dumont [1981].
12. Yarma Çay, rivulet near Aksehir, 26.7.1973.
13. Seyhan Reservoir on coastal plain of Adana, South Anatolia, 29.7.1973.
14. Ditch along road between Tarsus and Pozanti at foot of Cilician gates, South Anatolia, 29.7.1973.
15. Swamp 20 km East of Tarsus, South Anatolia, 31.7.1973.
16. Pool at Bozova, East Anatolia, 2.8.1973.
17. Euphrates River near Samsat, East Anatolia, 2.8.1973.
18. Adjandida, village pool near Adiyaman, East Anatolia, 2.8.1973.
19. Meandering river in woodlands 10 km E. of Kirklareli, Thracia, 17.8.1975.

Species collected

An alphabetical list of species is given below. The numbers following the species names refer to the localities (see above). The infrasubspecific category 'forma' is applied only to identify a particular phenotype, not a taxon. Underlined taxa are illustrated in the text. Species marked with an asterisk are new records for Turkey.

* *Anuraeopsis coelata* (de Beauchamp): 13
* *Anuraeopsis fissa* (Gosse): 6, 18
Asplanchna sp.: 6
Asplanchna priodonta Gosse: 7 (+ 0)
Asplanchnella brightwelli (Gosse): 7, 19
* Asplanchnella girodi (De Guerne): 7
Bdelloidea: 2, 5, 7, 12, 17
Brachionus angularis Gosse: 13
Brachionus budapestinensis Dad.: 14, 17
Brachionus calyciflorus Pallas: 4, 7, 17
f. *anuraeiformis*: 4, 7, 17
Brachionus plicatilis O.F.M.: 6 (20–30 m. level only), 13
Brachionus rubens Ehrbg. 6, 7, 19
Brachionus quadridentatus
f. brevispinus (Ehrbg.) : 19
f. cluniorbicularis (Skorikov)x 17
* *Cephalodella exigua* (Gosse): 19
* *Cephalodella gibba* (Ehrbg.): 6
Cephalodella sp.: 17
Colurella adriatica Ehrbg.: 17, 18
* *Colurella colurus* (Ehrbg.): 5
* *Colurella obtusa* Gosse: 3
* *Colurella uncinata* (O.F.M.): 7
* *Dicranophorus forcipatus* (O.F.M.): 7
* *Dicranophorus grandis* (Ehrbg.): 7
* *Eosphora ehrenbergi* Weber: 7
* *Epiphanes brachionus spinosa* (Rousselet): 16, 19
Epiphanes macroura Barr. & Dad.:
* *Euchlanis oropha* Gosse: 17
* *Filinia brachiata*: 7
* *Filinia cornuta* (Weissl): 16
f. *passa*: 7
Filinia longiseta (Ehrbg.): 10, 11
f. *limnetica* [Zacharias] : 2, 10
* *Filinia saltator*: 2
Filinia terminalis (Plate): 7, 9
* *Gastropus hyptopus* (Ehrbg.): 15
Hexarthra fennica (Lev.): 6, 10, 11 (0–50 m:only ♀; 50–70 m: ♂, ♀), 13, 17
Hexarthra mira (Hudson & Gosse): 1, 2 (♂), 15 (♂), 19
* *Hexarthra oxyuris* (Sernov): 6 (only between 5 and 20 m).
Hexarthra polyodonta (Hauer): 10
Keratella cochlearis (Gosse): 6, 10
Keratella quadrata (O.F.M.): 7
Keratella tecta (Gosse): 7
Keratella testudo (Ehrbg.): 7 (dominant)
* *Keratella tropica* (Apstein): 7, 10, 13, 15, 17, 18
f. *reducta* (Fadeew): 7, 13, 17
f. *brehmi* (Klausener): 13
Keratella valga Ehrbg.: 7
* *Lecane bulla* (Gosse): 1, 6, 17
* *Lecane closterocerca* (Schmarda): 1
* *Lecane hamata* (Stokes) : 1, 7
* *Lecane ludwigi* (Eckstein): 1
Lecane luna (O.F.M.): 1, 5, 6, 7, 17, 19
* *Lecane lunaris* (Wulf.): 1, 5, 6
f. *perplexa*: 1,7
* *Lecane pyriformis (Dad.) : 7*
Lecane quadridentata (Ehrbg.): 7
* *Lecane ungulata* (Gosse): 13, 16
Lecane sp. : 1
* *Lepadella acuminata* (Ehrbg.): 6

* *Lepadella ovalis* (O.F.M.): 6
* *Lepadella patella* (O.F.M.): 1, 5, 6
* *Lepadella persimilis* de Ridder: 10
* *Lepadella quinquecostata* (Lucks): 7
  Mytilina ventralis (Ehrbg.): 7
  Notommata cyrtopus Gosse: 6
  Notholca acuminata Ehrbg.: 6 (only between 40 and 50 m)
* *Platyias quadricornis* (Ehrbg.): 1
  Polyarthra dolichoptera Idels.: 7, 13, 16, 17, 19
  f. *aptera*: 19
* *Polyarthra euryptera* (Wierzejski): 7, 8
* *Polyarthra remata* (Skorikov): 7, 9
  Proalides verrucosus (Barr, & Dad.): 16
* *Rotaria neptunia*: 19
* *Synchaeta oblonga* Ehrbg.: 7
* *Synchaeta stylata* Wierzejski: 7, 17
  Synchaeta sp. : 16
* *Testudinella emarginula* (Stenroos): 1
  Testudinella patina Ehrbg.: 1, 16
* *Trichocerca collaris (Rousselet): 17*
  Trichocerca taurocephala Hauer: 9
* *Trichocerca tenuior* (Goose): 6, 17
* *Trichotria curta* (Skorikov): 17
* *Trichotria pocillum* (O.F.M.): 7
* *Trichotria tetractis* (Ehrbg.): 17
* *Tripleuchlanis plicata* (Levander): 17
* *Wolga spinifera* (Western): 7, 8

Discussion

The present contribution almost doubles the number of rotifer species known from Turkey, but the majority are common and widespread species. The results suggest that only a fraction of the fauna

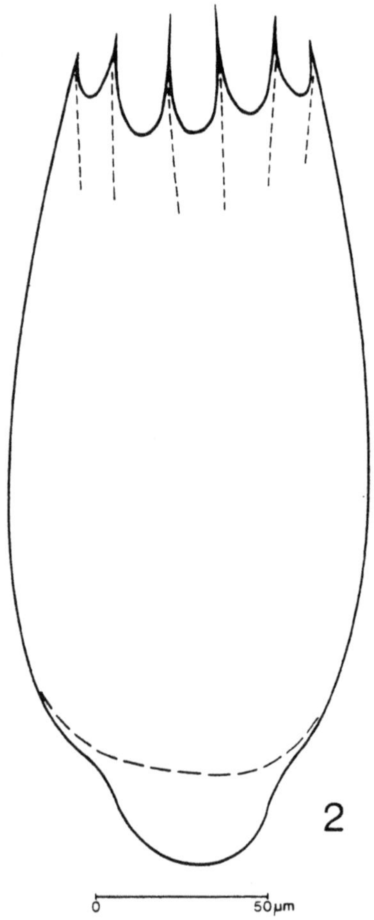

2

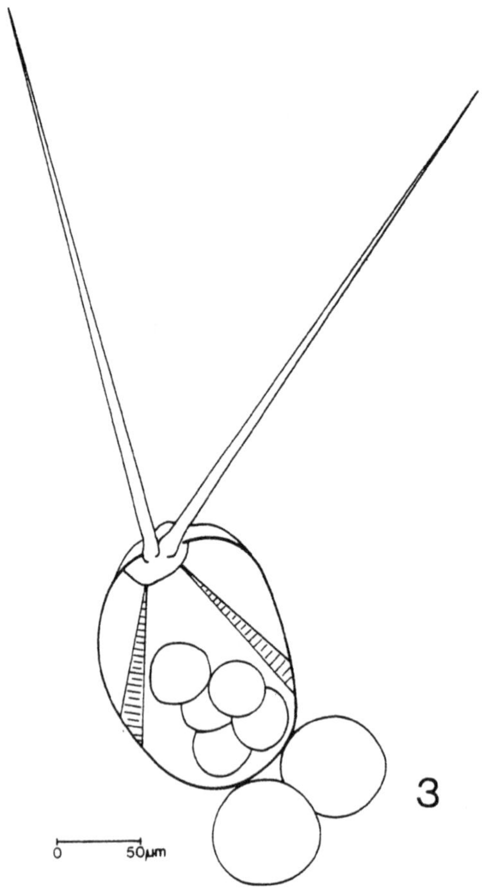

3

Fig. 2–3. 2. *Notholca acuminata* from Burdur; 3. *Filinia saltator* from Kanle Çesme.

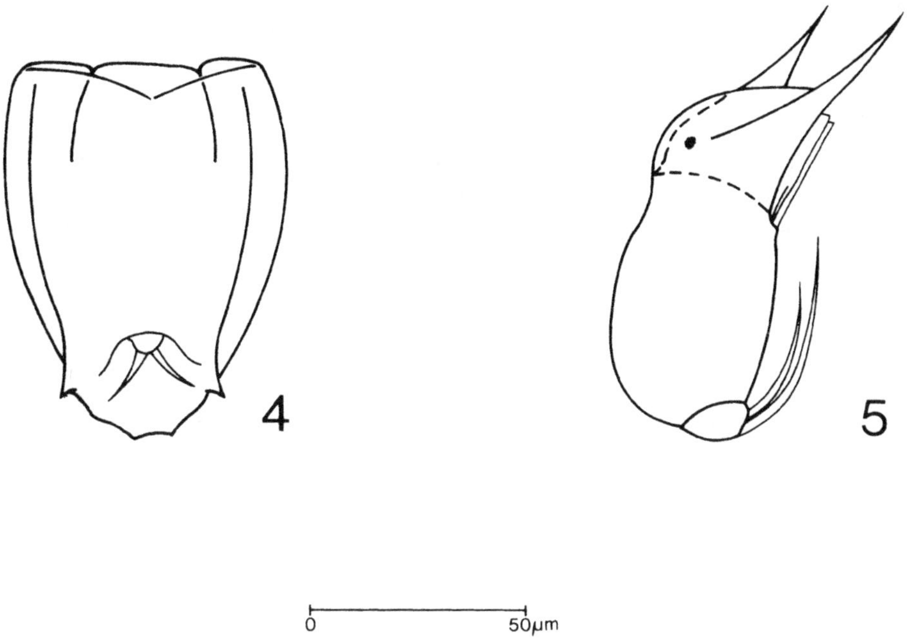

Fig. 4–5. 4. *Wolga spinifera* from Lake Beysehir; 5. *Trichocerca taurocephala* from Lake Egridir.

have been recorded to date. Information from other animal groups suggests that Turkey is an addition-substraction zone for faunas of southern, northern, and eastern origin, supplemented by some endemic species. This also appears to be true for rotifers.

Anuraeopsis coelata de Beauchamp, limited to Africa and India [Wulfert, 1966], has not been cited north of Sennar Reservoir in the Nile valley [de Ridder, 1984]. Its presence in Seyhan reservoir, on the hot coastal plain of Adana, now virtually eliminates the apparent gap in its range. Circumtropical taxa are also found here, including *Filinia saltator* (Fig. 3) and *Keratella tropica*.

Similarly, circumsubtropical taxa are represented by *Epiphanes brachionus spinosa* (Fig. 6) and several *Lecane* and *Lepadella*.

Northern species which used Anatolia as a stepping stone to extend southwards into Africa during the colder and more humid periods of the late Pleistocene, were also found. Relicts of these still survive on the Ethiopian highlands, and some species occasionally descend into the Nile valley in affluents of the Blue Nile, e.g. *Notholca acuminata* (Fig. 2), *Wolga spinifera* (Fig. 4), and *Keratella serrulata*. The first two of these species are now also known from the intermediate site of Anatolia.

The commonness of *Keratella testudo* (Fig. 9) on the Anatolian plateau is also significant. This northern species, typical of weedy pools, was first discovered in Lake Apolyont by Daday [1903] and was a dominant feature of the plankton of Lake Beysehir.

A similar surprise is the presence of *Trichotria curta* (Fig. 7) in the Upper Euphrates, in the steppes of East Anatolia. This species, predominantly boreal in occurrence, may have been carried down by the river current from the high altitude, continental zone of north-east Anatolia.

As usual among rotifers, the majority of taxa recorded were found to be wide-ranging, ubiquitous 'species', tolerant of considerable variations in pH, salinity, and temperature. *Lecane luna* (Fig. 8), *Keratella cochlearis*, *K.quadrata*, and many others, fall into this category.

Endemism is difficult to decide upon, for reasons explained in the first paragraphs of this paper. *Proalides verrucosus* (Fig. 13), which has only been recorded from the Levant, Anatolia, and the Wolga valley (districts of Vladimirsk, Sartovsk, Charkov and Astrachan) [Skorikov, 1914] is a possible candi-

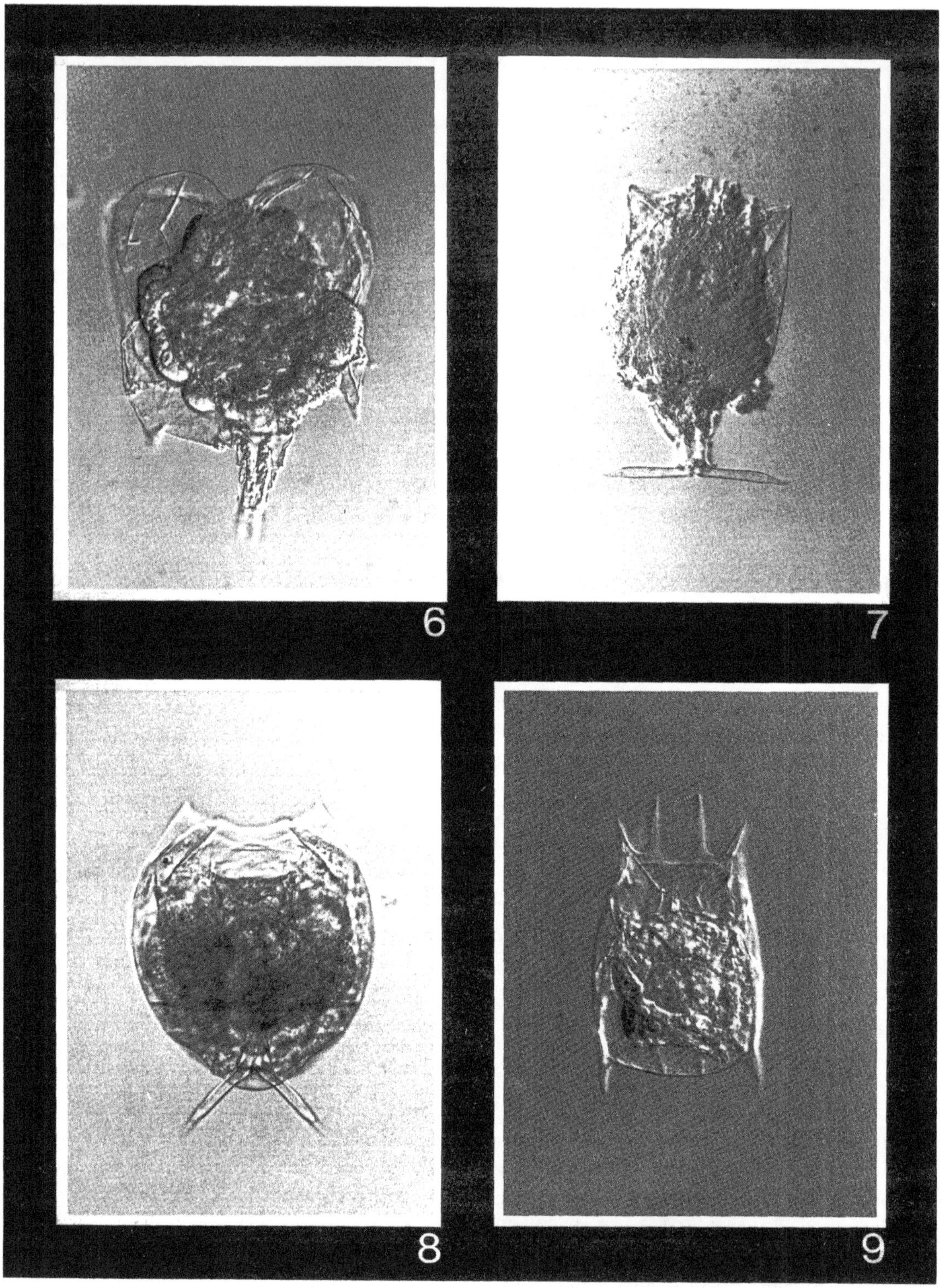

Fig. 6–9. 6. *Epiphanes brachionus spinosa* from Bozova; 7. *Trichotria curta* from the Euphrates River; 8. *Lecane luna* from Görëme; 9. *Keratella testudo* from Lake Beysehir.

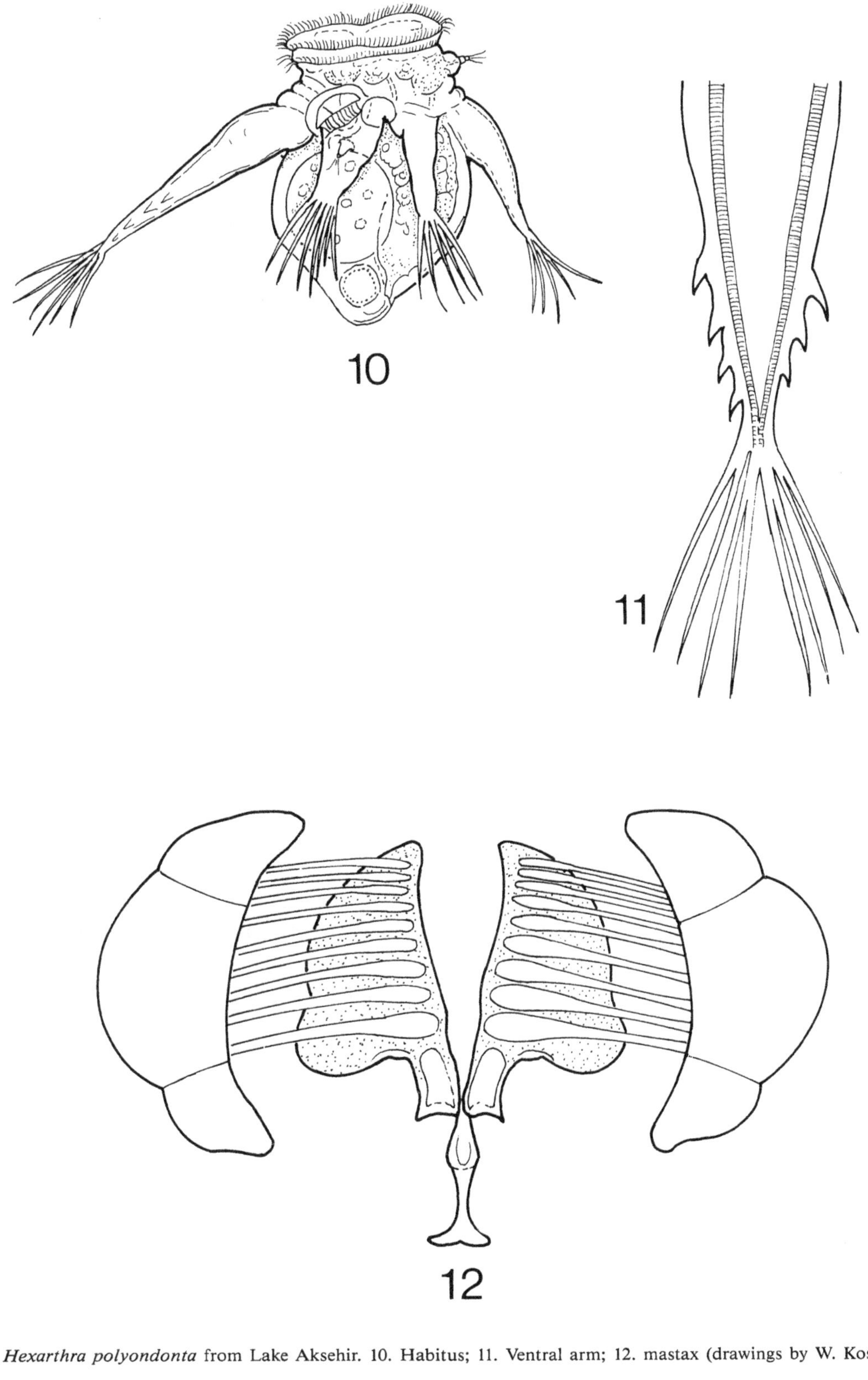

10

11

12

Fig. 10–12. Hexarthra polyondonta from Lake Aksehir. 10. Habitus; 11. Ventral arm; 12. mastax (drawings by W. Koste).

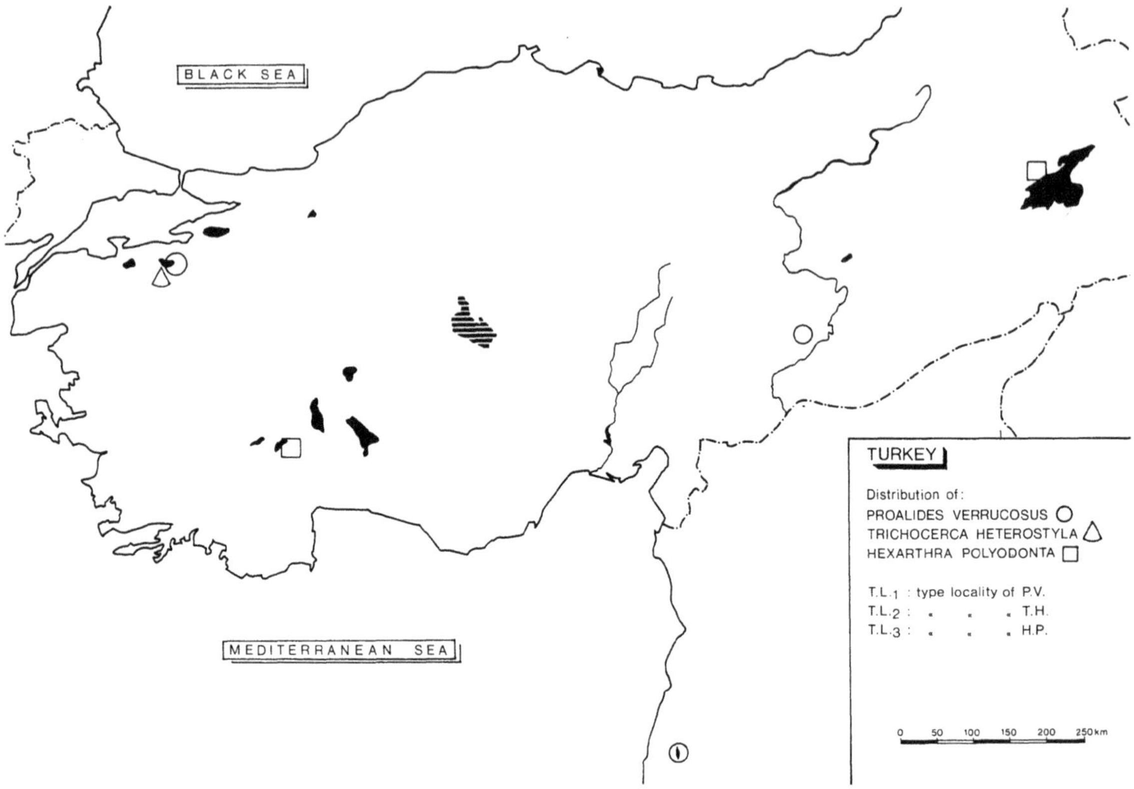

Fig. 13. Distribution in Anatolia and the Levant of three rotifer species with possibly restricted ranges.

date. Similarly, *H. polyondonta* and the little-known *Trichocerca heterostyla*, so far recorded only from Lake Apolyont, its type-locality, may also qualify for this status.

Finally, salinity is an important factor in the distribution of rotifer species. *Hexarthra fennica*, *H. polyodonta* (Figs. 10–12), *H. oxyuris*, and *Trichocerca taurocephala* (Fig. 5) are almost totally restricted to saline waters. A remarkable vertical zonation of these rotifer species occurred in Lake Burdur. *H. fennica*, by far the commonest form, was distributed throughout the water column, while *H. oxyuris* was restricted to depths of 5–20 m, and *Notholca acuminata* was found at 40–50 m depth. In Kratergöl, where *H. fennica* was the only rotifer present, the population was separated vertically in two parts [Dumont, 1981], and the males were restricted to the 50–70 m strata.

Acknowledgements

We are grateful to G. Daems, I. Miron, D. Somers for help in collecting the material studied. We thank W. Koste for Figures 10–12, and Mr. M. Bruyneel for help with the artwork.

Appendix

List of rotifers from Turkey (present paper not included)
Abbreviations:

D	= Daday, 1903
Dt	= Dumont, 1981
G	= Geldiay, 1949
H	= Hauer, 1957
V	= Vavra, 1903
T	= Tokat, 1976
M	= Mann, 1940
MC	= Margaritora & Cottarelli, 1970
MSM	= Margaritora *et al.*, 1977

Asplanchna priodonta Gosse: Ankara (G); Tatvan, Yaysin (MSM); L. Emir, L. Mermere (M).
Asplanchna sieboldii Leydig: Ankara (G); Aci Gol (Nevsehir) (MSM)
Asplanchnella brightwelli (Gosse): Ankara (G); L. Apolyont (D) (M); Lake Abant (MC).
Brachionus angularis Gosse: Lake Van (H), L. Apolyont (D)
Brachionus budapestinensis Dad.: L. Apolyont (D)
Brachionus calyciflorus Pallas: L. Hazar (T); Ankara (G); Bingol, Yaysin (MSM).
Brachionus diversicornis (Daday): L. Apolyont (D) (M).
Brachionus falcatus Zacharias: Ankara (G); Belgrade forest lakes (M).
Brachionus forficula Wierz.: L. Apolyont (D)
Brachionus leydigi f. *tridentatus*: Ankara (G)
Brachionus plicatilis O.F.M.: L. Van (H), L. Hazar (T).
Brachionus quadridentatus Hermann: L. Hazar (T); Bingol (MSM) f. *brevispinus*: Ankara (G)
*Brachionus q.*f. *ancylognathus* (Schmarda): Ankara (G).
Brachionus rubens Ehrbg: L. Iznik (D).
Brachionus urceolaris (O.F.M.): L. Iznik, L. Apolyont, L. Manyas, L. Burdur, L. Ilgaz, L. Emir, Guzel Yali, Cubuk Dam, Ankara (M).
Colurella adriatica Ehrbg.: L. Iznik (D).
Conochilus unicornis Rousselet: L. Abant (MC).
Epiphanes macroura (Barr. & Dad.): Ankara (G), L. Apolyont (D).
Euchlanis sp.: Ankara (G).
Euchlanis dilatata (Ehrbg): L. Sary (V).
Filinia longiseta (Ehrbg): Ankara (G); L. Apolyont (D) (M); Bingol, Yaysin, Tatvan (MSM); L. Ilgaz, L. Egridir (M).
Filinia l. f. *passa*: Ankara (G).
Filinia maior (Colditz): L. Van (H).
Filinia terminalis (Plate): L. Abant (MC).
Hexarthra fennica (Lev.): L. Hazar (T); Kratergol (DT); L. Sary (V) (almost certainly not this species but *H. bulgarica*).
Hexarthra mira (Hudson & Gosse): Ankara (G), L. Apolyont & Iznik (D).
Hexarthra polyodonta (Hauer): L. Van (H.).
Kellicottia longispina (Kell.): L. Hazar (T); L. Abant (MC).
Keratella c. cochlearis (Gosse): L. Van (H); L. Apolyont (D) (M); L. Abant (MC); L. Egridir, L. Burdur, Golbasi (M).
Keratella quadrata (O.F.M.): L. Van (H.); L. Hazar (T); Ankara (G); L. Egridir, L. Ekys (M).
Keratella q. frenzeli Carlin: L. Abant (MC).
Keratella tecta (Gosse): L. Van (H).
Keratella testudo (Ehrbg.): L. Apolyont (D).
Keratella valga Ehrbg.: Ankara (G.).
Lecane luna (O.F.M.): Ankara (G.), L. Sary (V).
Lecane quadridentata (Ehrbg): Ankara (G).
Lophocharis oxysternon (Gosse): Ankara (G).
Mytilina ventralis (Ehrbg): Ankara (G).
Notholca acuminata Ehrbg: Ankara (G).
Philodina roseola Ehrbg: Ankara (G).
Polyarthra sp.: Ankara (G), L. Apolyont (D).
Polyarthra dolichoptera Idels.: Ankara (G), L. Apolyont (D).
Proalides verrucosus (Barr. & Dad.): L. Apolyont (D).
Rotaria sp.: L. Iznik (D).
Synchaeta pectinata Ehrbg: Ankara (G).
Testudinella elliptica Ehrbg: Ankara (G).
Testudinella patina Ehrbg: Ankara (G).
Trichocerca elongata (Gosse): L. Apolyont (D) (M); Golbasi (M).
Trichocerca heterostyla (Dad.): L. Apolyont (D).
Trichocerca similis (Wierz.): L. Apolyont (D).
Trichocerca taurocephala (Hauer): L. Van (H).

References

Barrois, T. & E. Daday, 1893. Contribution à l'étude des Rotifères de Syrie et description de quelques espèces nouvelles. Rev. biol. n. Fr. 6: 391–409.

Daday, E., 1903. Mikroskopische Süsswassertiere aus Kleinasien. Sitz. Ber. k. Akad. Wiss. Wien. mathem.-naturh. Kl. 112: 139–167.

De Ridder, M., 1984. A review of the rotifer fauna of the Sudan. Hydrobiologia 110: 113–130.

Dumont, H. J., 1981. Kratergöl, a deep hypersaline crater-lake in the steppic zone of western Anatolia (Turkey), subject to occasional limnometeorological perturbations. Hydrobiologia 82: 271–279.

Dumont, H. J. 1983. Biography of Rotifers. Hydrobiologia 104: 19–30.

Geldiay, R., 1949. Etude comparée sur les faunes macroscopique et microscopique du barrage de Tchoubouk et du lac d'Emir (Voisinage d'Ankara). Comm. Fac. Sci. Ankara 2: 151–199.

Hauer, J., 1957. Rotatorien aus dem Plankton des Van-Sees. Arch. Hydrobiol 53: 23–29.

Mann, A. K., 1940. Über pelagische Copedpoden türkischer Seen (Mit Berücksichtigung des übrigen Planktons). Int. Revue ges. Hydobiol. Hydrograph. 40: 1–87.

Margaritora, F. G. & V. Cottarelli, 1970. Le biocenosi planctoniche estive del lago Abant (Turchia Asiatica, Regione del Mar Nero). Rendic. Accad. lombardo Sci. Lett., Cl. Sci B 104: 170–190.

Margaritora, F. G. & E. Stella & L. Mastrantuono, 1977. Contributo allo studio della fauna ad entomostraci delle acque temporanee della Turchia Asiatica. Riv. Idrobiol. 16: 151–172.

Skorikov, A. S., 1914. Über das Plankton im Unterlauf der Wolga mit Beiträgen zur Frage 'Potamoplankton'. Tr. Astrach. ichtyol. Lab. 3(5): 33 pp.

Tokat, M., 1976. Rotatortia of Lake Hazar (Golçuk) and their distribution (Turkish). Hydrobioloji, Istanbul 18: 13 pp.

Vavra, V., 1903. Ergebnisse einer naturwissenschaftlichen Reise zum Erdschias-Dag (Kleinasien). Rotatorien und Crustaceen. Arb. k.k. naturhist. Hofmus. 22: 1–7.

Wulfert, K., 1966. Rotatorien aus dem Stausee Ajwa und der Trinkwasser-Aufbereitung der Stadt Boroda (Indien). Limnologica 4: 53–93.

Zederbauer, E. & V. Brehm, 1907. Das Plankton einiger Seen Kleinasiens. Arch. Hydrobiol. Planktonk. 3: 92–99.

Hydrobiologia 147: 75–81 (1987)
May, L., Wallace, R. & Herzig, A. (eds), Rotifer Symposium IV

Distribution of *Brachionus* species in Spanish mediterranean wetlands

Maria R. Miracle,[1] Manuel Serra,[1] Eduardo Vicente[2] and Carmen Blanco[3]
Departments of [1]Ecology, [2]Microbiology, and [3]Zoology, University of Valencia, Burjassot (Valencia), Spain

Keywords: Rotifera, Spanish Mediterranean, wetlands, water chemistry, *Brachionus*, discriminant analysis

Abstract

In this study 200 zooplankton samples were collected (1979–1980), from 57 different wetlands in coastal Mediterranean Spain (CMS) and examined for the occurrence of *Brachionus* species. Data on 17 separate physical and chemical features of these water bodies were obtained from samples collected at the same time. Ten different *Brachionus* species were found in these wetlands, but only six occurred frequently enough to allow further examination of their distributional patterns using multivariate discriminant analysis. To separate these species, three analyses were performed using the 17 physical and chemical parameters, or their ratios. Three discriminant functions accounted for 80% or more of the among species variance. Separation of species occurred on each discriminant function which could be attributed to a reduced number of parameters. These were mainly temperature, sulphate, chloride, alkalinity and alkaline earths.

Introduction

Coastal wetlands in Mediterranean Spain frequently have large rotifer populations whose biomass can exceed that of other zooplankton. *Brachionus* and *Synchaeta* are two genera characteristically found in this area. Many species of the genus *Brachionus* are cosmopolitan [Pejler, 1977] and some have genetically different ecotypes, which apparently permits the occupation of a variety of environments [Serra & Miracle, 1983, 1985]. Nevertheless, a differential distribution of species belonging to the same genus is likely to occur [Hutchinson, 1978].

The physical and chemical parameters of wetlands in coastal Mediterranean Spain (CMS) have been described by López [1983]. Such abiotic information could prove useful in separating closely allied species and demonstrating how a genus can occupy a large number of sub-environments within a limited geographical area. The purpose of this paper is to analyse the distribution of the most abundant species of *Brachionus* in CMS wetlands and identify the more important ecological factors that affect species distribution.

The study area

The study area was a narrow band of brackish wetlands along the Mediterranean Spanish coast consisting of coast marshes, lagoons, springs and salt fields (Fig. 1). The wetlands were categorized into seven distinct groups (wetland types I to VII) by Lopez [1983], using physical and chemical parameters. These wetland types could be recognised on the basis of mineralization (freshwater inflow, marine influence, evaporation) and eutrophication (Table 1).

Material and methods

During four different expeditions (one per season during 1979–1980), 200 zooplankton samples

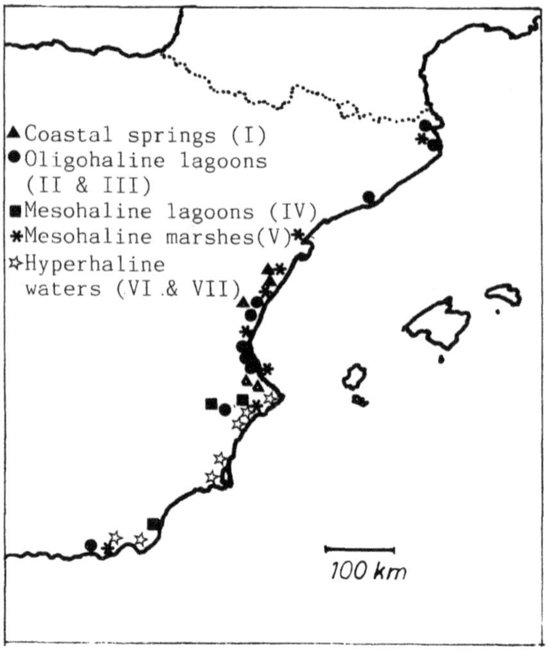

Fig. 1. Outline of the Spanish Mediterranean coast indicating the main sampling stations and their wetland type (I–VII). Only one sampling station of each type is indicated when they lie close together.

(50 μm mesh net) were collected from 57 different locations in CMS. Samples for the determination of 17 physical and chemical parameters were taken at the same time [López, 1983]. Zooplankton samples were counted with an inverted microscope and relative frequency was determined for each species.

Three multidiscriminant analyses [Cooley & Lohnes, 1971; Green, 1971] were performed with six groups of samples. These groups consisted of those samples containing each of six *Brachionus* species that occurred in >5% of cases. The samples were defined by the logarithmic transforms of 16 physical and chemical parameters and pH, which was not transformed (Table 3). The first analysis (no. 1) was made with groups consistingin all samples in which the *Brachionus* species defining the group was present. The second analysis (no. 2) was similar to the first, except that a few samples were excluded from the three groups defined by the more common species, *B. angularis, B. plicatilis* and *B. quadridentatus.* Samples were excluded when the species which defined the group was rare in that sample (i.e. < 2%), and other *Brachionus* species were dominant. The third analysis (no. 3) was like the second, except that the six major chemical parameters and total Fe and Mn were replaced by five ratios derived from the original data (Table 4).

Results and discussion

General distribution

Ten species of *Brachionus* were found in CMS wet-

Table 1. Occurrence and frequency of *Brachionus* species as a function of total zooplankton in CMS wetlands characterised by their mineralization parameters and trophic state (i.e. reactive + acid labile phosphorus). Wetland types (I – VII) are arranged along a gradient of increasing influence of marine waters as follows: I- Subterranean freshwater springs near the coast, with high relative concentrations of Ca^{2+}; II- Hypereutrophic freshwater coastal lagoons, with an important sewage inflow; III- Eutrophic freshwater coastal lagoons; IV- Sulphated coastal lagoons of intermediate salinity, with significant evaporation rates; V- Brackish water marshes; VI- Hyperhaline salt marshes and abondoned salt fields; VII- Artificially regulated hyperhaline salt fields.

Wetland type	Occurrence (percent)	Mean frequency (percent)	Conductivity (μS cm^{-1})	Alkalinity (meq l^{-1})	Ca^{2+} (meq l^{-1})	$Cl^-:SO_4^=:Al$	$Na^+:Mg^{2+}:Ca^{2+}$	PO_4^{3-} (μmol l^{-1})
I	47.60	4.17	3372	4.1	10.7	8: 2:1	2:0.6:1	0.17
II	73.68	21.96	2820	2.7	4.4	7: 4:1	4:2.0:1	2.06
III	57.10	23.04	3211	4.7	6.9	4: 2:1	3:1.0:1	0.25
IV	71.88	12.99	13704	5.4	24.8	22: 9:1	5:1.3:1	0.34
V	42.20	0.74	13415	4.4	23.4	35: 5:1	7:1.2:1	0.13
VI	10.70	0.83	42803	3.7	51.7	221:37:1	14:3.0:1	0.51
VII	11.10	2.08	48797	3.4	51.0	317:41:1	18:4.0:1	0.33

lands. They are, in order of importance, *B. plicatilis*, *B. angularis*, *B. quadridentatus* (mainly *brevispinus* and *cluniorbicularis* forms, but also *quadridentatus*, *rhenanus* and *fulleborni*), *Brachionus calyciflorus* (mainly *anuraeiformis* and *pala* forms, but also *dorcas* and *amphiceros*), *Brachionus urceolaris*, *Brachionus leydigi* (forms *leydigi* and *tridentatus*), *Brachionus bidentatus*, *Brachionus rubens*, *Brachionus bennini* and *Brachionus variabilis*.

Percentage occurrence of the genus *Brachionus* in wetlands I to VII are shown in Table 1. In general, the genus showed clear preference for coastal lagoons (wetland types II, III and IV). They occurred in springs and marshes (I, V), but tended to avoid hyperhaline systems (VI, VII).

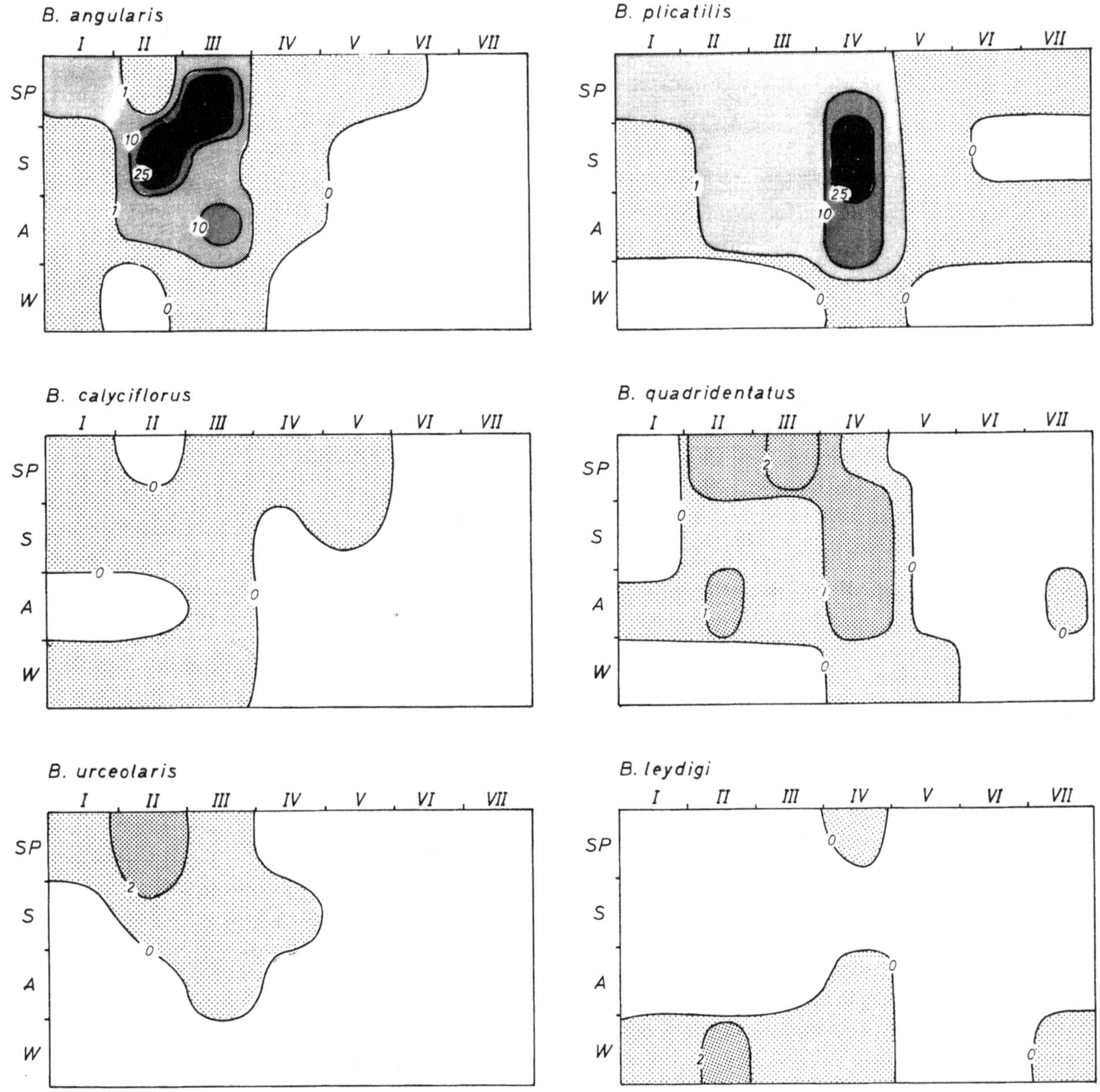

Fig. 2. Distribution of six *Brachionus* species from CMS. Isolines represent mean percentages with respect to the total zooplankton community as a function of wetland type (I–VII) and season. SP = spring; S = summer; A = autumn; W = winter.

As a group, the ten *Brachionus* species showed a slight preference for spring and avoided winter. *Brachionus* were found in 47% of spring samples, 42% of summer samples, 43% of autumn samples and 32% of winter samples. However, percent occurrence in summer samples may have been overestimated because some water bodies had dried up and the total number of samples was smaller than in other seasons. The six main species showed rather different distributions in relation to wetland type and season (Fig. 2). *Brachionus angularis*, for example, occurred most often in oligohaline coastal lagoons where freshwater inflow was important (II, III), although it tolerated a wide range of salinities (0.5 to 24‰). On the other hand, *B. plicatilis* occurred most often in waters of intermediate salinity and showed a clear preference for sulphated coastal lagoons (IV) whose ionic ratios were more similar to those of inland endoreic lagoons than brackish waters. However, *B. plicatilis* is the most widespread of these species, having been found in both oligo- and hyperhaline waters ranging from 0.5 to 88.6‰ salinity. Two species, *B. calyciflorus* and *B. urceolaris*, occurred mainly in oligohaline coastal waters (I, II & III), but in relatively low numbers. *B. urceolaris* showed the most well defined salinity limits, from 0.5 to 8.6‰. *Brachionus quadridentatus* had a similar distribution to that of *B. plicatilis*, except that it was much more restricted to lagoons (II, III and IV). *Brachionus leydigi* resembled *B. quadridentatus* in its distribution, but it was almost exclusively a winter resident, except among lagoons of group IV where it was more persistent.

Discriminant analysis

Analyses one and two

Only six species (Table 2) occurred frequently enough (5% of samples) to be used in the discriminant analysis. Results from the first two discriminant analyses (Nos. 1 and 2) using the 17 environmental parameters (Table 3) were very similar. Analysis No. 2, in which a few samples were eliminated from consideration (see Material and methods), had slightly more conclusive results, and these are reported here (Table 3, Fig. 3). The first three discriminant functions accounted for slightly more than 80% of the among-species variance. Discriminant function I accounted for 34.3% of the among-species variance and was largely a function of SO_4^{2-} concentration. Discriminant function I was well correlated with all the major constituents of water salinity and separated the three oligohaline species (*B. angularis, B. urceolaris. B. calyciflorus*) from the other three species (*B. quadridentatus, B. plicatilis* and *B. leydigi*). The latter three species appear to prefer those lagoons of group IV where

Table 2. Mean values of several physical and chemical parameters at which six species of *Brachionus* occurred in CMS wetlands.

Parameters	*Brachionus* species					
	angularis	*plicatilis*	*calyciflorus*	*urceolaris*	*leydigi*	*quadridentatus*
Conductivity (mS cm^{-1})	4.76	15.81	8.31	3.92	9.63	10.10
Temperature (°C)	20.83	20.46	21.12	23.87	15.32	20.15
Alkalinity (meq l^{-1})	3.95	4.40	3.72	3.59	4.08	4.71
Cl^- (meq l^{-1})	36.99	221.96	77.61	18.58	168.35	81.77
SO_4^{2-} (meq l^{-1})	15.47	46.20	20.29	14.20	36.23	36.69
Ca^{2+} (meq l^{-1})	9.33	26.23	10.16	6.93	18.66	15.75
Mg^{2+} (meq l^{-1})	12.02	48.39	15.98	7.29	40.33	22.63
Na^+ (meq l^{-1})	37.58	214.45	69.33	23.48	142.45	83.57
O_2 (ml l^{-1})	7.68	6.02	7.50	9.03	7.15	6.98
NO_3 (μmol l^{-1})	47.90	22.35	93.11	25.30	145.21	32.77
PO_4^3 (μmol l^{-1})	0.77	0.67	0.79	1.46	1.40	0.87
Cl^-/SO_4^{2-}	2.08	3.38	2.57	1.10	2.60	2.16
Alk/Ca^{2+} + Mg^{2+}	0.34	0.16	0.28	0.34	0.20	0.19

Table 3. The first three discriminant function coefficients from analysis No. 2, showing the relative contribution of 17 physical and chemical parameters to the separation of *Brachionus* species in CMS wetlands. The structure coefficients, or correlations between the discriminant functions and these physical and chemical parameters, are indicated in parentheses. Coefficients over ±0.3 and ±0.5, respectively, have been underlined.

	Discriminant functions					
	I		II		III	
Percentage of variance	34.3		26.5		19.5	
Parameters:						
Conductivity	0.304	(−0.69)	−0.271	(−0.22)	0.172	(0.13)
Temperature	−0.039	(0.23)	−0.729	(−0.92)	0.344	(0.65)
pH	0.162	(0.21)	0.107	(−0.28)	−0.098	(0.43)
Alkalinity	−0.064	(−0.43)	0.031	(0.20)	0.034	(−0.25)
Cl^-	−0.293	(−0.73)	0.407	(−0.03)	−0.605	(−0.04)
SO_4^{2-}	−0.800	(−0.79)	−0.040	(−0.09)	0.482	(0.18)
Ca^{2+}	−0.073	(−0.71)	−0.115	(0.14)	−0.040	(−0.19)
Mg^{2+}	−0.013	(−0.71)	0.248	(−0.04)	−0.056	(−0.02)
Na^+	0.203	(−0.69)	−0.177	(−0.02)	0.278	(0.00)
K^+	0.114	(−0.55)	−0.198	(−0.09)	−0.068	(−0.09)
O_2	0.062	(0.43)	−0.068	(−0.05)	0.309	(0.52)
NO_2^-	−0.009	(−0.06)	0.058	(0.32)	−0.020	(−0.12)
NO_3^-	−0.005	(−0.04)	−0.002	(0.43)	0.098	(−0.31)
Si	−0.072	(−0.34)	−0.148	(−0.40)	0.005	(0.43)
PO_4^{3-}	−0.069	(0.07)	0.172	(−0.14)	0.065	(0.46)
Total Fe	0.186	(−0.14)	0.019	(0.14)	−0.114	(−0.25)
Total Mn	−0.206	(−0.55)	−0.100	(0.19)	0.185	(−0.14)

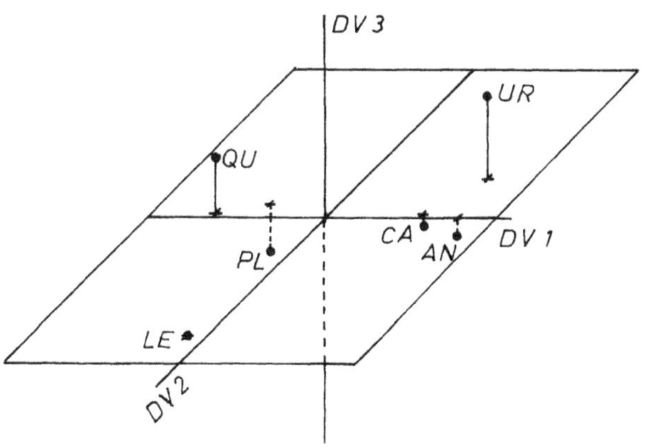

Fig. 3. Schematic diagram of discriminant analysis number two. The relative position of the mean values for each *Brachionus* species in the space are dimensioned by the first three discriminant functions or variates (DV1, DV2, DV3). QU – *B. quadridentatus*; PL – *B. plicatilis*; LE – *B. leydigi*; UR – *B. urceolaris*; CA – *B. calyciflorus*; AN – *B. angularis.*

evaporation rate and continental influence is high. The continental influence on CMS wetlands means that bicarbonate, calcium and sulphate levels are high, especially in the southern part. So, when evaporation rates are high, calcite and gypsum are precipitated; conductivity increases, but the relative proportion of sulphates and carbonates decreases. This phenomenon differentiates the lagoons of group IV from those of groups II and III, which have smaller Cl^-/SO_4^{2-} and Cl^-/alkalinity ratios (Table 1).

Discriminant function II (26.5% of the variance), which separates *B. leydigi* from the other species (Fig. 3), is largely a function of temperature and is also correlated with NO_3^- (Table 3). Discriminant function II has also a rather high influence of chloride, but is not well correlated with it. The most contrasting species to *B. leydigi* on this axis is *B. urceolaris*, which showed a preference for less saline and more calid waters (Table 2).

Discriminant function III, which accounted for

19.5% of the variance, is more difficult to interpret. It is related to an inverse relationship between Cl^- and SO_4^{2-}, but temperature and surface oxygen are also involved (Table 3). It is also positively correlated with phosphate (reactive plus acid labile phosphorus). The most eutrophic lagoons (group II) showed the smaller Cl^-/SO_4^{2-} ratios are the higher surface oxygen and phosporus concentrations (Table 1). Oligohaline waters in CMS wetlands, with a higher rate of inflow than evaporation, have small Cl^-/SO_4^{2-} ratios. This ratio becomes higher as evaporation rates increase. The Cl^- and SO_4^{2-} relationship may separate *B. urceolaris* from other species, especially from *B. plicatilis* (Fig. 3). Moreover, *B. urceolaris* was the species which showed the greatest preference for less mineralized and more eutrophic waters.

Analysis three

The first three functions of the third discriminant analysis (No. 3) accounted for 86% of the among-species variance (Table 4). The results of this analysis were similar to that of the second. Discriminant function I (40.85% of variance) is largely a function of the Cl^-/SO_4^{2-} ratio, but temperature was also important.

This is to be expected from the evaporation and precipitation phenomena described above for sulphate rich waters. This function accounts for separation of the same three species (i.e., *B. angularis, B. urceolaris* and *B. calyciflorus*) from the other three, as the first discriminant function of analysis 2.

The second discriminant function (II), accounted for 26.70% of the variance and separated *B. leydigi* from the other species, as in analysis No. 2. Once again temperature was important, but the ratio of alkalinity to $Ca^{2+} + Mg^{2+}$, which related to calcite precipitation, also had a strong influence. Throughout its winter occurrence, *B. leydigi* preferred those oligo-mesohaline lagoons with low alkalinity/$Ca^{2+} + Mg^{2+}$ ratios.

As in analysis No. 2, discriminant function III (18.45% of variance) also separated *B. urceolaris* from the other species. Both Cl^- and SO_4^{2-} were involved again, but as the ratio Cl^-/SO_4^{2-}. Temperature, oxygen and phosphate also showed rather high positive correlations.

Table 4. The first three discriminant function coefficients from analysis No. 3, showing the relative contribution of 13 physical and chemical parameters and ratios to the separation of *Brachionus* species in CMS wetlands. The structure coefficients, or correlations between the discriminant functions and these physical and chemical parameters, are indicated in parentheses. Coefficients over ±0.3 and ±0.5, respectively, have been underlined.

	Discriminant functions					
	I		II		III	
Percentage of variance	40.85		26.70		18.45	
Parameters:						
Conductivity	$\underline{-0.392}$	(−0.04)	−0.167	(0.36)	0.079	(−0.42)
Temperature	$\underline{0.482}$	$\underline{(\ 0.52)}$	$\underline{-0.508}$	$\underline{(-0.51)}$	$\underline{0.550}$	$\underline{(\ 0.64)}$
pH	0.100	(0.17)	0.038	(0.23)	−0.063	(0.29)
O_2	0.060	(0.10)	0.022	(0.11)	0.275	(0.43)
NO_2^-	0.015	(−0.18)	0.104	(−0.18)	−0.038	(−0.06)
NO_3^-	−0.050	(−0.13)	0.019	(−0.05)	0.090	(−0.24)
Si	−0.089	(−0.19)	−0.170	(−0.25)	0.113	(0.35)
PO_4^{3-}	−0.100	(−0.17)	0.222	(0.17)	0.013	(0.44)
Cl^-/SO_4^{2-}	$\underline{0.723}$	$\underline{(\ 0.49)}$	0.132	(0.42)	$\underline{-0.759}$	$\underline{(-0.81)}$
SO_4^{2-}/Ca^{2+}	0.027	(0.05)	−0.001	(0.00)	−0.024	(−0.06)
Fe/Mn	−0.130	(0.05)	−0.041	(−0.02)	0.111	(−0.08)
Alk/$Ca^{2+} + Mg^{2+}$	0.203	(0.08)	$\underline{-0.777}$	$\underline{(-0.76)}$	−0.006	(0.26)
$Ca^{2+} + Mg^{2+}/Na^+ + K^+$	−0.005	(0.33)	0.003	(−0.22)	0.002	(0.04)

Discussion

On the basis of the results described above, there appears to be a real separation among the six *Brachionus* species studied, and a relatively small number of ecological factors are sufficient to account for their differential distribution. These are, in order of importance, sulphate, temperature, chloride, alkalinity relative to total alkaline earths, conductivity and O_2.

The CMS wetlands studied represent only a fraction of the geographical range of the genus *Brachionus* and the physical and chemical factors examined in this study are not exhaustive. Biotic factors such as trophic relationships, competition and predation were not considered. However, the genus seems to have evolved to match the abiotic environmental heterogeneity and a preference for specific conditions is evident. *B. plicatilis* is known to attain peak densities in waters of high alkalinity [Walker, 1981]. The alkalinities of waters in the present study are far less than those considered high by Walker [1981], but we have recorded a tendency for *B. plicatilis* to inhabit lagoons of group IV, which have relatively higher alkalinities than the other waterbodies studied (Table 1).

When other geographical areas are considered, similar distributions are seen. For example, on the Iberian peninsula as a whole *Brachionus* species appear to be important only in steppe lagoons or shallow highly mineralized endoreic waters. In such habitats the main species are *B. plicatilis* and *B. quadridentatus*, the former being much more frequent than the latter [Miracle, 1982]. In the present study, these two species were the most closely related species, showing a clear preference for lagoons of group IV, i.e., those most closely related to inland endoreic waters. In other environments (e.g., Karst and alpine lakes and reservoirs) *Brachionus* is only a minor component of the zooplankton community. Only *B. calyciflorus* and *B. urceolaris* are occasionally found in reservoirs Margalef *et al.*, 1976.

The preference of *Brachionus* species for coastal lagoons with continental influence and athalassic saline lakes, supports the hypothesis that this genus has evolved from a freshwater ancestry and contrasts with the hypothesis of a marine origin proposed by Epp and Winston [1977]. Furthermore, the latitudinal distribution of the genus *Brachionus*, centred in the tropics [Green, 1972; Pejler, 1977], indicates their ancient origin. In a southern temperate region, such as the Iberian peninsula, the genus *Brachionus* is important in older environments such as coastal and steppe wetlands and it plays only a minor role in post-glacial or recent man-made aquatic systems.

Acknowledgement

We are very grateful to Dr. R. L. Wallace for his useful criticism and revision of the original manuscript.

References

Cooley, V. W. & P. R. Lohnes, 1971. Multivariate Data Analysis Wiley, N.Y.: 364 pp.

Epp, R. W. & P. W. Winston, 1977. Osmotic regulation in the brackish water rotifers Brachionus plicatilis (Müller). J. Exp. Biol. 68: 151–156.

Green, J., 1972. Latitudinal variation in associations of planktonic rotifera. J. Zool. 167: 31–39.

Green, R. H., 1971. A multivariate statistical approach to the hutchinsonian niche: bivalve molluscs of central Canada. Ecology 52: 543–556.

Hutchinson, G. E., 1978. An introduction to population ecology. Yale Univ. Press, New Haven, 260 pp.

López, P., 1983. Aguas salinas epicontinentales próximas a la costa mediterranea españa. Estudio del medio. Ph.D. Thesis University of Barcelona (Spain), 331 pp.

Margalef, R., D. Planas, J. Armengol, A. Vidal, N. Prat, A. Guiset, J. Toja & M. Estrada, 1976. Limnología de los embalses españoles. Ministerio de Obras Publicas, Madrid, 422 + 85 pp.

Miracle, M. R., 1982. Biogeography of the freshwater zooplanktonic communities of Spain. J. Biogeogr. 9: 455–467.

Pejler, B., 1977. On the global distribution of the family Brachionidae (Rotatoria). Arch. Hydrobiol. Suppl. 53: 255–306.

Serra, M. & M. R. Miracle, 1983. Biometric analysis of Brachionus plicatilis ecotypes from Spanish lagoons. Hydrobiologia 104: 258–271.

Serra, M. & M. R. Miracle, 1985. Isozyme polymorphism in Brachionus plicatilis populations from several Spanish lagoons. Verh. int. Ver. Limnol. 22: 2991–2996.

Walker, K. F., 1981. A synopsis of ecological information on the saline lake rotifer Brachionus plicatilis Müller 1786. Hydrobiologia 81: 159–167.

Hydrobiologia 147: 83–89 (1987)
May, L., Wallace, R. & Herzig, A. (eds), Rotifer Symposium IV

Biometric variation in three strains of *Brachionus plicatilis* as a direct response to abiotic variables

Manual Serra and Maria R. Miracle
Dep. Ecologia, Universidad de Valencia, Burjasot (Valencia), Spain

Keywords: Rotifera, *Brachionus plicatilis*, biometry, size, temperature, salinity, genotype

Abstract

Morphometric variation in three clones of *Brachionus plicatilis* cultured at 20 °C, 25 °C and 30 °C, and 9‰, 12‰ and 24‰ salinity was analysed. Size appeared to be largely under genetic control and this defined the narrow limits within which variation due to abiotic factors could occur. Temperature had a significant effect on size, but affected the three clones differently. The most general effect of temperature was a reduction in size which levelled off as the temperature rose. Of the measurements taken, only distance between the median spines was affected by salinity. An important inter-relationship between the effects of temperature and salinity was also detected.

Introduction

Ecologists have long been interested in environmentally induced morphological variation in rotifers. Of the possible factors involved, the effects of temperature and salinity on size have been studied most extensively. Serra & Miracle [1983] observed that morphological variation among *Brachionus plicatilis* (Müller) clones, obtained from a series of Spanish water bodies, had an important genetic component. This was later confirmed using isozyme analysis [Serra & Miracle, 1985]. This paper examines the biometrical response of three genetically distinct clones of *Brachionus plicatilis* to changes in temperature and salinity.

Materials and methods

Three clones of *Brachionus plicatilis* (CU, SPO, FCA) were grown from individuals obtained from natural environments and the Spanish Fisheries Research Institute ([Serra & Miracle, 1983]; N.B. SPO = SP in previous papers). Clone CU inhabited an endorreic continental lagoon, while clone SPO came from a coastal lagoon and clone FCA was collected from a coastal marsh. Electrophoretic analysis has shown that these clones belong to genetically different clusters [Serra & Miracle, 1985].

The clones were fed with the alga *Tetraselmis* sp. and cultured in sea water diluted to the experimental salinity, using the method of Guillard & Ryther [1962]. The clones were maintained under the experimental conditions for 10 days prior to the experiments.

The experimental design included three factorial fixed effects: clone (see above), temperature (20 °C, 25 °C, 30 °C) and salinity (9‰, 12‰ and 24‰). Two replicate flasks (blocks) were prepared for each set of experimental conditions. While in the exponential phase of growth, 25 females were randomly selected from each flask and measured (Fig. 1). The body measurements were determined with a Wild M40 microscope under 40 × (length and width) and 100 × (DLS, DMS and LMS) magnification.

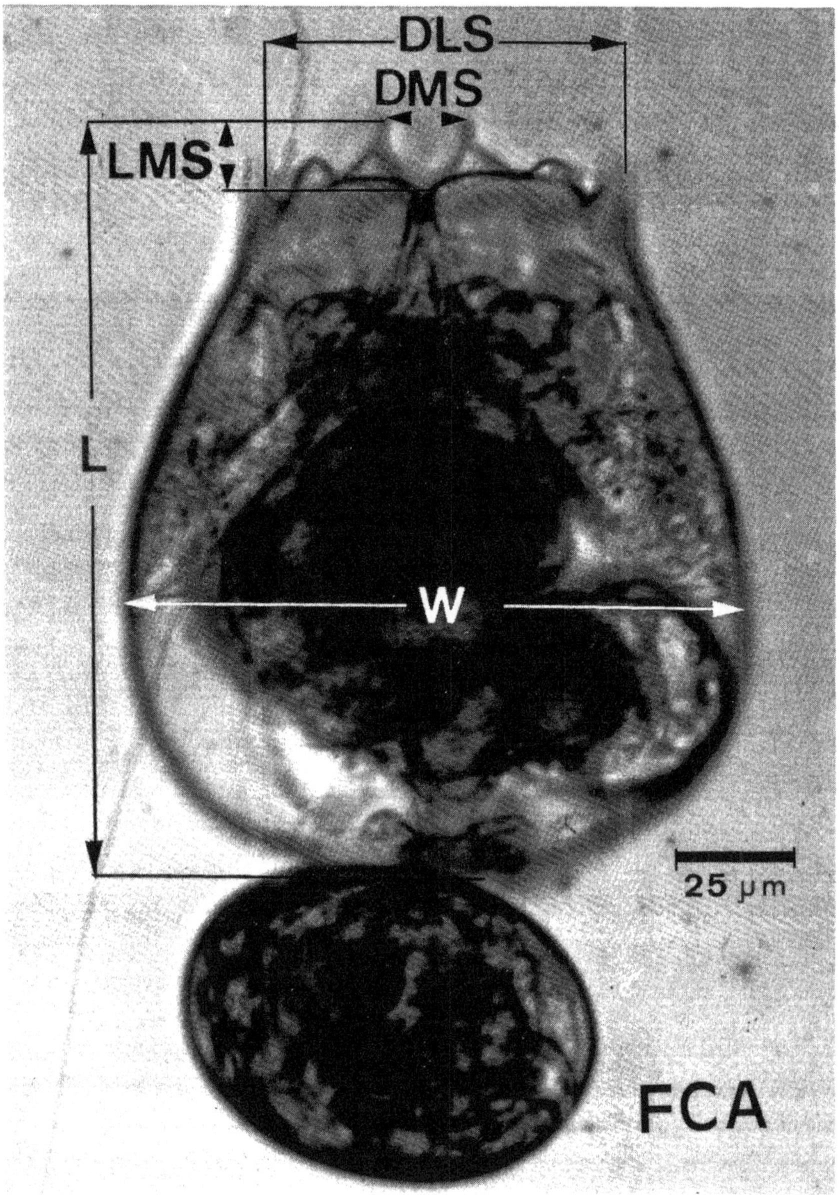

Fig. 1. *Brachionus plicatilis* (clone FCA) showing the body measurements analysed: L, length; W, width; DLS, distance between lateral spines; DMS, distance between median spines; LMS, length of median spines.

Statistical analyses were carried out on the logarithmic transformation of the measurements recorded. A multivariate analysis of variance (MANOVA) was performed on the data and the overall significance of each effect was tested with Wilks' criterion [Cooley & Lohnes, 1971; Kres, 1983]. The significance of each effect on the individual measurements was determined with a univariate analysis of variance (ANOVA) as described by Cooley & Lohnes [1971].

The relationships between measurements were explored by discriminant analysis, which transforms the original variables into linear combinations that maximise the separation between aprioristic groups. The computed analysis used 27 groups corresponding to the culture of the three clones under 9 different conditions (3 salinities × 3 temperatures × 3 clones). The variables defining the groups were the morphological characteristics of the rotifers (5 lorica measurements – see Fig. 1) in each of the 54 duplicate culture flasks.

Results

Two important non-biometric differences were observed among the experimental cultures. Firstly, reproductive behaviour differed markedly among clones. Clone CU deposited its resting eggs outside the body, while the resting eggs of clone SPO remained inside the animal after death. In contrast, clone FCA did not produce males or resting eggs at all. Secondly, the three clones showed different growth rates under the experimental conditions. Clone CU grew well at all salinities and temperatures tested, while clones SPO and FCA showed a poor tolerance of low salinities at high temperatures, and of high salinities at low temperatures.

Figure 2 shows the morphological characteristics measured in relation to temperature and salinity, for each clone. The most striking differences in

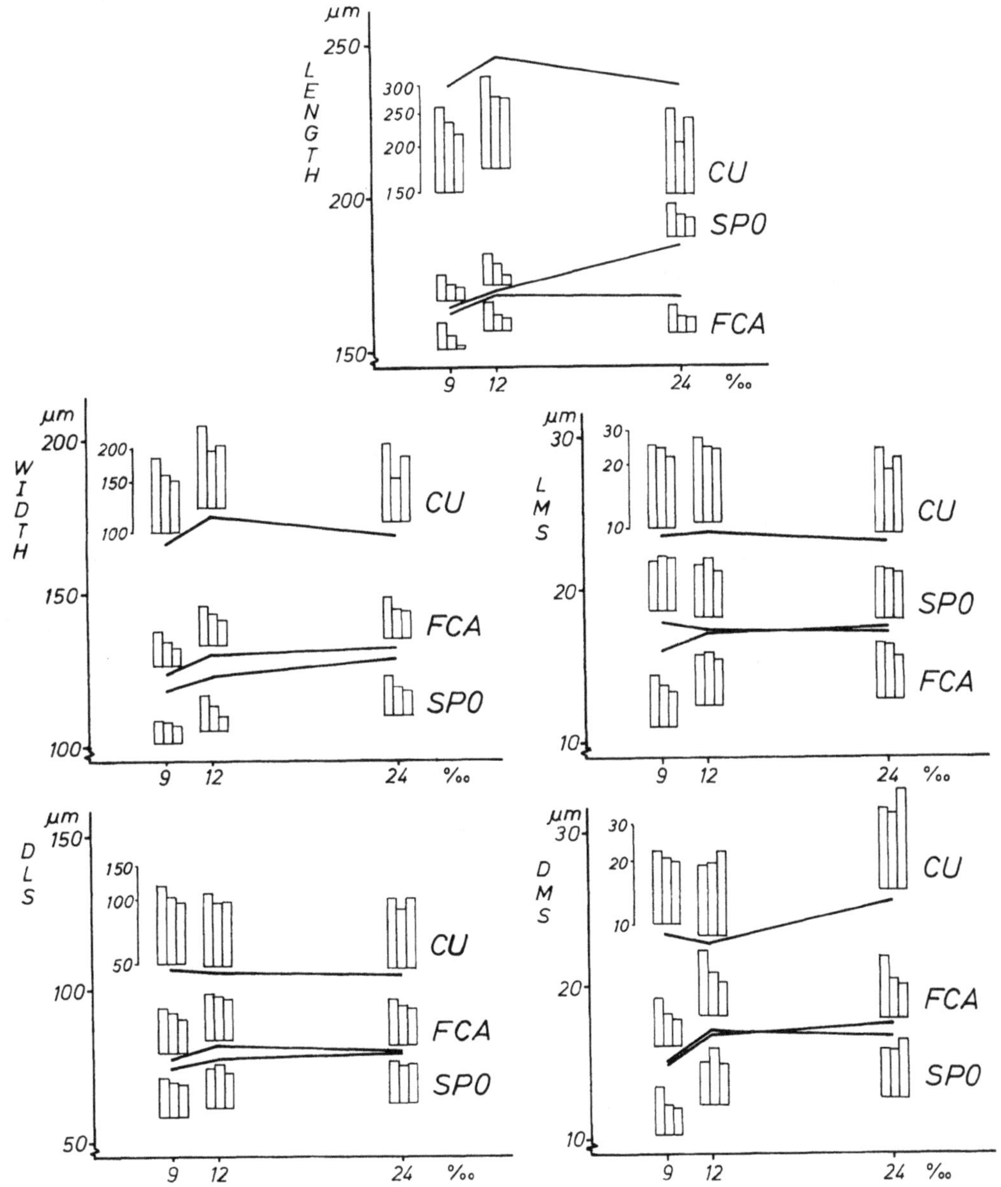

Fig. 2. Variation of five biometric characters (see Fig. 1) in three clones of *Brachionus plicatilis* in relation to incubation temperature (vertical bars correspond to 20 °C, 25 °C and 30 °C, respectively) and salinity (solid lines indicate mean values for each clone). The logarithmic scale relating to the vertical bars is indicated to the left of the first set of vertical bars in each panel.

these measurements occurred among clones, suggesting that genetic variation had a significant effect on size (Table 1). For each clone, variation due to temperature and salinity occurred within limits which seemed to be genetically determined. Moreover, differences in clone body shape were maintained.

Within clones, salinity appeared to have little effect on morphology. Only DMS was significantly affected by salinity (Table 1). Nevertheless, it was possible to recognise some trends. At the highest temperature, body length and width in all clones tended to increase slightly with increasing salinity (Fig. 2). Body measurements also increased slightly with increasing salinity at the two lower temperatures tested, except for length and width in clone CU, and length in clone FCA (Fig. 2). In the latter cases we observed an increase in these measurements, followed by a decrease when salinity rose.

In contrast, temperature had a significant effect on all measurements, except DMS (Table 1). Increasing temperature resulted in a reduction in length, width and DLS at all salinities in the smaller clones (SPO, FCA) and at low and intermediate salinities in the larger clone (CU) (Fig. 2). This reduction was most marked at the lower temperatures. However, at the higher salinity, length, width and DLS in clone CU decreased with increasing temperature between 20 °C and 25 °C, and increased again between 25 °C and 30 °C.

LMS and body length showed an allometric pattern of growth in clone CU ($K = 0.85$, $r^2 = 0.92$). In clone FCA, this relationship was isomorphic ($K = 1.09$, $r^2 = 0.53$). No significant relationship between these parameters was recorded in clone SPO. LMS also shows a slight inverse relationship with DMS.

The location of the 27 group centroids in the space defined by the first three discriminant functions (Fig. 3) helps to describe the nature of these group differences. The first discriminant function, accounting for 66% of the discriminating power, separates clones or genetic components. The larger clone, CU, is clearly discriminated from the other two by this function. The smaller coastal clones, FCA and SPO, lie closer together and are separated by the first discriminant function and also, to a lesser extent, by the second.

Hierarchically, within clones, there is a clear separation between cultures at different temperatures. In clone CU, the second discriminant function represents the effect of temperature. In the two smaller clones, the effect of temperature is seen in the first and second discriminant functions but more clearly in the latter, in this case.

The effect of salinity is associated with the third discriminant function; low salinities are located mainly on the positive side of the vertical axis and high salinities on the negative side.

From the first three discriminant function coefficients (Table 2), the groups are separated mainly by length and width. Length has more weight in the first discriminant function, which mainly separates clones (see above). Width has more weight in the

Table 1. MANOVA and ANOVA results of three factorial fixed effects (clone, temperature and salinity) and a random nested effect (blocks) on five body measurements ($N = 25$; * is $p < 0.05$ and ** is $p < 0.01$).

Source	Wilks' λ	df		F ratio				
		n_1	n_2	Length	Width	DLS	DMS	LMS
Clone (C)	0.0101**	2	27	165.7**	104.4**	108.1**	79.5**	93.5**
Temperature (T)	0.3451**	2	27	15.1**	17.3**	5.6**	2.5	6.1**
Salinity (S)	0.3525**	2	27	1.0	2.5	0.5	5.6**	0.4
C×T	0.0992**	4	27	0.6	1.7	0.8	6.4**	1.8
C×S	0.3268	4	27	0.5	0.6	0.6	1.5	1.6
T×S	0.4055	4	27	0.8	0.9	1.1	1.4	0.5
C×S×T	0.2272	8	27	0.5	0.5	0.5	1.1	0.9
Blocks	0.5364**	27	1296	7.8**	8.7**	10.4**	7.0**	7.7**

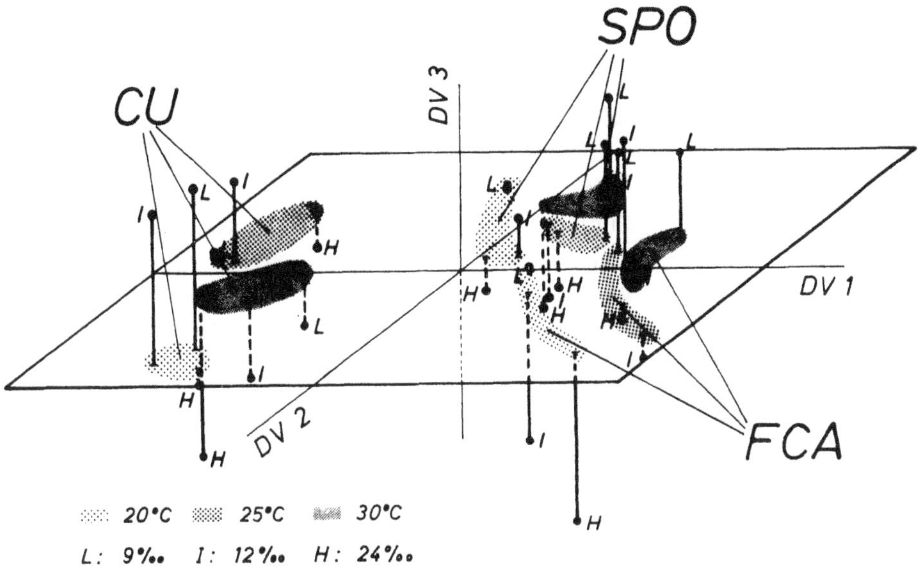

Fig. 3. Position of the centroids of each group of cultures of the same clone, in relation to combinations of clone, temperature and salinity, in the space defined by the first three discriminant functions (DV1, DV2, DV3). Centroids of the groups or cultures of the same clone, cultured at the same temperature, are grouped within the areas shown by different levels of shading. Salinities are indicated as follows: L = low; I = intermediate; H = high. Clones are named as in the text.

Table 2. Coefficients and proportion of discriminating power accounted for the three first discriminants variables.

	Discriminants variables		
	1	2	3
Body measurements			
Length	−0.796	−0.683	0.461
Width	0.596	0.713	−0.453
DLS	0.034	0.072	0.594
DMS	−0.037	−0.034	−0.475
LMS	0.090	0.138	−0.066
Proportion of discriminating power	0.66	0.16	0.11

second discriminant function, which mainly separates temperature conditions (Fig. 3). The other measurements are only important in the third discriminant function, which accounts for a small proportion of the discriminating power, and is associated with the effect of salinity. This function appears to measure the variation of length and DLS compared to width and DMS, i.e. there is variation in the shape of the anterior part of the lorica. These results are consistent with the analysis of variance (Table 1), in which length has a greater significance in the differences among clones, width in the differences due to temperature, and DMS in the effect of salinity.

Discussion

The variance and discriminant analyses used in the present study were tested against blocks (replicate culture flasks; Table 1). The blocks were significant, indicating that, even under rigorously controlled conditions, there are some uncontrolled factors, e.g. small differences in food or population density, that affect the results. A model with blocks is therefore necessary in a study of this kind to establish the level of significance of the effects; in this case, not only to lower the significance of the clone × temperature × salinity interaction, but also to make univariate effects more obvious.

The results show that body size and shape in *Brachionus plicatilis* is strongly affected by genotype. This has also been found for other *Brachionus* species [Pourriot, 1973; Yufera, 1982; Snell & Carrillo, 1984]. Serra & Miracle [1983] described geographical variation and cyclomorphosis in *B. plicatilis* from a range of Spanish lagoons. It now seems likely, from isoenzymatic studies [Serra & Miracle, 1985; Snell & Winkler, 1985], that an important part of this variation can be attributed to genotypic differences among populations of *B. plicatilis*.

The inverse relationship between size and temperature found in the present study has been confirmed for a variety of organisms. The same relationship was also found in *Brachionus* spp. [Halbach, 1970; Pourriot, 1973] and *Brachionus plicatilis* [Snell & Carrillo, 1984], although the effect was not significant in the statistical analyses carried out by those authors. They considered strain to be a random effect, while we consider strain to be a fixed effect, because the clones were not randomly selected. Furthermore, genetic variation among the strains studied resulted in different responses to changing temperature. This is evident from the work of Snell & Carrillo [1984] and from our own results (High significance of clone-temperature interaction; Table 1).

The inverse relationship between size and temperature is supported by hypotheses which state that the rate of biological processes are exponentially related to temperature. Data on the embryonic development time and population dynamics of rotifers show the dependence of development times and population growth rates on temperature [e.g. Herzig, 1983; Pascual & Yufera, 1983]. Furthermore, some studies have established an inverse relationship between these rates and size [Pourriot, 1973]; this relationship can also be derived from the data given by other authors [e.g. King & Miracle, 1980].

Examples in the literature of size-salinity relationships are contradictory [Margalef, 1974; Brand, 1971; Remane & Schleiper, 1971] and a general rule cannot be stated. This relationship is highly dependent on the experimental organisms which seem to have different growth optima. All clones followed the general trend of size reduction (length and width) which levelled off as temperature increased, with the exception of clone CU. This clone showed an increase in size after the expected decrease, at the highest salinity tested. This complex trend may be explained by the greater adaptability of clone CU to variations in temperature-salinity conditions. Its size at high salinities and high temperature is much greater than would be expected due to temperature alone.

Our observations on the viability of the three experimental clones of *B. plicatilis* are consistent with the general hypothesis that many species are more tolerant of low salinity at low temperatures, and of high salinity at high temperatures. These are the most likely combinations of salinity and temperature in the natural environment, and the species seems to be adapted to these conditions.

References

Brand, G. W., 1981. Responses of a calanoid copepod to existence in saline waters. Hydrobiologia 81: 145–149.

Cooley, V. W. & P. R. Lohnes, 1971. Multivariate Data Analysis. John Wiley & Sons, N.Y., 364 pp.

Guillard, R. R. L. & J. H. Ryther, 1962. Studies of marine planktonic Diatoms. I. Cyclotella nana Hustedt and Detomula confervacea (Cleve) Gran. Can. J. Microbiol. 8: 229–239.

Halbach, U., 1970. Die Ursachen des Temporalvariation von Brachionus calyciflorus Pallas (Rotatoria). Oecologia 4: 262–318.

Herzig, A., 1983. Comparative studies on the relationship between temperature and duration of embryonic development of rotifers. Hydrobiologia 104: 237–246.

King, C. E. & M. R. Miracle, 1980. A perspective on aging in rotifers. Hydrobiologia 73: 13–19.

Kinne, O., 1964. The effects of temperature and salinity on marine and brackish water animals. II. Salinity and temperature-salinity combinations. Oceanogr. Mar. Biol. Ann. Rev. 2: 281–293.

Kres, H., 1983. Statistical Tables for Multivariate Analysis. Springer-Verlag, N.Y., 504 pp.

Margalef, R., 1974. Ecologia. Omega, Barcelona. 951 pp.

Pascual, E. & M. Yufera, 1983. Crecimiento en cultivo de una cepa de Brachionus plicatilis O. F. Müller en function de la température y la salinidad. Inv. Pes. 47: 151–159.

Pourriot, R., 1973. Rapports entre la temperature, la tailles des adultes, la longueur des oeufs et taux de développement embryonnaire chez Brachionus calyciflorus Pallas (Rotifère). Ann. Hydrobiol. 4: 103–115.

Remane, A. & C. Schlieper, 1971. Biology of Brackish Waters. John Wiley & Sons, N.Y., 372 pp.

Serra, M. & M. R. Miracle, 1983. Biometric analysis of Brachionus plicatilis ecotypes from Spanish lagoons. Hydrobiologia 104: 271–279.

Serra, M. & M. R. Miracle, 1985. Enzyme polymorphism in Brachionus plicatilis populations from several Spanish lagoons. Verh. int. Ver. Limnol. 22: 2991–2996.

Snell, T. W. & K. Carrillo, 1984. Body size variation among strains of rotifer Brachionus plicatilis. Aquaculture 37: 359–367.

Snell, T. W. & B. Winkler, 1984. Isozyme analysis of rotifer proteins. Biochem. Syst. Ecol.

Yufera, M., 1982. Morphometric characterization of a small-sized strain of Brachionus plicatilis in culture. Aquaculture 27: 55–61.

Hydrobiologia 147: 91–95 (1987)
May, L., Wallace, R. & Herzig, A. (eds), Rotifer Symposium IV

Taxonomic and ecological notes on *Hexarthra bulgarica* from high mountain lakes and ponds in the Sierra Nevada (Spain)[1]

L. Cruz-Pizarro & R. Morales
Departamento de Ecología, Facultad de Ciencias, Universidad de Granada, Granada, Spain

Keywords: *Hexarthra bulgarica*, Rotifera, taxonomy, Sierra Nevada, Spain

Abstract

Hexarthra bulgarica was found in 16 of the 37 high mountain lakes (over 2500 m) studied in the Sierra Nevada (Southern Spain). These populations resemble closely the subspecies *H. b. canadensis*. Data on the morphology and ecology of the Sierra Nevada populations are given, and these are compared to published information on the three other subspecies of *H. bulgarica*.

Introduction

Taxonomy of the genus *Hexarthra* is still a matter of some uncertainty. Ruttner-Kolisko [1974] has divided the genus into two main groups of species: the *fennica-jenkinae* group and the *mira-intermedia* group and suggested that *Hexarthra bulgarica* occupies an intermediate position between them. However, Koste [1978] prefers to include the species in the 'formenkreis' *propinqua-mollis-bulgarica*.

Dumont *et al.* [1978] suggested that *H. bulgarica* can be divided into three subspecies: *H. b. bulgarica, H. b. nepalensis* and *H. b. canadensis*. Although these authors note that the only morphological feature common to all these subspecies is the presence of a pair of spines on the dorsolateral arm, certain ecological characteristics help to define the species.

Hexarthra bulgarica has already been reported from several high altitude lakes: the Rila Mountains, Bulgaria [Wiszniewski, 1933], the Italian Alps [Tonolli & Tonolli, 1951; Ferrari, 1971], the Himalayas [Edmonson & Hutchinson, 1934], the Khumbu area (east Nepal) in lakes within the subnival belt [Loffler, 1969; Daems & Dumont, 1974], the Rocky Mountains (Grouse Lake, Alberta, Canada) [Dumont *et al.*, 1978], and probably the Andes [Ruttner-Kolisko, 1974; citing Löffler, pers. comm.]. This is the first detailed description of the species in Southern Europe. In this paper we describe the general taxonomic and ecological characteristics of several populations of *H. bulgarica* collected from high mountain lakes in the Sierra Nevada (Southern Spain).

[1] Research supported by CAICYT Project No. 3069/83.

Materials and methods

During the summer of 1981 a survey was carried out of 37 oligotrophic ponds and small lakes located above 2500 m in the Sierra Nevada (Granada, Spain). Most of these lakes lie on siliceous rocks, but some basins contain small calcium carbonate deposits in amphibolites and micaschists. All the lakes remain frozen for at least 8 months of the year. During the ice-free period, the water level in most of the lakes is subject to wide fluctuations, and some of the smallest waterbodies dry up completely in summer. A description of the sites and their precise geographical position are given in Morales [1985a]. The different rotifer communities found in these lakes was discussed by Morales [1985b].

Net samples were collected from a variety of sites within each lake. A total of 137 qualitative samples were obtained from the littoral by sweeping a conical plankton net 25 cm diameter, 90 cm in length with a mesh size of 45 μm. All samples were preserved in 4% formalin. Individual specimens were mounted in glycerine and examined under a light microscope. Mastax preparations were made of several specimens from every population, the soft tissues having been dissolved by sodium hypochlorite.

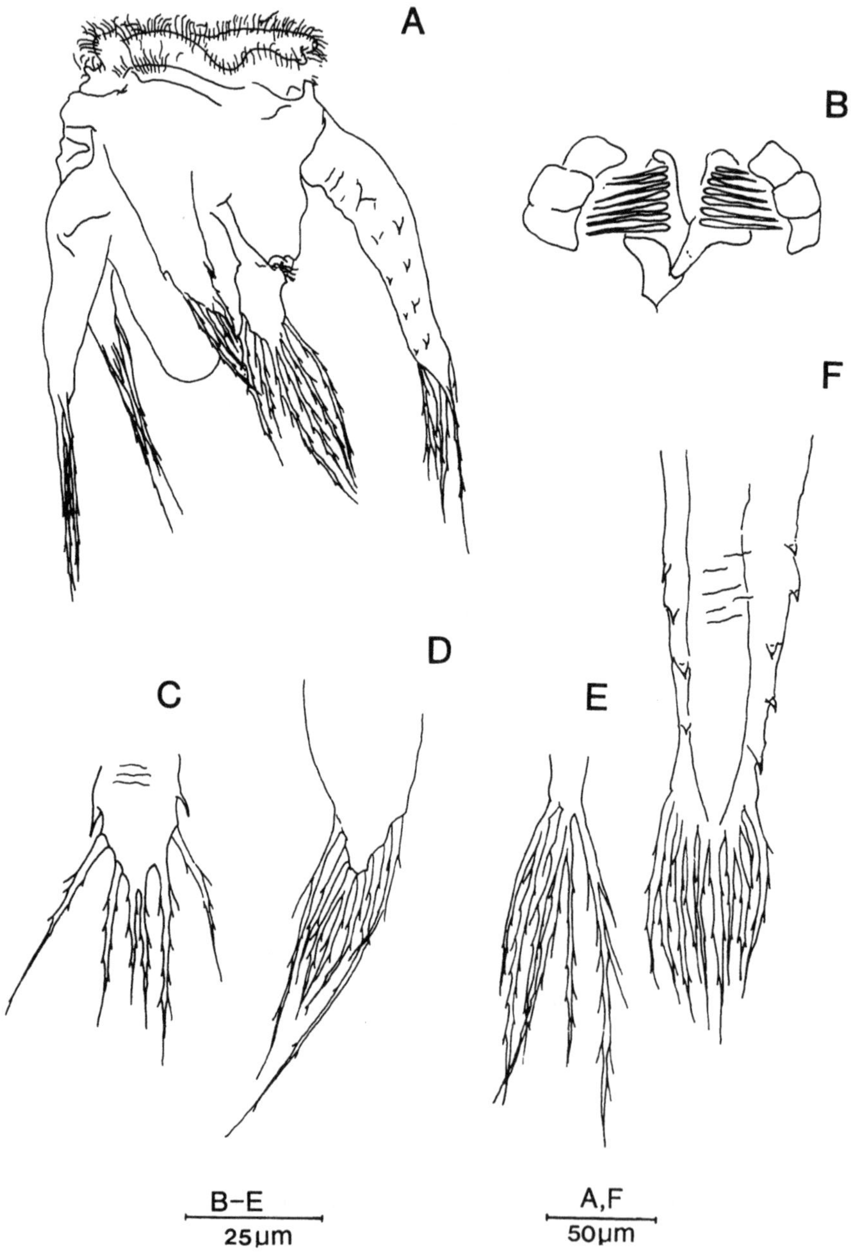

Fig. 1. Hexarthra bulgarica canadensis from the Sierra Nevada. A: whole animal; B: trophi; C: dorsolateral arm; D: ventrolateral arm; E: dorsal arm; F: ventral arm.

Results and discussion

Morphological characteristics

Of the 37 ponds and small lakes studied, 16 contained populations of *H. bulgarica*. All of the specimens collected (ca 200) had three common characteristics: (1) absence of ventral appendages, (2) absence of a lower lip, and (3) presence of spines on the dorsolateral arms (Fig. 1). Most individuals had a ventral arm to body length ratio greater than one. However, there was considerable intra- and inter-population variation. The intra-population variation may be demonstrated by examining the number of teeth on the unci of specimens from a single lake. For example, in Laguna Cuadrada the proportion of the population having teeth on the unci of 7/7, 7/8, 8/8, and 8/9 was 14%, 24%, 19%, and 43%, respectively. The inter-population variation may be seen by comparing these data to that of the overall population. A greater percentage of the Laguna Cuadrada population had more teeth on the unci than has the Sierra Nevada population as a whole (Table 1). This deviation from the prevailing trend has also been described for the population of lake La Caldera, where most of the individuals had 3/4 spines and 10 bristles on the ventral arm [Cruz-Pizarro, 1981].

Table 1 summarizes the range of variation in a number of key features for the Sierra Nevada populations, and compares them with data published for the other three subspecies [Dumont *et al.*, 1978]. Despite this great morphological variation, the presence of a pair of spines on the dorsolateral arms strongly suggests that this species is indeed *H. bulgarica*. Overall our material is most similar to *H. b. canadensis*, although more than half of the individuals from Sierra Nevada have a smaller number of spines on the ventral arm than the Grouse Lake (Canada) population. Further, many specimens have fewer bristles on the dorsolateral, ventrolateral, and dorsal arms than in the other subspecies.

Table 1. Comparison of the main features of the subspecies of *Hexarthra bulgarica*.

	H.b. bulgarica[a]	*H.b. nepalensis*[a]	*H.b. canadensis*[a] (Canada)	*H.b. canadensis*[b] (Sierra Nevada)
Body length (μm)	220–340	220–310	240–270	170–280
Ventro-anterior lip	Absent	Present	Absent	Absent
Teeth on unci	6/6	5/5, 5/4	7/7, 7/8, 8/8	7/7, 7/8, 8/8, 8/9 (19%) (45%) (30%) (6%)
Ventral arm:				
spines	3/3, 4/4	3/3	5/5, 5/4	3/4, 4/4, 5/4, 5/5 (9%) (55%) (29%) (7%)
bristles	10–12	10	10–13	10 12 (45%) (55%)
Dorsal arm:				
bristles	7, 8	7	6, 7, 8	4 5 6 7 (7%) (40%) (44%) (9%)
Ventrolateral arm:				
bristles	7	7	7, 8	5 6 7 8 (3%) (17%) (71%) (9%)
Dorsolateral arm:				
bristles	6	6	6, 7	5 6 7 (21%) (74%) (5%)
spines	2	2	2	1 2 (15%) (85%)

[a] Data from Dumont *et al.* [1978]; [b] this study.

Ecological characteristics

Hexarthra bulgarica is one of the most common rotifers inhabiting freshwaters in the Sierra Nevada and has been found both in typically planktonic habitats (i.e. Lake La Caldera) and in the littoral zone of small and shallow ponds (i.e. Rio Seco Inferior). Very often it is the dominant species in such communities, reaching population densities of over 10000 individuals m^{-3} in lake La Caldera [Cruz-Pizarro, 1982] and up to 70000 individuals m^{-3} in Rio Seco Inferior [Morales, 1985a].

Another notable characteristic of many of the Sierra Nevada populations is the presence of mictic females very early in the season. This has already been described in detail by Cruz-Pizarro [1982] for the population in lake La Caldera. According to King [1980] resting eggs produced early in the season should represent a '...reasonably complete sample of the genetic variation of the original cohort'. This may explain the great intra-population morphological variation observed in our study.

Hexarthra bulgarica occurs in lakes that differ widely in their physical and chemical conditions

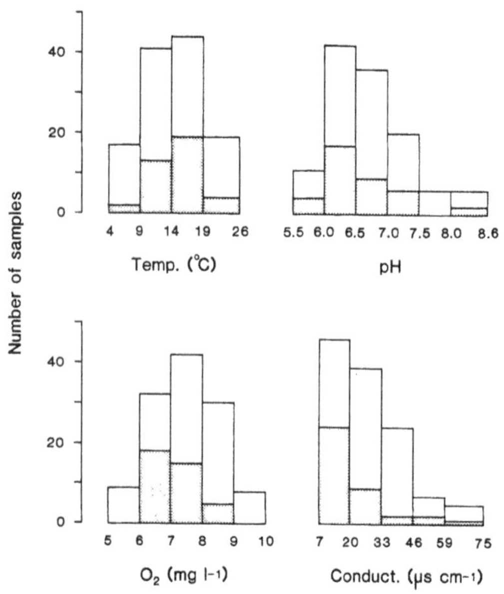

Fig. 2. The distribution of *Hexarthra bulgarica canadensis* in lakes of different physical and chemical characteristics. Open histograms show the proportion of samples falling into each defined category and shaded histograms the proportion containing *H. b. canadensis*.

(Fig. 2). However, this species does display a preference for the more oligotrophic waters of low conductivity. In fact, more than 70% of the lakes in which this species was found had conductivities ranging between 7 and 20 $\mu S\ cm^{-1}$.

Conclusions

Morphological analyses suggest that the populations of *H. bulgarica* found in the Sierra Nevada (Granada, Spain) are probably *H. b. canadensis*. The presence of *H. bulgarica* in lakes of the Sierra Nevada agrees with the general conclusion that it is truly a high mountain rotifer. All the reports of this species are from lakes of about 2200 m or greater. This study illustrates the need to consider ecological information in defining species [Ruttner-Kolisko, 1963; Dumont, 1980; Cruz-Pizarro, 1981].

Acknowledgements

We thank Dr. D. G. George and Miss D. Hewitt for reading and improving our manuscript. We also thank Mr. T. Furnass for drawing part of the figures and Mrs. J. Hawksford for typing the manuscript.

References

Cruz-Pizarro, L., 1981. Rotiferos planctónicos de un lago de alta montaña (La Caldera, Siarra Nevada, Granada). Necesidad de una taxonomía ecológica. Trab. Monogr. Dep. Zool. Univ. Granada, (N.S.) 4: 1–40.

Cruz-Pizarro, L., 1982. Ciclo de viva de Hexarthra bulgarica en la laguna de La Caldera (Sierra Nevada, Granada). Consideraciones sobre la re producción en Rotiferos Monogonontes. Foro Ciencias y Letras 3/4: 95–104.

Daems, G. & H. J. Dumont, 1974. Rotifers from Nepal, with the description of a new species of Scaridium and a discussion of the Nepalese repre sentatives of the genus Hexarthra. Biol. Jaarb. 42: 61–81.

Dumont, H. J., 1980. Workshop on taxonomy and biogeography. Hydrobiologia 73: 205–206.

Dumont, H. J., M. Coussement & R. S. Anderson, 1978. An examination of some Hexarthra species (Rotatoria) from western Canada and Nepal. Can. J. Zool. 56: 440–445.

Edmondson, W. T. & G. E. Hutchinson, 1934. Report on

Rotatoria. Mem. Conn. Acad. Arts Sci. 10: 153–186.
Ferrari, I., 1971. Notes on the dynamics of the reproductive activity of Arctodiaptomus bacillifer in high altitude alpine lakes. Boll. Zool. 38: 221–235.
King, C. E., 1980. The genetic structure of zooplankton populations. In W. C. Kerfoot (ed.), Evolution and Ecology of Zooplankton Communities. The University Press of New England, Hanover (N.H.); Lond.: 315–328.
Koste, W., 1978. Rotatoria, 2 Vols. Borntraeger Berlin. 673 pp., 234 plates.
Löffler, H., 1969. High altitude lakes in Mt. Everest region. Verh. int. Ver. Limnol. 17: 373–385.
Morales, R., 1985a. Estudio de las comunidades de Rotiferos Monogonontes de las lagunas de alta montaña de Sierra Nevada. Tesis Doctoral. Universidad de Granada, 296 pp.
Morales, R., 1985b. Rotifer fauna of lakes and ponds above 2500 m in the Sierra Nevada, Spain, with a description of a new subspecies. In this volume.
Ruttner-Kolisko, A., 1963. The interrelationships of the Rotatoria. In E. C. Dougharty (ed.), The Lower Metazoa. University of California Press, Berkeley: 263–272.
Ruttner-Kolisko, A., 1974. Plankton Rotifers, biology and taxonomy. Binnengewässer 26(1):146 pp.
Tonolli, V. & L. Tonolli, 1951. Osservazioni sulla biologia ed ecologia di 170 popolamenti zooplanctonici di laghi italiani de alta quota. Mem. Ist. Ital. Idrobiol. 6: 53–136.
Wiszniewski, J., 1933. Un nouveau rotifère du genre Pedalia habitant les lacs des hautes montagnes. Int. Revue ges. Hydrobiol. Hydrogr. 29: 229–236.

Hydrobiologia 147: 97–101 (1987)
May, L., Wallace, R. & Herzig, A. (eds), Rotifer Symposium IV

Rotifer fauna of lakes and ponds over 2500 m above sea level in the Sierra Nevada, Spain, with description of a new subspecies[1]

R. Morales-Baquero
Departmento de Ecologia, Fac. de Ciencias, Universidad de Granada, Granada, Spain

Keywords: Rotifera, Spain, alpine lakes, taxonomy

Abstract

The littoral zone of lakes in the high mountains of Sierra Nevada, Spain, were sampled during the ice-free periods of 1980 and 1981. A species list is given and a new subspecies, *Lepadella quinquecostata nevadensis*, subsp. nov. is described. Notes on the geographical distribution of some species are also included.

Introduction

High mountain lakes are of interest to biologists because of the extreme environmental conditions which they present. The rotifer fauna of lakes in several mountain ranges near Sierra Nevada have already been described: Pyrenees [Margalef, 1948; De Ridder, 1964; Miracle, 1978], Alps [Tonolli & Tonolli, 1951; Thomasson, 1956], Atlas [Coussement & Dumont, 1980]. However, in the Sierra Nevada, rotifers have been studied from only one mountain lake, La Caldera [Cruz-Pizarro, 1981a, b]. This paper presents new information on the rotifer fauna of other mountain lakes in this area.

Study site

Sierra Nevada (36°55′N –37°15′N, 2°31′–3°40′E) is the most southerly mountain range in Europe, reaching a maximum altitude of 3482 m. About 50 small post glacial lakes, on siliceous bedrock, are found above 2500 m. These lakes range from a few meters to about 200 m in length, and are generally less than 2 m deep, although one has a maximum depth of more than 12 m. Littoral vegetation is sparse along the shores and, where it does exist, consists of Bryophyta and filamentous green algae. There are no aquatic vertebrates. These lakes are generally ice-covered from October to May.

[1] Research supported by CAICYT Project No. 3069/83.

Materials and methods

Samples were collected from the littoral zone of 37 mountain lakes (>2500 m above sea level) in Sierra Nevada, during the ice-free periods of 1980 and 1981. A conical Nytal plankton net (45 μm mesh) was swept repeatedly through the open water. Samples were preserved in 4% formaldehyde solution immediately after collection.

Results and discussion

Table 1 lists the rotifer species collected during the study and the frequency with which they were found. For comparison, a list of the most common species in the Pyrenees and Alps is also included. Table 1 shows that the most common species of the Alpine and Pyrennean lakes are very similar. With the exception of *Euchlanis dilatata*, these species are much less common in lakes of Sierra Nevada.

Table 1. Rotifer species recorded in Sierra Nevada in order of frequency of occurrence in the 37 lakes sampled. The most common species of the Alpine and Pyrennean lakes are also shown [after Tonolli & Tonolli, 1951; Miracle, 1978]. Frequency is expressed as the percentage of lakes in which the species occurred.

SIERRA NEVADA		ALPS	
Euchlanis dilatata	81%	*Polyarthra vulgaris*	29%
Trichocerca relicta	78%	*Asplanchna priodonta*	11%
Hexarthra bulgarica canadensis	40%	*Keratella quadrata*	11%
Lepadella patella	40%	*Euchlanis dilatata*	10%
Trichocerca bicristata	38%	*Conochilus unicornis*	9%
Lecane furcata	38%	*Lecane luna*	8%
Trichocerca tenuior	32%		
Cephalodella gibba microdactyla	32%	PYRENEES	
Trichotria tetractis	30%	*Asplanchna priodonta*	24%
Colurella obtusa	30%	*Kellicottia longispina*	21%
Lecane flexilis	22%	*Polyarthra vulgaris*	15%
Lecane lunaris	19%	*Ascomorpha ovalis*	4%
Trichocerca rattus	16%	*Keratella quadrata*	3%
Lecane closterocerca	16%	*Euchlanis dilatata*	3%
Notholca squamula	13%		
Polyarthra dolichoptera	8%		
Lecane kluchor	8%		
Lepadella quinquecostata nevadensis	8%		
Trichocerca cavia	5%		
Lepadella acuminata	5%		
Lepadella triptera	3%		
*Keratella cochlearis**	3%		
*Filinia longiseta/terminalis**	3%		
*Polyarthra remata**	3%		
*Polyarhtra major/vulgaris**	3%		

(*) from Cruz-Pizarro (1981).

Hexarthra bulgarica, considered to be a typical high mountain rotifer species [Dumont, 1980], has a wide distribution in Sierra Nevada [see also Cruz-Pizarro & Morales, 1987].

Species of particular interest

Lepadella quinquecostata nevadensis subsp. nov. (Fig. 1A-D)

Lepadella quinquecostata [Lucks, 1912] is characterised by a low median keel on the dorsal plate of the lorica, flanked by 2 or 3 pairs of symmetrical lateral ribs, and a short characteristic crest at the posterior end of the dorsal plate. This species has never been recorded in the Iberian peninsula before. Specimens collected from Sierra Nevada are morphologically different from the type form of this species which strongly suggests that it is a new subspecies. These differences are summarised below:

1) Median keel of the dorsal lorica very low and hardly visible, with two pairs of symmetrical lateral ribs. The more internal of these lateral ribs reaches, but does not joiin, the posterior of the lorica.

2) Lorica high in cross-section, with flat median keel and two protuberances on the ventral surface.

3) Dorsal lorica terminates without crest.

The 9 specimens measured were also considerably smaller than the type form described by Koste [1978], see Table 2. Specimens were found in 5 of the lakes sampled, always in small numbers. All of these lakes had an abundance of moss along their shores.

Lecane (Heminostyla) kluchor [Tarnogradsky, 1930] (Fig. 1E, F)

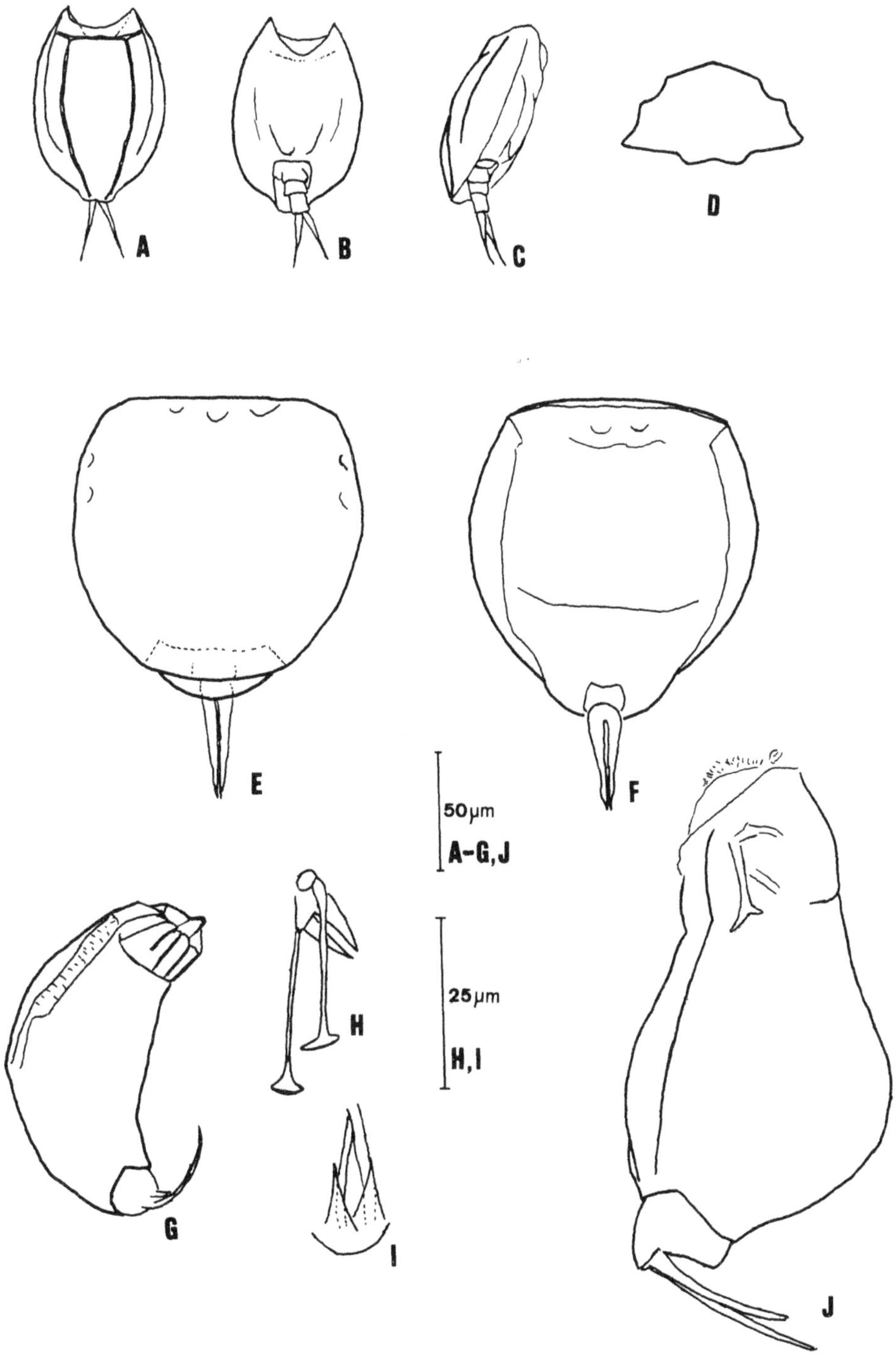

Fig. 1.A-D. Lepadella quinquecostata nevadensis subsp. nov. (A: dorsal; B: ventral; C: ventro-lateral; D: cross-section). E-F *Lecane*(H)*kluchor* (E: dorsal; F: ventral). G-I *Trichocerca*(D)*relicta* (H: trophi; I: toes). J *Cephalodella gibba microdactyla.*

Table 2. Size (in μm) of *Lepadella g. quinquecostata* (Lucks) (from Koste 1978) and *L.q. nevadensis.*

	L.q. quinquecostata	*L.q. nevadensis*
Length lorica	92–112	64–84
Breadth lorica	60– 70	44–61
Length toes	22– 30	22–25

This rare species was found in three of the lakes studied, although in very low numbers. The measurements of 10 specimens fell within the following ranges: length dorsal lorica 95–108 μm; width dorsal lorica 94–100 μm; length ventral lorica 102–111 μm; width ventral lorica 78–82 μm; toes 36–41 μm; claws 3.5–4.6 μm.

This species has a restricted distribution in Europe. It has been found in the Danube Valley, the Black Forest and the Subalpine Region [Berzins, 1978]. Recently, Francez and Pourriot [1984] recorded it for the first time in France in a peat bog of Puy de Dome (French Sub-alpine mountains). This is the first record in Spain.

Cephalodella gibba microdactyla [Koch-Althaus, 1963] (Fig. 1J)
This subspecies was thought to be restricted to central Europe, where it is frequently found [Koste, 1978; Berzins, 1978]. This is the first record from Spain.

The species was found frequently in the lakes studies, always associated with periphytic habitats. The specimens examined were usually contracted, due to the formalin preservative. Therefore, only measurements of the toes and trophi are given: toes 58–72 μm; trophi 52 μm (fulcrum 32 μm, manubrium 34 μm, uncus 17 μm).

Trichocerca (Diurella) relicta [Donner, 1950] (Fig. 1G–I)
This species was widely represented in Sierra Nevada. Its morphological characteristics correspond well to the description of Koste [1978] and its measurements were within the following ranges: lorica length 75–115 μm; left toe 27–33 μm; right toe 17–27 μm; trophi-fulcrum 28 μm, manubrium 23 μm.

Individuals from different populations were measured (Table 3). The results show a considerable variability in lorica length between populations. In many cases the animals were smaller than the size range given by Koste [1978] for this species (95–105 μm).

Table 3. Mean length (μm ± SE) of *Trichocerca relicta* in 20 populations from lakes in the Sierra Nevada.

Lakes and dates	Sample size	$\bar{X} \pm$ SE
A. Verdes (14-7-81)	9	93±1.2
A. Verdes (7-8-81)	3	93±0.8
Majano (30-6-81)	4	91±0.8
Majano (14-7-81)	10	81±0.7
Majano (2-9-81)	4	94±4.8
Gemelas (30-6-81)	10	85±2.0
Gemelas (7-8-81)	8	93±2.6
Mosca (14-8-81)	10	100±0.3
L. L. Larga (11-9-81)	4	103±3.8
P. Negro (6-6-81)	7	95±2.2
Siete L.7 (16-7-81)	10	101±1.2
Siete L.6 (11-8-81)	4	103±1.7
Siete L.5 (16-7-81)	10	85±0.3
P. Col. (7-7-81)	10	92±2.0
P. Col. (28-7-81)	10	82±0.5
Lanjaron 3 (7-7-81)	6	106±1.9
Lanjaron (7-7-81)	10	104±2.7
Lanjaron (28-7-81)	5	87±1.6
V. media (10-7-81)	10	85±1.9
V. sup. (25-7-81)	5	99±1.9

The geographical distribution of this species is also restricted to Europe. It has been found in the Alps, central Sub-alpine mountains, Pontic Province, central lowlands and Baltic Province [Berzins, 1978]. Our record is the most southerly location found to date and is the first record for Spain.

Acknowledgements

The author is deeply grateful to Dr Walter Koste for his kind help with taxonomic problems.

References

Berzins, B., 1978. Rotatoria. In: J. Illies (ed.) Limnofauna Europaea, 2nd. ed. Fischer, Stuttgart: 54–91.

Coussement, M. & H. J. Dumont, 1980. Some peculiar elements in the rotifer fauna of the atlantic Sahara and of the Atlas mountains. Hydrobiologia 73: 249–254.

Cruz-Pizarro, L., 1981a. Estudio de la comunidad zooplanctónica de un lago de alta montaña (LaCaldera, Granada). Unpubl. Doctoral thesis, Univ. Granada, 181 pp.

Cruz-Pizarro, L., 1981b. Rotiferos planctónicos de un lago de alta montaña (La Caldera, Granada). Trab. Monogr. Dep. Zool. Univ. Granada 4: 1–40.

Cruz-Pizarro, L. & R. Morales, 1987. Taxonomic and ecological notes on Hexarthra bulgarica canadensis from high mountain lakes and ponds in the Sierra Nevada (Spain). Hydrobiologia, this volume.

De Ridder, M., 1964. Enkele raderdiertjes uit de Spaanse Pyreneeën. Biol. Jaarb. Dodonaea 32: 185–201.

Dumont, H. J., 1980. Workshop on taxonomy and biogeography. Hydrobiologia 73: 205–206.

Francez, A. J. & R. Pourriot, 1984. Remarques taxonomiques sur quelques rotiferès des tourbières avec la description d'une espèce et d'une sous-espèce nouvelles. Hydrobiologia 109: 125–130.

Koste, W., 1978. Rotatoria. Die Rädertiere Mitteleuropas, begründet von Max Voigt. Vols. 1 & 2. Gebrüder Borntraeger, Berlin, Stuttgart, 673 & 234 pp.

Margalef, R., 1948. Flora, fauna y communidades bióticas de las aguas dulces del Pirineo de la Cerdaña. C.S.I.C. Zaragoza. Monografias de la estación de estudios pirenaicos: 4–226.

Miracle, R., 1978. Composición específica de las communidades zooplanctónicas de 153 lagos de los Pirineos y su interés biogeográfico. Oecologia aquatica 3: 167–191.

Thomasson, K., 1956. Reflections on Artic and Alpine lakes. Oikos 7: 117–143.

Tonolli, L. & V. Tonolli, 1951. Ossevazioni sulla biologia ed ecologia di 170 popolamenti zooplanctonici di laghi italiani di alta quota. Mem. Ist. ital. Idrobiol. 6: 53–136.

Hydrobiologia 147: 103–105 (1987)
May, L., Wallace, R. & Herzig, A. (eds), Rotifer Symposium IV

Rotifer fauna in the periphyton of Karst rivers in Croatia, Yugoslavia

Radovan Erben
Department of Zoology, Faculty of Natural Sciences, University of Zagreb, Yugoslavia

Keywords: Rotifera, periphyton, Karst, river, Yugoslavia

Abstract

Rotifers from periphyton communities of rivers in the Karst area of western Yugoslavia were examined. Thirty-two species, belonging to 10 families, were found. Rotifers were absent from locations without periphyton. Low densities of animals were noted at stations near urban centres and in the estuaries of the Mirna, Răša, Rječina and Krka rivers. Several species were found in all of the rivers studied, but always in low numbers. Most of these species are generally considered to be euryvalent and cosmopolitan.

Introduction

The western part of Yugoslavia, particularly the Republic of Croatia, belongs to the Karst region. Qualitative and quantitative descriptions of the flora and fauna of this area are incomplete, although the structure of the macro-zoobenthos, with respect to the physical, chemical and hydographic characteristics of these waters, has been described by Matoničkin & Pavletić [1961, 1964], Matoničkin *et al.* [1966], Matoničkin *et al.* [1966], Matoničkin [1969] and Pavletić & Matoničkin [1972]. These investigations did not consider the micro-zoobenthos of the periphyton, which is predominantly rotifers. This paper gives a brief account of the rotifers found in this region.

Methods

Seven Karst rivers were investigated, Mirna, Raša, Rječina, Korana, Mrežznica, Gacka and Krka. Four to ten sites were selected on each river, giving a total of 47 sampling sites. Samples were taken in April, June, September and October, between 1978 and 1984.

Samples were collected by removing the periphyton from the substrata, e.g. stones, blocks of wood or branches, with tweezers or a special scraping tool; the samples were not preserved. The rotifers were identified according to Kutikova [1970] and Koste [1978]. Relative frequency of occurrence was expressed on a scale of 1–7, after Knöpp [1954].

Results and discussion

In general, the rivers studied were characterized by fast flow, a predominantly stony bottom and low water temperature. The periphyton consisted mainly of mosses, with some green and blue-green algae, and diatoms. The sampling stations differed in type of periphyton, water discharge and current.

The rotifers of Karst rivers are well adapted to life in periphyton communities by their habit and type of movement. Priphyton provides a suitable habitat and nourishment for these animals; sufficient oxygen is available and the rotifers are protected against strong currents. High oxygen concentrations were found in the upper courses of the rivers where mosses dominated the periphyton. Bryophilic species including the genera *Cephalodella, Colurella, Lepadella* and *Philodina* were found here (Table 1). The low temperatures

Table 1. Systematic account of the rotifers in the periphyton of the Karst rivers with frequency values after Knöpp (1954).

Species	Type of periphyton			
	mosses	blue-green algae and mosses	green algae	green algae and diatoms
Superorder Monogononta				
Euchlanis dilatata	2		6	
Trichotria tetractis var. *similis*			1	
T. pocillum			2	
Colurella uncinata		4	5	3
Lepadella ovalis			6	
L. patella			4	4
Lecane ungulata			3	2
L. luna		2		1
Proales sp.		1	1	
Cephalodella catellina	1	2		
C. hoodi	3		3	3
C. pachydon	3	5	7	
C. gracilis			2	1
C. nodosa	2	2	3	
C. gibba	1	2	2	3
C. incila		5	6	
C. exigua		2	2	1
C. bryophila	2	2		
C. sterea f. *minor*			3	
Cephalodella sp.	1	2		1
Taphrocampa annulosa		2		
Dicranophorus forcipatus		1	1	1
D. grandis	1	2	4	
Encentrum mucrenatum			4	2
Encentrum sp.		1	1	
Testudinella patina		2		
Testudinella sp.		1		
Superorder Bdelloidea				
Philodina roseola	7	7		
Rotaria citrina	6	6		
R. tardigrada	3	2	1	
R. rotatoria		6	7	6

characteristic of these waters did not seem to hinder development of the genera *Cephalodella, Colurella, Dicranophorus* or *Rotaria.* Thirty-two species were found, the most numerous being *Colurella uncinata, Lepadella patella* and *Philodina roseola.* Many species appeared only once, preferring warmer periods of the year when periphyton was well developed and flow rates low. In general, a rich periphyton community and slow currents revealed a high number of species and an abundance of individuals. In areas where travertine barriers frequently traversed the rivers, fast currents developed and mosses dominated the periphyton. Here, rotifers occurred in low numbers only.

Many of the species found were cosmopolitan forms [Illies 1978] which are also euryvalent, according to Kutikova [1970]. All of the species recorded occurred in the periphyton only, with two exceptions. *Rotaria tardigrada* and *R. rotatoria* were also found in leaf litter. Rotifers were notably absent from the river estuaries, where high salt concentrations probably limited their distribution. Besides cosmopolitan forms, a number of rare forms were also collected, i.e. *Cephalodella pachydon, C. gracilis, C. nodosa, C. bryophila, Taphrocampa annulosa.* Only a few species, *C. uncinata, P. roseola* and *R. citrina,* appeared at most stations along the whole length of a river.

According to Sladeček [1973] most of the rotifer species found in the present study are characteristic of unpolluted waters. However, many of these species were also found at polluted sites. This suggests that some of these species can tolerate higher levels of pollution than was first thought and that their use as indicatoirs of pollution is questionable.

References

Illies, J., 1978. Limnofauna Europaea. Gustav Fischer Verlag, Stuttgart.

Knöpp, H., 1954. Ein neuer Weg zur Darstellung Biologischer vorflueruntersuchungen erläutert an einem Gütelängsschnitt des Mains. Die Wasserwirtschaft 45: 9–15.

Koste, W., 1978. Rotatoria. Die Rädertiere Mitteleuropas, begründet von Max Voigt. Überordnung Monogonota. 2 Vols. Gebrüder Borntraeger, Berlin; Stuttgart, 673 pp., 234 figs.

Kutikova, A. S., 1970. Kolovratki fauni SSSR. ANSSR, Leningrad.

Matoničkin, I. & Pavletić, Z., 1961. Le mond vivant des cascades travertines en rivières karstiques. Biološki glasnik 14: 105–128.

Matoničkin, I. & Pavletić, Z., 1964. Studi ecologici delle bi-

ocenose sulle deposizioni travertinose. Estratto Dall' Archivio di Oceanografia e Limnologia X111, 2: 197–205.
Matoničkin, I., Pavletić, Z. & Tavčar, V., 1966. Speed of water as ecological factor in calacareous running waters. Biološki glasnik 19: 51–63.
Matoničkin, I., 1969. Problems concerning the biology of waste waters in karst regions. Krš Jugoslavije 6: 433–442.
Pavletić, Z. & Matoničkin, I., 1972. Biocenological structure in the river Korana as a reflection of the quality of water. Ekologija 7, 1–2: 59–79.
Sladeček, V., 1973. System of Water Quality from the Biological Point of View. Arch. Hydrobiol. Beih. 7: 1–218.

Hydrobiologia 147: 107–116 (1987)
May, L., Wallace, R. & Herzig, A. (eds), Rotifer Symposium IV

Rotifer occurrence in relation to pH

Bruno Bērziņš[1] & Birger Pejler[2]
[1]*Limnological Institute, Box 65, 221 00 Lund, Sweden*
[2]*Limnological Institute, Box 557, 751 22 Uppsala, Sweden*

Keywords: Rotifera, pH, indicator species, acidification

Abstract

Information on the distribution of more than 225 species of planktonic, periphytic, and benthic rotifers from diverse waters in south and central Sweden was analyzed for pH preference. No particularly strong correlation was found between pH and any other environmental factor, with regards to rotifer distribution. However, species indicating oligotrophy generally have their pH optima at or below 7.0, while those indicating eutrophy occur at or above this level. Rotifers found in acidic waters are often non-planktonic or semiplanktonic.

Introduction

The use of bio-indicators has sometimes been regarded with some scepticism. However, when based on a substantial amount of survey information the method seems to be a very useful one [Pejler, 1981 and references therein; Karabin, 1983].

Between 1945 and 1982, a large amount of data was collected from various types of water bodies in southern and central Sweden. The data include information on temperature, about 20 chemical factors, and the abundance of microzoans (planktonic, periphytic, and benthic) for each water body. The material is probably the largest of its kind ever collected. This paper discusses the pH preferences of rotifers from the study area.

Materials and methods

Planktonic, periphytic and benthic samples were collected from about 350 lakes, 50 ponds, 20 pools, 150 running waters, and 15 bogs in southern and central Sweden, between 1945 and 1982 [Bērziņš, 1978]. Most samples were collected during the ice-free period, but some winter samples were also taken.

Samples (1–5 l) for physical, chemical, and biological analysis were taken in open water at various depths, including near the bottom, and in the littoral zone.

Water samples collected for biological analysis were filtered through a 20 μm mesh net and the concentrate fixed in formalin. In most cases net samples also were collected and examined without fixation to facilitate identification of soft-bodied species.

Periphyton was collected in the following way. Moist macrophyte material (biovolume >0.5l) was squeezed into a bottle, dipped and squeezed again: this procedure was repeated several times. Subsamples were examined without fixation until no additional species were encountered. Density was estimated on a scale of 1–10, 1 indicating rare and 10 abundant.

To investigate the epifauna of stones and logs an area of 100–200 cm^2 was scraped off and the material sucked directly into a bottle from the knife. This was done by slowly letting air out of the collection bottle and water in, while keeping the bottle next to the knife blade.

The benthos was sampled by slowly scraping a small vessel over the bottom, collecting a volume of sediment a few millimeters thick and ca. 100–200 cm^2 in area. Subsamples of the sediment were analyzed without fixation and the density of organisms estimated using the scale of 1–10.

The taxonomy of Koste [1978] was followed for identification of monogonont rotifers and that of Donner [1965] for bdelloids. However, in some cases we deviated from these works e.g. concerning the genus *Dissotrocha* [see Bērziņš, 1982]. Furthermore we agreed with Donner [1965: 17–18], who considered it premature to distinguish between 'geographical, ecological and cytological races, microraces, clines, etc.', and accordingly used only one infraspecific category [see also Bērziņš, 1967; Ruttner-Kolisko, 1972; but compare Koste, 1978]. The infraspecific taxa discussed here should be regarded as forms, though the designation 'f.' has been omitted for simplicity.

Data were analyzed with the aid of a computer, in which the scaled value of species abundance was correlated with various environmental factors as single items. Here we report only rotifer occurrence as a function of pH for those species which occurred in ≥50 samples. The remaining information has been archived at the Limnological Institute in Lund.

Results and discussion

pH is, of course, related to many other chemical parameters in freshwaters, which in themselves are correlated with zooplankton distribution [e.g., Lindström, 1983]. Nevertheless, pH appears to be important in explaining distribution of rotifers [Edmondson, 1944; Skadowsky, 1923].

Most species examined in this study (N = 227) tolerated a broad range of pH values (Fig. 1a–f). Many species (N = 168, 74%) had their peak abundance within 1.0 units of pH 7.0. Others had their median value at quite low pH-values, two species at 4.4. The corresponding maximum on the alkaline side was 8.5.

Several rotifers were found only in acidic environments, never or only very rarely being found above pH 7.0, and with their peak occurrence well below pH 6.0. In the literature these species are said to be restricted to bogs, often in wet *Sphagnum*. Some of these rotifers are classified as 'acidophil' by Wulfert [1951] or as 'acidic' by Bateman & Davis [1980]. Many of these species were encountered in acid zones of a bog pool investigated by Skadowsky [1923] and/or in swamps studied by Francez [1981]. According to Bateman & Davis [1980: 150] they are 'mainly particle feeders or grazers'.

It is more difficult to find a common denominator for rotifers present in alkaline waters. However, *Euchlanis d. deflexa* is said to exhibit mass development in decomposing, drifting vegetation [Koste, 1978], and some of the planktonic forms with alkaline preferences (Fig. 1f) are known as indicators of eutrophic conditions: *Keratella cochlearis tecta* [Pejler, 1962], *Brachionus angularis, Brachionus calyciflorus, Filinia longiseta* and *Trichocerca pusilla* [Pejler, 1965, 1983, references therein]. However, this is not the case of the three *Synchaeta* species at the alkaline end of the pH gradient (*Synchaeta longipes, Synchaeta tremula kitina, Synchaeta stylata*). *Synchaeta longipes* is even said to occur 'in the plankton of oligotrophic lakes...' [Koste, 1978].

Although there are several planktonic rotifers whose peak abundance occurs on the alkaline end of the scale (pH > 8.0), there are none of the acid extreme (pH < 5.5). Thirty species of eulimnoplanktonic rotifers occur with their peak abundance values below or equal to pH 7.0 and 23 species above pH 7.0 (see Fig. 1, species marked with a * and a %, respectively). It is interesting that many species which have been reported to prefer oligotrophic conditions [Pejler, 1965, 1983] are found in the first group of these 54 species (i.e., pH ≤ 7.0), though not especially concentrated there. The 'most alkaline' species of this group, *Asplanchna herricki*, has its abundance maximum at pH 7.0, which is also the case of the 'most acid' indicator of eutrophy, *Polyarthra euryptera*. All other indicators of eutrophy [according to Pejler, 1965, 1983] have their peak abundance clearly in alkaline waters, and *Keratella cochlearis tecta* is

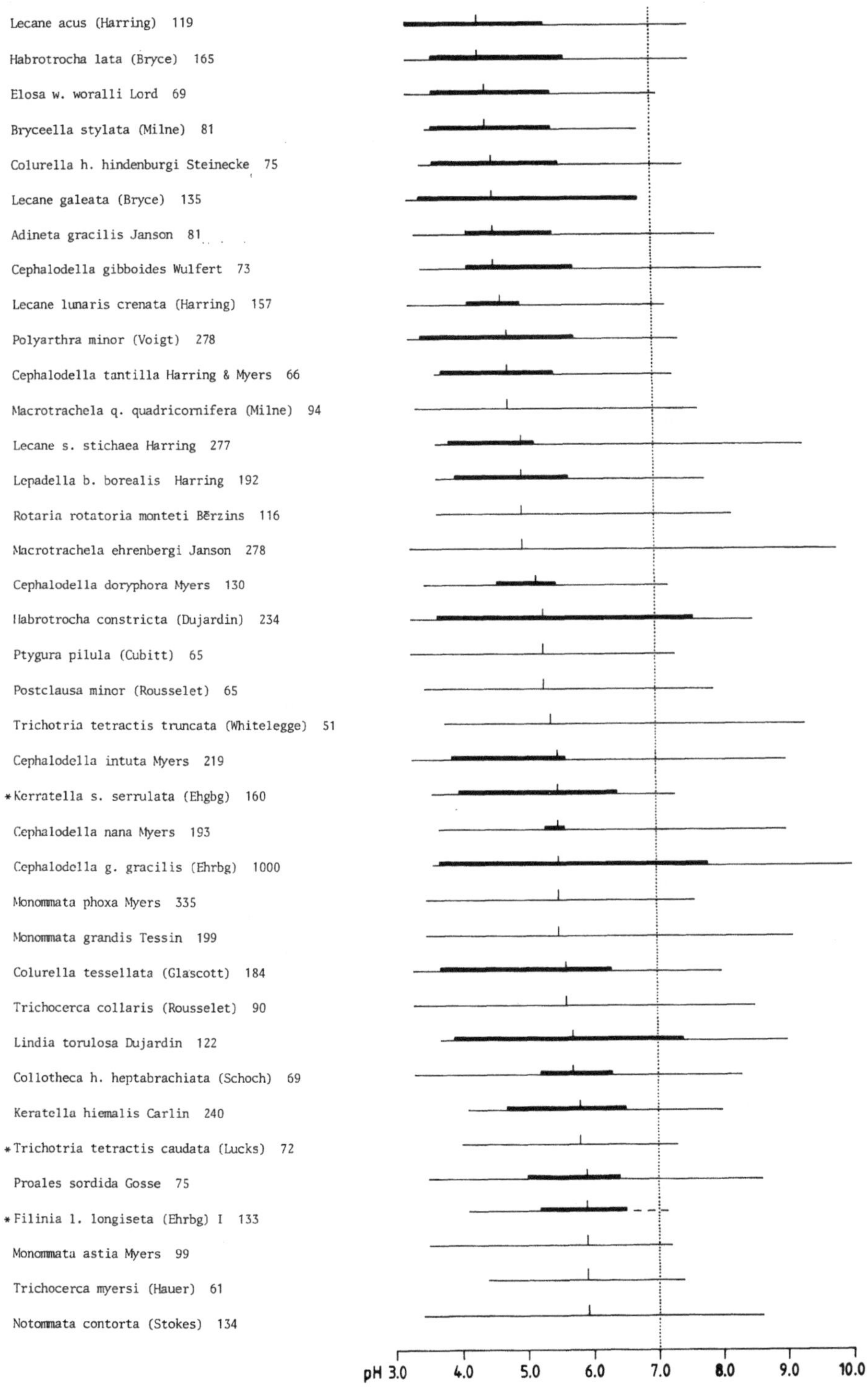
Lecane acus (Harring) 119
Habrotrocha lata (Bryce) 165
Elosa w. woralli Lord 69
Bryceella stylata (Milne) 81
Colurella h. hindenburgi Steinecke 75
Lecane galeata (Bryce) 135
Adineta gracilis Janson 81
Cephalodella gibboides Wulfert 73
Lecane lunaris crenata (Harring) 157
Polyarthra minor (Voigt) 278
Cephalodella tantilla Harring & Myers 66
Macrotrachela q. quadricornifera (Milne) 94
Lecane s. stichaea Harring 277
Lepadella b. borealis Harring 192
Rotaria rotatoria monteti Bērzins 116
Macrotrachela ehrenbergi Janson 278
Cephalodella doryphora Myers 130
Habrotrocha constricta (Dujardin) 234
Ptygura pilula (Cubitt) 65
Postclausa minor (Rousselet) 65
Trichotria tetractis truncata (Whitelegge) 51
Cephalodella intuta Myers 219
*Kerratella s. serrulata (Ehgbg) 160
Cephalodella nana Myers 193
Cephalodella g. gracilis (Ehrbg) 1000
Monommata phoxa Myers 335
Monommata grandis Tessin 199
Colurella tessellata (Glascott) 184
Trichocerca collaris (Rousselet) 90
Lindia torulosa Dujardin 122
Collotheca h. heptabrachiata (Schoch) 69
Keratella hiemalis Carlin 240
*Trichotria tetractis caudata (Lucks) 72
Proales sordida Gosse 75
*Filinia l. longiseta (Ehrbg) I 133
Monommata astia Myers 99
Trichocerca myersi (Hauer) 61
Notommata contorta (Stokes) 134
pH 3.0
4.0
5.0
6.0
7.0
8.0
9.0
10.0

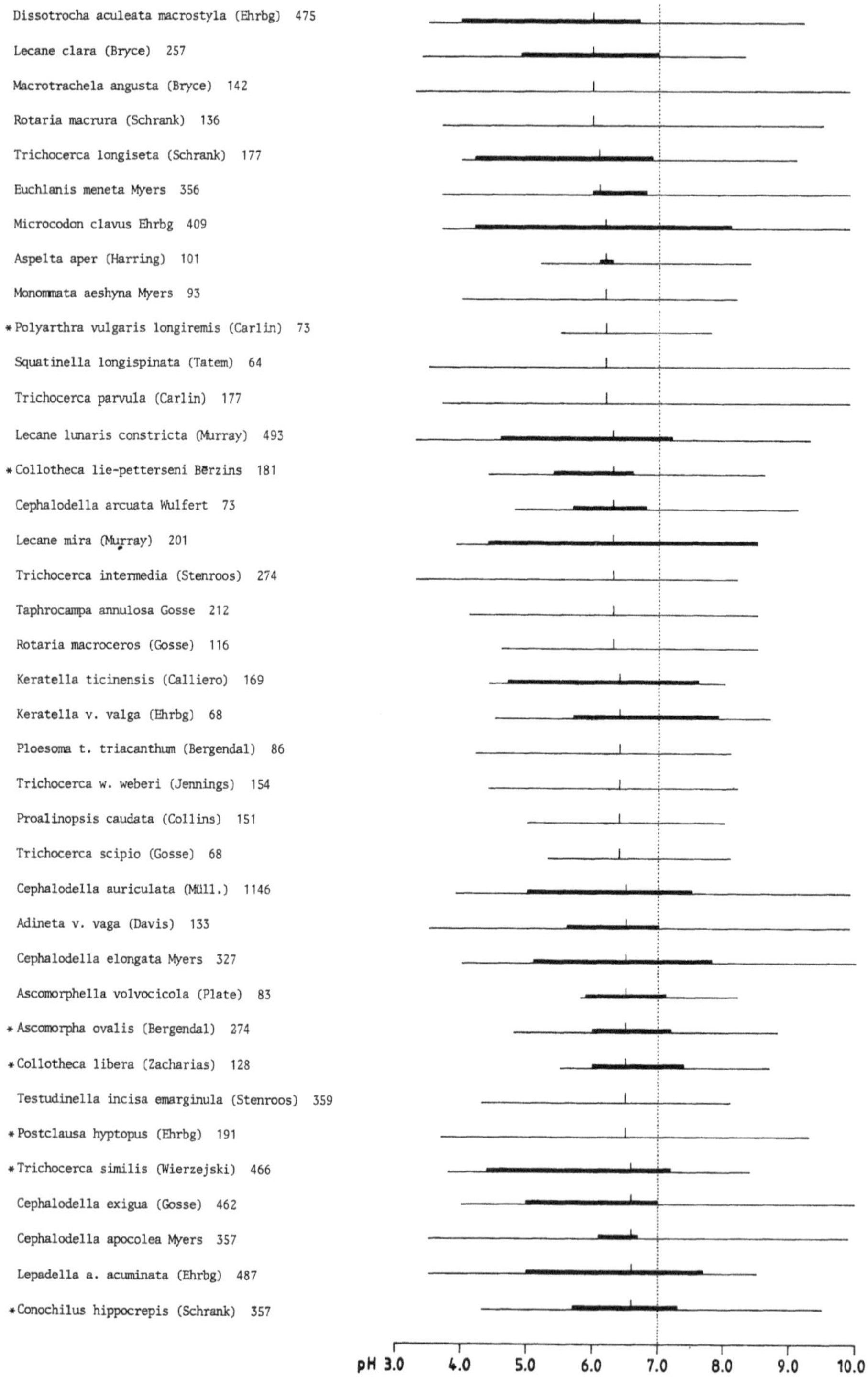
Dissotrocha aculeata macrostyla (Ehrbg) 475
Lecane clara (Bryce) 257
Macrotrachela angusta (Bryce) 142
Rotaria macrura (Schrank) 136
Trichocerca longiseta (Schrank) 177
Euchlanis meneta Myers 356
Microcodon clavus Ehrbg 409
Aspelta aper (Harring) 101
Monommata aeshyna Myers 93
* Polyarthra vulgaris longiremis (Carlin) 73
Squatinella longispinata (Tatem) 64
Trichocerca parvula (Carlin) 177
Lecane lunaris constricta (Murray) 493
* Collotheca lie-petterseni Bērzins 181
Cephalodella arcuata Wulfert 73
Lecane mira (Murray) 201
Trichocerca intermedia (Stenroos) 274
Taphrocampa annulosa Gosse 212
Rotaria macroceros (Gosse) 116
Keratella ticinensis (Calliero) 169
Keratella v. valga (Ehrbg) 68
Ploesoma t. triacanthum (Bergendal) 86
Trichocerca w. weberi (Jennings) 154
Proalinopsis caudata (Collins) 151
Trichocerca scipio (Gosse) 68
Cephalodella auriculata (Müll.) 1146
Adineta v. vaga (Davis) 133
Cephalodella elongata Myers 327
Ascomorphella volvocicola (Plate) 83
* Ascomorpha ovalis (Bergendal) 274
* Collotheca libera (Zacharias) 128
Testudinella incisa emarginula (Stenroos) 359
* Postclausa hyptopus (Ehrbg) 191
* Trichocerca similis (Wierzejski) 466
Cephalodella exigua (Gosse) 462
Cephalodella apocolea Myers 357
Lepadella a. acuminata (Ehrbg) 487
* Conochilus hippocrepis (Schrank) 357
pH 3.0
4.0
5.0
6.0
7.0
8.0
9.0
10.0

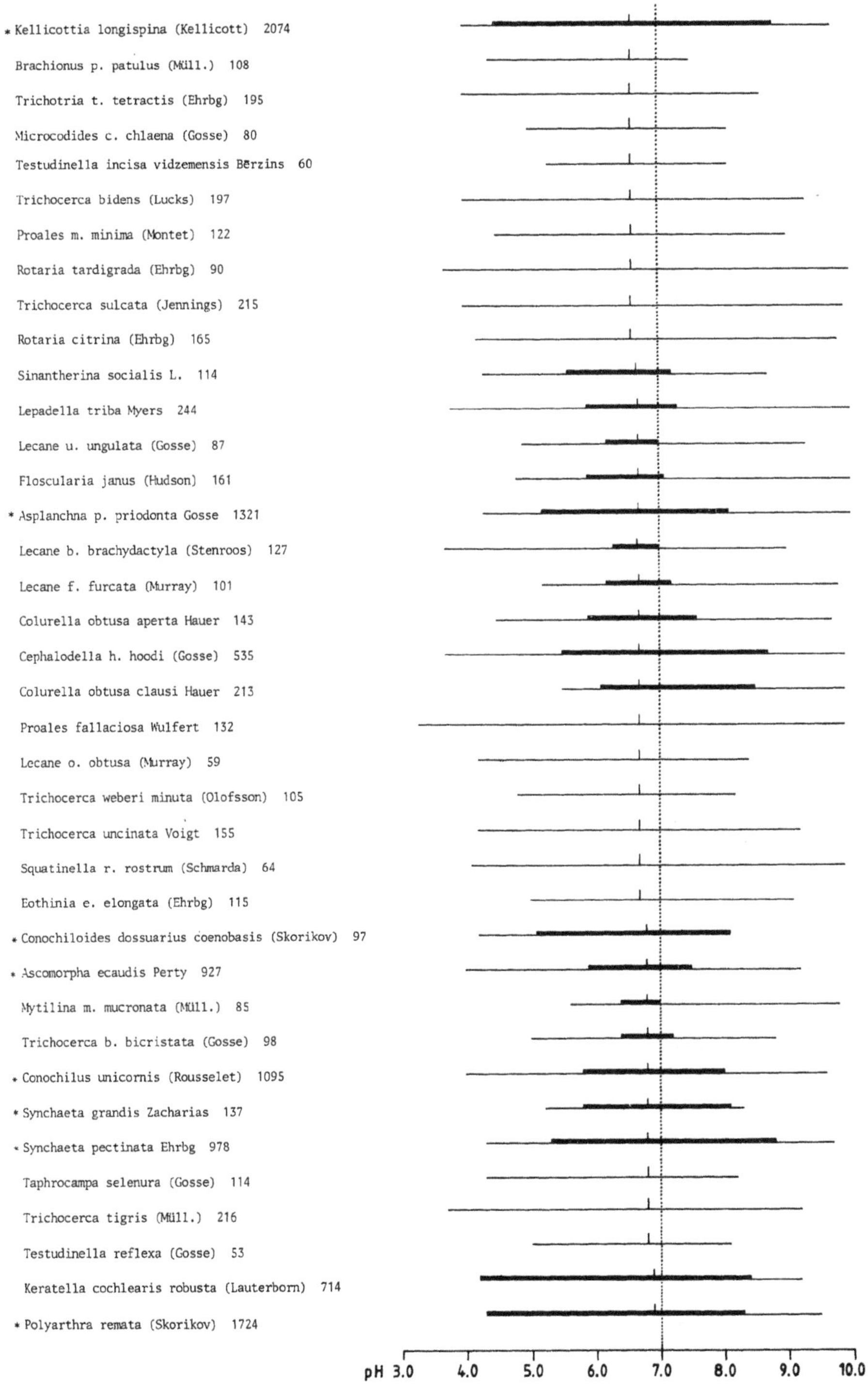
* Kellicottia longispina (Kellicott) 2074
Brachionus p. patulus (Müll.) 108
Trichotria t. tetractis (Ehrbg) 195
Microcodides c. chlaena (Gosse) 80
Testudinella incisa vidzemensis Bērzins 60
Trichocerca bidens (Lucks) 197
Proales m. minima (Montet) 122
Rotaria tardigrada (Ehrbg) 90
Trichocerca sulcata (Jennings) 215
Rotaria citrina (Ehrbg) 165
Sinantherina socialis L. 114
Lepadella triba Myers 244
Lecane u. ungulata (Gosse) 87
Floscularia janus (Hudson) 161
* Asplanchna p. priodonta Gosse 1321
Lecane b. brachydactyla (Stenroos) 127
Lecane f. furcata (Murray) 101
Colurella obtusa aperta Hauer 143
Cephalodella h. hoodi (Gosse) 535
Colurella obtusa clausi Hauer 213
Proales fallaciosa Wulfert 132
Lecane o. obtusa (Murray) 59
Trichocerca weberi minuta (Olofsson) 105
Trichocerca uncinata Voigt 155
Squatinella r. rostrum (Schmarda) 64
Eothinia e. elongata (Ehrbg) 115
* Conochiloides dossuarius coenobasis (Skorikov) 97
* Ascomorpha ecaudis Perty 927
Mytilina m. mucronata (Müll.) 85
Trichocerca b. bicristata (Gosse) 98
* Conochilus unicornis (Rousselet) 1095
* Synchaeta grandis Zacharias 137
* Synchaeta pectinata Ehrbg 978
Taphrocampa selenura (Gosse) 114
Trichocerca tigris (Müll.) 216
Testudinella reflexa (Gosse) 53
Keratella cochlearis robusta (Lauterborn) 714
* Polyarthra remata (Skorikov) 1724
pH 3.0
4.0
5.0
6.0
7.0
8.0
9.0
10.0

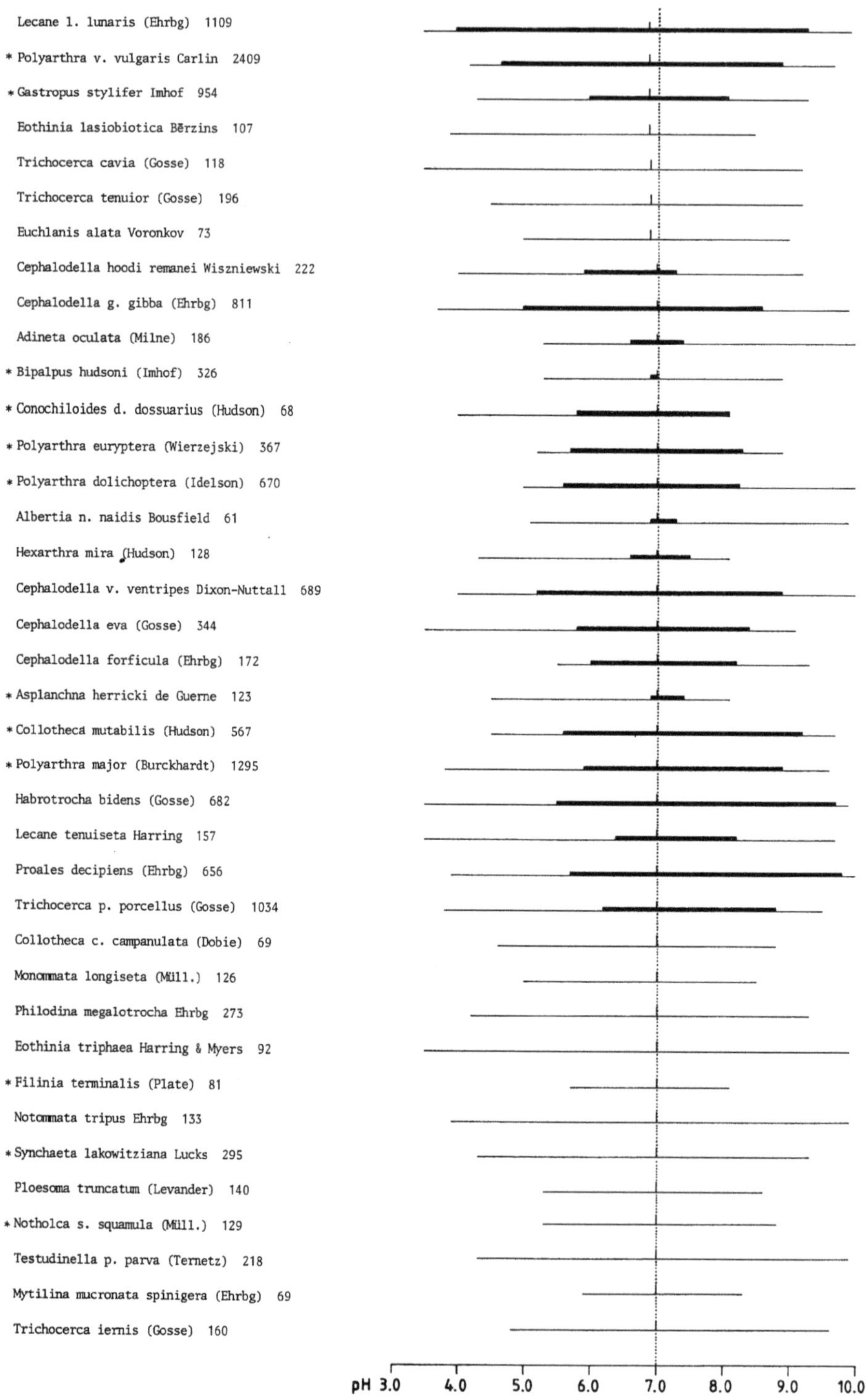
Lecane l. lunaris (Ehrbg) 1109
* Polyarthra v. vulgaris Carlin 2409
* Gastropus stylifer Imhof 954
Eothinia lasiobiotica Bērzins 107
Trichocerca cavia (Gosse) 118
Trichocerca tenuior (Gosse) 196
Euchlanis alata Voronkov 73
Cephalodella hoodi remanei Wiszniewski 222
Cephalodella g. gibba (Ehrbg) 811
Adineta oculata (Milne) 186
* Bipalpus hudsoni (Imhof) 326
* Conochiloides d. dossuarius (Hudson) 68
* Polyarthra euryptera (Wierzejski) 367
* Polyarthra dolichoptera (Idelson) 670
Albertia n. naidis Bousfield 61
Hexarthra mira (Hudson) 128
Cephalodella v. ventripes Dixon-Nuttall 689
Cephalodella eva (Gosse) 344
Cephalodella forficula (Ehrbg) 172
* Asplanchna herricki de Guerne 123
* Collotheca mutabilis (Hudson) 567
* Polyarthra major (Burckhardt) 1295
Habrotrocha bidens (Gosse) 682
Lecane tenuiseta Harring 157
Proales decipiens (Ehrbg) 656
Trichocerca p. porcellus (Gosse) 1034
Collotheca c. campanulata (Dobie) 69
Monommata longiseta (Müll.) 126
Philodina megalotrocha Ehrbg 273
Eothinia triphaea Harring & Myers 92
* Filinia terminalis (Plate) 81
Notommata tripus Ehrbg 133
* Synchaeta lakowitziana Lucks 295
Ploesoma truncatum (Levander) 140
* Notholca s. squamula (Müll.) 129
Testudinella p. parva (Ternetz) 218
Mytilina mucronata spinigera (Ehrbg) 69
Trichocerca iernis (Gosse) 160
pH 3.0
4.0
5.0
6.0
7.0
8.0
9.0
10.0

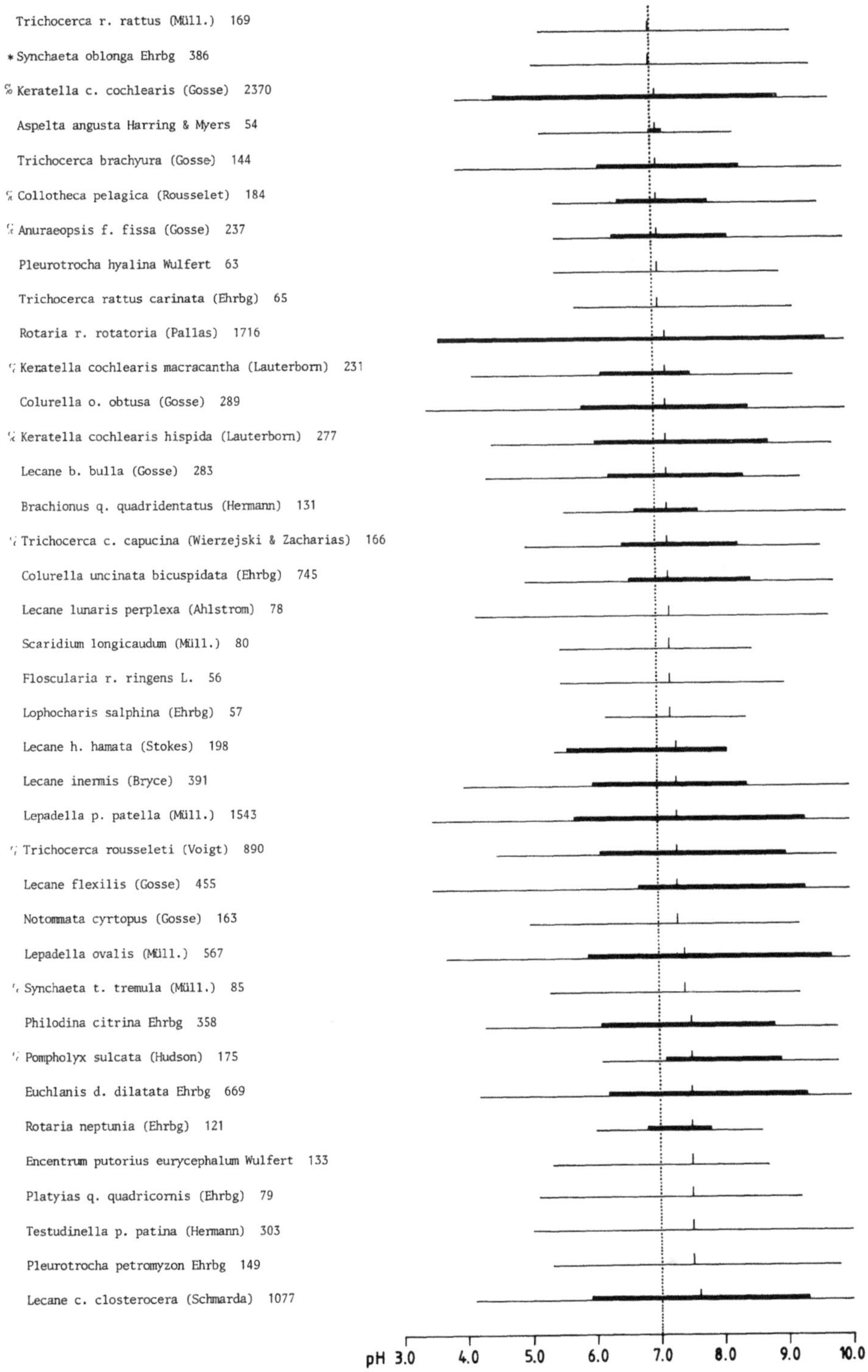
Trichocerca r. rattus (Müll.) 169
* Synchaeta oblonga Ehrbg 386
Keratella c. cochlearis (Gosse) 2370
Aspelta angusta Harring & Myers 54
Trichocerca brachyura (Gosse) 144
Collotheca pelagica (Rousselet) 184
Anuraeopsis f. fissa (Gosse) 237
Pleurotrocha hyalina Wulfert 63
Trichocerca rattus carinata (Ehrbg) 65
Rotaria r. rotatoria (Pallas) 1716
Keratella cochlearis macracantha (Lauterborn) 231
Colurella o. obtusa (Gosse) 289
Keratella cochlearis hispida (Lauterborn) 277
Lecane b. bulla (Gosse) 283
Brachionus q. quadridentatus (Hermann) 131
Trichocerca c. capucina (Wierzejski & Zacharias) 166
Colurella uncinata bicuspidata (Ehrbg) 745
Lecane lunaris perplexa (Ahlstrom) 78
Scaridium longicaudum (Müll.) 80
Floscularia r. ringens L. 56
Lophocharis salphina (Ehrbg) 57
Lecane h. hamata (Stokes) 198
Lecane inermis (Bryce) 391
Lepadella p. patella (Müll.) 1543
Trichocerca rousseleti (Voigt) 890
Lecane flexilis (Gosse) 455
Notommata cyrtopus (Gosse) 163
Lepadella ovalis (Müll.) 567
Synchaeta t. tremula (Müll.) 85
Philodina citrina Ehrbg 358
Pompholyx sulcata (Hudson) 175
Euchlanis d. dilatata Ehrbg 669
Rotaria neptunia (Ehrbg) 121
Encentrum putorius eurycephalum Wulfert 133
Platyias q. quadricornis (Ehrbg) 79
Testudinella p. patina (Hermann) 303
Pleurotrocha petromyzon Ehrbg 149
Lecane c. closterocera (Schmarda) 1077
pH 3.0
4.0
5.0
6.0
7.0
8.0
9.0
10.0

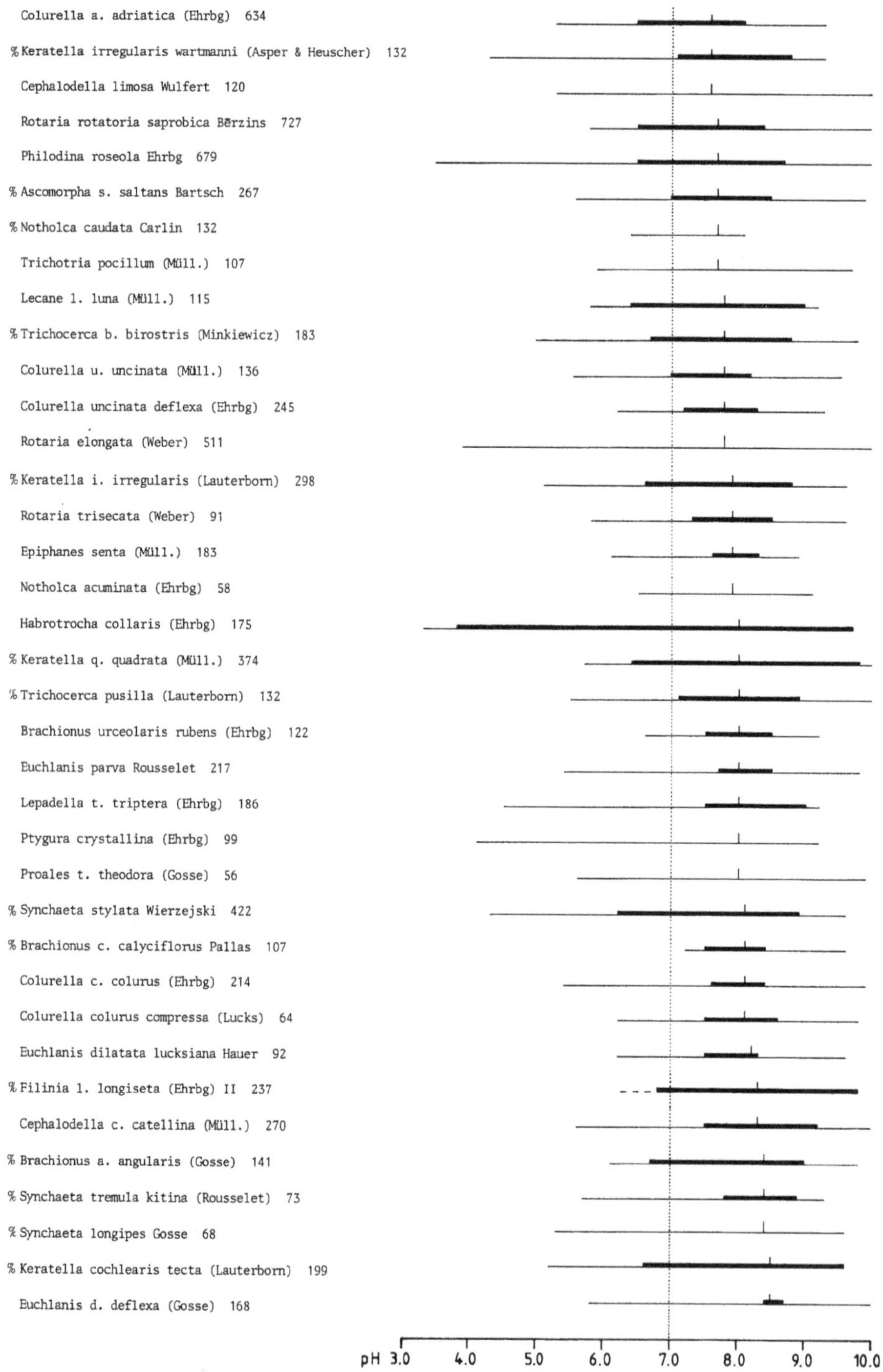

Fig. 1. Occurrence of rotifers in relation to pH in Swedish waters. The range (horizontal thin lines), high abundance (horizontal thick lines) and maximum abundance (vertical lines) of 227 rotifers are given. Species are ranked according to pH preference, beginning with those having their peak abundance at pH < 6.0 (a) through those having their peak abundance at pH ≥ 7.6 (f). Numbers refer to the number of samples in which each species was observed. Planktonic species whose peak abundance occurs at pH ≤ 7.0 (*) and pH > 7.0 (%) are indicated. *Filinia longiseta* showed a bi-modal pH tolerance and is recorded twice.

even the 'next-most alkaline' of all rotifer species included: pH ca. 8.5 at peak abundance in this study (Fig. 1f). However, the fairly wide scattering of the trophic indicators along the pH scale shows that there is no absolute link between pH and trophic status.

A comparison should be made with the recent results concerning acidified Scandinavian lakes [Hobaek & Raddum, 1980; Hörnström & Ekström, 1983; Morling, 1984]. These workers showed that the planktonic rotifers dominating these lakes were common species with their peak abundance at ≤pH 7.0. These species included: *Keratella cochlearis, Kellicottia longispina, Trichocerca capucina, Gastropus stylifer, Ascomorpha ecaudis, Asplanchna priodonta, Polyarthra vulgaris, Polyarthra remata, Conochilus unicornis, Conochilus hippocrepis* and the cold stenothermal *Keratella hiemalis.* The only 'specialist' of acid water reported is *Keratella serrulata.*

Stenothermal forms are fairly well dispersed along the pH-scale [Carlin, 1943; Pejler, 1957b; May, 1983]. Cold-water forms may achieve their peak abundance at various levels (e.g., *Keratella hiemalis*, pH 5.8; *Collotheca lie-petterseni*, 6.3; *Polyarthra dolichoptera, Filinia terminalis, Synchaeta lakowitziana, Synchaeta oblonga*, and *Notholca squamula*, 7.0; *Notholca caudata*, 7.7). The pattern is similar for warm-stenothermal forms (e.g., *Collotheca libera*, 6.5; *Polyarthra remata*, 6.9; *Polyarthra euryptera*, 7.0; *Pompholyx sulcata*, 7.5; *Trichocerca birostris*, 7.8; *Trichocerca pusilla*, 8.0; *Synchaeta stylata*, 8.1; *Keratella cochlearis tecta*, 8.5). Though peak abundance of warm-water forms tends to occur at higher pH values than cold-stenothermal forms, there is no good evidence suggesting a positive correlation between pH of peak abundance and temperature.

Conclusions

As a group, rotifers exhibit a very wide range of pH tolerance. All 227 species in this study were found in waters with pH values spanning at least 2.0 units, and several were found in waters which differed by as much as 5.0 units. A few species even exhibited high abundance over an extremely broad pH range, e.g., *Rotatoria rotatoria* and *Habrotrocha collaris* (Fig. 1e, f). Pejler [1957a, 1977] has suggested that bdelloids have the potential for such extensive ranges because their mode of reproduction (obligatory parthenogenesis) favors development of strains uniquely adapted to a particular environment. Ruttner-Kolisko [1983] suggested that a similar situation may exist for many monogonont rotifers. Clearly further research is needed here.

At present we have not determined any correlation between peak rotifer abundance, pH, and other chemical factors. We intend to return to this problem in future papers.

Acknowledgements

Data analysis was supported by the Environmental Protection Board of Sweden. Thanks are due to Lars Norling, Leif Robertsson and Mike Coveney for help in analysing these data. We are grateful to Ellinor Barringe and Janis Prieditis for their help with a variety of routine work.

References

Bateman, L. & C. Davis, 1980. The Rotifera of hummock-hollow formations in a poor (mesotrophic) fen in Newfoundland. Int. Revue ges. Hydrobiol. 65: 127–153.

Bērziņš, B., 1967. Rotatoria. In J. Illies (ed.), Limnofauna Europaea, Fischer, Stuttgart, pp. 35–38.

Bērziņš, B., 1978. Ekologisk valens. Koder. Zookomponenter, substrat och lokaler. Internal report, Limnol. inst., Lund. ISSN 0348-0798. 86 pp. (Swedish only).

Bērziņš, B., 1982. Die Unterarten von Dissotrocha aculeata (Macrostyla). Limnol. inst., Lund. 10 pp.

Carlin, B., 1943. Die Planktonrotatorien des Motalastrom. Zur Taxonomie und Ökologie der Planktonrotatorien. Medd. Lunds univ. Limnol. inst., 5. 256 pp.

Donner, J., 1965. Ordnung Bdelloidea (Rotatoria, Rädertiere). Akademie-Verlag, Berlin, 297 pp.

Edmondson, W. T., 1944. Ecological studies of the sessile Rotatoria. Part I. Factors affecting distribution. Ecol. Monogr. 14: 31–66.

Francez, A. J., 1981. Rotifères de quelques tourbières d'Auvergne. Ann. stat. biol. Besse-en-Chandesse 15: 276–287.

Hobaek, A. & G. Raddum, 1980. Zooplankton communities in acidified lakes in South Norway. SNSF-prosj., Oslo – As. IR 75/80, 132 pp.

Hörnström, E. & C. Ekström, 1983. pH-, närings- och aluminiumeffekter pa plankton i västkustsjöar. SNV PM 1704, Solna (Sweden). 124 pp. (Swedish, with English summary).

Karabin, A., 1983. Ecological characteristics of lakes in north-eastern Poland versus their trophic gradient. VII. Variations in the quantitative and qualitative structure of the pelagic zooplankton (Rotatoria and Crustacea) in 42 lakes. Ekol. pol. 31: 383–409.

Koste, W., 1978. Rotatoria. Die Rädertiere Mitteleuropas. Ein Bestimmungswerk begr. von Max Voigt. Überordnung Monogononta. Vol. 1–2, 673 pp. + 234 pl.

Lindström, K., 1983. Changes in growth and size of Keratella cochlearis (Gosse) in relation to some environmental factors in cultures. Hydrobiologia 104: 325–328.

May, L., 1983. Rotifer occurrence in relation to water temperature in Loch Leven, Scotland. Hydrobiologia 104: 311–315.

Morling, G., 1984. Acidification and plankton in west-Swedish lakes 1966–1983. Acta univ. upsal. 727, 22 pp.

Pejler, B., 1957a. On variation and evolution in planktonic Rotatoria. Zool. Bidr. Upps. 32: 1–66.

Pejler, B., 1957b. Taxonomical and ecological studies on planktonic Rotatoria from northern Swedish Lapland. K. svenska Vetensk. Akad. Handl., ser. 4, bd. 6, no. 5, 68 pp.

Pejler, B., 1962. On the variation of the rotifer Keratella cochlearis (Gosse). Zool. Bidr. Upps. 35: 1–17.

Pejler, B., 1965. Regional-ecological studies of Swedish fresh-water zooplankton. Zool. Bidr. Upps. 36: 407–515.

Pejler, B., 1977. General problems on rotifer taxonomy and global distribution. Arch. Hydrobiol. Beih. 8: 212–220.

Pejler, B., 1981. On the use of zooplankters as environmental indicators. In M. Sudzuki (ed.), Some approaches to saprobiological problems, Sanseido, Tokyo, pp. 9–12.

Pejler, B., 1983. Zooplanktic indicators of trophy and their food. Hydrobiologia 101: 111–114.

Ruttner-Kolisko, A., 1972. Rotatoria. In Die Binnengewasser, 26. Das Zooplankton der Binnengewässer, I: 99–234.

Ruttner-Kolisko, A., 1983. The significance of mating processes for the genetics and for the formation of resting eggs in monogonont rotifers. Hydrobiologia 104: 181–190.

Skadowsky, S. N., 1923. Hydrophysiologische und hydrobiologische Beobachtungen über die Bedeutung der Reaktion des Mediums für die Süsswasserorganismen. Verh. int. Ver. Limnol. 1: 341–358.

Wulfert, K., 1951. Das Naturschutzgebiet auf dem Glatzer Schneeberg. Die Rädertiere des Naturschutzgebietes. Arch. Hydrobiol. 44: 441–471.

Hydrobiologia 147: 117–127 (1987)
May, L., Wallace, R. & Herzig, A. (eds), Rotifer Symposium IV

Ecology of bdelloids: how to be successful

Claudia N. Ricci
Dipartimento di Biologia, Sez. Ecologia, Via Celoria 26, 20133 Milano, Italy

Keywords: bdelloids, rotifers, biogeography, ecology, parthenogenesis, anhydrobiosis

Abstract

Bdelloids inhabit many different environments. The entire taxon is an order belonging to the phylum Rotifera. In spite of its wide distribution, it has a very uniform morphology, suggesting that natural selection has had almost no effect on its morphological characteristics.

This paper reviews the geographical and ecological distributions of bdelloids and their ability to tolerate different conditions is discussed. Two characteristics account for the wide distribution of bdelloids, parthenogenetic reproduction and their ability to withstand unfavorable conditions through anhydrobiosis. The former is an apomictic thelytoky which may ensure genetic homogeneity within clones. However, evidence which suggests variability among and within parthenogenetic clones is discussed. Some recent experimental evidence indicates that anhydrobiosis does not affect the life history of the surviving individuals. The effects of anhydrobiosis on bdelloid populations are considered.

Introduction

Bdelloid rotifers are ubiquitous animals, capable of dwelling in lakes, streams and in the water film covering terrestrial mosses and soil. Two biological phenomena distinguish them from other classes of Rotifera: 1) obligatory parthenogenetic reproduction and 2) ability to withstand harsh periods through anhydrobiosis. Both characteristics make them pioneering animals, able to colonize new environments. According to Maynard-Smith [1978], obligatory unisexuality represents an evolutionary dead end, and those groups in which it is found are incapable of creating new phylogenetic groups. Nevertheless, the parthenogenetic class Bdelloidea contains four orders and 18 genera. Most bdelloids have a very similar basic morphology. Thus it appears that natural selection influences their morphology, maintaining little difference in gross morphological characteristics. Therefore one could suggest either that (1) sibling species can occur more commonly than it is generally thought or (2) the entire class has a very plastic 'general purpose' genotype, in accord with the recent hypothesis of Lynch [1984].

This paper considers the geographical and ecological distribution of bdelloids, in an attempt to assess the adaptability of the class. The major factors responsible for their ability to survive in very extreme and unstable environments are discussed.

Geographical distribution

Studies on the geographical distribution of rotifers have dealt chiefly with the Monogononta, e.g. Green, 1972; Pejler, 1977; De Ridder, 1981; Dumont, 1983. These authors have indicated that, despite the assumption of generalized cosmopolitism, distributional barriers for rotifers do exist and many species show a very distinct geographic distribution [Dumont, 1983]. Several genera are confined to distinct latitudinal zones, and many species are considered to be endemic to certain regions.

A total of 363 species of Bdelloidea are known throughout the world. Of these, as many as 146 (40%) have been found only in Europe [Berzins, 1978]. An additional 42% have been reported from Europe and one or more other continents, and are assumed, here, to be cosmopolitan. The remaining 18% have been recorded from other continents. A few of these occurred on two widely separated continents, such as Africa and New Zealand, while the others seem to be truly endemic (Table 1).

Information on bdelloids from many localities is very fragmentary. For example, bdelloids from North America have rarely been studied apart from the work of Jennings [1901], Burger [1948] and, more recently, Berzins [1984] and Chengalath [1984]. Clearly the lack of data does not necessarily mean the lack of bdelloids in any particular area, but simply reflects the distribution of bdelloid students. To what extent are bdelloids really endemic? I will deal in some detail with a few regions, as examples of the contradictory patterns of bdelloid distribution.

Antarctic

The Antarctic is a very harsh environment for freshwater animals, since water and soil surfaces are covered with ice for long periods. As expected, this habitat has a low species diversity. Anhydrobiosis or resistant stages, coupled with parthenogenesis and certain physiological properties are characteristics necessary for the success of bdelloids in this region [Everitt, 1981]. In spite of these considerations, however, of 19 species of bdelloids reported from the Antarctic [José de Paggi & Koste, 1984], only four (21%) are endemic (*Habrotrocha angularis, Philodina gregaria, P. alata, P. antarctica*). The others are cosmopolitan species, occurring also in north and south America, Europe, Africa and New Zealand.

Table 1. List of the genera of Rotifera Bdelloidea.

Genus	Number of endemic species				
	Total number of species	Europe	Outside Europe	Cosmopolitan	Comments
Habrotrocha	103	45	14	44	9 endemic to Africa
Otostephanos	10	5	1	4	1 endemic to Africa
Scepanotrocha	10	3	1	6	1 common to N.Z., S. Am.
Anomopus	2	1	1		
Ceratotrocha	4	3		1	
Didymodactilus	1			1	
Dissotrocha	6	4	1	1	1 endemic to Africa
Embata	5	2		3	
Macrotrachela	78	25	22	31	9 endemic to Africa
Mniobia	48	25	8	15	7 endemic to Austr.
Pleuretra	10	3	1	6	1 endemic to Africa
Philodina	43	15	13	15	7 endemic to Africa
Rotaria	24	10	1	13	1 endemic to N.Z., Austr.
Adineta	13	3	1	9	1 endemic to N.Z.
Bradyscela	2	1		1	
Abrochta	1	1			
Henoceros	2		1	1	1 endemic to Sumatra
Philodinavus	1			1	
Total	363	146	65	152	

Africa

Studies of African bdelloids began in the first decades of the 20th century, concentrating mostly on tropical and southern Africa [Murray, 1908, 1911a, b; Milne, 1916a, b]. In addition, a few data from northern regions are given by Murray [1911b]. Later, a list of rotifers from the Belgian Congo was compiled by Gillard [1957], without adding significantly to the species list. A total of 104 bdelloid species are now known from Africa; of which 24 are endemic (23%). However, many of these have not been found since their first description, probably due to the lack of recent studies. One such species, *Macrotrachela plicatula* Murray, was found in Transvaal and described by Murray [1911b]. It was not collected again until Ricci [1980] found specimens in a moss sample collected in Kampala, Uganda. This species is probably endemic to Africa, although its distribution ranges some thousand kilometers. No entomologist would consider this endemism!

Australia and New Zealand

146 species of bdelloid rotifers have been recorded in Australia and New Zealand [Haigh, 1963, 1965, 1966, 1968, 1970; Shiel & Koste, 1979]. Twenty-two of these species (15%) are endemic.

The rates of endemism reported here for the bdelloids are similar to those values reported for monogonont rotifers by Dumont [1983]. However, the scattered reports of bdelloid fauna make it difficult to attribute any genus to one particular region. More information is needed before any attempt can be made to trace a pattern of biogeography on the basis of phylogenesis. Nevertheless, the conclusions reached by Dumont [1983] for monogononts also fit bdelloids.

Species and varieties

A bdelloid species is recognized on the basis of morphology alone, since these animals reproduce exclusively by parthenogenesis and the standard biological meaning of 'species' cannot be applied. Burger's [1948] useful remarks on the identification of bdelloids can be summarized as follows:

1) The rules of classification are often arbitrary, depending on the importance an author ascribes to different characters. In this regard, Pejler [1977] adds that 'the systematics has to be based on the subjective decision of experienced specialists.'

2) Morphological variations between species, or subspecies, could be temporary variations. The size and shape of animals, or parts of them, is generally deemed important information, but various factors such as starvation and age can modify the shape and the size of animals. Moreover 'the relative size of certain organs vary during growth.' [Burger 1948].

3) The mastax is a useful diagnostic character for monogononts, and has a definitive size from birth in all rotifers. The ramate type of mastax is common to most bdelloids, and the number of teeth can be a variable character, 'most variable in the species with most teeth' [Burger, 1948].

According to the third edition of the International Code of Zoological Nomenclature [1985], any form or variety must be considered as a 'subspecific rank', or, if established before 1961, as an 'infrasubspecific rank'. When using the subspecific rank, all bdelloid students, aware of the many morphological variations that most bdelloid species exhibit, split the species into infrasubspecific groups, making use of forms, varieties or, more properly, subspecies. However, on some occasions, the splitting goes so far as to use not only trinomina but also tetranomina: e.g. *Habrotrocha pusilla textrix* f. *longilabris* Donner, 1950 versus *H. pusilla textrix* f. *brevilabris* Donner, 1950.

According to Pejler [1977], it is advisable to restrict the use of subspecies to groups of animals inhabiting 'a geographical area separate from the one occupied by related subspecies.' For this purpose, I will consider the distribution of *Macrotrachela quadricornifera* Milne and its six subspecies (*ligulata* Berzins, *loricata* Donner, *rigida* Milne, *quadricorniferoides* Bryce, de Koning, *scutellata* Schulte, *vanoyei* Schepens) (Fig. 1). None of these is limited to any particular region or habitat, except *M. quadricornifera vanoyei* which has been reported only

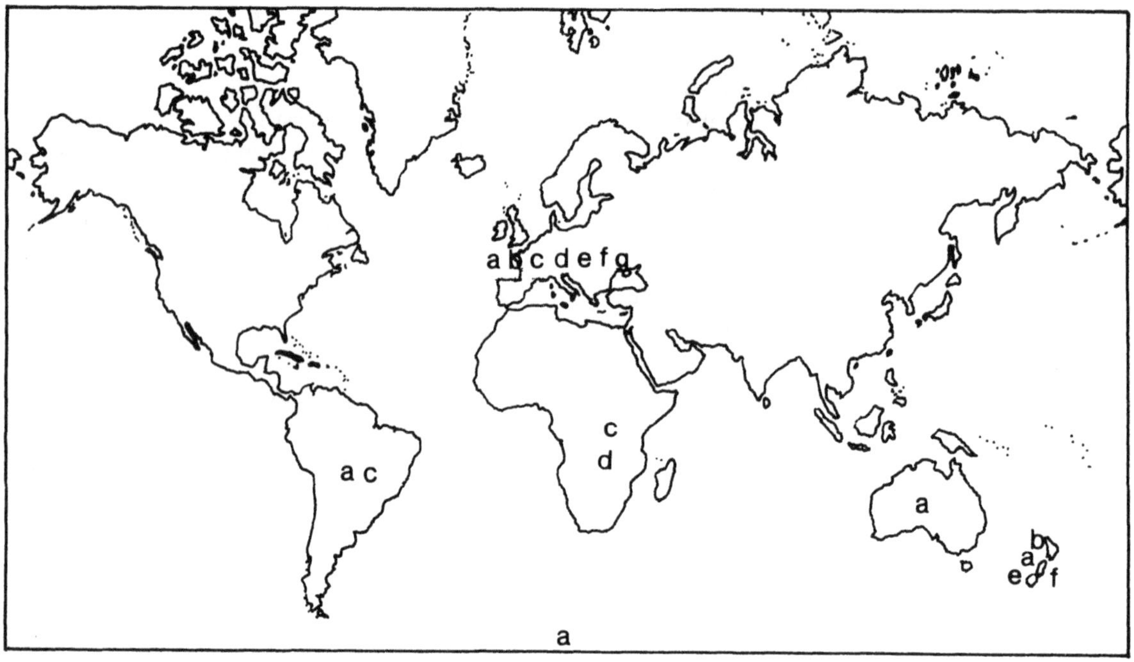

Fig. 1. Distribution of *Macrotrachela quadricornifera* and of its six subspecies: a, *M. quadricornifera;* b, *M. q. ligulata;* c, *M. q. loricata;* d, *M. q. rigida;* e, *M. q. quadricorniferoides;* f, *M. q. scutellata;* g, *M. q. vanoyei.*

from Belgium. On the basis of its large variability, Donner [1965] considers *M. quadricornifera* to be a 'complex species' and Koste [1970] to be a 'superspecies'. Moreover, Koste [1970] points out quite large variability among both populations collected from different habitats (geographical or ecological races) and, more significantly, individuals from the same habitat.

A study was carried out with five populations of *M. quadricornifera* to distinguish them on the basis of parameters other than morphological ones, which except for size, are very similar. Rotifers were collected from four different environments (spring water, wet soil, cold stream and dry mosses) in the geographic area of Lombardia, Northern Italy. Evidence for differences among these populations was obtained from electrophoretic patterns of three enzymes (PGI, EST and MDH) [Bisol & Ricci, in preparation] and from their life histories. At present, only data on three populations are available [Ricci & Bombonati, 1985]. The life spans and fertilities of the considered populations differed and temperature influenced the growth parameters (r and R_0 in particular) of each population in a different way (Fig. 2). Mass culture experiments indicated their food requirements were different:

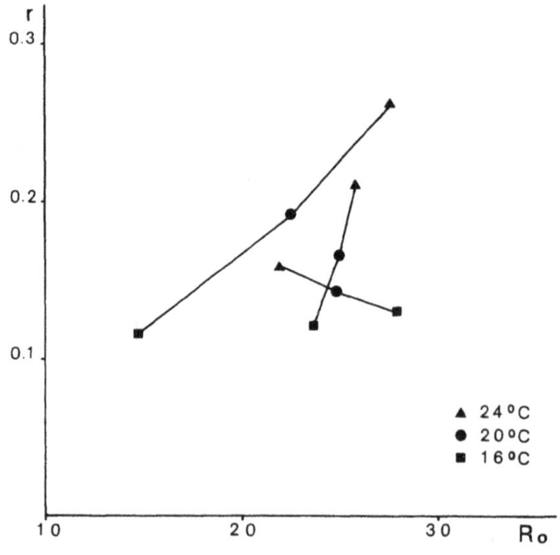

Fig. 2. Effects of temperature on intrinsic rate of natural increase (r) and on reproductive rate (R_0) in three populations of *M. quadricornifera.*

some populations grew better when fed algae, others when fed yeasts.

These preliminary results suggest that the different populations of *M. quadricornifera* are ecospecies or, perhaps, sibling species, rather than morphospecies. Clearly, there is no simple solution to the problems of bdelloid taxonomy. So, I shall continue using the term 'species', keeping in mind how limited its meaning is for this group of animals.

Ecological distribution

It is generally considered that the major factor affecting the distribution of bdelloid rotifers is the availability of water [e.g. Dobers, 1915; Bartos, 1951; Donner, 1965].

Bdelloids can be divided into two ecological groups, hydrophile and xerophile species. Dobers [1915] and Bartos [1951] suggested that only the latter group is capable of anhydrobiosis and can dwell successfully in unstable environments. Donner [1956, 1975] points out that 95% of rotifers inhabiting terrestrial mosses, soil and lichens are bdelloids. Most species of soil and moss rotifers belong to the *Habrotrocha* genus, while the majority of the animals (30%) belong to the *Macrotrachela* genus.

Many authors have attempted to relate types of bdelloid species to types of moss. Dobers [1915] states that there are such relationships, but Bartos [1951], Francez [1980] and Burger [1948] recognize only humidity as the regulating factor. Burger [1948] in particular states that bdelloid occurrence in moss is regulated by the following three conditions.

1) As in island biogeography the age of the moss at the site affects the length of time colonization has been possible and therefore the number of bdelloid species present.

2) Presence of water and its osmotic potential affects bdelloid activity.

3) The condition of food supply is also important. Although Burger [1948] states that food selection is based on size, food quality also seems to be important to moss dwelling and aquatic bdelloids [Ricci, 1984a].

According to Donner [1953, 1956] moss dwelling bdelloids have certain morphological peculiarities. They tend to have narrower wheel-organs, a more rigid cuticle covered with cuticular spines or knobs, and smaller toes. The latter feature, in particular, varies within a species according to its habitat, and is considered to be the result of adaptation to a terrestrial environment. Such morphological variability in bdelloids adds to the problems of the systematic identification.

The ratio of bdelloids to total rotifers (Monogononta + Bdelloidea) depends on water availability. In moss and soil 95% of the rotifers are bdelloids while around the shores of freshwater environments, 20–30% of rotifers are bdelloids [Donner, 1975]. The dominant species are *Mniobia* in dry mosses, *Scepanotrocha* and *Ceratotrocha* in humus, *Philodina nemoralis* and *Habrotrocha constricta* in the 'Aridale' (=arid zone) 'Geniste' (=spray zone) 'Gespülse' (=rinse zone) and *P. flaviceps* at the banks of water courses (Fig. 3).

Bdelloid communities are generally stable, throughout all seasons [Madalinski, 1961]. Koste [1976] recorded only slight differences between winter and summer fauna. In contrast, Zullini & Ricci [1980] observed seasonal changes in the abundance of *P. flaviceps*, which was most abundant in early spring.

Similar soils are inhabited by similar bdelloid communities. Donner [1975] suggested that lignin plays an important role in determining the distribution of bdelloids. In particular, he reported an uncommon richness of bdelloid population in a stream in which sawdust is abundant [Donner, 1970, 1972].

Most authors assume that bdelloids evolved from a wholly aquatic existence. Perhaps they abandoned this habitat as mosses colonized the terrestrial environment, and adapted to survival in dry periods. Some bdelloids have remained aquatic, but most drift with heavy rains from one habitat to another. Although some species remain in wholly aquatic environments, there is no evidence that any bdelloid species is completely incapable of cryptobiosis, and therefore bdelloids can generally be assumed to be able to withstand dry or frozen periods through cryptobiosis. Moreover, in freshwater

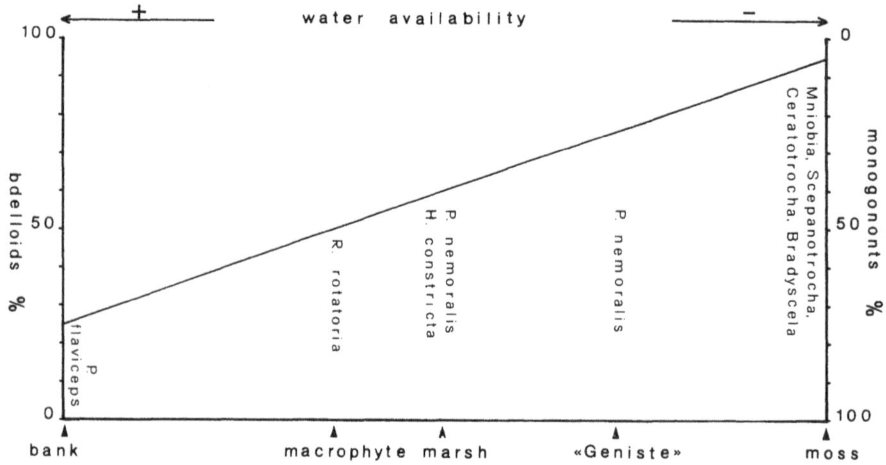

Fig. 3. Percentage of Bdelloidea (of the total Rotifera) in environments with different amounts of water: the species and genera reported are the dominant ones [after Donner, 1975].

they seem to be less successful: they are generally not as well adapted as monogononts, their life spans are longer and fertility rates lower [Ricci, 1983].

A few species (e.g. *Rotaria macrura* and *Philodina megalotrocha*) seem to be pelagic, on the basis of scattered reports [Zacharias cit. Dobers, 1915]. On the whole, aquatic bdelloids seem to tolerate a wide range of ecological conditions. Acid waters, with pH values from 4.2 to 5.5 shelter only one species, *Dissotrocha macrostyla*, [Wulfert, 1960], but in thermal water (up to 45 °C), as many as 7 species were reported [Pax & Wulfert, 1941]. Of these, *Adineta glauca* occurred only when the temperature rose above 40 °C. It may be considered to be a real thermobiont. Other very common and widely distributed species, including *H. constricta, P. flaviceps, P. acuticornis, P. roseola,* and *Rotaria rotatoria*, are considered to be eurytherms. The depth of the water seems to be inversely correlated with bdelloid distribution. Klimowicz [1972] observes that the number of species and population densities decreased with increasing depth.

Donner [1964] states that water current is an important factor for bdelloid communities, as revealed by a comparison of the rotifer community in a slow branch of the river Danube and in rapid streams in the High Tatra mountains. Bdelloids generally inhabit tufts of mosses, where they are protected from drifting with the current [Madalinski, 1961; Koste, 1976].

Indicator species

Slàdecek [1983] studied the use of rotifers, including bdelloids, as indicators of water quality. Nine of the 13 species of *Macrotrachela* listed by the author, were classified as indicators of oligosaprobity

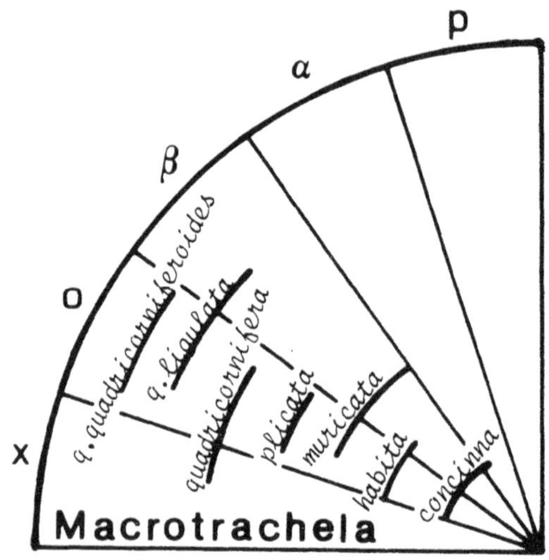

Fig. 4. Arrangement of 7 species of *Macrotrachela* within limnosaprobity: x, xenosaprobity; o, oligosaprobity; b, betamesosaprobity; a, alfamesosaprobity; p, polisaprobity [after Sladecek 1983].

and scored equal to the maximum indicative weight (=5) (Fig. 4); *M. quadricornifera*, a very common species occurring in springs, streams, ponds etc., belongs to this list. However, as we have already seen, different populations of this 'morphospecies' can differ in their tolerance to temperature, their food requirement, and their enzymatic pattern. If we assume that the different populations have adapted to different environmental conditions, we must also assume that the populations have different needs, depending on the conditions they have had to face, invalidating their use as indicators of water quality. However, some very good studies based on experimental evidence have shown that bdelloids do indicate water quality quite well [Donner, 1978; Koste, 1976].

Parthenogenesis

Two features that are important for the success of bdelloids in unstable environments are parthenogenesis and anhydrobiosis. The group as a whole has forfeited chromosome homology [Hsu 1956a, b; Plasota & Plasota 1980], although homologous pairs have been recognised in at least one species [Robotti 1987]. In any case, it is generally recognised that ordinary meiosis does not take place. This kind of reproduction is termed apomictic thelytoky. Since gene recombination is not possible, all members of a population should be identical to their ancestors, except for any mutational events which have occurred.

Lynch [1984] states that parthenogenesis is favored in extreme environments because it is coupled to highly generalized genotypes that allow habitat requirements to be less restrictive. Many studies, mostly based on electrophoretic results, deal with variability among apomictic thelytokous clones and support the hypothesis that these populations maintain substantial amounts of genetic variability, e.g. insects [Atchley 1977; Suomalainen & Saura 1973; Saura *et al.* 1976], crustacea [Hebert & Crease 1983] and mollusca [Stoddard 1983]. Many authors [e.g. Cuellar 1977; Saura *et al.* 1976; King 1972] have advanced the hypothesis that different clones exploit the same habitat in a season succession. According to this hypothesis the best adapted clone increases when conditions are favourable for its genotype. Few data are available on intraclonal variability of apomictic populations. There is some evidence that intraclonal variability does exist [Birky & Gilbert 1971; Beranek & Berry 1974; Badino & Robotti 1975; Ricci 1984b] but its significance and origins are not understood.

What genetic mechanisms are responsible for variability in apomictics? Lynch [1984] suggests that 'in addition to mutation, a variety of other variance-generating mechanisms are available to parthenogens. These include mitotic recombination, intra-cistronic recombination, mitotic non-disjunction, endomeiosis and endomitosis, transposition of regulatory genes, heterocromatization and polyploidization.' Perhaps to this list could be added mitochrondrial DNA and cytoplasmic heredity; the latter being well known in rotifers [Lansing 1954].

The capacity for generating variability in parthenogens is much lower than in sexually reproducing individuals [Suomalainen *et al.* 1976; Lynch 1985]. However, parthenogenetic rather than bisexual reproduction can be more successful under some circumstances:

a) When a genotype assures advantages in a given environment, it could be useful to maintain its integrity: a parthenogenetic clone is a complete linkage group and selection can operate on the entire group of properties of its genotype [Lynch 1984].

b) Any population started by one or few founder organisms would face a significant drop in levels of genetic variation because of small population size and strong inbreeding, and few populations will survive the 'genetic revolution' [Mayr 1954]. Thelytokous individuals would escape any genetic revolution during colonization.

c) Parthenogens can persist and reproduce at low densities since they do not require mating encounters [Gerritsen 1980].

Bdelloids often dwell in fluctuating and unpredictable environments, where populations have to face low densities conditions or new colonization is required.

Anhydrobiosis

Cryptobiosis, or more properly anhydrobiosis, was first observed in bdelloids by Leeuwenhoek in 1702. Jacobs [1909] studied the morphology, physiological resistance and survival of dried bdelloids and stressed the importance of slow drying for greatest survivorship. Dobers [1915] and Hickernell [1917] found an increased fecundity in rotifer populations after drying and suggested that cryptobiosis had a positive influence on bdelloid populations. Jacobs [1909], however, did not agree with this finding.

The fine structure modifications of dried rotifers were described by Dickson & Mercer [1967] and Ricci & Melone [1984]: tissues and cells became condensed, maintaining their identity, and no cyst was secreted by the drying animal, at least in *Macrotrachela quadricornifera*. The increased resistance of anhydrobiotic stages to extreme temperatures was considered extensively by Gilbert [1974], but since his review of dormancy in rotifers no information has been added.

Loss of water by the anhydrobiont is high, as demonstrated by Jacobs [1909] and Hickernell [1917], but is not complete, since part of the chemically bound water is retained. It is still not known how membranes of dried rotifers are stabilised in the absence of water. Recently, high concentrations of trehalose, a non-reducing disaccharide of glucose, have been found in many organisms which dry out (e.g. spores, yeasts, brine shrimps and nematodes). It is thought that trehalose may act as an important factor in stabilizing dry membranes [Crowe *et al.* 1984], and studies of trehalose content in bdelloids are therefore advisable.

There is no information in the literature concerning the effect of anhydrobiosis on bdelloid populations. Degeneration of populations not submitted to anhydrobiosis has been reported by Dobers [1915] and Hickernell [1917], but this is not supported by experiments. When the effect of anhydrobiosis on a population of *Philodina roseola* was evaluated, it appeared that individuals from a culture derived from animals surviving cryptobiosis had longer life spans and higher reproductive rates than closely related populations kept under conditions of continuous reproduction (Fig. 5). It would

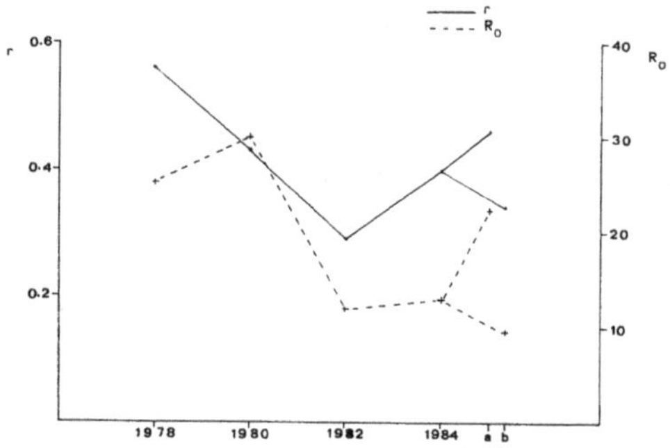

Fig. 5. Variations of r and R_0 for a clonal population of *P. roseola* in different years. a: population after anhydrobiosis. b: population under continuous hydration.

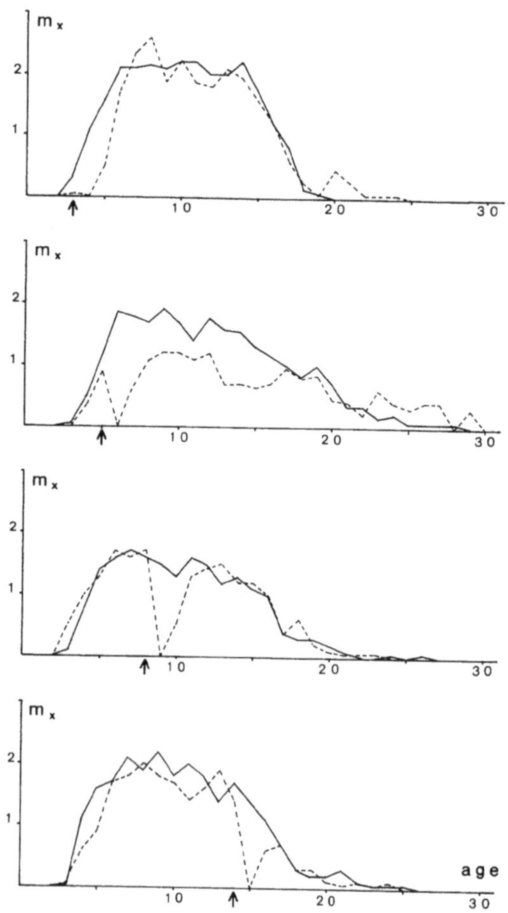

Fig. 6. Effects of anhydrobiosis on age-specific fecundity rates (m_x) at four ages (arrows). Continuous line: cohort under continuous culture conditions. Dashed line: anhydrobiont cohort.

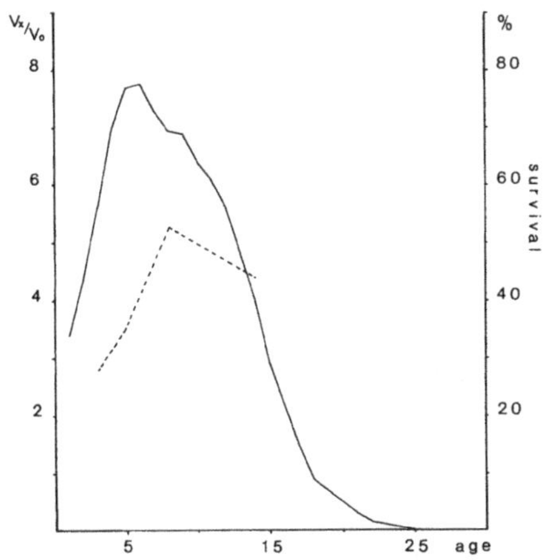

Fig. 7. Percentage of survival after cryptobiosis at different ages versus the specific reproductive value (Fisher value). Continuous line: age-specific reproductive value. Dashed line: survival percentages.

be interesting to see if this effect can be repeated in a variety of species indicating that life table parameters may be improved by anhydrobiosis. However, how anhydrobiosis acts on the animals is still unknown.

To evaluate the effect of anhydrobiosis on the life cycle of *Macrotrachela quadricornifera* the following experiment was carried out [Ricci, unpublished]. Anhydrobiosis was induced in cohorts of individuals at four different ages. Age-specific survivorship (l_x) and fecundity (m_x) were calculated for each cohort. The results indicated that anhydrobiosis influenced neither life span nor reproductive activity, but represented a lag period in the lives of the animals (Fig. 6). However, survival after anhydrobiosis was very low, ranging from 20% to 50% depending on the age of the animals. It seems, therefore, that anhydrobiosis has a very strong selective effect on some characteristics that we have failed to detect, and maximizes the survival of animals who are at the age of their maximum reproductive value (Fisher value) (Fig. 7). Mortalities were highest in young animals.

Anhydrobiosis may be supposed to assure the survival and reproduction of the animals best fitted to the environment, while the least well fitted individuals are eliminated.

Conclusion

The analysis of geographical and ecological distributions of bdelloids provides evidence for their ability to adapt and survive under a variety of environmental conditions. Some species can be recognised as endemic to certain regions, a few species appear to be limited to specific habitats, but generally most bdelloids are able to inhabit both terrestrial and aquatic environments. However, bdelloids can be used as ecological indicators of water quality.

The success and wide distribution of bdelloids is based on two features, parthenogenetic reproduction and anhydrobiosis. Parthenogenesis, which is the most profitable reproductive strategy in unpredictable environments, appears to be able to generate genetic variability and so provides the conditions on which environmental selection (e.g. anhydrobiosis) can act. If this is true, bdelloid populations dwelling in unstable habitats have undergone continual selection, and different clones with slightly different genotypes should be expected.

Acknowledgements

I wish to thank Luisella Vaghi and Lucia Manzini for their help in the experimental work. Financial support came from a CNR grant, CT 8401201.

References

Atchley, W. R., 1977. Evolutionary consequences of parthenogenesis: Evidence from the Warramaba virgo complex. Proc. natn. Acad. Sci. U.S.A. 74: 1130–1134.

Badino, G. & C. Robotti, 1975. Selection in parthenogenetic lines of Asplanchna sieboldi (Rotatoria). Experientia 31: 298–299.

Bartoš, E., 1951. The Czechoslovak Rotatoria of the Order Bdelloidea. Věstník Čs. Zool. spol. 15: 241–500.

Beranek, A. P. & R. J. Berry, 1974. Inherited changes in enzyme patterns within parthogenetic clones of Aphis fabae. J. Entomol. (Sez. A) 48: 141–147.

Berzins, B., 1978. Rotatoria. In: Illies, J. (ed.). Limnofauna Europaea. 2nd ed. Fisher, Stuttgart: 54–91.

Berzins, B., 1984. Some rotifers from North America: a reminiscence. Limnol. Institutionen Lunds Universitet. 15 pp.

Birky, C. W. & J. J. Gilbert, 1971. Parthenogenesis in rotifers: the control of sexual and asexual reproduction. Am. Zool. 11: 245–266.

Burger, A., 1948. Studies on the moss dwelling bdelloids (Rotifera) of Eastern Massachusetts. Trans. am. microsc. Soc. 67: 111–142.

Chengalath, R., 1984. Bibliographia Invertebratorum Aquaticorum Canadensium. Vol. 3. Rotifera. National Museums of Canada. 102 pp.

Crowe, J., L. Crowe & D. Chapman, 1984. Preservation of membranes in anhydrobiotic organisms: the role of trehalose. Science 223: 701–703.

Cuellar, O., 1977. Animal Parthenogenesis. Science 197: 837–843.

De Ridder, M., 1981. Some considerations on the geographical distribution of rotifers. Hydrobiologia 85: 209–225.

Dickson, M. & E. Mercer, 1967. Fine structural changes accompanying desiccation in Philodina roseola (Rotifera). J. Micr. 6: 331–348.

Dobers, E., 1915. Über die Biologie der Bdelloidea. Int. Revue ges. Hydrobiol. Hydrol. Suppl. zu Bd. 7: 1–128.

Donner, J., 1953. Einige Fälle von Reduktion der Zehen zu einer Haftscheibe bei bdelloiden Rotatorien des Bodens. Österr. Zool. Ztschr. 4: 346–353.

Donner, J., 1956. Über die Mikrofauna, besonders die Rotatorienfauna des Bodens. VI Congrès Int Science du Sol. III, 20: 121–124.

Donner, J., 1964. Die Rotatorien. – Synusien submerser Makrophyten der Donau bei Wien und mehrer Alpenbäche. Arch. Hydrobiol. Suppl. 27: 227–324.

Donner, J., 1965. Ordnung Bdelloidea. Akademie Verlag. 297 pp.

Donner, J., 1970. Die Rädertierbestände submerser Moose der Salzach und anderer Wasser-Biotope des Flussgebietes. Arch. Hydrobiol. Suppl. 34: 109–254.

Donner, J., 1972. Die Rädertierbestände submerser Moose und weiterer Merotope im Bereich der Staueräume der Donau an der deutsch-österreichischen Landesgrense. Arch. Hydrobiol. Suppl. 36: 109–254.

Donner, J., 1975. Randbiotope von Fliessgewässern als orte der Anpassung von Wasserorganismen an Bodenbedingungen, gezeigt an Rotatorien der Donau und Nebenflüsse. Verh. Ges. Ökol.: 231–234.

Donner, J., 1978. Material zur saprobiologischen Beurteilung mehrerer Gewässer des Donau-Systems bei Wallsee und in der Lobau, Österreich, mit besonderer Berücksichtigung der litoralen Rotatorien. Arch. Hydrobiol. Suppl. 52: 117–228.

Dumont, H. J., 1983. Biogeography of rotifers. Hydrobiologia 104: 19–30.

Everitt, D. A., 1981. An ecological study of an Antarctic freshwater pool with particular reference to Tardigrada and Rotifera. Hydrobiologia 83: 225–237.

Francez, A. J., 1980. Quelques Bdelloides (Rotifères) des tourbieres d'Auvergne. Rev. Sc. Nat. d'Auvergne 46: 21–26.

Gerritsen, J., 1980. Sex and parthenogenesis in sparse populations. Am. Nat. 115: 718–742.

Gilbert, J. J., 1974. Dormancy in rotifers. Trans. am. Microsc. Soc. 93: 490–513.

Gillard, A., 1957. Katalog der Radertieren van Belgish Kongo. Natuurwet. Tijdschr. 38: 113–141.

Green, J., 1972. Latitudinal variations in associations of planktonic rotifera. J. Zool., Lond. 167: 31–39.

Haigh, S. B., 1963. The Bdelloid Rotifers of New Zealand. Part 1. J. Quekett micr. Club 29: 161–170.

Haigh, S. B., 1965. The Bdelloid Rotifers of New Zealand. Part 2. J. Quekett micr. Club 30: 7–13; 36–41.

Haigh, S. B., 1966. The Bdelloid Rotifers of New Zealand, Part 3. J. Quekett micr. Club 30: 193–201.

Haigh, S. B., 1968. The Bdelloid Rotifers of New Zealand, Part 4. Microscopy 31: 33–41.

Haigh, S. B., 1970. The Bdelloid Rotifers of New Zealand. Part 5. Microscopy 31: 326–330.

Hebert, P. D. N. & T. J. Crease, 1983. Clonal diversity in populations of Daphnia pulex reproducing by obligate parthenogenesis. Heredity 51: 353–369.

Hickernell, L. M., 1917. A study on desiccation in the rotifer Philodina roseola, with special reference to cytological changes accompanying desiccation. Biol. Bull. 32: 343–407.

Hsu, W. S., 1956a. Oogenesis in the Bdelloidea rotifer Philodina roseola. La Cellule 57: 283–296.

Hsu, W. S., 1956b. Oogenesis in Habrotrocha bidens. Biol. Bull. 111: 364–374.

International Code of Zoological Nomenclature, 1985. Published for the International Commission on Zoological Nomenclature, London.

Jacobs, M., 1909. The effects of desiccation on the rotifer Philodina roseola. J. Exp. Zool. 6: 207–263.

Jennings, H. S., 1901. Synopsis of North-American Invertebrates. XVII The Rotatoria. Am. Nat. 35: 725–777.

José de Paggi, S. & W. Koste, 1984. Checklist of the rotifers recorded from Antarctic and Subantarctic areas. Senckenbergiana biol. 65: 169–178.

King, C. E., 1972. Adaptation of rotifers to seasonal variations. Ecology 53: 404–418.

Klimowicz, H., 1972. Rotifers of the near bottom zone of lakes Mikolajskie and Taltowisko. Pol. Arch. Hydrobiol. 19: 167–178.

Koste, W., 1970. Das Rädertier-Porträt. Macrotachela quadricornifera, ein moosbewohnendes bdelloides Rädertier. Mikrokosmos 59: 328–332.

Koste, W., 1976. Über die Rädertierbestände (Rotatoria) der oberen und mittleren Hase in den Jahren 1966–1969. Osnabrücker Naturw. Mitt. 4: 191–263.

Lansing, A. I., 1954. A nongenic factor in the longevity of Rotifers. Ann. N.Y. Acad. Sci. 57: 455–464.

Lynch, M., 1984. Destabilizing hybridation, general-purpose genotypes and geographic parthenogenesis. Quat. Rev. Biol. 59: 257–290.

Lynch, M., 1985. Spontaneous mutations for life-history characters in an obligate parthenogen. Evolution 39: 804–818.

Madaliński, K., 1961. Moss dwelling Rotifers of Tatra streams. Pol. Arch. Hydrobiol. 9: 243–263.

Maynard-Smith, J., 1978. The evolution of sex. Cambridge University Press, 222 pp.

Mayr, E., 1954. Change of genetic environment and evolution. In: Huxley, J., A. C. Hardy & E. B. Ford (eds), Evolution as a process. Allen & Unwin, London: 157–180.

Milne, W., 1916a. On the bdelloid Rotifera of South Africa. Part I. J. Quekett Micr. Club 13: 47–83.

Milne, W., 1916b. On the bdelloid Rotifera of South Africa. Part II. J. Quekett Micr. Club 13: 149–184.

Murray, J., 1908. Some African Rotifers. J. r. microsc. Soc. 28: 665–670.

Murray, J., 1911a. Some African Rotifers: Bdelloidea of Tropical Africa. J. r. microsc. Soc. 31: 1–18.

Murray, J., 1911b. Bdelloid Rotifera of South Africa. Ann. Transvaal Mus. 3: 1–19.

Pax, F. & K. Wulfert, 1941. Die Rädertiere der deutschen Thermen. Lotos 88: 246–262.

Plasota, K. & M. Plasota, 1980. The determination of the chromosome number of Habrotrocha rosa Donner. Hydrobiologia 73: 43–44.

Pejler, B., 1977. General problems on rotifer taxonomy and global distribution. Arch. Hydrobiol. Beih. 8: 212–220.

Ricci, C., 1980. Rotiferi bdelloidei da muschi dell'Uganda. In Ricerche Zoologiche in Uganda. Accad. Naz. Lincei, quad. n. 250: 17–21.

Ricci, C., 1983. Life histories of some species of Rotifera Bdelloidea. Hydrobiologia 104: 175–180.

Ricci, C., 1984a. Culturing of some bdelloid rotifers. Hydrobiologia 112: 45–52.

Ricci, C., 1984b. Variabilità in popolazioni clonali a riproduzione partenogenetica. Boll. Zool. Atti XLX Convegno UZI 51: 93.

Ricci, C. & D. Bombonati, 1985. Ciclo vitale e produttività in Macrotrachela quadricornifera. S.IT.E. Atti, 5: 259–262.

Ricci, C. & G. Melone, 1984. Macrotrachela quadricornifera (Rotifera, Bdelloidea); a SEM study on active and cryptobiotic animals. Zoologica Scripta 13: 195–200.

Robotti, C., 1987. Chromosome number of some Bdelloidea rotifer species. Hydrobiologia, this volume.

Saura, A., J. Lokki, P. Lankinen & E. Suomalainen, 1976. Genetic polymorphism and evolution in parthenogenetic animals. III. Tetraploid Otiorrhynchus scaber (Coleoptera: Curculionidae). Hereditas 82: 79–100.

Shiel, R. J. & W. Koste, 1979. Rotifera recorded from Australia. Trans. r. soc. South Australia 103: 57–68.

Slàdecek, V., 1983. Rotifers as indicators of water quality. Hydrobiologia 100: 169–201.

Stoddard, J. A., 1983. The accumulation of genetic variation in a parthenogenetic snail. Evolution 37: 546–554.

Suomalainen, E. & S. Saura, 1973. Genetic polymorphism and evolution in parthenogenetic animals. I. Polyploid Curculionidae. Genetics 74: 489–508.

Suomalainen, E., A. Saura & J. Lokki, 1976. Evolution of parthenogenetic insects. Evol. Biol. 9: 209–257.

Wulfert, K., 1960. Die Rädertiere saurer Gewässer der Dübener Heide. Arch. Hydrobiol. 56: 311–333.

Zullini, A. & C. Ricci, 1980. Bdelloid rotifers and nematodes in a small Italian stream. Freshwat. Biol. 10: 67–72.

Hydrobiologia 147: 129–133 (1987)
May, L., Wallace, R. & Herzig, A. (eds), Rotifer Symposium IV

A bdelloid rotifer living as an inquiline in leaves of the pitcher plant, *Sarracenia purpurea*

Lois E. Bateman
Biology Department, Sir Wilfred Grenfell College, Memorial University of Newfoundland, Corner Brook, Newfoundland, Canada A2H 6P9

Keywords: Rotifera, pitcher plant, inquiline, bdelloid

Abstract

A new form of *Habrotrocha rosa* Donner 1949 and aspects of its life as an inquiline of the northern pitcher plant, *Sarracenia purpurea* L. are described.

Introduction

It has long been known that the northern pitcher plant, *Sarracenia purpurea* L. is not only a carnivorous plant, but also host for a variety of organisms which thrive in the fluid contents of its leaves, or pitchers. The most thoroughly studied of the inquilines are two dipterans, a tendipedid midge, *Metriocnemus knabi* Coq. and a culicid mosquito, *Wyeomyia smithii* (Coq) [See Evans, 1971; Judd, 1959; Paterson & Cameron, 1982; and Swales, 1972 for references]. Larval development of both insects occurs in the pitchers. Also recorded in this unusual habitat have been acarine mites, sarcophagid larvae, dorylaimoid nematodes, and harpacticoid copepods [Judd, 1959; Fish & Hall, 1978; Swales, 1972]. Some aspects of the community structure within the leaves have been studied [Addicott, 1974; Bradshaw & Creelman, 1984; Fish & Hall, 1978], but little attention has been paid to rotifers. Their presence has been noted, however, by Swales [1972], Addicott [1974], and Bateman [1975].

During a study of the rotifers of Newfoundland bogs and fens, the contents of pitcher plant leaves were examined. A previously unreported bdelloid rotifer of the genus *Habrotrocha* was discovered to be present in over 90% of the leaves. This rotifer is a variety of the variable and difficult species *Habrotrocha rosa*, Donner, 1949 (Dr. W. Koste pers. comm.). This new form of *H. rosa* is described and compared to previous descriptions of the species. Results of a study undertaken to examine the life history of this form of *H. rosa* are described.

Methods and materials

From July 1984 until July 1985, pitcher plant leaves were obtained from various localities in western Newfoundland (Fig. 1). Due to snow cover, leaves were not obtained from December to March. Leaves were cut from the plant at their base, to ensure that the fluid-filled cavity remainded intact. Each leaf was stored upright in a separate plastic bag or bottle, transported to the laboratory, and stored in a refrigerator at 60 °C until examined, usually within one or two days. The contents of each leaf were flushed into a finger bowl with distilled water and examined with a dissecting microscope at 10× – 70× magnification. Individual rotifers were identified to species using a compound microscope.

Single rotifers and eggs were placed in distilled

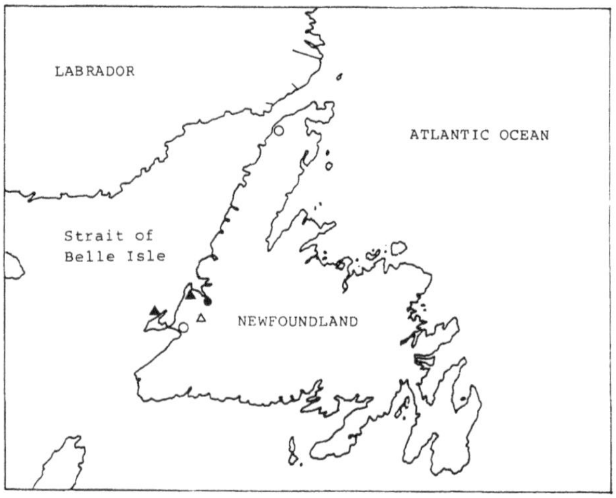

Fig. 1. Map of Newfoundland showing sites from which samples of *Sarracenia purpurea* were obtained.

- ● samples obtained monthly from July–November, 1984 and April–July, 1985
- △ samples obtained August, 1984 and June, 1985
- ○ samples obtained November, 1984
- ▲ samples obtained June, 1985

water-filled depression slides or the cavities of staining plates and covered with glass coverslips or plates. They were maintained at room temperature (21 °C) and examined at least daily for time of egg hatching, changes in size and colour, age at first and subsequent egg productions, and age at death. Each newly hatched rotifer was provided with a few grains of commercial fish food [Spirulina 20, Aquatrol Inc., Anaheim, CA, U.S.A.] which had been ground to a powder.

Long-term cultures (6–12 months) of the rotifers were maintained in finger bowls by adding distilled water to the contents of a pitcher plant leaf. Approximately 0.5 cm^3 of the ground fish food was added monthly.

Results

All of the 54 pitcher plant leaves examined contained insect remains and 22 contained larvae of *W. smithii.* Mites were present in only two leaves. Larvae of *M. knabi* and specimens of *H.* cf *rosa* occurred in fifty and fifty-one of the leaves, respectively. A single specimen of *Macrotrachela quadricornifera quadricorniferoides* (Bryce) de Koning was found in one leaf, and one of *Adineta steineri* Bartos in another.

The number of rotifers per leaf varied from between 10 and 20 in May, 1985 to more than 200 individuals in August and September, 1984. The contents of leaves obtained during November and April were frozen, but yielded 8 to 12 mature specimens within the first three days after thawing. Immature rotifers were not seen in the contents of these leaves. In other leaves, populations consisted of rotifers of various sizes from 200 μm to 340 μm. Many of the larger animals contained eggs. Twenty-eight individuals were isolated and successfully raised for periods varying from 22 to 40 days. The following description results from observations of these individuals.

Newly hatched *H.* cf *rosa* are colourless and about 200 μm in length. When they are 6–7 days old they are about 250 μm in length and begin to produce eggs. At this time their stomach contents have become pale brown. Thereafter the rotifers become slightly darker in colour and reach a size of about 300 μm when they are 12 days old.

The egg is conspicuous, being medium brown in colour and darker than the rest of the animal. It appears first on one side of the animal and within the last two hours before release, moves to a lower central position. When deposited, the smooth elliptical egg adheres at one end to a substratum. The brown colour remains for the first 24 h, gradually becoming lighter as the embryo develops. By the second day the egg is colourless. At about 52 h the mastax may be seen and by 65 h it is moving. Eggs hatch in 72–73 h at 20 °C. For two hours preceeding hatching, the young rotifer twists and turns at irregular intervals. During this period, coronal cilia can be observed moving (400× magnification). After hatching, the rotifer crawls away within two minutes, leaving a very thin egg shell which collapses when probed gently. Eggs are released at the rate of 0.8 day^{-1}. During daily examinations of single rotifers, it is customary to see an egg in the chamber and another developing within the adult. Rotifers were observed to produce eggs over periods varying from 25 to 40 days. The ovaries of a rotifer which has ceased egg production are dark brown in colour and very conspicuous next to the lighter-coloured stomach. Rotifers died three or four days after the colour change of the ovary was noticed. The stomach is yellow-brown in colour unless there is a shortage of food, when it is colourless.

Single rotifers which were introduced into small beakers with distilled water and ground fish food gave rise to over two hundred descendents within one month.

Rotifers were capable of moving freely throughout the culture dish. They crawled and fed on the bottom, on the underside of the water surface, and on any detritus present. Swimming individuals were present at all depths.

Description of the pitcher plant form of *Habrotrocha* cf *rosa* and comparisons with other forms.

The pitcher plant form of *H.* cf *rosa* conforms to the characteristics of the species as described by Donner [1965], with the exception of the size of the corona. In Donner's key, the corona is described as usually narrower than the head. In the pitcher plant form, it is slightly wider than the head (Fig. 2). (However, a form with a wider corona has been reported by Donner, 1972). Of the several forms originally described by Donner [1949] the pitcher plant form does not conform completely to any single form, but, while creeping, most closely resembles the general shape of Donner's form B. The body retains a smooth outline, which is slightly wider in the posterior half of the trunk. While feeding, the foot usually remains extended, and while swimming the extended foot is held at an angle of approximately 135° to the body.

As described by Donner [1949, 1965] *H. rosa* has several variable features including number of teeth of the trophi, size of mastax, shape of spur and form of head, corona, upper lip and egg. This form has a dental formula of 3/3 in a large mastax. The spurs are conical and diverge slightly with an interspace of 6 μm. The rostrum has two lobes. There is a distinct sulcus extending half the depth between the pedicels. The upper lip is rounded, with no

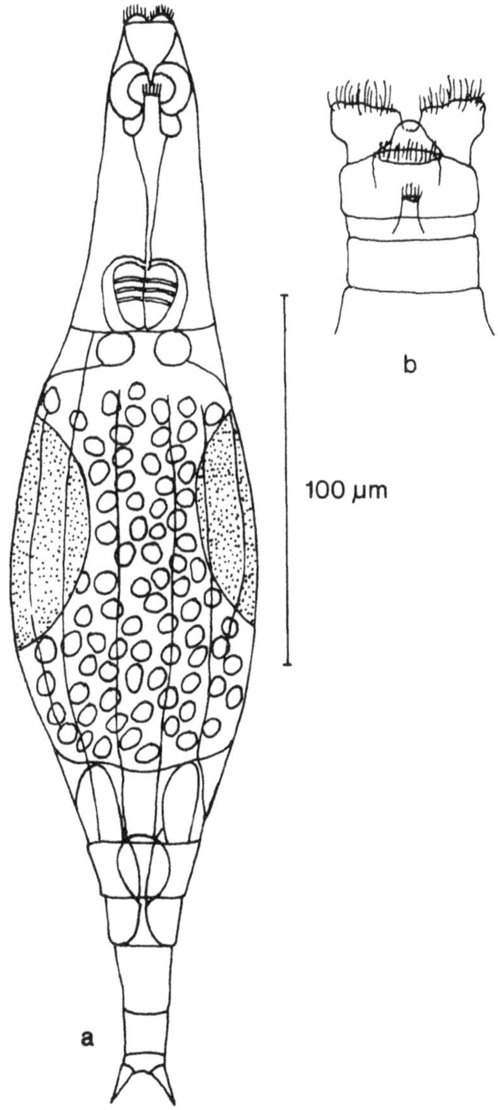

Fig. 2. Pitcher plant form of *Habrotrocha rosa.*
a) creeping
b) corona

projections and extends far enough to cover the bottom of the sulcus, but does not reach the top of the corona (Fig. 2). Figure 3 shows forms described by Donner [1949] which have similar features.

The pitcher plant form is larger than specimens of *H. rosa* described by Donner [1949, 1965] (Fig. 4). Measurements are as follows (with those in [ ] from Donner [1965] except where indicated): creeping length 200–340 μm, most about 300 μm [185–322 μm, most about 250 μm], corona

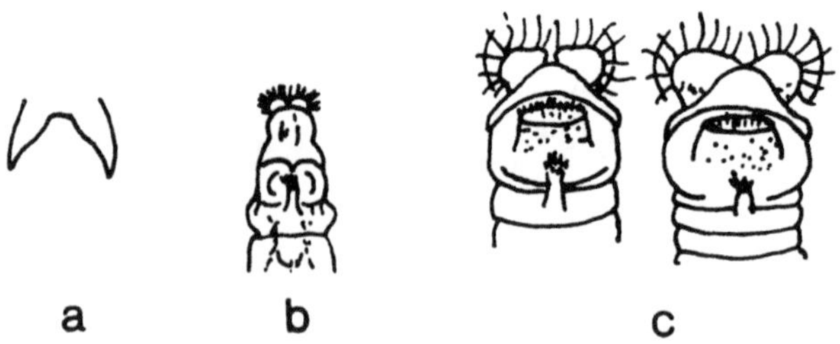

Fig. 3. Some features of *Habrotrocha rosa,* from Donner, 1949.
a) spurs, form E.
b) rostrum, form B.
c) corona (2 versions), form F.

29–39 μm [13–26.5 μm], spurs 9 μm [4.5–6 μm], pellets 4–6 μm [3.3–6 μm], mastax 17–19 μm × 7–9 μm [12–18 μm × 6–8 μm [Donner, 1949]], egg 90 μm × 45 μm [usually about 60 μm].

Habrotrocha rosa has previously been reported from a variety of habitats, including leaf litter, soil, and moss. This appears to be the first record for North America, previous records occurring in Central and Southern Europe and New Zealand [Donner, 1965].

Discussion

The list of inquilines of *S. purpurea* can be expanded to include the bdelloid rotifer *H. rosa.* The presence of young, of individuals of various sizes, and of egg-bearing individuals is evidence that the species is reproducing within the pitcher. The fluid contents of the plant leaf are rich in bacteria and particulate matter, probably the result of decomposition of the insect victims and the feces of other inquilines, chiefly the larvae of the midge, *M. knabi.* These materials are potential food sources for this suspension-feeding rotifer, which represents a fourth level in the pitcher-plant community as described by Fish and Hall [1978]. Although the rotifers spatially overlap the strata occupied by *M. knabi* (the bottom of the leaf chamber) and *W. smithii* (the body of the water volume) competition is unlikely since they probably feed on smaller particles than are used by either insect. Additional investigation is necessary to determine the specific food source of this rotifer and to determine its role in the pitcher plant community.

Habrotrocha cf *rosa* has been found in leaves at every site on the west coast of Newfoundland where pitcher plants have been sampled, but there is no information yet to help delineate its geographic range on the North American continent. *Sarracenia purpurea* is found in sphagnum bogs and other wet areas from the Northwest Territories of Canada to the Gulf of Mexico. Previous records of rotifers associated with the plant include bdelloids in Quebec, Canada [Swales, 1972] and unidentified illoricate and loricate rotifers in Michigan, U.S.A. [Addicott, 1974]. *Habrotrocha* cf *rosa* may therefore occur widely throughout the pitcher plant range. The possibility that there are other rotifer inquilines of the pitcher plant is suggested by the presence of numbers of loricate rotifers in Michigan and of several specimens of *Philodina proterva* Milne with eggs in eastern Newfoundland [Bateman, 1975]. The presence of one or two individuals of a rotifer species in a pitcher may be the result of accidental introductions of eggs or of dessicated individuals.

H. cf *rosa* appears to overwinter in the frozen pitcher plant leaf as an adult, since no eggs were found in frozen leaves, nor did young individuals appear in advance of mature adults. The capability of bdelloid rotifers to withstand freezing is well known [Gilbert, 1974].

It has yet to be determined how *H.* cf *rosa* is transmitted from one leaf to another. Leaves of the pitcher plant survive for approximately one year, gradually decaying during June and July. When decay is advanced, the contents of the leaves are spilled into surrounding vegetation, often *Sphagnum* moss. No specimens of *H.* cf *rosa* have been found here or in any other habitats in bogs where *S. purpurea* normally occurs. New leaves remain sealed until they are approximately 8 cm long which is taller than much of the moss surrounding them. Many new leaves are well-developed, with *H.* cf *rosa* present in them, before the old leaves are fully decayed. Thus, it is unlikely that rotifers enter leaves directly from surrounding vegetation.

The ease and rapidity with which *H.* cf *rosa* reproduces indicates that the introduction of a single rotifer is probably adequate to populate a pitcher. Swales [1972] has suggested that the midge *M. knabi* might act as a vector for the transmission of mites. The high coincidence of *M. knabi* larvae and *H.* cf *rosa* in pitcher plant leaves suggests that the insect could also be the vector for the rotifer. However, it remains to be seen whether rotifers could be transmitted by newly emerged midges. It is also possible for rotifers to be transmitted between pitcher plants by attaching to the ovipositor of a female midge as she visits successive pitchers. Paterson and Cameron [1982] note that a generation of *M. knabi* emerging in July in New Brunswick, Canada, can oviposit in both overwintered and newly opened leaves. The same authors also note that midge larvae from decaying leaves can migrate to younger leaves. Perhaps a migrating larva could carry rotifers and other small inquilines in the moisture clinging to its cuticle.

Acknowledgements

The author thanks Dr. Walter Koste for advice on classification of the pitcher plant form of *Habrotrocha* cf *rosa*, and for useful diagrams and measurements. Appreciation is expressed to Warwick Hewitt for the drawings.

References

Addicott, J. F., 1974. Predation and prey community structure: an experimental study of the effect of mosquito larvae on the protozoan communities of pitcher plants. Ecology 55: 475–492.

Bateman, L. E., 1975. An ecological study of hummock-hollow formations on certain peatlands of the Avalon Peninsula, Newfoundland, with special emphasis on the rotifera. M.Sc. Thesis, Memorial University of Newfoundland, 131 pp.

Bradshaw, W. E. & R. A. Creelman, 1984. Mutualism between the carnivorous purple pitcher plant Sarracenia purpurea and its inhabitants. Am. Midl. Nat. 112: 294–304.

Donner, J., 1949. Rotatorien der Humusböden, Österreich zool. Zeitschr. 2: 117–151.

Donner, J., 1965. Ordnung Bdelloidea. Akademie-Verlag, Berlin, 297 pp.

Donner, J., 1972. Die Rädertierbestände submerser Moose und weiterer Merotope im Bereich der Stauräume der Donau an der deutsch-österreichischen Landesgrenze. Arch. Hydrobiol. Suppl. 44: 49–114.

Evans, K., 1971. The bionomics of Wyeomyia smithii (Coquillett) the pitcher plant mosquito (Diptera: Culicidae: Sabethiui). Msc Thesis, Univ. of Manitoba, 279 pp.

Fish, D. & D. W. Hall, 1978. Succession and stratification of aquatic insects inhabiting the leaves of the insectivorous pitcher plant Sarracenia purpurea. Am. Midl. Nat, 99: 172–183.

Gilbert, J. J., 1974. Dormancy in rotifers. Trans. am. microsc. soc. 93: 490–513.

Judd, W. W., 1959. Studies of the Byron Bog in southwestern Ontario. X. Inquilines and victims of the pitcher plant, Sarracenia purpurea L. Can. Entomol. 91: 171–180.

Paterson, C. G. & C. J. Cameron, 1982. Seasonal dynamics and ecological strategies of the pitcher plant chironomid Metriocnemus knabi Diptera Chironomidae in southeast New Brunswick, Canada. Can. J. Zool. 60: 3075–3083.

Swales, D. E., 1972. Sarracenia purpurea L. as host and carnivore at Lac Carre, Terrebonne Co., Quebec. Part II. Nat. Can. 99: 41–47.

Hydrobiologia 147: 135–139 (1987)
May, L., Wallace, R. & Herzig, A. (eds), Rotifer Symposium IV

Specificity of the α-tocopherol (Vitamin E) effect on lifespan and fecundity of bdelloid rotifers

James R. Litton, Jr
Department of Biology, Saint Mary's College, Notre Dame, IN 46556

Keywords: Rotifera, lifespan, fecundity, Bdelloida, α-tocopherrol, Vitamin E

Abstract

Experiments were undertaken to determine molecular specificity of Vitamin E-effects on lifespan and fecundity in four bdelloid rotifers (*Habrotrocha* sp., *Philodina* sp., *Pleuretra* sp., and *Rotaria* sp.). Results indicate that lifespan and fecundity could be significantly increased by addition of any one of three tocopherol compounds (*d*-α-, β-, and δ-tocopherol) to the rotifer medium. Life table functions were increased the most by the *d*-α-tocopherol form. Improvement of these life table functions was not achieved by substitution of tocopherol analogs or other antioxidants in the rotifer medium.

Introduction

Several studies have demonstrated extended life span and increased fecundity of a few invertebrates when α-tocopherol (Vitamin E) or other antioxidants are added to the culture media. Such work is often considered to support the free-radical hypothesis of aging [Comfort, 1979; Green, 1972]. For example, α-tocopherol has been reported to extend the lifespan of an insect (*Drosophila*, 10% increase), two nematodes (*Turbatrix aceti* and *Caenorhabdites elegans*, 34% and 23% increase, respectively), and a rotifer (*Philodina*, 9% increase) [Enesco & Verdone-Smith, 1980]. Enesco and Verdone-Smith [1980] found that addition of α-tocopherol (as DL-α-tocopherol) to the culture medium (10^{-4} M) significantly extended average lifespan and increased number of offspring per female in *Philodina*. Some of my recent investigations also have demonstrated both increased lifespan and fecundity following addition of *d*-α-tocopherol (10^{-4} to 10^{-6} M) to the culture medium of four bdelloid genera (*Habrotrocha, Philodina, Pleuretra,* and *Rotaria* [Litton, unpubl. data].

In any Vitamin E-mediated response, there is always the question of its specificity to *d*-α-tocopherol. For example, tocopherol-induced polymorphism of *Asplanchna* is very specific; only *d*-α-tocopherol and several close analogs initiate and maintain the response [Gilbert, 1980]. In the case of life table functions, it is not known if the same extension of lifespan and increase in fecundity will be observed if *d*-α-tocopherol analogs or general antioxidants are used. The purpose of this study was to determine specificity of the Vitamin E effects on lifespan and fecundity in four bdelloid rotifers.

Materials and methods

Cultures

Clones of four bdelloids (*Habrotrocha* sp., *Philodina* sp., *Pleuretra* sp., and *Rotaria* sp.) were established from individuals isolated from Spicer Lake (Saint Joseph County, Indiana, USA) and maintained in mass culture on a diet of *Enterobacter* (=*Aerobacter aerogenes*) (10^5-10^8 cells ml^{-1}). The basic rotifer medium was prepared by boiling 0.75 g of cerophyll [Ward's, Rochester, NY]. 0.06 g $CaCO_3$, 7 ml of 1 M phosphate buffer (pH 6.5) in 993 ml of distilled water. Stock cultures of each species were kept in the dark at 23 °C in 250 ml Erlenmeyer flasks and transferred to fresh culture medium every three days. At the start of each experiment, cultures were placed in an incubator (constant darkness) at 21 °C, the reported optimum temperature for *Philodina* [Enesco & Verdone-Smith, 1980].

After five days, 32 individuals were placed in separate wells of a 24-well tissue culture dish (# 3524, Costar, Cambridge, MA) containing ≅ 2.5 mls of rotifer medium. The offspring of these individuals were used as experimental animals, since their birth date was precisely determined. Experimental animals were transferred into the appropriate media (see below) and every 24 hours the number of viable offspring were determined, and the parent (experimental animal) transferred to a new experimental dish containing previously prepared, fresh rotifer medium with 10^5-10^8 cells ml^{-1} *Enterobacter.*

No regular study of bacterial ingestion was done, but visual inspection of the medium at the time of changing showed a substantial quantity of bacteria still present in each culture well.

Rotifers were considered dead when gentle prodding with a micropipette failed to induce any motion. A one-way analysis of variance (ANOVA) and *post hoc* Tukey test were used to determine whether there were statistically significant differences in lifespan and fecundity among treatments [Sokal & Rohlf, 1981].

Experimental compounds

The following 14 compounds were tested for ability to affect lifespan or fecundity: *d*-α-tocopherol, *l*-α-tocopherol, 2, 5, 7, 8-tetramethyl-2-(4′, 8′-dimethylnonyl)-6-hydroxychroman, 5, 5′ bi-α-tocopherol, α-tocopherol quinone, ethyoxyquin, Menadione (Distillation Products Industries and Eastman Organic Chemicals, Rochester, NY); β-tocopherol, δ-tocopherol, δ-tocopherol, α-tocotrienol (Eisai Co., Tokyo, Japan); *dl*-α-tocopheramine, tocopherolactone (Hoffman-La Roche, Inc., Nutley, NJ); sodium selenate and *dl*-methionine (Nutritional Biochemicals Corp., Cleveland, OH). Most of these compounds are insoluble in the aqueous rotifer medium, but each was dissolved using Tween-80 as outlined by Enesco and Verdone-Smith [1980]. The solubilized mixtures were added to the culture medium at concentrations of 10^{-4} or 10^{-6} M by dilution with aqueous rotifer medium. Control medium contained neither test compound nor Tween-80. An additional control medium, containing 10^{-4} or 10^{-6} M Tween-80 was used.

Samples of the experimental culture media (100 ml) in each treatment were extracted for lipids with chloroform:methanol (2:1) using a procedure outlined by Litton (1978). These lipid fractions were chromatographed on silica gel plates and compared with plates of known mixtures of the twelve lipid compounds being tested. Sodium selenate and the selenate/methionine mixture were not tested in this procedure. Samples taken at 24 h, 3 day, and 5 day intervals of storage at 23 °C showed that all twelve of the compounds were present in the experimental culture media. Thus the compounds under investigation were all present in the media during the course of the study using the Tween-80 solubilization procedure.

Results and discussion

Effects of test compounds (10^{-4} and 10^{-6} M) on lifespan for the four bdelloids is shown in Table 1. One-way analysis of variane showed significant differences within groups ($p < 0.001$). Subsequent *post hoc* analysis (Tukey test, $\alpha = 0.01$) showed that rotifers maintained in *d*-α-tocopherol, β-tocopherol, or δ-tocopherol media (at both concentrations) had a significantly longer lifespan that the two control groups. All other compounds tested, including Tween-80, apparently had no effect on lifespan.

Table 2 shows the data obtained on fecundity for the four bdelloids. A significant difference in num-

Table 1. Effect of Vitamin E compounds (d-α-tocopherol, tocopherol analogs, and other antioxidants) on average lifespan (days) in four bdelloids.

Treatment	*Philodina* sp.		*Pleuretra* sp.		*Rotaria* sp.		*Habrotrocha* sp.	
	10^{-4} M	10^{-6} M	10^{-4} M	10^{-6} M	10^{-4} M	10^{-6} M	10^{-4} M	10^{-6} M
Control	18.7	18.6	18.6	18.5	18.6	18.5	18.5	18.5
Tween-80 control	18.6	18.5	18.6	18.5	18.5	18.6	18.7	18.7
d-α-tocopherol	20.6*	20.6*	21.0*	21.7*	21.3*	21.5*	20.4*	21.1*
β-tocopherol	19.2*	19.0*	19.1*	19.5*	19.3*	19.3*	19.2*	19.0*
γ-tocopherol	19.1*	19.1*	19.2*	19.3*	19.2*	19.5*	19.3*	19.5*
δ-tocopherol	18.7	18.6	18.5	18.6	18.6	18.1	18.7	18.3
α-tocotrienol	18.6	18.4	18.7	18.4	18.6	18.5	18.5	18.6
l-α-tocopherol	18.7	18.5	18.6	18.0	18.7	18.6	18.6	18.5
2, 5, 7, 8-tetramethyl-2-(4′, 8′-dimethyl-nonyl)-6-hydroxy-chroman	18.6	18.4	18.5	18.3	18.4	18.6	18.5	18.5
dl-α-tocopheramine	18.5	18.4	18.5	18.3	18.6	18.5	18.5	18.1
α-tocopherol quinone	18.7	18.6	18.5	18.7	18.6	18.6	18.6	18.4
tocopherolactone	18.6	18.6	18.7	18.1	18.8	18.7	18.6	18.2
ethoxyquin	18.5	18.6	18.6	18.6	18.5	18.8	18.7	18.3
menadione	18.6	18.5	18.5	18.7	18.6	18.7	18.8	18.7
sodium selenate	18.6	18.6	18.6	18.6	18.6	18.3	18.7	18.1
sodium selenate and *dl*-methionine	18.6	18.6	18.0	18.5	18.5	18.3	18.6	18.5

* = $P<0.001$

Table 2. Effect of Vitamin E compounds (*d*-α-tocopherol, tocopherol analogs, and other antioxidants) on average number of offspring in four bdelloids.

Treatment	*Philodina* sp.		*Pleuretra* sp.		*Rotaria* sp.		*Habrotrocha* sp.	
	10^{-4} M	10^{-6} M	10^{-4} M	10^{-6} M	10^{-4} M	10^{-6} M	10^{-4} M	10^{-6} M
Control	19.0	20.1	20.5	20.0	20.0	20.3	20.1	20.3
Tween-80 control	20.5	20.3	20.5	20.0	19.5	20.0	19.7	19.6
d-α-tocopherol	24.5*	26.1*	25.6*	24.4*	24.3*	24.5*	26.0*	25.3*
β-tocopherol	22.9*	22.5*	22.0*	23.0*	22.6*	22.7*	23.0*	22.6*
γ-tocopherol	22.1*	22.7*	22.8*	22.5*	22.6*	22.5*	22.5*	22.8*
δ-tocopherol	20.1	19.9	20.3	19.8	19.9	20.1	20.6	20.1
α-tocotrienol	20.3	20.6	19.8	20.3	20.4	20.3	20.4	20.3
l-α-tocopherol	19.5	19.6	20.6	20.6	20.5	20.0	20.3	20.1
2, 5, 7, 8-tetramethyl-2-(4′, 8′-dimethyl-nonyl)-6-hydroxy-chroman	20.7	20.5	20.7	20.5	20.3	20.3	20.5	20.2
dl-α-tocopheramine	20.6	20.9	20.8	19.8	20.1	20.6	19.7	20.4
α-tocopherol quinone	20.5	20.1	20.1	19.6	20.1	20.0	19.6	20.0
tocopherolactone	19.8	20.3	20.2	20.0	20.2	20.3	20.5	20.4
ethoxyquin	20.2	20.4	20.2	20.4	20.4	20.1	20.6	20.3
menadione	20.3	20.8	20.5	20.5	20.3	20.4	19.9	20.5
sodium selenate	20.6	20.1	20.3	20.1	20.6	20.2	20.3	19.7
sodium selenate and *dl*-methionine	20.5	20.1	20.4	20.0	19.5	20.1	20.4	20.1

* = $P<0.001$

ber of viable offspring was observed among groups ($p < 0.001$). Subsequent *post hoc* analysis (Tukey test, $\alpha = 0.01$), indicated that *d*-α-tocopherol, β-tocopherol, and δ-tocopherol (at both concentrations) had a significantly higher number of offspring as compared to either control groups. This may be attributed, in part, to the increased length of the reproductive period in rotifers receiving these compounds.

No significant differences in maximum longevity was noted among any of the groups at either concentration. Maximum longevity was approximately 22 days. However, the mean length of the reproductive period showed significant differences among groups ($p < 0.001$). The *post hoc* analysis (Tukey test, $\alpha = 0.01$) showed a significantly longer reproductive period ($\bar{X} = 5.8$ days) for those rotifers exposed to *d*-α-tocopherol, β-tocopherol, and δ-tocopherol when compared to control groups or other treatments ($\bar{X} = 4.5$ days). Each species showed a remarkably similar increase in mean reproductive period (*Habrotrocha* – 5.9 days, *Philodina* – 5.6 days, *Pleuretra* – 6.0 days, *Rotaria* – 5.8 days), when exposed to these three tocopherol compounds as compared to mean length of reproductive period for the controls (*Habrotrocha* – 4.4 days, *Philodina* – 4.4 days, *Pleuretra* – 4.6 days, *Rotaria* – 4.6. days).

The bdelloids chosen for these experiments have similar life history patterns under these culture conditions [Litton, unpubl. obs.]. They all show evidence of reproductive activity early in life, with a sharp increase in egg production that peaks and then rapidly declines over a 4–6 day period. The reproductive period occurred at similar times in all four species (*Habrotrocha:* days 2.5–8.5; *Philodina:* days 3–9; *Pleuretra:* days 3–9; *Rotaria:* days 4–10).

Although lifespan and fecundity were increased by *d*-α-, β-, and δ-tocopherol, there was a noticeable difference in the magnitude of the response (Tables 1 and 2). While the effect of β-tocopherol and δ-tocopherol on mean lifespan and mean number of offspring produced were significantly different from each other ($p < 0.001$, they did differ significantly from *d*-α-tocopherol ($p < 0.001$). Thus, *d*-α-tocopherol is more effective in increasing mean lifespan ($\simeq 8\%$ greater than β- or δ-tocopherol) and average offspring ($\simeq 9\%$ greater than β- or δ-tocopherol) than either of the other two active forms of Vitamin E.

It should be noted that preliminary experiments, conducted as outlined above with these four genera, using concentrations of *d*-α-tocopherol at 10^{-10} M, gave almost identical results to those in Tables 1 and 2 for this compound.

Conclusions

My results confirm and extend earlier findings which showed that rotifers placed in media containing Vitamin E had a significantly longer mean lifespan, as well as increased mean length of reproductive period and increased mean offspring per rotifer [(Enesco & Verdone-Smith, 1980; Litton, unpubl. data].

Results of these experiments also show that lifespan and fecundity effects in these bdelloids are fairly tocopherol specific. The lack of effect of a particular compound is probably bot due to its failure to enter the rotifer. This assumption is based on the fact that these rotifers indiscriminantly ingest any fine particulate matter in their medium. Thus, the test compounds must certainly enter the rotifer stomachs during feeding as well as impact the outer surface of the rotifer while in its aqueous medium. However, it is impossible to determine availability of any of these compounds at the cellular level without radiotracer studies.

While structurally unrelated antioxidants and tocopherol analogs often have Vitamin E like activity in other animal systems, they could not substitute for tocopherol in this study. The antioxidants ethoxyquin, menadione, and selenium (as sodium selenate or sodium selenate with *dl*-methionine) did not effect a response. The compound 2, 5, 7, 8-tetramythyl-2 (4′, 8′-dimethylnonyl)-6-hydroxychroman, a molecule identical to *dl*-α-tocopherol except for one less isoprenoid unit in the side chain, suggests that the response requires the entire tocopherol side chain. Lack of activity of *dl*-α-tocopheramine showed that replacement of the hydroxyl group on the α-tocopherol molecule by an amino group is significant in certain rotifers. This compound is just as active in mammalian systems as α-tocopherol [Bieri & Prival, 1967]. Tocopheronolactone, a metabolite of α-tocopherol, shown to be active in mammals [Corwin & Schwarz, 1960] was also without effect in these rotifer responses. An oxidation product of α-tocopherol with some activity in mammals, α-tocopherol quinone, [Bunyan *et al.,* 1963] was also inactive.

The total inactivity of *l*-α-tocopherol is surprising. The compound, which is levarotatory at the C-2 center and dextrarotatory at the C-4′ and C-8′ centers, contrasts with the *d*-α-tocopherol, which is dextrarotatory at all three centers. This lack of activity in the *l*-form may signal an unusual specificity in these responses for the dextrarotatory configuration at the C-2 center.

Lack of saturation at three sites on the aliphatic side chain in a α-tocotrienol may be responsible for absence of activity in this molecule. The remaining four naturally occurring tocopherols tested (α-, β-, γ- and δ-) differ only in number and position of methyl groups on the aromatic ring of 2-metyl-6-chromanol. Evidently the presence of methyl groups at C-5, C-7 and C-8 makes α-tocopherol most active. Having two methyl groups at C-5 and C-8 for β-tocopherol; at C-7 and C-8 for δ-tocopherol, seems to confer some activity. The presence of only one methyl group (C-8) does not confer activity in δ-tocopherol. Although at variance with the activity reported for β- and δ-tocopherol in mammalian systems [reviewed by Kasparek, 1980] and even somewhat for the rotifer *Asplachna* [Gilbert & Birkey, 1971] for these naturally occurring series of tocopherol compounds, these findings point to the apparent specific nature of these age-related Vitamin E-effects in bdelloid rotifers. The preliminary results showing the lifespan and fecundity responses at 10^{-10} M also indicate the rather sensitive nature of this response, at least so far as *d*-α-tocopherol is concerned. In many ways, the finding that *d*-α-tocopherol is the most active compound in this response is not surprising; α-tocopherol is the most prevalent form of Vitamin E in organisms and in the environment, and it usually has the greatest activity of all tocopherol compounds [Litton, 1978]. Further investigation of the specificity of the response, concentrating on specific deviations from α-tocopherol on the chroman ring and side chain and the threshold concentration for these responses, seem warranted.

Acknowledgements

Special thanks are due to R. L. Wallace for reading and improving the manuscript. Funds for rotifer culture materials were provided by grants from Saint Mary's College and the Indiana Academy of Science.

References

Bieri, J. G. & E. L. Prival, 1967. Vitamin E activity and metabolism of N-methyltocopheramines. Biochemistry 6: 2153–2158.

Bunyan, J., D. McHale & J. Green, 1963. The Vitamin E activity and α-tocopheryl-quinone and α-tocopherylhydroquinone in the rat. Brit. J. Nutr. 17: 391–398.

Comfort, A., 1979. The Biology of Senescence 3rd edition, Elsevier, NY, 377 pp.

Corwin, L. M. & K. Schwartz, 1960. Maintenance of α-ketoglutarate and succinate oxidation in E-deficient liver homegenates by α-tocopherol, α-tocopherol metabolite, menadione, and dephenylphenylinediamine. Nature 186: 1048.

Enesco, H. E. & C. Verdone-Smith, 1980. α-tocopherol increases lifespan in the rotifer Philodina. Exp. Gerontol. 15: 335–338.

Gilbert, J. J., 1980. Developmental polymorphism in the rotifer Asplanchna sieboldi. Amer. Sci. 68: 636–646.

Gilbert, J. J. & C. W. Birkey, Jr., 1971. Sensitivity and specificity of the Asplanchna response to dietary α-tocopherol. J. Nutr. 101: 113–126.

Green, J., 1972. Vitamin E and the biological antioxidant theory. Ann. N.Y. Acad. Sci. 203: 29–44.

Kasparek, S., 1980. Chemistry of tocopherols and tocotrienols, pp. 7–65 in Vitamin E: A Comprehensive Treatise. Marcel Dekker, Inc., NY, 660 pp.

Litton, J. R., Jr., 1978. The ole of tocopherol (Vitamin E) in rotifer ecology and measurement of the availability of lipid-soluble vitamins in a freshwater ecosystem. Ph.D. dissertation. Dartmouth College, Hanover, NH, 323 pp.

Sokal, R. R. & F. J. Rohlf, 1981. Biometry. 2nd ed., W. H. Freeman and Co., NY, 859 pp.

Hydrobiologia 147: 141–155 (1987)
May, L., Wallace, R. & Herzig, A. (eds), Rotifer Symposium IV

Coloniality in the phylum Rotifera

Robert Lee Wallace
Biology Department, Ripon College, Ripon, WI 54971-0248, USA

Keywords: Rotifera, colony, behavior, feeding, clearance rate, predator-prey, evolution

Abstract

Coloniality in the phylum Rotifera is defined and reviewed. Only two families of rotifers contain truly colonial forms: Flosculariidae and Conochilidae (order Gnesiotrocha, suborder Flosculariacea). Most species form intraspecific colonies ranging in size from a few to about 200 individuals, but species that produce enormous colonies (>1 000 individuals) are also known. All seven genera of the Flosculariidae contain species that form colonies to widely varying degrees (*Beauchampia, Floscularia, Lacinularia, Limnias, Octotrocha, Ptygura, Sinantherina*). All four species of the monogenetic Conochilidae (*Conochilus*) are colonial, but two species form colonies of only an adult and a few young. At least one other family (Philodinidae; Bdelloida) contains a species that exhibits a form of coloniality (*Philodina megalotrocha*). Two hypotheses that attempt to explain the adaptive significance of coloniality (Energetic Advantage and Predatory Avoidance) are reviewed and new information concerning the former is presented. Evolution of coloniality is discussed briefly.

Introduction

A colony may be defined as a relatively stable group of animals found in numbers greater than expected based on environmental factors, such as food, shelter, light, and temperature (Buskirk, 1975). Colonies may be further distinguished from aggregations or populations by the high level of coordination among individuals (modular units) that make up the colony [Rosen, 1979: Table I]. Research has shown that a colony may provide certain advantages to its members, including increased feeding ability, improved defense, increased interspecific competitive ability, and enhanced breeding capacity [Buskirk, 1975; Buss, 1981; Brown & Orians, 1970].

Although colony formation is widespread among invertebrates, particularly social insects, most rotifer species do not form colonies. Generally, individual rotifers have little to do with one another, except as potential food or as mates. An interesting aspect of coloniality in rotifers is that it exists at all. Given the close proximity of individuals in a colony, one might predict competition for food to be so intense that colony formation would never have evolved [see Buss, 1981]. Yet it is clearly an evolutionary stable strategy [*sensu* Davies & Krebs, 1978: 3] for members of Conochilidae and Flosculariidae [Canella, 1952]. The purposes of this paper are to (1) define coloniality as it relates to rotifers and review its development within the phylum, (2) consider hypotheses that attempt to explain the adaptive significance of coloniality, and (3) suggest some ideas which may lead to a plausible explanation for the evolution of coloniality within the phylum.

Characteristics of colonies

1. Definition

A colony is a group of individuals found in numbers greater than expected, based on environmental factors. Members of that colony are in direct contact with one another either bodily, by tubes, or by gelatinous secretions. The colony is usually formed by intraspecific accumulation or recruitment of young, and not of mature individuals.

This definition excludes species that occasionally form dense populations when large numbers of larvae settle on a substrate [Kolisko, 1939]. For example, neither *Ptygura beauchampi* [Wallace, 1978] nor *Collotheca gracilipies* [Wallace & Edmondson, 1986] are colonial species, although both have been found so densely packed that individuals touch one another. In these species, larvae settle in large numbers because they are responding to specific cues associated with the substrate [Wallace, 1980: 189], and not to previously settled adults. Also excluded are adults that form temporary aggregations. In the laboratory, small clusters of *Asplanchna herricki* have been seen to adhere to one another by their pedal glands [Dr. R. Stemberger, pers. commun.], but this species is not considered to be colonial. An exception to my definition of coloniality may be *Philodina megalotrocha* (Philodinidae; Bdelloida). This species forms small temporary colonies (aggregations) of adults while attached to a variety of substrates (abiotic surfaces, stalked sessile protozoa, sessile rotifers, Isopoda, and Bryozoa) [Burwitz, 1977]. This activity may be widespread in bdelloids [Burwitz, 1977; Kolisko, 1939].

2. Common features

Eight genera, including ≃25 species (or subspecies), of rotifers form permanent colonies (Table 1). However, colony formation is unevenly distributed among these genera. All species of the genera *Conochilus, Lacinularia, Octotrocha* and *Sinantherina* usually form colonies, while in *Ptygura* most species are solitary. Of the remaining three genera, *Floscularia* and *Limnias* have solitary and colonial species, while the sole species of the genus *Beauchampia* occasionally forms interspecific colonies.

Three important features are shared by all colonial rotifers. (1) They are all microphagous; no predatory species form colonies [Canella, 1952: 180]. The intense competition for prey would probably outweigh any benefits gained by colonial existence. I have excluded the predator *Acyclus inquietus* from the list of colonial species, even though it is found in colonies of *Sinantherina ariprepes* and *Sinantherina socialis*, because *Acyclus* is simply responding to an abundant food source: larvae of its prey [Edmondson, 1940]. (2) Most colonial

Table 1. Selected features of rotifers that commonly form colonies. Except where indicated, all species form intraspecific colonies. The species listed here are those reviewed by Koste [1978] unless otherwise indicated.

Species	Features				
	Coloniality	Planktonic or sessile	Tube type	Colony formation type	Selected references
		Permanent colonies			
Flosculariidae					
Beauchampi					
crucigera	rare	sessile	gelatinous secretions + detritus	Type I	Tiefenbacher [1972] Wallace [1977a]
Floscularia[a]					
conifera	solitary or colonial to >50	sessile	pellets	Type I	Edmondson [1945] Wallace [1977a]
ringens	solitary, rarely colonial	sessile	pellets	Type I	Tiefenbacher [1972]

Table 1. (Continued).

Species	Features				
	Coloniality	Planktonic or sessile	Tube type	Colony formation type	Selected references
Lacinularia					
causeyae[b,c]	30 – 60	planktonic	gelatinous matrix	Type II	Vidrine *et al.* [1985]
elliptica	very large up to 500	planktonic	gelatinous matrix	Type II	Bhardwaj & Dattagupta [1984] Shephard [1897] Vidrine *et al.* [1985]
elongata[c]	colonial, apparently large	sessile	gelatinous matrix	Type II (?)[d]	Shephard [1896]
flosculosa	solitary to large (100 – 300, or more)	sessile[e]	gelatinous matrix	Type II	Bhardwaj [1985] Bhardwaj & Dattagupta [1984] Edmondson [1959] Vidrine *et al.* [1985]
ismailoviensis	200 – 300	planktonic	gelatinous matrix	Type II	Western [1891] Vidrine *et al.* [1985]
megalotrocha[c]	solitary or colonial	sessile	gelatinous matrix	Type II (?)	Thorpe [1893]
pedunculata[c]	colonial > 20	sessile with a long peduncle	gelatinous matrix	Type II (?)	Anderson & Shephard [1892] Hudson & Gosse [1889] Shephard [1899]
reticulata[c]	a few to apparently quite large	sessile	gelatinous matrix	Type II	Anderson & Shephard [1892] Shephard [1899][f]
striolata[c]	very large colonies (> 3500!)	sessile with a long peduncle	gelatinous matrix	Type II	Shephard [1899]
Limnias					
ceratophylli	solitary or colonial to about 50	sessile	gelatinous secretions + detritus	Type I	Hudson & Gosse [1889] Tiefenbacher [1972]
Octotrocha					
speciosa	solitary to very large colonies	sessile	gelatinous matrix	Type II	Edmondson [1959] Thorpe [1893]
Ptygura					
crystallina	rarely colonial	sessile	gelatinous secretions + detritus	Type I intra- and interspecific	Tiefenbacher [1972] Wallace [1977a]
longicornis	few	sessile	gelatinous secretions + detritus	Type I (?)	Koste [1970]
pilula	rarely colonial	sessile	gelatinous + fecal pellets	Type I interspecific	Wallace [1977a]
tihanyensis[c]	small colonies	sessile	gelatinous + detritus	Type II (?)	Edmondson [1949]
Sinantherina					
ariprepes	solitary or 5 – 20, up to ≃ 200	sessile	no tube	Type I	Edmondson [1940] (but see Canella 1952: 220)

Table 1. (Continued).

Species	Features				
	Coloniality	Planktonic or sessile	Tube type	Colony formation type	Selected references
procera	40 – 50 perhaps larger	sessile	no tube	Type II	Bhardwaj & Dattagupta [1984] Thorpe [1893]
semibullata	25 – 100 occasionally solitary	planktonic, temporary attachment	no tube	Type II	Canella [1952] Gunter & Knight [1978] Thorpe [1889] Vidrine *et al.* [1985]
spinosa	50 – 100	planktonic, temporary attachment	no tube	Type II	Bhardwaj & Dattagupta [1984] Thorpe [1893] Vidrine *et al.* [1985]
socialis	moderate to large (ca 30 – 200)	sessile, planktonic larval colony	no tube	Type II	Champ & Pourriot [1977b] Pourriot *et al.* [1972][g] Surface [1906] Vidrine *et al.* [1985] Wallace [1980]
Conochilidae					
Conochilus[h]					
dossuarius	adult with 1 or 2 young	planktonic	gelatinous matrix	Type II	Hudson & Gosse [1889] Gilbert [1980]
hippocrepis	25 to 200	planktonic	gelatinous matrix	Type II	Hudson & Gosse [1889] Wesenberg-Lund [1930]
natans	adult with 1 or 2 young	planktonic	gelatinous matrix	Type II	Wesenberg-Lund [1930]
unicornis	usually 5 – 20 occasionally many > 150	planktonic	gelatinous matrix	Type II	Burckhardt [1944] Kolisko [1939] Rousselet [1892] Wallace & Starkweather [1985]
			Temporary colonies		
Philodinidae					
Philodina					
megalotrocha	small groups of adults	planktonic, but may attach to various surfaces	no tube	intra- and interspecific grouping of adults	Burwitz [1977]

[a] Although they have only minor differences of anatomy, I have separated *F. conifera* and *F. ringens* because of their contrasting propensities to form colonies [see Koste 1978: 541].

[b] *Lacinularia causeyae* is a new species described by Vidrine *et al.* [1985].

[c] Several species of *Lacinularia* and one *Ptygura*, not reviewed by Koste [1978], are included here for the sake of completeness. Their designation may be considered tentative until they can be properly reviewed [see also Rousselet, 1900].

[d] (?) = Colony type assumed.

[e] Vidrine *et al.* [1985] reports that small colonies (ca 30) are free-swimming, while large colonies are sessile and contain thousands of individuals.

[f] Short reference to colony size and colony formation.

[g] Originally identified as *L. flosculosa*, but changed to *S. socialis* by Champ & Pourriot [1977b].

[h] The species described by Michael [1966] as *C. madurai* is probably *L. elliptica*.

rotifers ($\simeq 70\%$) are sessile. Three genera containing eight species have planktonic forms (*Conochilus, Lacinularia, Sinantherina*), but species in the latter two genera probably inherited coloniality from their sessile ancestors. Four species attach to plants by a tethering peduncle of colony origin: *Lacinularia pedunculata* and *Lacinularia striolata* (permanently attached) and *Sinantherina semibullata* and *Sinantherina spinosa* (temporarily attached). (3) Colonies do not reproduce by budding or branching, and there are no specialized zooids as in other colonial invertebrates, e.g., Bryozoa [Kolisko, 1939; Surface, 1906]. Further, there are no connections between individuals; consequently, resources are not redistributed within the colony.

3. *Colony size*

Colony size varies from 2 to well over 1000 individuals (Table 1). Some species form only small colonies (usually <5 individuals) (e.g., *Floscularia ringens* and *Ptygura crystallina*). Of those that are capable of forming larger colonies, the number in a colony may vary considerably. For example colonies of *Floscularia conifera* range from 2 to >50 individuals, *S. semibullata* from 30 to $\simeq 100$, *Conochilus hippocrepis* from $\simeq 25$ to >200, and *Conochilus unicornis* from 5 to >150. Two species have been reported to produce enormous colonies: *Lacinularia flosculosa* (thousands of individuals, Vidrine *et al.* [1985] and *Lacinularia striolata*, (>3500 individuals, Shephard [1899]!

Variance in colony size may be partly due to variable reproduction rates [Ruttner-Kolisko, 1974: 123]. In *F. conifera* the greatest proportion of colonial animals was found when population level was high [Wallace, 1977a]. However, the case of *C. unicornis* is more complex. This species is generally reported as having from 5 to 20 individuals per colony [Edmondson, 1959: 475; Kolisko, 1939; Ruttner-Kolisko, 1974: 123]. However in Pleasant Lake, (Merrimack Co., NH, USA), most colonies were very large ($\bar{x} = 84.4 \pm 40.3$, 1 SD), some having >150 individuals [Wallace & Starkweather, 1985, Table 1]. Kolisko [1939, Figs. 9, 10] found that small ('typical') colonies of *C. unicornis* (6–10) were formed so that the distal ends of the feet of all colony members met at a common center or focus. However, in larger colonies (>10), no single center existed ('atypical') and clusters of individuals each had a separate focus to which their feet extended. The same feature was present in the very large Pleasant Lake colonies, except that there were many clusters of individuals and consequently many foci. As Kolisko [1939: 89] pointed out, this colony architecture is probably related to the formation of new colonies by separation or division of the gelatinous matrix between clusters in the colony. A similar method of division takes place in *C. hippocrepis* and probably other species [Hudson & Gosse, 1886; Wesenberg-Lund, 1930: 176]. *Conochilus hippocrepis* was also present in Pleasant Lake, but it did not produce unusually large colonies for this species ($\bar{x} \simeq 92$) [Wallace & Starkweather, 1985].

4. *Tube production*

Tubes and tube building activities have been investigated for only a few species. There appears to be no correlation between coloniality and presence, absence, or composition of a tube (Table 1, but see below Predator Avoidance Hypothesis). Some species do not produce any tubes (*Sinantherina*) and those that do use several materials: cement and detritus (e.g., *Limnias ceratophylli*), detritus formed into pellets (e.g., *Floscularia*), and gelatinous secretions (*Conochilus, Lacinularia, Octotrocha*).

5. *Colony formation*

Detailed observations of colony formation are rare in the literature. Based on the information available I have classified colony formation methods into two distinct types (Table 1). Type I colony formation is found in species possessing free-swimming larvae which select tubes of previously settled adults as substrata (*Beauchampia, Floscularia, Limnias,* some *Ptygura,* and possibly one *Sinantherina*). Larvae of Type I colonies may be said to have been recruited to the colony, implying active substrate selection [see Wallace, 1980: 181]. In Type II colony formation, larvae accumulate directly into their parent colony as young individuals without a free-swimming phase, but daughter colonies may also separate from a large parent colony by splitting or as an aggregation of larvae (*Conochilus, Lacinularia, Octotrocha*, and *Sinantherina*). A complicated variation of Type II colony formation is found in *S. socialis* [Surface, 1906]. In this species colony formation is accomplished by synchronous egg hatching and formation of larvae into a planktonic colony, which later selects a substrate [Champ, 1978; Champ & Pourriot, 1977a, b; Surface, 1906; Wallace, 1980]. Release of daughter colonies by a similar process apparently takes place in *Lacinularia elliptica, L. flosculosa, L. striolata, Sinantherina procera* and *S. semibullata* [Bhardway & Dattagupta, 1984; Huxley, 1853; Shephard, 1899; Western, 1891]. However, daughter colonies of *C. hippocrepis, C. unicornis* and *S. spinosa* are made up of both young and adults [Bhardway & Dattagupta, 1984; Kolisko, 1939; Wesenberg-Lund, 1930].

In most Type II colonies, young appear to be randomly dispersed within the colony, however, in some species young are located at one end, e.g., *L. elliptica* and *Lacinularia ismailoviensis* [Vidrine *et al.*, 1985] and *S. semibullata* [Canella, 1952: Figs. II & III]. In Type I colonies, young tend to be distributed closer to the tops of the tubes of older colony mates, e.g., *F. conifera* [Edmondson, 1945: Fig. 7].

Genotypic diversity (genotypic relatedness of individuals) within colonies is probably radically different depending on colony formation type. Type I colonies probably have lower genotypic relatedness than Type II colonies, unless there are some unknown mechanisms that affect distribution and/or substrate selection by larvae. Individuals of Type II colonies no doubt are highly related, because newborn individuals tend to stay in the parental colony. The significance of high genotypic relatedness in colonies is unknown, but would be a very interesting matter to investigate, especially considering the amount of interest this topic has generated recently (see Other Ideas: Sexuality).

Most colonial species produce multigeneration colonies, but *S. socialis* is a notable exception. Because of its unique mode of colony formation, nearly all individuals in a colony are the same age (± 3 h) [Wallace, 1980: 182, Fig. 1]. During the growing season some Type II colonies, are in a certain sense, deathless or continuous. In *C. unicornis*, new individuals are added by accumulation, but once the colony becomes large, it splits into two or more smaller daughter colonies [Kolisko, 1939]. All Type I and a few Type II colonies have a finite life span within the growing season. That is, *F. conifera* colonies do not grow continuously; eventually the individuals die and larval recruitment ceases. *Sinantherina socialis* colonies also have a limited life span, during which time they may produce 5 or more larval colonies [Wallace, unpubl. obs.].

Formation of interspecific (mixed species) colonies is not common (Table 1). Of six species that have been reported to form them, all are Type I colonies [Tiefenbacher, 1972; Wallace, 1980]. The colonists are *Beauchampi crucigera, F. conifera, F. ringens, L. ceratophylli, P. crystallina*, and *Ptygura pilula*. They usually attach to *F. conifera* (a frequent nuclear species), *F. ringens, L. ceratophylli*, or *P. crystallina. Philodina megalotrocha* has been reported to attach to *Ptygura* and *Collotheca* species [Burwitz, 1977]. As previously considered, presence of *Acyclus* in *Sinantherina* colonies is not an example of interspecific colony formation. Unfortunately, reports of interspecific colonies are so scattered and infrequent that few useful conclusions may be drawn.

Adaptive significance of coloniality

Two main hypotheses have been proposed to explain the adaptive significance of coloniality: the Energetic Advantage hypothesis [Edmondson, 1945; Wallace, 1977b, 1980] and the Predatory Avoidance hypothesis [Gilbert, 1980; Wallace, 1980]. However, they are not mutually exclusive and together may provide considerable explicative power concerning variation of colony size in certain species.

1. Energetic advantage hypothesis

Edmondson [1945] clearly showed that several life table functions of *F. conifera* are strongly affected (improved) by colonial life. When solitary animals were compared with colonies consisting of two or more individuals, Edmondson found that rate of maturation, rate of reproduction, longevity, and tube length at death were all significantly increased by grouping. (However, tube elongation rates were not elevated.) Thus, colonial animals somehow gain an advantage over solitary ones by virtue of their colonial existence. This seems counter intuitive; one might expect colonial animals to suffer from intraspecific competition [Buss, 1981]. One hypothesis which has been developed from Edmondson's data is as follows. A solitary animal can remove (clear) food particles from the water at a particular rate. If two animals are close enough together, their filtering currents do not interfere, but reinforce each other in a nonlinear manner. As a result, the volume cleared by the two is more than doubled [see also Mackie, 1963, but compare Bishop & Bahr, 1973; Buss, 1981]. Therefore, rate of food intake per animal is increased and the extra energy is used in improved maintainance (repair) and reproduction. An alternative hypothesis is that clearance rate (CR) is not elevated, but the amount of energy expended per unit food captured is decreased (i.e., increased feeding efficiency). Of course both mechanisms may be in operation [Edmondson, 1945; Wallace, 1977a].

I have been able to perform three different tests of the elevated CR hypothesis. In two sets of experiments I examined CRs of several colonial species using radioactively labeled food particles [Wallace & Starkweather, 1983, 1985]. In the first of these, I determined CR as a function of colony size for *C. hippocrepis*, *C. unicornis*, and *S. socialis*. In the second set of experiments, I determined CR of solitary and colonial *F. conifera* as a function of tube length. In a third set of experiments I used latex microspheres (Sigma Chemical Co.) ($\simeq 5.7 \pm 1.5$ SD μm diameter) instead of radioactively labeled yeast to determine CR in solitary and colonial *S. socialis*. By counting the number of latex microspheres in each animal's stomach, I was able to determine a CR for every individual in the colony. The results of these three sets of experiments are discussed below.

a) Colony size and CR: Conochilus *and* Sinantherina. There was no statistically significant correlation between CR and colony size for naturally occurring colonies of between $\simeq 25$ to 225 individuals/colony in the three species tested. This is indicated by the Pearson-product-moment correlation coefficients [Sokal & Rohlf, 1981: 565] which were calculated for each species: *C. hippocrepis* $r = 0.06$, *C. unicornis* $r = -0.15$, *S. socialis* $r = -0.19$). However, the elevated CR hypothesis should not be rejected based on these results alone. Edmondson [1945: 165, Fig. 23] showed that mean length at death for *F. conifera* in colonies of only two individuals was significantly increased over solitary individuals. Further, there was only a slightly greater (but regular) increase in length at death for those in colonies of three to seven individuals, but little increase in CR beyond that [Edmondson, 1945: 166, Fig. 23]. None of the colonies used in my experiments had fewer than 20 individuals, therefore, a positive correlation between CR and colony size might not be expected with these large colonies. That is, individual CR may increase as a function of colony size, but only in colonies that have between two and ten individuals.

b) Clearance rates of solitary and colonial F. conifera. There is a positive relationship between CR and tube-length in both solitary and colonial *F. conifera* (Fig. 1). However, there was no significant difference ($P < 0.05$, Mann-Whittney U-Statistic, one-tailed test) found between CR of solitary and colonial *F. conifera* in the four size ranges available for testing ($\simeq 250$, $\simeq 600$, $\simeq 950$, $\simeq 1200$ μm). The substantial variance in CR seen for colonial animals >1500 μm may be due to animal senescence [Wallace & Starkweather, 1985]. This experiment strongly suggests that there is no elevation in CR of colonial rotifers.

c) Clearance rates of solitary and colonial Sinantherina socialis. To further test the tentative conclusion that there is no elevation in CR of

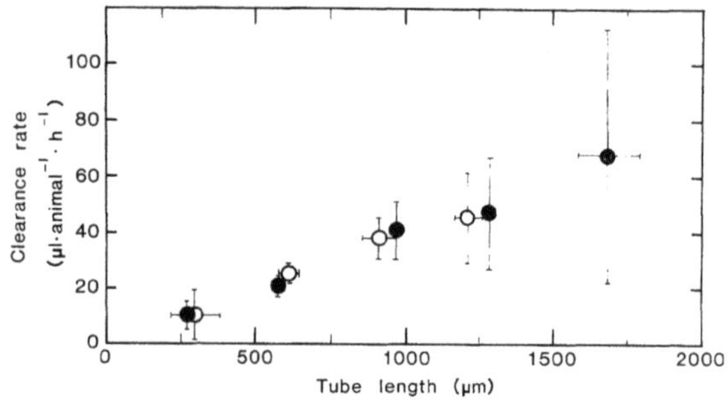

Fig. 1. Clearance rates (μl animal^{-1} h^{-1}) of solitary (open circles) and colonial (closed circles) *Floscularia conifera*, fed yeast (*Rhodotorula*), as a function of tube length. Error bars are $\pm$1.0 SD of the mean. Error bars associated with tube length are also given because several animals were used for each determination. [See Wallace & Starkweather 1985 for procedures].

colonial rotifers, a slightly different experiment was performed using *S. socialis*. This species was used because all members of a colony are approximately the same age (see above). The general procedures of Wallace & Starkweather [1985] were used in these experiments, except that I substituted latex microspheres for radioactively labeled yeast. For certain purposes use of microspheres may be superior to radioisotopes, because it avoids the problem of ^{32}P loss from fixed animals [Wallace & Starkweather, 1985].

Plants colonized by *S. socialis* were collected and returned to the laboratory (without refrigeration). There the colonies were bisected, using dental tools. One half of the colony was then left alone, while individuals were systematically removed from the other half until a single, undamaged animal remained. Both the colony and the solitary individual were immediately transferred to a 100 ml beaker filled with pond water [Wallace & Starkweather, 1985]. After an acclimatization period of at least 1 h, latex microspheres were added to the beaker and the suspension stirred using a magnetic spin bar ($<$10 s). The final concentration of microspheres was either 1 or 10 spheres μl^{-1}. After 15 minutes the experiment was terminated by adding 100% formalin to achieve a final concentration of $\simeq$5%. Animals were then removed from their substrate using dental tools and micropipets and washed three times in rotifer medium. The number of microspheres in each animal's gut was determined by direct observation.

Clearance rates were calculated for all individuals and a Z-score (standard deviate) was calculated for each solitary animal using the mean and standard deviation from the corresponding colony half [Sokal & Rohlf, 1981: 105]. These Z-scores should be normally distributed, if there was no significant difference between CRs of the colonial animals and their corresponding solitary individuals. The Z-scores for 18 separate experiments showed considerable variance (Fig. 2), but the null hypothesis (no difference in distribution of Z-scores from a normal distribution) could not be rejected when a t-test was performed on these data (N = 18, t = 0.71, $P > 0.49$). Therefore, I concluded that there was no statistically significant difference in CRs of colonial and solitary individuals in this species.

d) Increased feeding efficiency hypothesis. At present, it appears the Elevated CR hypothesis can not be accepted, as none of the three experiments described above demonstrated an elevation in CR by colonial rotifers. In fact, one may argue that increased CRs should not be expected, since it would mean a decrease in gut passage time and food

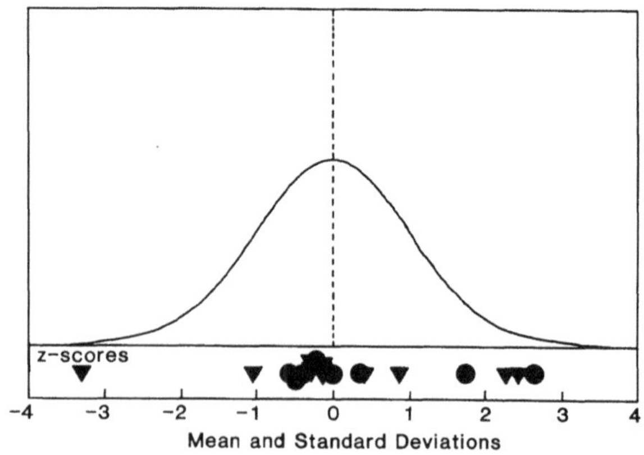

Fig. 2. Clearance rates, plotted as Z-scores (standard deviates), of solitary *Sinantherina socialis* fed latex microspheres. The solid line is a standard normal curve [Sokal & Rohlf, 1981: 101, equation 6.1]. No vertical scale is implied in the distribution of Z-scores; co-incident data points were displaced slightly for plotting purposes. (Circles = 10 microspheres μl^{-1}; diamonds = 1 microsphere μl^{-1}.) (See text for details).

would be less well digested and absorbed [Calow, 1981: 43]. However, further work needs to be done on this problem, especially at low food levels. For the present, it seems reasonable to assume the energetic benefit afforded colonial life in *F. conifera* is due to an increased feeding efficiency. That is, the cost of producing feeding currents is energetically less expensive in colonial animals. A direct test of the Feeding Efficiency Hypothesis probably will be difficult to undertake. Perhaps determining ciliary beat frequency (using strobe lighting) and oxygen consumption rates (using Cartesian diver microrespirometery, see Epp & Lewis, [1979] for solitary and colonial animals would provide useful information on this question.

I have recently made some observations on the pattern of current flow around colonies of *S. socialis* that relate to the question of feeding efficiency. To make these observations I used the following techniques. Colonies of *S. socialis* which were still attached to their substrate were collected from Buffalo Lake (Marquette Co., WI, USA) (21 °C) and returned to the laboratory without refrigeration. They were then positioned in a small petri dish filled with lake water (≃25 ml) using locking forceps that were attached to a micromanipulator. Colonies were held at least 3 – 5 mm above the bottom of the petri dish and below the water surface. This was done to minimize the potential for interference of feeding currents from those surfaces [(see Alcaraz *et al.*, 1980: 243]. Small amounts (< 0.5 ml) of a pond-water and carmine suspension were periodically added to the petri dish near the colony to aid in determining current position. Observations of the feeding currents were made using a Zeiss SR stereomicroscope with darkfield illumination. Video tapes were recorded for later analysis using a Panasonic (WV-3890B) color video camera and a NEC (VC-N40EU) video cassette recorder.

Using the procedure described above I have observed that individuals in adult *S. socialis* colonies, do not act independently in producing their feeding currents. The coronae of a large number of animals face in the same direction (an array). This results in a feeding current that has a specific pattern around the colony and is more-or-less stable for several minutes. Incurrent water usually flowed to the colony in one or two dominant streams, the water then flowed completely around the colony and left it from another point, usually at the side opposite the inflow. Several authors have reported a similar phenomenon in other species. Shephard [1899] described the feeding currents of *L. striolata*, as follows: '... action of the ciliary wreaths of a large colony produces a very distinct current in the water ... all the individuals over a large portion of the colony are operating to produce a current in the same direction'. Kolisko [1939: 84, Fig. 6] and Burckhardt [1944: 368] depicted the flow of currents around *C. unicornis* as following definite paths which approached the colony tangentially. Further, their figures seem to indicate that the feeding currents of adjacent individuals interact with each other. Canella [1952: Fig. II] described individuals in colonies of the planktonic species *S. semibullata* as all facing in the same direction while the colony was tethered to a substrate. Therefore colonies have an anterior and posterior, but this changes somewhat when the colony is swimming freely [Canella, 1952: Fig. III].

These reports of current flow resemble those described by Banta *et al.* [1974] for certain colonial Bryozoa. These authors described specific regions, called chimneys, where excurrent water was released from a colony. Zooids in these regions had lophophores that were slightly different from normal. They have longer necks and a few of the tentacles bordering the chimney are longer too. Bryozoan chimneys are blank areas in the colony where the lophophores of local zooids lean away from the chimney. Banta *et al.* [1974] suggest that chimneys represent specific avenues for excurrent water flow which create more efficient feeding currents for colonial zooids.

I propose that a similar mechanism is in operation in colonial rotifers. The excurrent area in colonial rotifers functions as a chimney, providing an efficient flow of water over a group of closely spaced coronae and then away from the colony (Fig. 3). Chimneys in rotifer colonies are easily recognized by tracing current flow around the colony or by determining where two or more arrays meet. At that point the ventral regions of rotifers in different arrays oppose one another (Fig. 3).

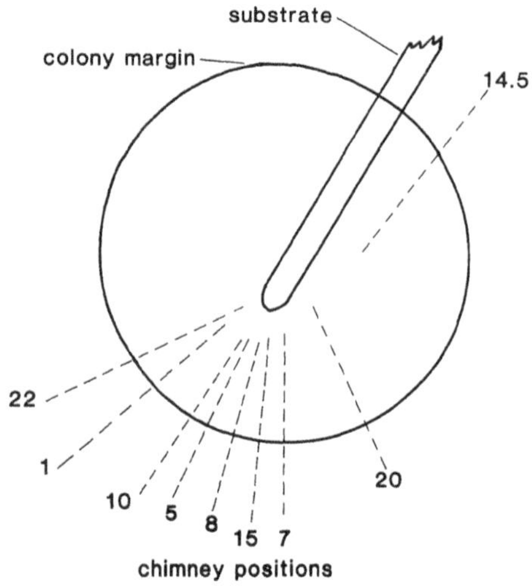

Fig. 3. Schematic diagram showing the position of chimneys in a *Sinantherina socialis* colony. Dashed lines indicate chimney positions, numbers indicate time in minutes. At 14.5 minutes a small, short-lived, secondary chimney was also seen.

There are three differences between bryozoan and rotifer chimneys. (1) In *S. socialis* there was usually only one chimney, on the side opposite from the incurrent flow. In bryozoans, there are many regularly-spaced chimneys. This difference probably results from the fact that bryozoans have a flat, encrusting growth form that may be several centimeters in diameter. In that case chimneys would be needed at regular intervals to permit excurrent water to escape. (2) Unlike bryozoans, there do not appear to be any anatomical modifications in the individual rotifers that occupy a chimney. However, coronae of individuals that make up a chimney face in the opposite direction from one another. (3) Rotifer chimneys appear to have a relatively stable position, although they may not be permanently fixed. Their location depends on the direction colony members face, which may change with time (Fig. 3). In contrast, bryozoan chimneys maintain a constant position. When bryozoan zooids retract and then elongate again, the chimneys are in the same place. This is no doubt a function of their anatomical modifications.

2. *Predator avoidance hypothesis*

It has been suggested that colony formation may provide some protection from predation. Surface [1906] implied that solitary *S. socialis* were very susceptible to predation, but he did not offer any support for his statement. Several invertebrate predators have been reported to feed on *Conochilus* [Williamson, 1983: Table 1, and refs. therein]. However, Gilbert [1980] showed that *C. unicornis* was protected from the predator *Asplanchna girodi* in two different ways. (1) The colony was much too big to be captured whole by *Asplanchna.* (2) The gelatinous matrix provided a refuge for individuals when *Asplanchna* attempted to engulf single animals. For *C. unicornis*, it appears that overall probability of ingestion after encounter is very low compared to more accessible species such as *Brachionus, Synchaeta,* and other small, soft-bodied prey [Stemberger & Evans, 1984; Dr. C. Williamson, pers. commun.].

A colony also may be able to detect predators more efficiently because of the larger number of individuals present. When disturbed, *F. conifera* individuals rapidly contract into the shelter of their tubes. Often this behavior is so forceful that other individuals are disturbed and the entire colony retracts rapidly.

Inability of invertebrate predators to handle larger prey (i.e., size-selective predation, Dodson, [1974] may explain the range of colony size reported for *C. unicornis* (see Characteristics of colonies). Larger colony size in the Pleasant Lake population may be a response to predation pressure from *Leptodora kindti* [see Edmondson & Litt, 1986]. *Conochilus unicornis* has a multigeneration colony and new colonies are produced during the growing season when a large colony separates [Kolisko, 1939]. In any given population, colony size may be adjusted by a modification in adhesive properties of the gelatinous matrix. A refractory matrix would result in larger colonies which are better able to withstand the attacks of invertebrate predators: a form of size-selective predation. Larger colonies, however, would be more vulnerable to predation by fish. Therefore, it is not surprising that Pleasant Lake *C. unicornis* ($\bar{x} \simeq 85$ individuals/colony) were located below $\simeq 10$ m [Wallace, unpubl. obs.]. Other selection pressures, such as energetic costs of swimming, should not be ignored as contributing factors to colony size in this species [Dr. R. Stemberger, pers. commun.].

All species of the genus *Conochilus* form colonies to some extent (Table 1), although in the *natans-dossuarius* group, only an adult and one or two offspring form a colony [Ruttner-Kolisko, 1974: 123]. Retaining young in the gelatinous matrix might simply be a holdover from an ancestral condition which produced a larger colony. Nevertheless, young may be protected from invertebrate predators by the adult's gelatinous matrix until they have secreted enough gelatinous material to provide their own protection. In the genus *Conochilus* there has been only one report that examined the physical nature of the gelatinous matrix and the configuration of animals within the colony [Kolisko, 1939]. It would be interesting to test my hypothesis that the *unicornis-hippocrepis* group (especially very large *unicornis* colonies) secrete a more refractory gelatinous material than that of the *natans-dossuarius* group (or small *unicornis* colonies).

Very large colonies in other genera with gelatinous secretions may be due solely to rapid accumulation of young during periods of favorable growth (*Lacinularia, Octotrocha*). However, a large gelatinous matrix may provide an important refuge from invertebrate predators (e.g., insect larvae, planaria).

Absence of a gelatinous matrix in the genus *Sinantherina* is puzzling, especially since members of this genus are very noticable to human observers: *S. ariprepes* [Edmondson, 1940: 450], *S. semibullata* [Canella, 1952; Gunter & Knight, 1978], and *S. socialis* [Surface, 1906; Wallace, pers. obs.]. Unpalatable animals are often very conspicuously colored which is usually considered to be an aposematic protective mechanism. Perhaps potential vertebrate predators find *Sinantherina* unpalatable. To my knowledge, there has been no known function attributed to the warts found in three species of *Sinantherina* (*procera, semibullata, socialis*). These structures are prominently exposed when the animal contracts [Thorpe, 1889; Wulfert, 1939] and may hold concentrations of chemicals obnoxious to predators [see Kerfoot *et al.*, 1980: 20]. One planktonic species (*S. spinosa*) has spines on its ventral body surface which may function in defense. When disturbed, the colony does not contract into a tight ball, as do other members of the genus. Instead, the animals tuck their coronae dorsally and the body takes on the shape of a question mark (?), exposing the spines [Thorpe, 1893: 151, Fig. 6c]. Although the spines are small, they may irritate the esophagus of a planktivorous fish and deter ingestion.

Sinantherina semibullata may avoid vertebrate predation in another way. Gunter and Knight [1978] have described an interesting behavior which they call avoidance contraction [see also Canella, 1952: 227]. When coming into contact with foreign objects all colony members contract and reduce the colony diameter by up to 50%. This permits the colony to sink rapidly. This behavior is very different from the spontaneous contraction of individuals that happens irregularly while the colony swims. Avoidance contraction may be similar to passive sinking ('dead-man response') exhibited by *Bosmina* when attacked by copepods [Kerfoot, 1977, 1978]. Also, the reduced colony diameter may allow it to pass through the gill rakers of planktivorous fish.

In general, no systematic investigations have proceeded on the question of vertebrate predation. I have observed fish preying on *Conochilus* in the laboratory, but these were only incidental observations [Wallace, 1980: 190]. Based on the observation that *Conochilus* clog plankton nets (210 μm), Beavers & Stavin [1975] concluded that this gelatinous rotifer can inhibit the feeding of certain planktivorous fish (i.e., threadfin shad, *Dorosoma petenense*) by clogging their gill rakers. This conclusion may not be warranted, as the authors did not perform an analysis of fish gut contents, nor did they observe the fish feeding.

3. Other ideas: Sexuality

Coloniality may strongly influence sexuality in two ways [see Wallace, 1977a: 550]. Crowding in colonies may induce sexuality as population density does in *Brachionus calyciflorus* [Gilbert, 1963]. Some supportive evidence for this idea was reported by Champ & Pourriot [1977b]. They concluded that occurrence of mictic female production in *S. socialis* was related to population density and photoperiod. Unfortunately, the influence of colony size alone apparently was not examined by these authors. Detailed research needs to be done which closely examines the relationship between population size, colony size, and mixis.

Secondly, grouping animals together in colonies may increase the probability that a male could fertilize many females from his own colony. This would increase probability of resting egg production, because solitary females would be more widely dispersed and difficult to locate before the male's energy reserve was exhausted [Vasisht & Dawar, 1970: 375]. This hypothesis could be tested by comparing rates of male and resting egg production in solitary and colonial *F. conifera* attached to the same plant. However, fertilization of colony mates may be possible only for colonies with low genotypic relatedness (colony Type I). Perhaps inability of males to fertilize colony mates which has been reported by Bhadwaj & Dattagupta [1984] for certain species of *Lacinularia* and *Sinantherina* (colony Type II) is due to high genotypic relatedness in those colonies [but compare Bogolovsky, 1958].

Another feature of colonies which merits more investigation is the disposition of resting eggs within the colony. Resting eggs are deposited in the gelatinous matrix of *L. flosculosa*, but apparently they are not held there very long in *C. hippocrepis*. Perhaps the added weight of resting eggs reduces the swimming ability in this planktonic species [Wesenberg-Lund, 1930] or their dark color makes the colony more visible and therefore more vulnerable to predation by fish.

Evolution of coloniality

A striking feature of true coloniality in rotifers is that it is limited to two microphagus taxa (Flosculariidae, Conochilidae). All eight genera of these two families are involved. The extent to which coloniality is developed in the Flosculariidae suggests that sessile species are preadapted for its development [Gilbert, 1980]. The planktonic examples of this family are considered to have evolved from sessile ancestors [Canella, 1952: 180; Gilbert, 1980; Kutikova, 1983]. However, presently it is difficult to understand how coloniality in the family Conochilidae evolved. Predation pressures from both invertebrate and vertebrate predators during early evolution of a colony may pose serious problems for the evolution of colonial species directly from planktonic forms. Also, it is interesting to note that there are very few planktonic solitary species in these two families (e.g., *Ptygura libera*). Perhaps predation has played an important role in the evolution of such forms as well.

How might have the sessile condition lead to coloniality? There must be strong selective pressure for attachment to a substrate as sessile rotifers no doubt enjoy considerable energy savings over planktonic forms [Fairchild, 1981: 1351]. The evolution of a true sessile condition probably involved several important events including permanent attachment, development of the larval stage, development of a sensory apparatus which permitted recognition of particular substrates [Wallace, 1980], and modification of the corona. Once a species becomes sessile, two paths are open to evolution of coloniality. (1) Adoption of a specific substrate selection cue which recruits conspecific larvae into a colony (colony Type I). (2) A change in life history so that young either remain with the adult, forming a colony by accumulation or form their own colonies as cohorts (colony Type II). The sessile condition was lost in some species, but note that some planktonic species have apparently retained the ability to temporarily attach to a surface (e.g., *S. semibullata*) [Canella, 1952; Gunter & Knight, 1978].

Conclusions

The principal conclusion of this review is that coloniality as an adaptive strategy in the phylum Rotifera presently may be explained both in terms of energetic efficiency and predator avoidance. Coloniality appears to increase the amount of usable energy per individual probably through increased feeding efficiency. Along with tube production, coloniality provides protection from certain invertebrate and perhaps some vertebrate predators.

However, much research remains to be done on all aspects of the biology of colonial rotifers. More work on the autecology of colonial species such as those efforts produced by Bhardwaj & Dattagupta [1984], Canella [1952], Edmondson [1945], Kolisko [1939], Tiefenbacher [1972] and Vidrine *et al.* [1985] are required. I suggest that future research pay attention to the following points. (1) Examine the development of colonies from resting eggs and determine how daughter colonies arise after the founding colony is established. It appears that colony formation Type II may be further differentiable. (2) The question of genotypic relatedness as a function of colony formation type should be inspected and its effect on sexual reproduction explored [Bhardwaj & Dattagupta, 1984]. Further, the distribution and significance of amphoteric female production in colonial rotifers requires additional research [Bhardway & Dattagupta, 1984; Bogoslovsky, 1958; Champ & Pourriot, 1977b]. (3) Additional experimentation testing the Elevated CR and the Increased Feeding Efficiency hypotheses using a wider variety of species and experimental techniques is required. (4) A detailed study of the hydrodynamics of colonial feeding currents merits investigation using high speed films or video tapes and Schlieren optics [Strickler, 1977]. (5) Additional information on predator-prey relationships should be obtained. (6) The significance of temporary colony formation in *P. megalotrocha* and other species needs to be investigated. Are these groupings fortuitous aggregations (just a matter of relief from swimming) or do they represent a type of colony formation? What is the adaptive significance of this grouping?

Answers to these and similar questions should help provide an evolutionary synthesis of colonial and sessile rotifers.

Acknowledgements

I wish to thank W. T. Edmondson, J. J. Gilbert, and P. L. Starkweather for their generous aid on aspects of the original research discussed here. I acknowledge, with many thanks, the discussions I had with J. J. Gilbert, R. Stemberger and C. Williamson. Thanks are also due to L. B. Wallace and D. Zei who greatly aided this work by providing several translations, W. Brooks and J. Quilter who read and improved the manuscript, and L. Toussaint who help type it. Support for research and publication was provided by several sources, including the US National Science Foundation (WTE and RLW – BMS 75-03175, DEB 75-03175 A01; JJG and PLS – DEB 78-02882; RLW – BSR 84-05157), the American Philosophical Society (RLW), and Ripon College (RLW).

References

Alcaraz, M., G.-A. Paffenhöfer & J. R. Strickler, 1980. Catching the algae: a first account of visual observations on filter-feeding calanoids. In W. C. Kerfoot (ed.), Evolution and Ecology of Zooplankton Communities. The University Press of New England, Hanover (N.H.); Lond.: 241–248.

Anderson, H. H. & J. Shephard, 1892. Notes on Victorian Rotifers. Proc. r. Soc. Victoria, Melbourne n. ser. 4: 69–80.

Banta, W. C., F. K. McKinney & R. L. Zimmer, 1974. Bryozoan monticules: excurrent water outlets? Science 185: 783–784.

Beavers, R. A. & R. H. Stavin, 1975. Seasonal changes in diaptomid copepod populations as a function of vertebrate predation. Verh. int. Ver. Limnol. 19: 2966–2975.

Bhardwaj, S. C., 1985. Studies on the morphology of five species of colonial rotifers. Uttar Pradesh J. Zool. 5: 6–13.

Bhardway, S. C. & A. K. Dattagupta, 1984. Life cycle and colony formation in a few species of rotifers of the order Flosculariacea. In A. K. Dattagupta & R. P. Maleyvar (eds.), Environment and Biotic-interaction. Proc. IV All India Symp. Envir. Biol.: 74–80.

Bishop, J. W. & L. M. Bahr, 1973. Effects of colony size on feeding by Lophopodella carteri (Hyatt). In R. S. Broadman, A. H. Cheetham & W. A. Oliver (eds.), Dowden, Huctinson, and Ross, Inc. Strondsburg, PA. 603 pp.

Bogolovsky, A. S., 1958. New data on the reproduction of heterogenous rotifers. Observations on the reproduction of Sinantherina socialis (Lin.). Zooh. Zh. 37: 1616–1623 (in Russian).

Brown, J. L. & G. H. Orians, 1970. Spacing patterns in mobile animals. Ann. Rev. Ecol. Syst. 1: 239–262.

Burckhardt, G., 1944. Hydrobiologische Studien an Schweizer Alpenseen. Zeit. Hydrol. 10: 354–384.

Burwitz, P., 1977. Eine sonderbare 'Koloniebildung' des Rädertieres Philodina megalotrocha. Mikrokosmos 66: 293–294.

Buskirk, R. E., 1975. Coloniality, activity patterns and feeding in a tropical orb-weaving spider. Ecology 56: 1314–1328.

Buss, L. W., 1981. Group living, competition, and the evolution of cooperation in a sessile invertebrate. Science 213: 1012–1014.

Calow, P., 1981. Invertebrate Biology. Halsted Press, N.Y., 183 pp.

Canella, M. F., 1952. Osservazioni su Sinantherina semibullata (Thorpe) e su altri Flosculariacea. Ann. d'Univ. Ferrera (N.S.), Sez III, 1: 171–257.

Champ, P., 1978. Dynamique d'une population d'un Rotifère épiphyte thermophile (Sinantherina socialis) en présence de pollution thermique. Arch. Hydrobiol. 83: 213–231.

Champ, P. & R. Pourriot, 1977a. Reproductive cycle in Sinantherina socialis. Arch. Hydrobiol. Beih. 8: 184–186.

Champ, P. & R. Pourriot, 1977b. Particularités biologiques et écologiques du Rotifère Sinantherina socialis (Linne). Hydrobiologia 55: 55–64.

Davies, N. B. & J. R. Krebs, 1978. Introduction: ecology, natural selection and social behaviour. In J. R. Krebs & N. B. Davies (eds.), Behavioural ecology an evolutionary approach. Blackwell Scientific Publications, Oxford: 1–18.

Dodson, S. I., 1974. Adaptive change in plankton morphology in response to size-selective predation: a new hypothesis of cyclomorphosis. Limnol. Oceanogr. 19: 721–729.

Edmondson, W. T., 1940. The sessile Rotatoria of Wisconsin. Trans. am. Microsc. Soc. 59: 433–459.

Edmondson, W. T., 1945. Ecological studies of sessile Rotatoria, 2. Dynamics of populations and social structure. Ecol. Monogr. 15: 141–172.

Edmondson, W. T., 1949. A formula key to the Rotatorian genus Ptygura. Trans. am. Microsc. Soc. 68: 127–135.

Edmondson, W. T., 1959. Rotifera. In W. T. Edmondson (ed.), Fresh-water Biology, 2nd edition, John Wiley & Sons, Inc., N.Y.: 420–494.

Edmondson, W. T. & A. H. Litt, 1986. Conochilus in Lake Washington. Hydrobiologia (this volume).

Epp, R. W. & W. M. Lewis, Jr., 1979. Sexual dimorphism in Brachionus plicatilis (Rotifera): evolutionary and adaptive significance. Evolution 33: 919–928.

Fairchild, G. W., 1981. Movement and microdistribution of Sida crystallina and other littoral microcrustaceans. Ecology 62: 1341–1352.

Gilbert, J. J., 1963. Mictic female production in the rotifer Brachionus calyciflorus. J. exp. Zool. 153: 113–123.

Gilbert, J. J., 1980. Observations on the susceptibility of some protists and rotifers to predation by Asplanchna girodi. Hydrobiologia 73: 87–91.

Gunter, D. D. & L. A. Knight, Jr., 1978. Observations of the rotifer Sinantherina semibullata (Thorpe) from Ross Barnett reservoir, Mississippi. Egypt. J. Microbiol. 13: 99–106.

Hudson, C. T. & P. H. Gosse, 1886. The Rotifera; or wheel-animalcules. Vol I, Longmans, Green and Co., London, 128 pp.

Hudson, C. T. & P. H. Gosse, 1889. The Rotifera; or wheel-animalcules, both British and foreign. Supplement, Longmans, Green and Co., London, 64 pp.

Huxley, T. H., 1853. Lacinularia socialis. A contribution to the anatomy and physiology. Trans. Micr. Soc., London, n. ser. 1: 1–19.

Kerfoot, W. C., 1977. Implications of copepod predation. Limnol. Oceanogr. 22: 316–325.

Kerfoot, W. C., 1978. Combat between predatory copepods and their prey: Cyclops, Epischura, and Bosmina. Limnol. Oceanogr. 23: 1089–1102.

Kerfoot, W. C., D. L. Kellogg, Jr. & J. R. Strickler, 1980. Visual observation of live zooplankters: evasion, escape, and chemical defenses. In W. C. Kerfoot (ed.), Evolution and Ecology of Zooplankton Communities. The University Press of New England, Hanover (N.H.); Lond.: 10–27.

Kolisko, A., 1939. Über Conochilus unicornis und seine Koloniebildung. Int. Revue ges. Hydrobiol. Hydrogr. 39: 78–98.

Koste, W., 1970. Über die Sessilen Rotatorien einer Moorblänke in Nordwestdeutschland. Arch. Hydrobiol. 68: 96–125.

Koste, W., 1978. Rotatoria. Die Rädertiere Mitteleuropas. Gebrüder Borntraeger, Berlin-Stuttgart, 2 vols., 673 pp., 234 plates.

Kutikova, L. A., 1983. Parallelism in the evolution of rotifers. Hydrobiologia 104: 3–7.

Mackie, G. O., 1963. Siphonophores, bud colonies and superorganisms. In E. C. Dougherty (ed.), The Lower Metazoa Comparative Biology and Phylogeny, Univ. Calif. Press, Berkeley: 329–337.

Michael, R. G., 1966. A new rotifer, Conochilus madurai sp. nov. from an astatic pool in Madurai, South India. Zool. Anz. 177: 439–441.

Pourriot, R., G. Rouyer & M. Peltier, 1972. Prolifération de rotifères epiphytes et pollution thermique dans la Loire. Bull. fr. Piscic. 224: 111–118.

Rosen, B. R., 1979. Modules, members and communities: a postscript introduction to social organisms. In G. Larwood & B. R. Rosen (eds.), Biology and Systematics of colonial Organisms. Academic Press, N.Y.: xiii–xxxv.

Rousselet, C. F., 1892. On Conochilus unicornis and Euchlanis parva – two new rotifers. J. Quekett Micr. Club, ser. 2, 4: 367–370.

Rousselet, C. F., 1900. Note on the genus Lacinularia. J. Quekett Micr. Club, ser. 2, 7: 313–314.

Ruttner-Kolisko, A., 1974. Planktonic Rotifers; Biology and Taxonomy. Binnengewässer Suppl. 26, 146 pp.

Shephard, J., 1896. A new rotifer – Lacinularia elongata. Vict. Nat. 13: 22–24.

Shephard, J., 1897. A new rotifer – Lacinularia elliptica. Vict. Nat. 14: 84–86.

Shephard, J., 1899. A new rotifer – Lacinularia striolata, with note on L. pedunculata. Proc. Roy. Soc. Vist. 12: 20–35.

Sokal, R. R. & F. J. Rohlf, 1981. Biometry. 2nd edition, W. H. Freeman and Company, N.Y., 859 pp.

Stemberger, R. S. & M. S. Evans, 1984. Rotifer seasonal succession and copepod predation in Lake Michigan. J. Great Lakes Res. 10: 417–428.

Strickler, J. R., 1977. Observation of swimming performances of planktonic copepods. Limnol. Oceanogr. 22: 165–170.

Surface, F. M., 1906. The formation of new colonies of the rotifer, Megalotrocha alboflavicans Ehr. Biol. Bull. 11: 182–192.

Thorpe, V. G., 1889. Description of new species of Megalotrocha. J. r. microsc. Soc. 1889: 613–616.
Thorpe, V. G., 1893. The Rotifera of China. J. r. microsc. Soc. 1893: 145–152.
Tiefenbacher, L., 1972. Beiträge zur Biologie und Ökologie sessiler Rotatorien unter besonderer Berücksichtigung des Gehäusebaues und der Regeneration-fähigkeit. Arch. Hydrobiol. 71: 31–78.
Vasisht, H. S. & B. L. Dawar, 1970. The Anatomy and Histology of the rotifer Lacinularia flosculosa Müller. Res. Bull. Panjab Univ. 21: 361–377.
Vidrine, M. F., R. E. McLaughlin & O. R. Willis, 1985. Free-swimming colonial rotifers (Monogononta: Flosculariacea: Flosculariidae) in southwestern Louisianna rice fields. Freshwat. Invertebr. Biol. 4: 187–193.
Wallace, R. L., 1977a. Distribution of sessile rotifers in an acid bog pond. Arch. Hydrobiol. 79: 478–505.
Wallace, R. L., 1977b. Adaptive advantages of substrate selection by sessile rotifers. Arch. Hydrobiol. Beih. 8: 53–55.
Wallace, R. L., 1978. Substrate selection by larvae of the sessile rotifer Ptygura beauchampi. Ecology 59: 221–227.
Wallace, R. L., 1980. Ecology of sessile rotifers. Hydrobiologia 73: 181–193.
Wallace, R. L. & W. T. Edmondson, 1986. Mechanism and adaptive significance of substrate selection by a sessile rotifer. Ecology 67: 314–323.
Wallace, R. L. & P. L. Starkweather, 1983. Clearance rates of sessile rotifers: In situ determinations. Hydrobiologia 104: 379–383.
Wallace, R. L. & P. L. Starkweather, 1985. Clearance rates of sessile rotifers: In vitro determinations. Hydrobiologia 121: 139–144.
Wesenberg-Lund, C., 1930. Contributions to the biology of the Rotifera. II: Periodicity and sexual periods. K. danske Vidensk. Selsk. Skr. 9: 1–230.
Western, G., 1891. Notes on rotifers: a free-swimming variety of Lacinularia and a new rotifer found at Guildford. J. Quekett Micro. Cl. ser. 2, 4: 254–258.
Williamson, C. E., 1983. Invertebrate predation on planktonic rotifers. Hydrobiologia 104: 385–396.
Wulfert, K., 1939. Beiträge zur kenntnis der Rädertierfauna Deutschlands. Teil IV. Die Rädertiere der Saale-Elster-Niederung bei Merseburg in ökologisch-faunistischer Beziehung. Arch. Hydrobiol. 35: 563–624.

Hydrobiologia 147: 157–162 (1987)
May, L., Wallace, R. & Herzig, A. (eds), Rotifer Symposium IV

Conochilus in Lake Washington

W. T. Edmondson & Arni H. Litt
Department of Zoology, NJ-15 University of Washington Seattle, Washington 98195 U.S.A.

Keywords: zooplankton, Rotifera, population dynamics, predation, community structure, *Conochilus*

Abstract

The rotifer *Conochilus unicornis* appeared in Lake Washington sporadically and usually in small numbers during a total of 28 years of observation since 1933. *Conochilus hippocrepis* was present even less frequently until the 3 year period 1977–1979, when it became extraordinarily abundant. The abundances of food organisms and known predators have been examined.

The only consistent correlation found was between the seasonal maximum of *Leptodora* and the decrease of *C. hippocrepis* from its peak abundance each year. *Leptodora* was generally less abundant during 1977–1979 than during most of the preceding years and the years following the *C. hippocrepis* episode. No such correlations were found with *Epischura, Cyclops* or edible algae.

In July 1985, there was a small resurgence of *C. hippocrepis* preceded by a small population of *C. unicornis. Leptodora* was less abundant than in the immediately preceding years. Observations were made on stomach contents of both animals and colony size of *C. hippocrepis.* Recommendations are made for programs of new field studies and laboratory experiments to establish the actual requirements for successful populations of *Conochilus.*

Introduction

Lake Washington has been sampled repeatedly during 28 of the 52 years since 1933. During this period, the rotifer *Conochilus unicornis* has occurred sporadically and generally in small numbers; only in 1957 was it conspicuous, approaching 1000 individuals 1^{-1}. Most occurrences since then have been brief and small, although in successive years 1970–1973 peak values were of the order of 100 or 200 1^{-1}. *Conochilus hippocrepis* has been even more sporadic and less conspicuous except for 1950 and the 3 year period 1977–1979 when it dominated the zooplankton because of the bulk of its gelatinous matrix [Edmondson & Litt, 1982: Fig. 2]. (N.B. that figure erroneously identifies the 1950 peak as *C. unicornis.*) The peak population in 1979 was about 1000 individuals 1^{-1}, approximately 13 colonies 1^{-1}. In 1985 another small resurgence of *C. hippocrepis* was preceded by a small peak of *C. unicornis.*

The purpose of the present paper is to describe the circumstances surrounding the 1977–1979 population maxima in an attempt to identify the conditions responsible for the success of *C. hippocrepis.* Recently we have observed living animals in an attempt to further our understanding of the requirements of the species.

On the basis of an analysis of the data, we propose some additional studies that should provide greater understanding of the mechanism of control of *Conochilus* populations. The studies would involve both additional kinds of observational data and experimental laboratory work. The aim is to identify and measure the processes that generate the observed correlations. This kind of approach has led to an identification of the probable complex of reasons for an unexpected outburst of *Daphnia* in 1976 [Edmondson & Litt, 1982; Infante & Abella, 1985].

Materials and methods

The preserved samples, taken mostly at weekly intervals, are the same as those used by Edmondson & Litt [1982]. The rate of population growth r was calculated by the conventional method of the difference between natural logarithms of two successive counts divided by the time interval. We used values of population density smoothed by a running mean of three.

We made a study of the seasonal course of population changes of *C. hippocrepis* and its potential predators and food during the five year period 1976–1980 looking for coordination of events. We also examined the data for correlations in general abundance of these organisms over a longer time (1950–1985).

We examined the three known invertebrate predators in Lake Washington that might be expected to affect *C. hippocrepis. Epischura nevadensis* is known to eat *C. hippocrepis* [Mesner, 1984]. As many as 12 trophi were found in a single gut during 1978. *Leptodora kindti* is known to be able to catch organisms larger than a *C. hippocrepis* colony (1 mm). *Cyclops* was examined because of the general predatory habits of the genus, but we did not expect the small *C. bicuspidatus thomasi* to have as much effect as the larger *C. vernalis* which, however, is very scarce in Lake Washington. We categorized the phytoplankton of the size range we expected *C. hippocrepis* to use into two size groups of noncolonial species without a gelatinous covering.

One assumption of this approach is that predatory activity is proportional to the abundance of predators, causing the death rate of the prey to vary accordingly. A second assumption is that the feeding rate of the prey is related to the abundance of food organisms. This has been shown to be true for a wide variety of zooplankton species over a wide range of food abundance, but can be modified by temperature and photoperiod.

Results and discussion

Rather unexpectedly, changes of *C. hippocrepis* are not correlated with the abundance of *Epischura*, either total copepodid stages or adults (Fig. 1). The rotifer consistently achieved major increases in population size during maximum or near maximum abundance of *Epischura*. Evidently the rate of capture was too small to deplete the rotifer population.

On the other hand, there is a fairly consistent relation between *C. hippocrepis* population and *Leptodora* (Fig. 1). In each of the three years the decrease in *C. hippocrepis* came during the peak of *Leptodora*. In 1978 *C. hippocrepis* increased irregularly in the presence of a population of *Leptodora* only about half as big as the population present during the subsequent decrease of the rotifer.

All population maxima of *C. hippocrepis* came during periods when *Cyclops* was scarce, and peak *Cyclops* populations were well separated temporally from *C. hippocrepis*. Thus, there is no way to ascertain from this material what effect a dense population of *Cyclops bicuspidatus* might have on *C. hippocrepis*.

The long term correlations are generally consistent with those just described. The three years of abundant *C. hippocrepis* were years of relatively scarce *Leptodora*. This followed a period of unusually abundant *Leptodora* and near absence of *C. hippocrepis. Leptodora* had not been detected in 1957 and was scarce in 1950, both years in which large populations of one of the species of *Conochilus* developed. Other striking long term correlations are shown in Fig. 2 of Edmondson & Litt [1982]. The years of abundant *C. hippocrepis* coincide with three years of more abundant *Epischura* and three years of scarcity of *Bosmina, Keratella cochlearis, Kellicottia longispina* and *Diaphanosoma. Daphnia* had become abundant the previous year, 1976, and made large populations in all years after that. During the period 1980–1984, the species that had been scarce during 1977–1979 remained relatively scarce with a few outstanding exceptions. *Kellicottia longispina* produced the largest population on record in 1981 and *Keratella cochlearis* produced populations in 1981 and 1982

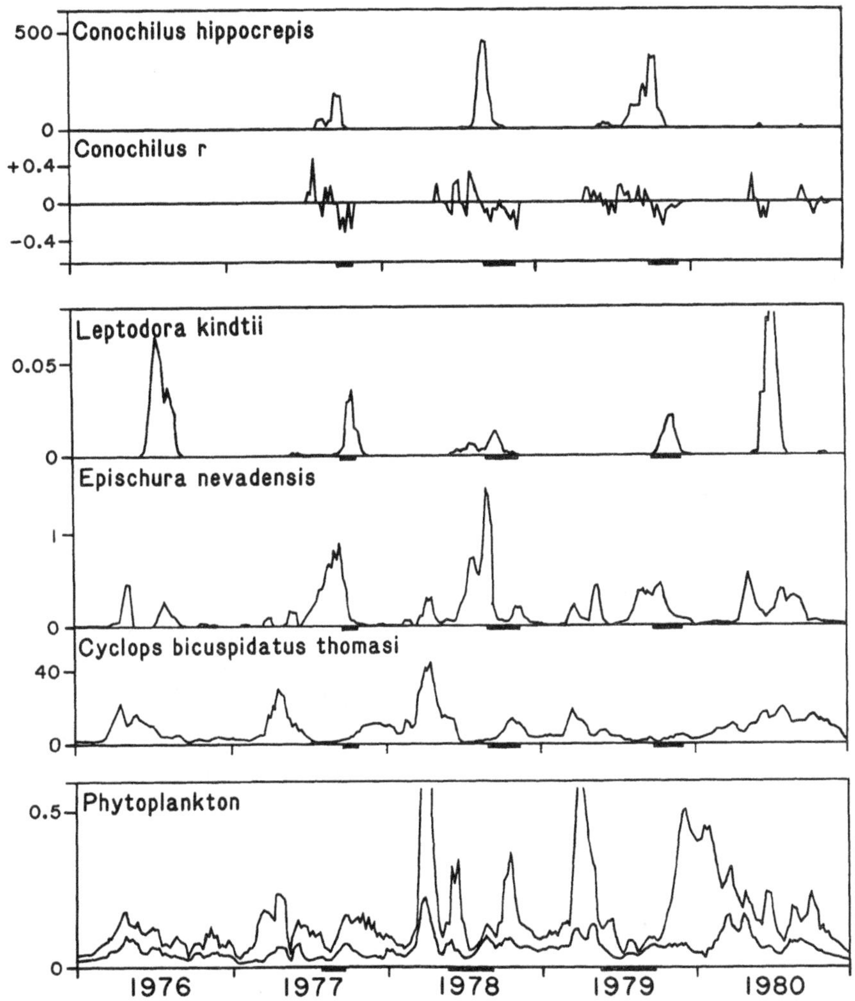

Fig. 1. Upper panel: Population density of *Conochilus hippocrepis* (organisms 1^{-1}) and population growth rate *r* (day^{-1}). Solid bars on the baseline show periods of decreasing abundance of *Conochilus.* Middle panel: Population density of three species of predatory crustaceans (organisms 1^{-1}). Total of all copepodid stages is shown for *Cyclops*, only adults (Stage VI) for *Epischura.* Total of all sizes is shown for *Leptodora.* Bars on the baseline are the same as in the upper panel. Bottom panel: Population abundance of potential food organisms for *Conochilus* expressed as total cell volume ($mm^3 \times 1^{-1}$). Top line: Organisms 18 m or less in longest dimension. Bottom line: Organisms 10 m or less. The species of algae selected for inclusion are solitary or form groups within the size indicated and do not have sheaths. Solid bars on the base line indicate periods of increasing abundance of *Conochilus.* All population curves have been smoothed by a running mean of three.

larger than any before except in the record year 1969. During this time, *Leptodora* was not notably abundant. At present we are unable to explain what prevented *Conochilus* from becoming abundant.

These rotifers and cladocerans can be regarded as competitors since to some extent they all can use some of the same food organisms. If food supply were involved in the coordination of the species it is hard to understand why *C. hippocrepis* did so well in the years when all the other rotifers were scarce, unless it has unusual dietary requirements. Further, the years of scarcity of the zooplankton mentioned, are not years of notably different populations of microalgae. *Daphnia* may have some influence since it is known to be able to interfere with *Keratella cochlearis* reproduction [Gilbert & Stemberger, 1985], but it did not prevent the large populations of 1981 and 1982.

Conclusions and recommendations

The results of the study so far are consistent with the idea that *Leptodora* is an effective predator on *C. hippocrepis*, although it is surprising that less than ten individuals in a cubic meter of water could prevent the growth of *C. hippocrepis* while much larger populations of *Epischura nevadensis* cannot. To understand fully the reasons for the changes observed in Lake Washington we require more information. Much of this information can be obtained only by experiment under controlled laboratory conditions or observations on freshly collected material. Since the individuals usually contract and the colonies fall apart in samples preserved with conventional methods, some important information is available only in fresh material. Following are some comments and recommendations about these needs.

1. We need to know more about the abilities of *Leptodora* and other plankton predators to find, catch, and consume *Conochilus*. *Leptodora* has a reputation for being primarily a 'fluid feeder', sucking out the contents of prey [Cummins *et al.*, 1969; Karabin, 1974] or reducing prey to small shreds [Weismann, 1874]. Nevertheless, we have seen the disassembled but unshredded components of two entire copepods in a single stomach. Parts of *Daphnia*, entire *Polyarthra*, and clumps of *Microcystis* have been seen. One stomach contained 48 *C. hippocrepis* trophi, each accompanied by two bright red eyes. When we pushed a *C. hippocrepis* colony into the mouth area of a *Leptodora*, it was grasped and held tightly for a while, then let go; we were unable to witness ingestion. However, the movements of the *Leptodora* seemed to be disturbed by confinement in a small, shallow dish. In small vessels *Leptodora* appears slow and clumsy, but it feeds much more effectively in vessels tens of centimeters across. Feeding rate experiments evidently must be done in relatively large vessels. Fortunately, a colony of *C. hippocrepis* is visible to the naked eye. *Leptodora* is difficult to see because it is so transparent, but can be observed with back lighting or made more visible when lightly stained with neutral red. This treatment makes it possible at least to observe the behavior of the animal in the presence of prey under minimal confinement, and possibly to determine the rate of capture and ingestion. Details of handling prey once caught might be better observed with tethered animals.

2. More information about *Conochilus* food is needed. The stomach content of *Conochilus* is usually dominated by very small, unrecognizable particles, possibly plastids and other organelles from broken soft-walled algae; however, one sometimes sees recognizable organisms. We have seen a pair of *Melosira distans* cells (5 $\mu m \times 13$ μm), an ovoid empty chlorphyte cell wall (13 $\mu m \times 8 \mu m$) and small green algae, 5 μm in diameter. An interesting feature of the *C. hippocrepis* stomach is an asymmetric median internal ridge that sometimes divides the food mass into two sections.

3. Large advances in zooplankton ecology have been based on the ability to calculate birth and death rates using the egg ratio method. By defining conditions existing during periods of high birth and death rates one can sometimes identify, with high probability, the cause of some event, or at least define the kind of information required to identify it. Unfortunately, *Conochilus* eggs are laid in such an advanced state of development that they may take less than an hour to hatch, so that one rarely sees an egg in the colony; the egg ratio method is inapplicable. Possibly one could develop the use of a count of embryos before laying that would serve the same purpose as a normal egg count. We were limited to evaluating the effect of predators on the basis of population decrease, not death rate.

4. Age distribution of a population is an informative property since it changes in response to changes in birth and death rates. When age cannot be determined, size distribution can sometimes be used. We observed a range of total length of individuals established in colonies from 290 to 840 μm. Newly hatched individuals were about 100 μm long. Sometimes individuals in a colony are of nearly uniform size. At other times there are several much smaller, presumably younger, animals.

Table 1. Size of colonies of *Conochilus hippocrepis* in Lake Washington.

Date	Range of colony size (inds./colony)	$\bar{X} \pm$ S.D.	Number counted	Population density (ind. $\times l^{-1}$)
14 May	15–161	65.5±35.0	45	0.8
21 May	12– 38	31.3±13.7	14	14.0
5 June	22–104	57.4±24.2	12	2.7
12 June	15– 44	32.0±12.1	6	1.4
18 June	12– 30	24.0±10.4	3	1.4
25 June	40– 90	64.4±18.2	8	10.9
2 July	28–134	83.5±33.9	24	16.2
19 July	29– 75	52.1±15.4	8	4.0
24 July	10– 91	41.4±21.9	26	0.8

5. When colonies of *Conochilus unicornis* contain more than about 10 individuals they usually split in two [Kolisko, 1939], although larger colonies are sometimes seen. Large colonies of *C. hippocrepis* split into two or three when compressed.

Colony size varies greatly, from 15 to 161 individuals in our material (Table 1). Sometimes one observes asymmetrical colonies, as if they had recently split. Possibly this condition could also result from attack by a predator. Observations on characteristics of colony and individual size during the growth and decline of a population might give insight into conditions affecting success.

6. Several kinds of symbiotic algae have been seen growing in the matrix of *Conochilus* colonies [De Graaf, 1953]. During the 1985 occurrence of *C. hippocrepis* in Lake Washington we saw two kinds. One was *Phormidium*, the other an unidentified chlorophyte that occurred in such density as to make the infected colonies very conspicuous. The cells were about 5 μm in diameter. On one occasion they were producing biflagellated zoospores about 3 μm long. We have seen vegetative cells, but not the spores in the stomach of *C. hippocrepis*. It would be useful to find out more about the nature and significance of the algae to the rotifers and the conditions that lead to massive development of algae.

7. There are still some taxonomic problems within the genus *Conochilus* [Pejler, 1956, 1957]. Most of our specimens could be rather securely assigned to one of the two species. Sometimes we had trouble because the character of the antennae was not clearly definable in some of the preserved specimens. Sometimes a large colony with the characteristic appearance of *C. hippocrepis* was found to consist of animals with a single antenna. With increasing frequency cases are being discovered in which morphological variations of zooplankton are induced by substances liberated by predators. In most of these cases the result is a structure that protects against predation. The variations in antennal structure in *Conochilus* hardly seem capable of anti-predator protection, but a tendency to form larger colonies might well have such an effect [Wallace, 1986]. In any case, it seems reasonable to review all cases of seemingly random morphological variation to see if they are consistently associated with some other species or condition.

Most of the observations mentioned above must be made on freshly collected, living material, but we have not taken the next step, an experimental study of feeding by *Leptodora*. Such a program can best be carried out in a lake that produces *Conochilus* more reliably than does Lake Washington.

Acknowledgements

The work has been supported by a long series of grants from the National Science Foundation. Sally E. B. Abella made a special compilation of phytoplankton data for Fig. 1. Mardi Varela made the counts of colony size.

References

Cummins, K. W., R. R. Costa, R. E. Rowe, G. A. Moshiri, R. M. Scanlon & R. K. Zajdel, 1969. Ecological energetics of a natural population of the predaceous zooplankter Leptodora kindtii Focke (Cladocera). Oikos 20: 189–223.

De Graaf, F., 1953. Commensal organisms in the colonies of Conochilus unicornis Rousselet with a note on Phormidium mucicola Huber et Naumann. Hydrobiologia 5: 390–399.

Edmondson, W. T. & A. H. Litt, 1982. Daphnia in Lake Washington. Limnol. Oceanogr. 27: 272–293.

Gilbert, J. J. & R. S. Stemberger, 1985. Control of Keratella populations by interference competition from Daphnia. Limnol. Oceanogr. 30: 180–188.

Karabin, An., 1974. Studies on the predator role of the cladoceran, Leptodora kindtii (Focke), in secondary production of two lakes with different trophy. Ekol. pol. 22: 295–310.

Kolisko, A., 1939. Über Conochilus unicornis und seine Koloniebildung. Int. Revue ges. Hydrobiol. Hydrogr. 39: 78–98.

Infante, A. & S. E. B. Abella, 1985. Inhibition of Daphnia by Oscillatoria in Lake Washington. Limnol. Oceanogr. 30: 1046–1052.

Mesner, N. O., 1984. The feeding ecology of Epischura nevadensis in Lake Washington. M.S. Thesis, University of Washington.

Pejler, B., 1956. Introgression in planktonic Rotatoria with some points of view on its causes & conceivable results. Evolution 10: 246–261.

Pejler, B., 1957. Taxonomical and ecological studies on planktonic Rotatoria from Northern Swedish Lappland. Kungl. Svenska Vetenskapsakademiens Handlingar. Fj. Ser. 6: 1–68.

Wallace, R. L., 1986. Coloniality in the phylum Rotifera. Hydrobiologia (in this volume).

Weismann, A., 1874. Über Bau and Lebenserscheinungen von Leptodora hyalina. Lilljeborg. Z. wiss. Zool. 24: 349–418.

Hydrobiologia 147: 163–180 (1987)
May, L., Wallace, R. & Herzig, A. (eds), Rotifer Symposium IV

The analysis of planktonic rotifer populations: A plea for long-term investigations

Alois Herzig
Biologische Station Neusiedlersee, A-7142 Illwitz, Austria

Keywords: Rotifer plankton, qualitative and quantitative aspects, short-term and long-term development, seasonality, physical – chemical limitations, competition, predation, parasitism

Abstract

Short-term species succession, and long-term community development, of planktonic rotifer populations of the temperate zone and factors influencing species' abundance (ie., physical and chemical limitations, food and exploitative competition, mechanical interference competition, predation, parasitism) are described and discussed. The long-term development of plankton communities in three European lakes is described and the major events are discussed in relation to ecological interactions. Frequently, the shortcomings of our knowledge about population ecology and ecophysiology of rotifers prevent explanations of short-term and, especially, of long-term developments of these plankton organisms. The need for qualitative and quantitative observations in the field and in the laboratory over long periods becomes obvious.

Introduction

There are three main components of freshwater zooplankton: Protozoa, Rotifera and Crustacea. Of these, usually only the Crustacea are analysed quantitatively because of their greater biomass. In terms of biomass, rotifers certainly cannot compete with the larger Crustacea, but their rapid reproduction and fast development rate means that rotifers can account for between 10% and 44% of the total zooplankton production.

Rotifers play an important role as grazers, suspension feeders and predators within the zooplankton community. Consequently, they are often important competitors of the larger Crustacea. They also serve as an essential food source for invertebrate and vertebrate predators. This paper examines the problem of rotifers 'living together' in a community. Short-term species succession and long-term community developments are described and discussed.

Short-term variation and seasonality

In water bodies of the temperate zone, seasonal changes in environmental factors result in an annual succession of rotifer species. Carlin [1943], in his study of planktonic rotifers from the Motala River (Sweden), was one of the first to demonstrate the influence of temperature and food on the occurrence of the various rotifer species. Carlin distinguished between perennial and seasonal species, and divided seasonal species into summer and winter forms. Frequently these summer and winter rotifers were associated with warm and cold temperature conditions.

Analysis of the rotifer communities of 16 water bodies from the temperate zone, with varying morphometric, trophic and climatic conditions [Carlin, 1943; Pejler, 1961, 1962; Nauwerck, 1963; Zimmermann, 1974; Schaber, 1975, 1976, 1977; de Bernardi & Soldavini, 1976; Elliott, 1977; Ruttner-Kolisko, 1977a; Larsson, 1978; Nauwerck, 1978; Herzig,

1979; Bangerter, *et al.* 1980; May, 1980a; Wappis, 1980; Eloranta, 1982; Einsle, 1983; Frank, *et al.* 1985], revealed the following seasonal classification of the dominant rotifer species:

Perennial species: *Keratella quadrata, Keratella cochlearis, Polyarthra vulgaris, Asplanchna priodonta, Conochilus unicornis;*

Winter species: *Keratella hiemalis, Notholca acuminata, Notholca caudata, Notholca squamula, Synchaeta lakowitziana, Polyarthra dolichoptera, Filina terminalis, Conochilus natans;*

Summer species: *Trichocerca capucina, Trichocerca porcellus, Trichocerca pusilla, Trichocerca rousseleti, Gastropus stylifer, Ascomorpha ecaudis, Ascomorpha saltans, Ascomorpha ovalis, Synchaeta tremula, Synchaeta oblonga, Synchaeta pectinata, Synchaeta grandis, Synchaeta stylata, Synchaeta kitina, Polyarthra major, Polyarthra remata, Ploesoma hudsoni, Pompholyx sulcata, Filinia longiseta.*

Perennial species can reach peak abundance at any time of year. They have low population levels until conditions are favourable and then quickly develop a dense population. *Keratella cochlearis* populations, for example, can reach their maximum abundance during spring or summer only, or in April/May/June and August/September/October, or several times during the year. To some extent these differences can be explained by the local climatic conditions which are dependent on the geographical location of the population, especially latitude [see Pejler, 1977; *K. cochlearis, K. longispina*].

Rotifers with a pronounced temperature preference tend to reach maximum population growth when temperature conditions are favourable; one or several peaks in abundance may occur within a relatively short period of time (e.g. *S. oblonga, S. grandis, P. dolichoptera, F. terminalis, F. longiseta, Brachionus angularis, Hexarthra fennica, K. valga, T. pusilla, G. stylifer*).

Within a single genus a succession of several species can occur within a year, a season, or even a month. In the genus *Synchaeta*, for example, the following succession of dominant species has been observed between April and September: *S. oblonga-S. pectinata – S. grandis – S. stylata – S. kitina + S. oblonga – S. pectinata – S. stylata – S. kitina – S. grandis* [Bangerter *et al.*, 1980; Berner-Fankhauser, 1986]. Similarly, within the genus *Polyarthra* the successional sequence *P. vulgaris – P. dolichoptera – P. vulgaris – P. major – P. remata– P. vulgaris* has been recorded [Carlin, 1943].

The seasonal succession of rotifer species at any given location often varies from year to year. Figures 9 and 10, for example, show the abundance of the dominant rotifer species from Neusiedlersee, Austria, for the years 1951 and 1968. In both years as many as 19 peaks of various populations developed. During the summer of 1951, the following species succession was observed: *Filinia (longiseta) – P. vulgaris – Anuraeopsis fissa – Hexarthra fennica – B. angularis – A. fissa – P. vulgaris.* In contrast, the summer 1968 was characterized by the following succession: *P. vulgaris – H. fennica – B. angularis – F. longiseta – P. vulgaris – F. longiseta – H. fennica – B. angularis – K. valga – F. longiseta – B. angularis – K. valga – F. longiseta – B. angularis – K. valga.*

Five factors seem to influence species' abundance and cause species succession:

(1) physical and chemical limitations, (2) food and exploitative competition, (3) mechanical interference competition, (4) predation and (5) parasitism.

1. Physical and chemical limitations

The main limiting factor for ectotherms is temperature and it is, therefore, not surprising that temperature is one of the main factors controlling the occurrence of rotifers in temperate waters. Carlin [1943] showed a correlation between the occurrence of various rotifer species and temperature. His data were more clearly evaluated by Edmondson [1946] with respect to the influence of temperature on the rate of population change. Nauwerck [1963] further describes the temperature tolerance and the optimum temperature for the growth of various rotifer populations of Lake Erken. More recently May [1983] examined the relationship between water temperature and the occurrence of certain rotifer species in Loch Leven, Scotland. These species showed a similar temperature preference when cultured in the laboratory.

Perennial species tend to be eurythermal. They can tolerate a wide range of temperatures within which they find optimum conditions for population growth. For example, *K. quadrata, K. cochlearis, K. longispina, A. priodonta, P. vulgaris* and *C. unicornis* tolerate temperatures from less than 1 °C to more than 20 °C and can reach peak abundance throughout this range. Some rotifers, however, need a very abundant food supply to maintain a succesful population at higher temperatures [e.g., *S. kitina* – May, 1983].

Winter-spring species usually reach maximum densities at temperatures below 10 °C [e.g., *K. hiemalis, S. lakowitziana, P. dolichoptera, N. squamula*). In general, the mechanism which restricts these species to cooler temperatures is not known. However, it has been shown that *N. squamula* cannot maintain a population at higher temperatures because they fail to produce viable eggs [May, 1983].

Species occurring in late spring – summer – early autumn need temperatures higher than 10 °C to maintain a population [e.g., *G. stylifer, P. sulcata, Trichocerca* spp., *P. major*). In the laboratory, *P. sulcata* was cultured successfully at 15 °C and 20 °C while at 10 °C the animals starved inspite of abundant food and rarely laid eggs. It was impossible to maintain a laboratory culture at this temperature [May, 1983]. Field observations suggest a similar situation for *T. pusilla* [May, 1983], *G. stylifer* [Herzig, unpublished results] and *P. major* [Carlin, 1943]. Most of these species reach their peaks in abundance at temperatures of 15 °C to 20 °C.

The genera *Synchaeta* and *Polyarthra* offer interesting examples of a succession of species within a genus due to different temperature preferences. The '*Polyarthra year*' (Fig. 1) normally begins with high numbers of *P. dolichoptera*. This is followed by a peak abundance of *P. vulgaris*, then *P. major*, *P. remata* during late summer/autumn and then the perennial form *P. vulgaris* during autumn and winter. The populations overlap to some degree and this overlap varies from year to year [Carlin, 1943]. In the genus *Synchaeta* a succession of *S. lakowitziana* – *S. oblonga* – *S. pectina* – *S. stylata* – *S. grandis* – *S. pectinata* – *S. tremula* can develop.

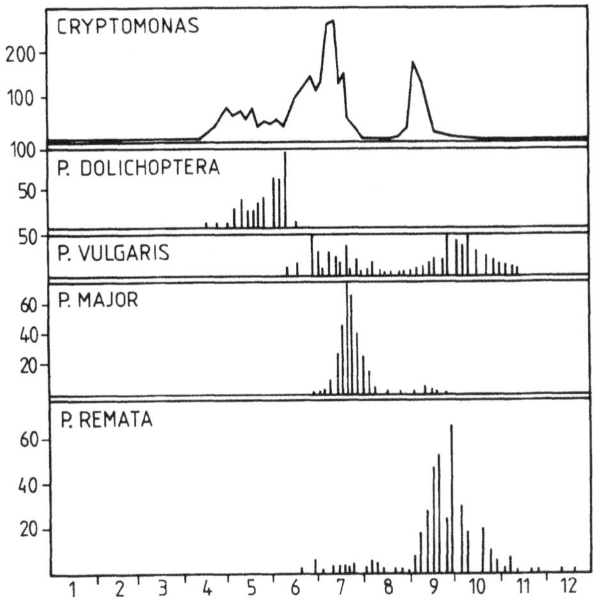

Fig. 1. Phenology of *Cryptomonas* and various species of *Polyarthra* in the year 1940 (data from Carlin 1943).

Occurrence at a distinct time of the year may well reflect adaptation to specific temperature conditions. This should be recognisable from the relationship between temperature and rate of development. There are clear differences between warm water species and cold water species [Herzig, 1983]; the former are more successful at high temperatures, the latter at low temperatures. However, differences between succeeding generations [seasonal temperature adaption – Edmondson 1960], as demonstrated for spring and summer generations of *Diaphanosoma brachyurum* [Herzig, 1984], cannot be ignored.

Obviously temperature is a key factor in restricting the occurrence of rotifers, but more laboratory work is needed to determine how this temperature limitation acts. The relationship between temperature and physiological parameters (enzyme activity, assimilation, respiration rate), and temperature and population parameters (fertility, rate of development, birth and death rates, male and resting egg production) should be studied in greater detail.

Salinity and dissolved oxygen concentration are examples of chemical limitation of rotifer communities, [Ruttner-Kolisko, 1971, 1977a]. Only the latter will be considered here. In many stratified water bodies, especially in eutrophic ones, hypolimnetic oxygen depletion may restrict rotifer occurrence. Generally, it is not oxygen requirement alone which limits the occurrence of rotifer species but the combined effect of temperature and oxygen concentration. Species can avoid high epilimnetic temperatures by migrating downwards to the cooler meta- or hypolimnion. But, here they may become trapped between the warm epilimnion and the deoxygenated hypolimnion.

Until now, only field observations recording the vertical distribution of rotifers and the variation in oxygen concentration with depth can contribute to this topic. Elliott [1977], for example, correlated occurrence and abundance of rotifers with temperature and oxygen concentration. Perennial eurythermal species like *K. cochlearis* and *K. longispina* tolerated low oxygen concentrations and could reach high densities under these conditions. Others, like *A. priodonta, C. unicornis,* or *Gastropus stylifer,* needed higher oxygen concentrations. *Filinia terminalis*, a cold stenothermal species, developed extremely well at very low oxygen concentrations. Ruttner-Kolisko [1977a] reports similar observations from Lunzer Obersee where *K. cochlearis* and *K. hiemalis* were found in abundance at very low oxygen concentrations ($0-2$ mg 0_2 l^{-1}). *Filinia terminalis* reached high numbers even at 0 to 0.5 mg 0_2 l^{-1}. Within this context, the role of food should not be ignored. In these layers of low or no oxygen, very large bacteria populations can develop that may form an important food source for the rotifers. Those species capable of tolerating low oxygen concentrations are able to take advantage of this additional food supply.

2. Food availability and exploitative competition

In simple terms, the role of food supply in the development of rotifer populations is indicated by comparing the time abundance graphs of edible algae with those of the rotifers. Frequently a correlation has been demonstrated between the succession of rofier species and that of nannoplankton (Cryptophyceae, Chrysophyceae) [Carlin, 1943; Pejler, 1962; Nauwerck, 1963; Zimmermann, 1974; Bangerter *et al.*, 1980]. For example, the growth phases of *Polyarthra* species (*dolichoptera, vulgaris, major, remata*) from River Motalaström, Sweden, closely followed the development of *Cryptomonas* sp. [Carlin, 1943] (Fig. 1). In Sempachersee, Switzerland, 85% of the phytoplankton during May and June consisted of nannoplankton. During that period ten different populations of rotifers reached an exponential phase of growth (*K. cochlearis, S. pectinata, P. dolichoptera* being dominant) [Zimmermann, 1974]. In Bielersee, Switzerland, two peaks of rotifer abundance occured, dominated by *Synchaeta* spp., both related to the development of Chrysophyceae and Cryptophyceae [Bangerter, *et al.*, 1980]. In contrast, the occurrence of other species (e.g., *F. terminalis* or *C. unicornis*) are independent of algal development and show a correlation to the abundance of detritus and bacteria [Pejler, 1962; Nauwerck, 1963].

Each suspension feeding rotifer in a community may have a different food preference and hence the impact on the ecosystem will be different according to its feeding habits. A qualitative review of feeding can be found in Pourriot [1977]. Comparative quantitative data are found in Bogdan *et al.* [1980], Starkweather [1980], Bogdan & Gilbert [1982] and Gilbert & Bogdan [1984]. These reviews clearly show that different species have different diets. In consequence a rich spectrum of edible particles of various sizes will favour the co-existence of numerous species and reduce competition. Exploitative competition will occur only if the resources are poor in assortment and/or quantity.

Gilbert & Bogdan [1984] give examples of resource partitioning even within genera. Cooccuring large and small *Keratella* (large morph: *K. crassa,* small morph: *K. c. typica*) ingested different sized particles with different efficiences, the smaller form feeding more efficiently on the smaller particles. Similarly, *P. vulgaris* and P. dolichoptera (small forms) had 5 to 8 times higher clearance rates when feeding on small *Chlamydomonas* than *P. euryptera* (a large form).

Gilbert & Bogdan [1984] also suggest that 'generalist' rotifers (e.g., *Keratella*) tend to occur over much longer time periods than 'specialist' rotifers, such as *Polyarthra* and *Synchaeta*. The 'generalist' seems to respond in its feeding habits to the available food source and this ability to adapt to changing conditions may be one of the reasons for their success. On the other hand, food specialization may reduce competition at least for the period of exponential growth. One example of food specialization is shown by *N. squamula* from Loch Leven which feeds almost exclusively on the colonial alga *Asterionella formosa* [May, 1980].

In addition, specialization for distinct temperature conditions, which is reflected in the relationship between grazing rate and temperature can lead to a succession of rotifer species. *Polyarthra dolichoptera* and *P. vulgaris* display such a difference in adaptation [Gilbert & Bogdan, 1984]. *Polyarthra dolichoptera* had as high a clearance rate at about 7 °C as *P. vulgaris* at 20 °C. Thus, *P. dolichoptera* becoming increasingly more common at lower temperatures is probably due to its more effective food uptake at lower temperatures.

Intraspecific competition for food is also an important factor in determining the structure of rotifer communities. Work on different clones of *Asplanchna* by Snell [1979] demonstrates that the superior competitor with respect to food (i.e., the one with the highest feeding rate) rapidly eliminates the other clone. Thus competition among genotypes could be an important force structuring rotifer populations. This topic needs further study.

The relationship between suspension feeding rotifers, cladocerans and copepods is also important as they compete for food resources of similar size. It is commonly thought that cladocerans will out-compete rotifers. This is not surprising since daphnids are more than 10 times larger (in length) than rotifers and their clearance rates can be 20 to 1300 times higher [rotifers: Bogdan & Gilbert, 1982; daphnids: Burns & Rigler, 1967; Jones *et al.* 1979]. Nevertheless, Bogdan & Gilbert [1982] and Gilbert & Bogdan [1984] showed that large *K. cochlearis* fed on *Chlamydomonas* reached clearance rates which were similar to those of *Bosmina longirostris*. Their results showed that *K. cochlearis* was responsible for 80% of the grazing activity during most of the year.

There is some evidence from field studies (see also Figs. 3, 6, 12) and from laboratory experiments supporting the idea that 'Cladocera will out-compete Rotifera' [Gilbert & Stemberger, 1985]:

From field studies

1) Rotifers often initially dominate zooplankton communities during colonization and are then largely displaced by cladocerans.
2) The removal of large cladocerans from zooplankton communities often permits smaller cladocerans and rotifers to dominate these communities.
3) Removal of *Daphnia* from in situ enclosures caused a marked increase in the fertility and density of rotifer populations [Neill, 1984].

From laboratory experiments

1) *Daphnia pulex* excludes the rotifer *Brachionus calyciflorus* from mixed-species cultures through exploitative competition.
2) *Daphnia galeata mendotae* and *Daphnia rosea* control *Keratella cochlearis* populations by mechanical interference competition.

3. Mechanical interference competition

Mechanical interference can also influence a rotifer population. Gilbert & Stemberger [1985] showed experimentally that the cladocerans *Daphnia galeata mendotae* and *D. rosea* inhibited the growth of *K. cochlearis* at natural population densities (i.e., 22–75 *Daphnia* l^{-1}, 220 – 1400 *Keratella* l^{-1}). *Keratella* were swept into the branchial chamber of these daphnids and were either killed instantly, mortally wounded, or lost their attached eggs. The degree of injury to the rotifer decreased with age and increased with retention time in the chamber. The susceptibility of a rotifer to mechanical interference depended upon the probability of being swept into the chamber and the degree to which the rotifer could be damaged. Nevertheless, one has to consider the probability of mechanical interference competition in future work.

4. Predation

Fish larvae probably feed on small food items, such as rotifers, for at least part of their life, but evidence from field and laboratory experiments is scarce (apart from information about the use of *Brachionus plicatilis* in aquaculture). The information available suggests that most planktivorous fish prefer microcrustaceans as food [e.g., *Daphnia*]. Consequently, it is unlikely that fish predation has any significant impact on planktonic rotifer populations. In contrast, invertebrate predators tend to concentrate on smaller prey and frequently include rotifers in their diet. This topic is reviewed by Williamson [1983]. Potential predators include some calanoid copepods, cyclopoid copepods, Mysidacea, Chaoborus larvae, and some rotifers.

Of the predatory rotifers, most data are available for the genus *Asplanchna* [reviewed by Gilbert, 1980]. *Asplanchna priodonta* is known to suppress the growth of *K. cochlearis* in the field [e.g., Zimmermann, 1974; Hofmann, 1983]. *Asplanchna girodi* easily ingested *K. cochlearis, S. pectinata* and single individuals of *Conochilus unicornis* in the laboratory, while *Kellicottia* sp. and colonies of *Conochilus* sp. were rarely captured. Experiments have also shown that *Polyarthra* sp. is an attractive prey for *Asplanchna*, but is protected from predation by its quick escape reaction.

Many observations on the food of cyclopoid copepods are derived from gut content analyses. Quantitative data are scarce and are mostly derived from field studies or laboratory enclosures [Brandl & Fernando, 1978; Gilbert & Williamson, 1978; Williamson, 1980; Williamson & Gilbert, 1980; Williamson, 1984; Zankai, 1984]. The available data suggests that *Polyarthra, Pompholyx, Brachionus, Asplanchna, Synchaeta, Filinia* and *Keratella quadrata* are a substantial component of the diet of cyclopoid copepods. Selection for smaller specimens of a given prey species has also been observed. Smaller *B. angularis* suffered a three times higher predation rate by *Cyclops vicinus* than the bigger specimens [Brandl & Fernando, 1978].

Rotifer communities are strongly influenced by predator – prey interactions which are sometimes very complex. Gilbert & Williamson [1978] showed experimentally that predator composition affects survival in *A. girodi, P. vulgaris* and *K. cochlearis*. The survival of *Polyarthra* and *Keratella* was significantly affected by the species of predator present. *Mesocyclops edax* fed on *Asplanchna* and *Polyarthra* but not on *Keratella*; *A. girodi* fed on *Keratella*, but failed to catch *Polyarthra*. When both predators were present *Asplanchna* was quickly eliminated, *Polyarthra* was heavily influenced by *M. edax*, but a high percentage of the *Keratella* population survived.

5. Parasitism

The effects of parasitism on rotifer populations have rarely been quantified. Nevertheless, field observations have shown that parasitism may lead to a collopse of rotifer populations [Eg., *C. unicornis* – Ruttner-Kolisko, 1977; *F. terminalis* – Miracle, 1977]. Clearly this subject needs further investigation.

Long-term community developments

The question now arises of whether the patterns of short-term variation are simply statistical events, partly reflecting sampling strategy, or the result of continuous interactions of populations with their environment. Ecological phenomena such as species succession, or the *concepts of 'normal' or 'typical'*, can be *understood or defined only on a long-term basis* [Likens, 1984]. In parallel the influence of the various factors on populations have to be studied in the laboratory. The long-term data then provide the opportunity to evaluate the results of experimental studies.

Three examples of long-term development of plankton communities of European lakes are described below: (1) a deep lake undergoing eutrophication: Lago di Mergozzo, Italy, (2) a small

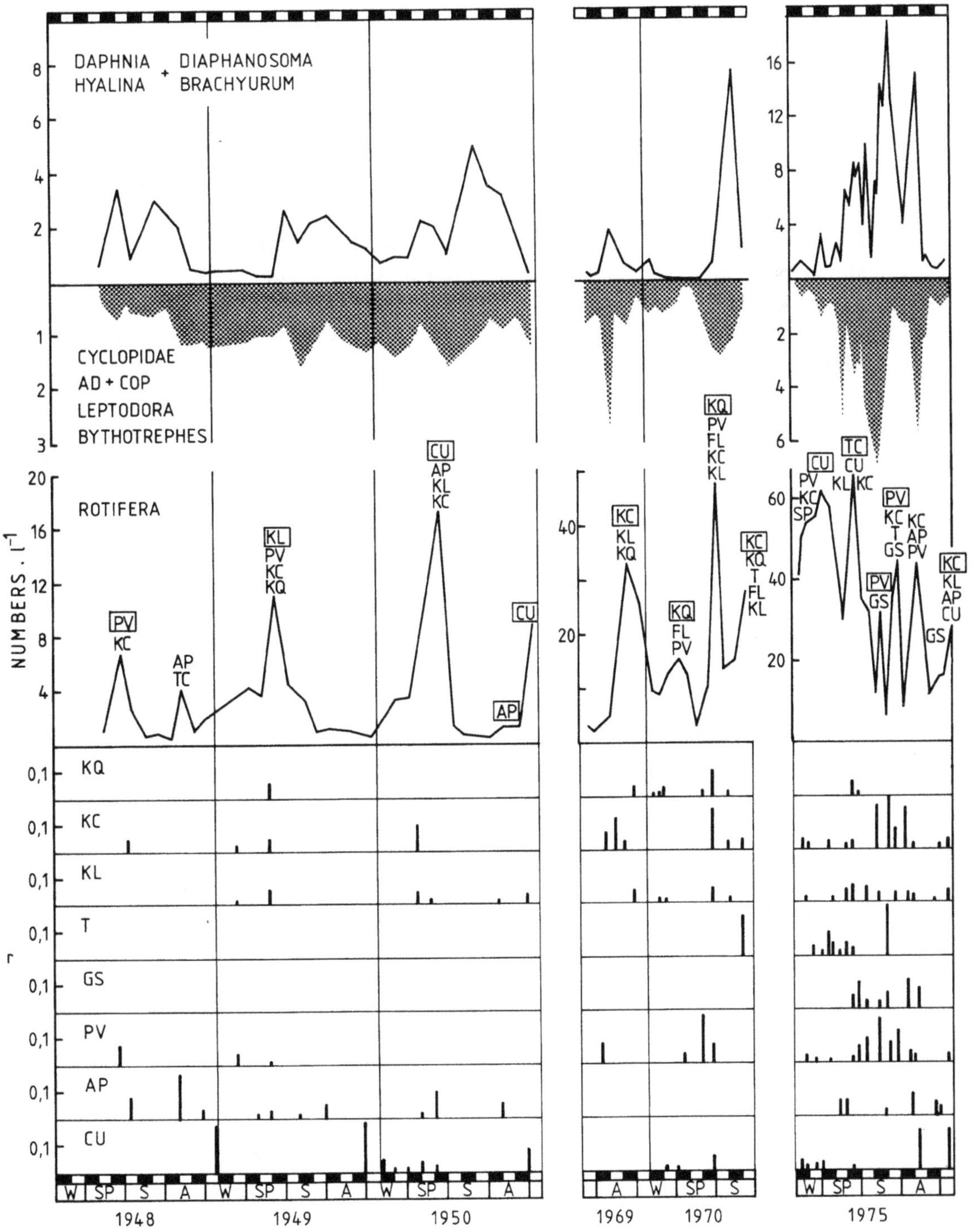

Fig. 2. Development of the plankton of Lago di Mergozzo. r = instantaneous rate of increase; PV – *Polyarthra vulgaris*; KC – *Keratella cochlearis*; AP – *Asplanchna priodonta*; TC – *Trichocerca capucina*; KL – *Kellicottia longispina*; KQ – *Keratella quadrata*; CU – *Conochilus unicornis*; FL – *Filinia longiseta*; T – *Trichocerca*; SP – *Synchaeta pectinata*; GS – *Gastropus stylifer*.

mesotrophic lake which was eutrophied in recent decades and is now in a phase of rehabilitation: Piburger See, Austria, and (3) a large shallow lake undergoing eutrophication: Neusiedlersee.

1. Lago di Mergozzo (45°57′N, 8°39′E; altitude = 194 m; A = 1.825 km^2; Z_{max} = 73 m; $\bar{Z}$ = 45.4 m). [de Bernardi & Soldavini, 1976; Saraceni, *et al.*, 1978].

Lake Mergozzo is a small, deep, subalpine lake in Northern Italy (close to L. Maggiore) which has steep slopes, a small watershed and a long theoretical water residence time (4.7 years). Data on the development of zooplankton (1948–1975) are summarized in Fig. 2. Rotifer numbers have increased gradually since 1948 as a result of changes in trophic conditions. The number of rotifer species changes throughout the time: 10 species – 1950; 22–1970; 14–1975. This is possibly another indication of eutrophication [see Pejler, 1983], but may be the result of very long sampling intervals (one month 1948–1950) within which populations growth and decline (e.g., of *G. stylifer, Trichocerca* spp., or *Synchaeta* spp.) could easily have taken place. Often rotifers show two peaks in abundance, late spring/summer and autumn, but with a higher sampling frequency the number of peaks increases [e.g., 1975].

Each year is characterized by a few dominant species but these species are different from year to year. In 1950, *C. unicornis* and *A. priodonta* were most abundant, during 1969/70 brachionid species (*K. cochlearis, K. quadrata, K. longispina*) were more common and 1975 is characterized by *C. unicornis, P. vulgaris* and *K. cochlearis.* Even three years in a sequence (1948–1950) show different rotifer dominance every year. 1948: *P. vulgaris (+ A. priodonta + K. cochlearis)*; 1949: *K. longispina (+ P. vulgaris + K. cochlearis*); 1950: *C. unicornis (+ A. priodonta).* In 1975, warm stenothermal rotifers occur in high numbers for the first time (*G. stylifer, T. capucina).* This 'lag-phase' in population development may well reflect the dependence of the size of a pelagic population on the inoculum size (= number of viable resting eggs) (Table 1).

Phytoplankton data are available only for 1975. The high winter/spring densities of rotifers do not appear to be correlated with algal development. The numerically dominant rotifer, *C. unicornis,* most probably was feeding on detritus at this time. From March until October/November peaks of rotifer populations coincide with the occurrence of Chlorophyceae and Chrysophyceae.

In Fig. 2, the lower panels show the periods of population growth for the numerically important species expressed as the instantaneous rate of population increase. As the populations will grow only when food is abundant and predators are not suppressing the population, 'r' is a good indicator of good growing conditions [Edmondson, 1965]. Perennial species like *K. cochlearis, K. longispina* and *P. vulgaris* show rates of increase over the whole year. These species are known to be eurythermal and, apart from *P. vulgaris,* are thought to be non-specialist feeders. *C. unicornis* grows mainly during autumn, winter, and spring. *Trichocerca* species prefer spring and summer, *G. stylifer* summer and autumn, *A. priodonta* spring, summer and autumn. *Keratella quadrata* has success during 1969/70 only. It has to be mentioned that 'r' was not calculated when the populations did not exceed an abundance of 100 animals m^{-3}.

Figures 3 and 4 show the relationship between rotifers and large suspension feeders (*Daphnia hyalina, Diaphanosomabrachyurum*) and potential predators, respectively. The rotifers react more quickly to the process of eutrophication than Cladocera and Cyclopoida. Rotifers dominate the zooplankton during winter and spring, whereas *D. hyalina* prevails in June and *D. brachyurum* in July/August/September. As soon as the daphnids occur, the rotifers decrease markedly which suggests competition. Until 1970, there was no obvious predation pressure on the rotifers (Fig. 4) but, in 1975 a pattern appears which resembles the classical predator – prey cycle. *Keratella cochlearis* and *G. stylifer* continuously show positive rates of population increase, but never attain high numbers. This is probably due to predation by cyclopoid copepods and partly by *Leptodora.*

2. *Piburger See* (47°12′N/10°53′E; altitude = 915 m; A = 0.134 km^2; Z_{max} = 24.7 m; $\bar{Z}$ = 13.0 m). [Schaber, 1975, 1976, 1977; Rott, 1981].

Piburger See is a mesotrophic lake, with a theoretical water renewal time varying between 1.9 and

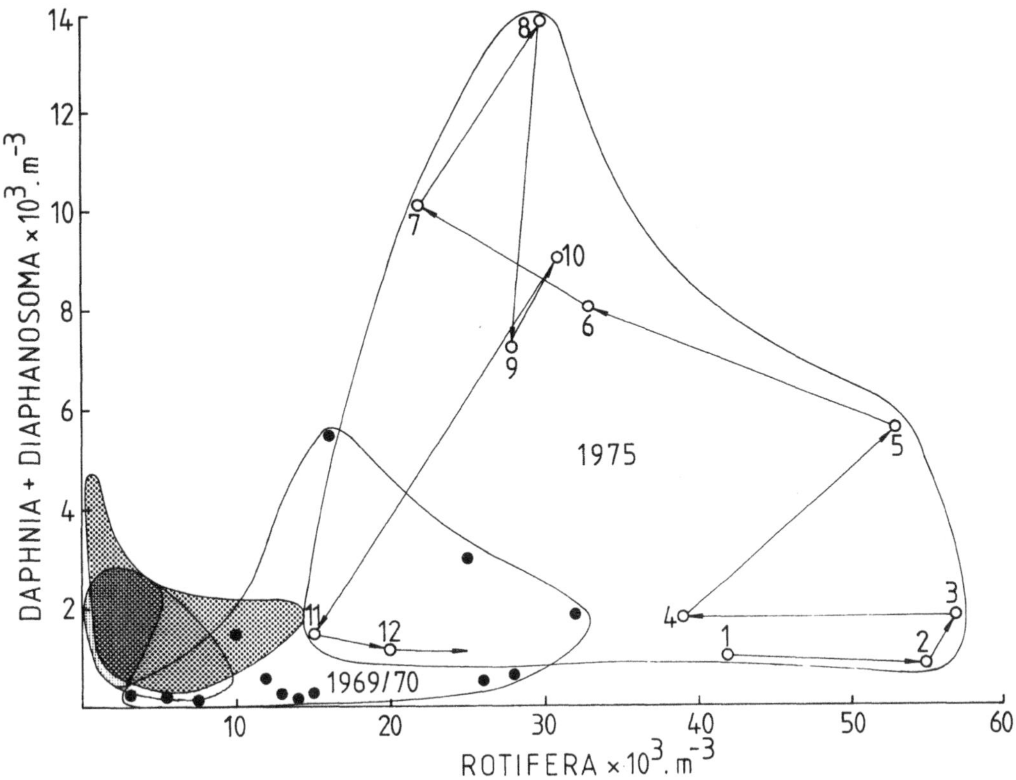

Fig. 3. Lago di Mergozzo. Development of Rotifera in relation to Cladocera (monthly averages). Shaded areas: 1948/49/50. Numbers 1–12 indicate the months.

3.2 years. The lake was eutrophied due to recreational activities and since 1979 several attempts have been made to improve this situation. The development of zooplankton, algal biomass and bacterial biomass is shown in Fig. 5. From 1972 through 1976, the zooplankton generally showed rather irregular variations in abundance. As in Lake Mergozzo, the rotifer zooplankton of Piburger See is characterized every year by a few species which differ from year to year. In 1972, *P. dolichoptera* was most abundant, whereas the peaks of 1973 were due to *Synchaeta* spp. 1974 was dominated by *Synchaeta* spp. and *F. terminalis*, and 1975 by *C. unicornis, K. hiemalis* and *F. terminalis.*

From 1972 to 1976, the peaks in rotifer abundance follow the development of the Chrysophyceae and Cryptophyceae. An exception is *F. terminalis*, which shows close correlation to the biomass of bacteria. The rates of population increase reveal *P. dolichoptera, C. unicornis,* and *A. priodonta* as perennial species which show population growth throughout the year (Fig. 5). The cold water forms *K. hiemalis* and *F. terminalis* show population growth at all seasons, too. They live in the cool hypolimnion of this lake, where they seem to use the bacteria feeding grounds very effectively.

Figure 6 presents the relationship between rotifers and cladocerans (small species are dominant: *Ceriodaphnia quadrangula, Bosmina longirostris*) in Piburger See. Compared to Mergozzo, where maximum rotifer abundance is 3 to 5 fold that of cladocerans, in Piburger See rotifers reach 10 to 50 times the number of cladocerans. The development of the zooplankton community differs from year to year. In 1972, the Cladocera seemed to compete with the rotifers very successfully, whereas in spring 1973 the faster growing rotifer population seems to have inhibited the development of a more numerous cladoceran population. In 1974, num-

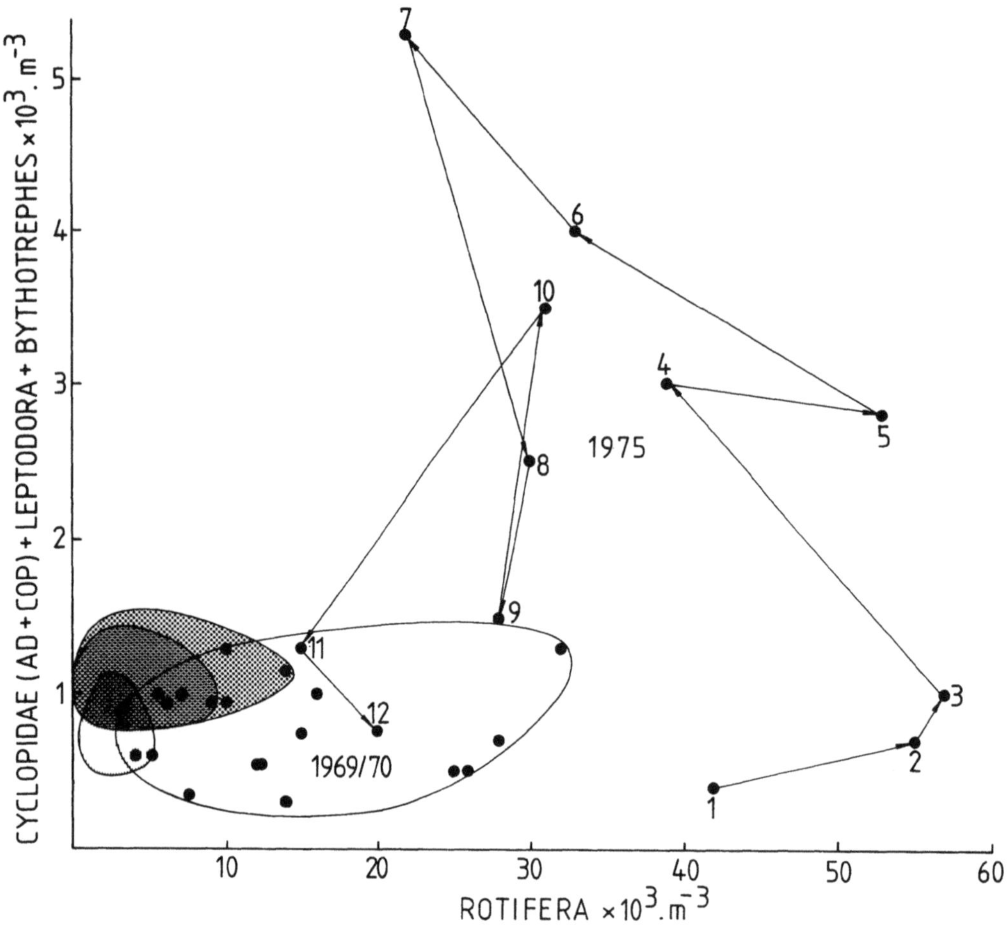

Fig. 4. Lago di Mergozzo. Development of Rotifera in relation to predators (monthly averages). Shaded areas: 1948/49/50. Number 1–12 indicate months.

bers of planktonic animals were much reduced, and during 1975 and 1976 the summer months were characterized by a succession of Rotifera and Cladocera.

3. *Neusiedlersee* (47°38′-47°51′N/16°41′-16°52′E; altitude = 115.5 m; A = 300 km^2, open water area = 170 km^2; Z_{max} = 2.0 m; $\bar{Z}$ = 1.3 m). [Dokulil, 1979, 1984; Herzig, 1979, present paper].

Neusiedlersee is a shallow, well mixed lake, which only occasionally stratifies. Alkalinity of 8.5 - 11.5 meq l^{-1}, pH of 8–8.5 and conductivity of about 1500 μS are characteristic chemical features. The open water region is always turbid due to the sediment particles which are stirred up from the bottom by wind action. Large surface area, shallow water and continuous mixing result in high water temperatures during summer and extremely low ones throughout winter. Since 1970 the lake has become increasingly eutrophic.

Figures 7 and 8 represent the summarized data on the long-term development of the zooplankton community. From 1968 through 1975 a gradual decrease in rotifer abundance occurred in parallel with an increase in crustacean abundance. The amount of algae available as food, expressed as total chlorophyll *a* concentration, increased from 1970 onwards.

Only one perennial species, *K. quadrata*, occurred every year. *Filinia terminalis, F. longiseta* and *H. fennica* (acc. to Koste, pers. comm.: *H. mira, H. fennica, H. jenkinae, H. polyodonta*) are typical seasonal rotifers which occurred through-

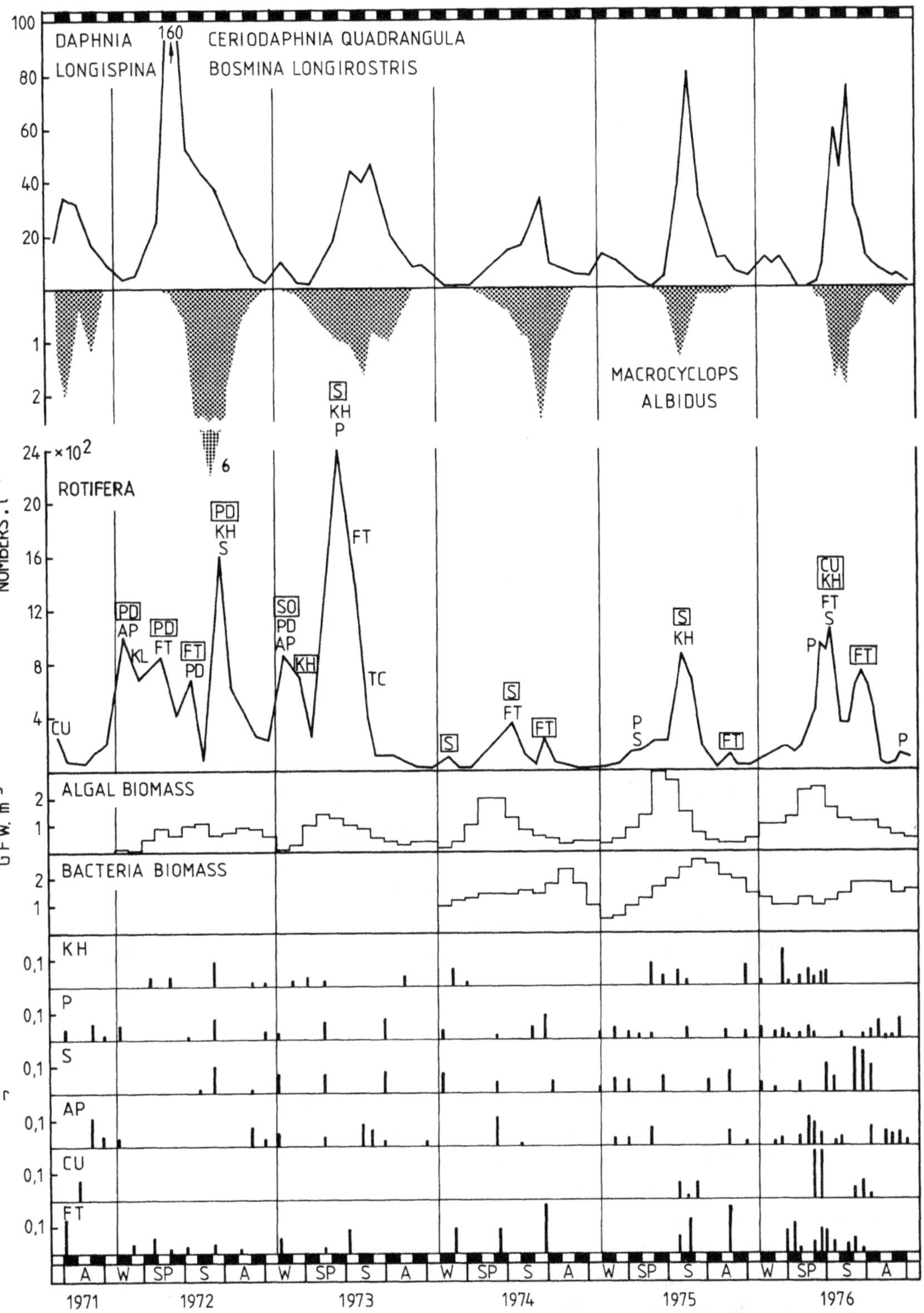

Fig. 5. Development of the plankton in Piburger See. Algal and bacteria biomass – monthly averages. r = inst.rate of increase. CU – *Conochilus unicornis*; PD – *Polyarthra dolichoptera*; AP – *Asplanchna priodonta*; KL – *Kellicottia longispina*; FT – *Filinia terminalis*; KH – *Keratella hiemalis*; SO – *Synchaeta oblonga*; S – *Synchaeta*; TC – *Trichocerca capucina*; P – *Polyarthra*.

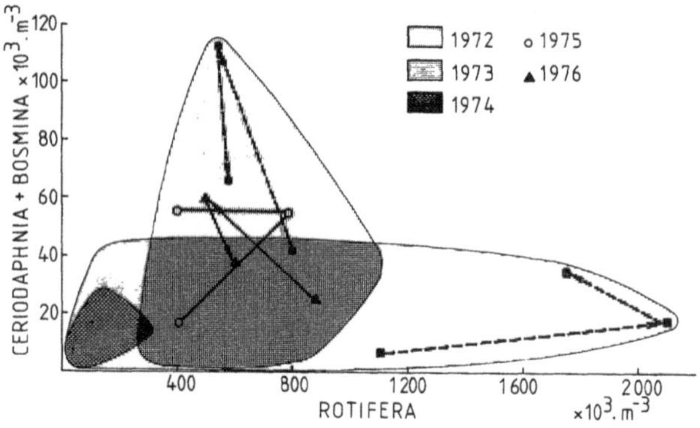

Fig. 6. Piburger See. Development of Rotifera in relation to Cladocera (monthly averages). Only the summer months are plotted for 1975 and 1976.

Table 1. Maximum population density of *Rhinoglena fertöensis* in relation to the number of resting eggs produced.

Year	N_{max}/l	Resting eggs produced/l
1972/73	78	130
1975/76	615	318
1976/77	1774	476
1977/78	1176	828
1978/79	1100	400
1979/80	1047	347

out the whole investigation period. Another perennial species, *P. vulgaris,* which was abundant in the summer and autumn plankton of the years 1950, 1968 and 1969, had disappeared almost completely ten years later. Two interesting examples are *K. cochlearis* and *Anuraeopsis fissa. Keratella cochlearis* was rare in 1950, reached peak abundance in the spring of 1968, and since then has declined to a point where it occurs in low numbers. *Anuraeopsis fissa,* a dominant species during the summer of 1951, has since declined to a point where it is rare.

Several other species almost completely disappeared following a period of population abundance: *B. angularis, B. calyciflorus, K. valga, N. acuminata* and *Synchaeta* spp. This may be the result of the fast rate of eutrophication, but this is not a satisfactory explanation of this phenomenon.

In contrast, *Rhinoglena fertöensis* was rarely observed in 1951, 1968, 1969 and 1970, but since then has gradually increased in numbers, and, since 1975, has become the dominant winter plankton organism. This seems to be due to optimum food conditions occurring during late autumn and winter, accompanied by the absence of any predator [Herzig, 1980]. The maximum number reached during winter is also dependent upon the number of resting eggs produced the previous year (Table 1); these numbers in turn are related to the maximum individual number reached by the population which produced them. Therefore it took 5 years until *R. fertöensis* reached high individual numbers.

The rates of increase of the numerically important species are found in Fig. 7. The summer species (*H.* spp., *B. angularis, F. longiseta, K. valga*) have their growth phases in the summer months, the perennial *P. vulgaris* shows growth phases during most months although it prefers spring and summer, the winter species *R. fertöensis* grows mainly during late autumn and winter and *B. calyciflorus* develops either a spring and an autumn growth phase, or only one of these. Independent of the actual individual numbers reached, the rotifers continue their periodicity of occurrence.

The abundance graphs of the numerically important rotifer species are shown in Figs. 9, 10, and 11, for the years 1951, 1968 and 1976. The most obvious difference between years is the change in the number of species present: 12 in 1951 and 1968, and only 5 in 1976. This may be explained by the eutrophication of Neusiedlersee. Furthermore, the number of succeeding peaks in abundance is quite different: 19 in 1951 and 1968, only 7 in 1976. In all three years one species reached very high numbers (2 to 20-fold the numbers of the other species); 1951 – *A. fissa*; 1968 – *K. cochlearis;* 1976 – *R. fertöensis*; these 'blooms' seem to be the result of optimum growth conditons which obviously occur very rarely.

If one compares the summer rotifer communities of 1951 and 1968, a great similarity in species composition is found (except that *K. valga* is abundant instead of *A. fissa* in 1968), but the quantitative development and the sequence of peaks are completely different. One recognises rapid changes of ups and downs which means the rotifer communities have to be analyzed species by species at least at

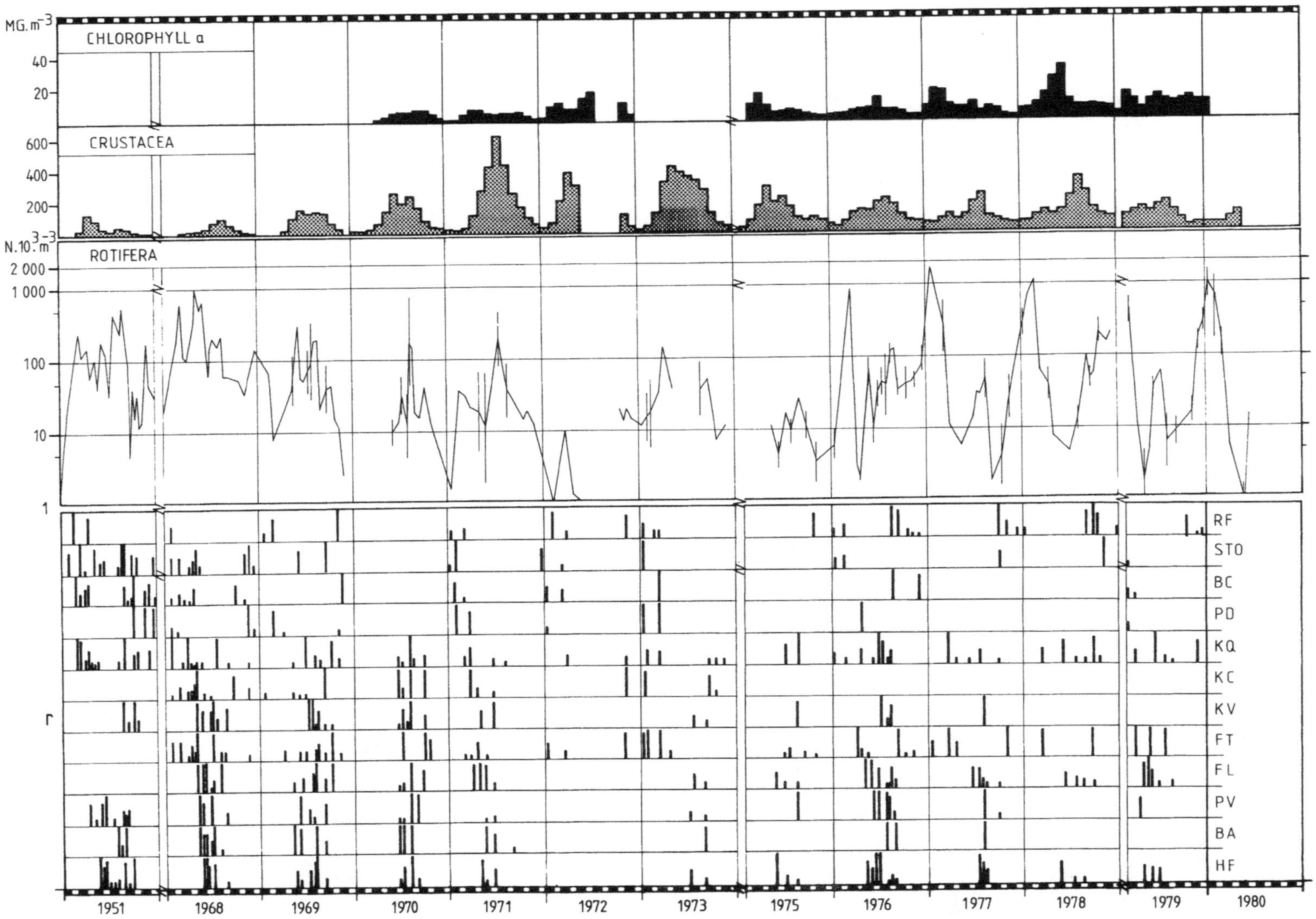

Fig. 7. Development of the plankton in Neusiedlersee. r = inst. rate of increase. Vertical lines in the rotifer panel indicate 95% confidence interval. Chlorophyll and crustacean abundance are monthly averages. RF – Rhinoglena fertöensis; STO – Synchaeta tremula-oblonga; BC – Brachionus calyciflorus; PD – Polyarthra dolichoptera; KQ – Keratella quadrata; KC – Keratella chochlearis; KV – Keratella valga; FT – Filinia terminalis; FL – Filinia longiseta; PV – Polyarthra vulgaris; BA – Brachionus angularis; HF – Hexarthra spp.

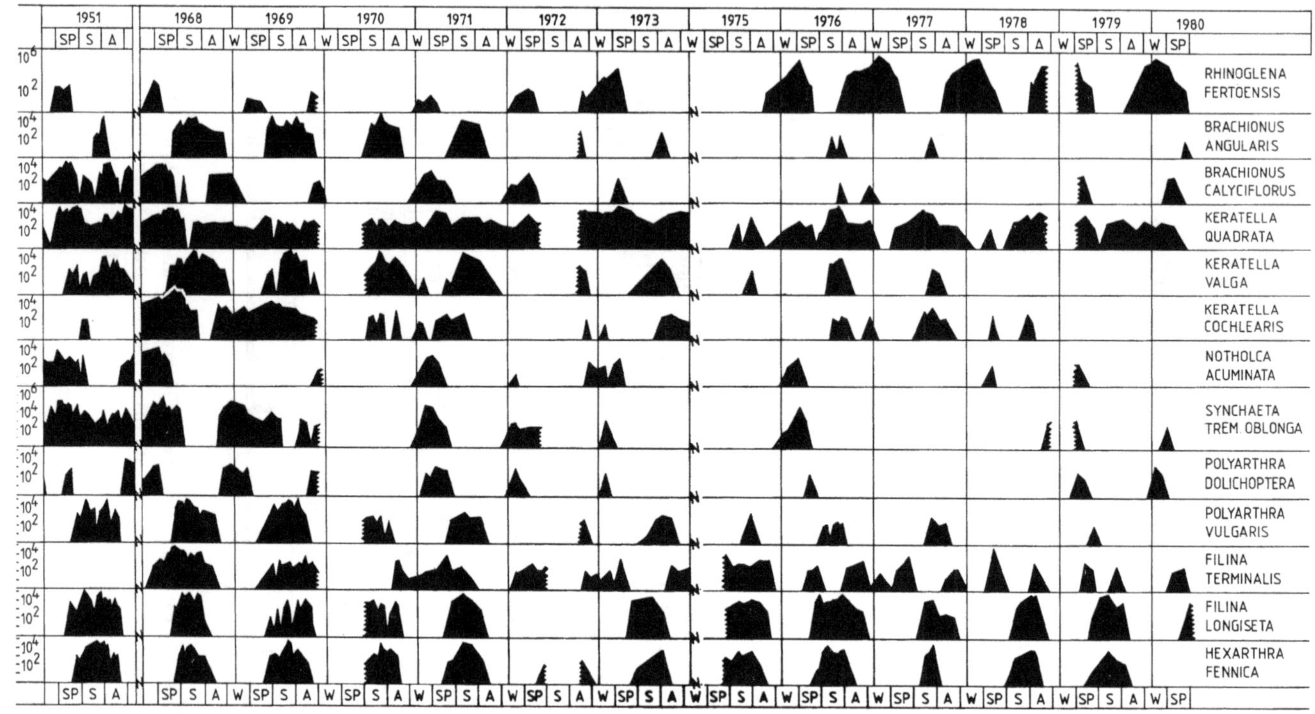

Fig. 8. Neusiedlersee: Phenologies of the most important rotifer species.

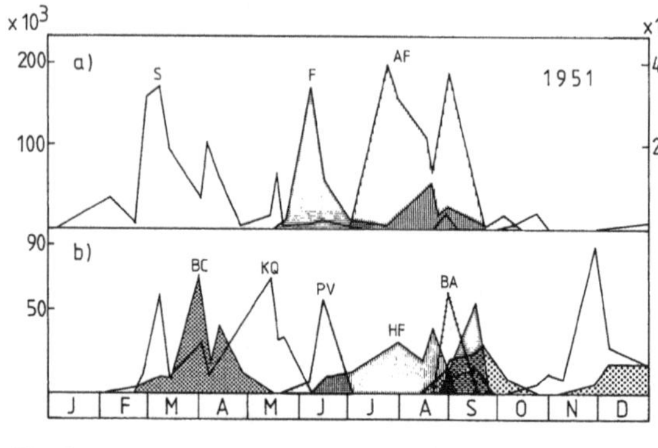

Fig. 9. Neusiedlersee: Species succession of rotifers in 1951.

a) S – *Synchaeta*; F – *Filinia*; AF – *Anuraeopsis fissa* (refer to the numbers at the right hand scale).

b) BC – *Brachionus calyciflorus*; KQ – *Keratella quadrata*; PV – *Polyarthra vulgaris*; HF – *Hexarthra* spp.; BA – Brachionus angularis.

weekly intervals, an interval of 3 days would be better.

Figure 12 summarizes the relationship between rotifers and crustaceans for all the years. From 1951 to 1968, rotifers developed numbers which were 6–8 times the crustacean numbers. This was fol-

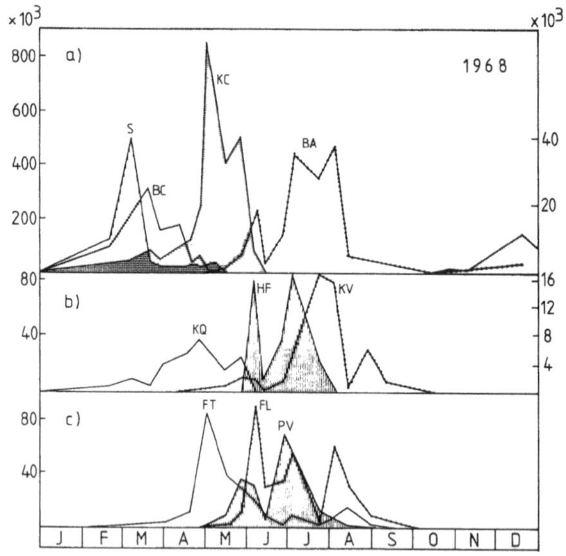

Fig. 10. Neusiedlersee: Species succession of rotifers in 1968.

a) S – *Synchaeta*; KC – *Keratella cochlearis* (refer to the numbers at the left hand scale). BC – *Brachionus calyciflorus*; BA – *Brachionus angularis.*

b) KQ – *Keratella quadrata*; KV – *Keratella valga* (refer to the numbers at the left hand scale). HF – *Hexarthra* spp.

c) FT – *Filinia terminalis*; FL – *Filinia longiseta*; PV – Polyarthra vulgaris.

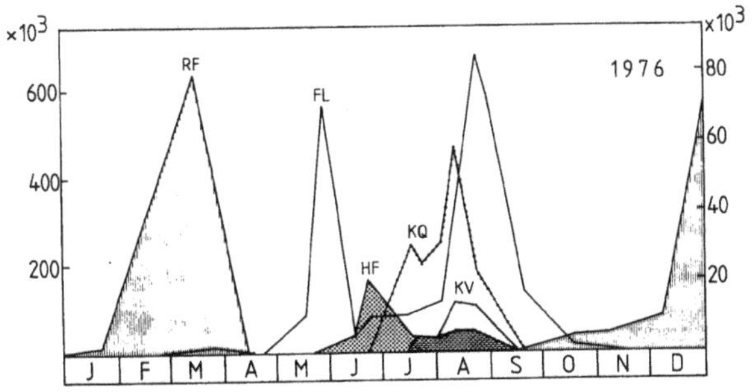

Fig. 11. Neusiedlersee: Species succession of rotifers in 1976. RF – *Rhinoglena fertöensis* (left hand scale); FL – *Filinia longiseta*; HF – *Hexarthra* spp.; KQ – *Keratella quadrata*; KV – *Keratella valga.*

lowed by an explosive growth of the crustacean populations (especially *Diaphanosoma brachyurum* during summer) and a dramatic drop in rotifer numbers, leading to the conclusion that the rotifers were outcompeted by the crustaceans (mainly Cladocera). Since 1976 another 'modus vivendi' developed: Late autumn/winter/spring are dominated by rotifers (*R. fertöensis*), the summer is dominated by crustaceans. This pattern of occurrence has been documented for 5 years.

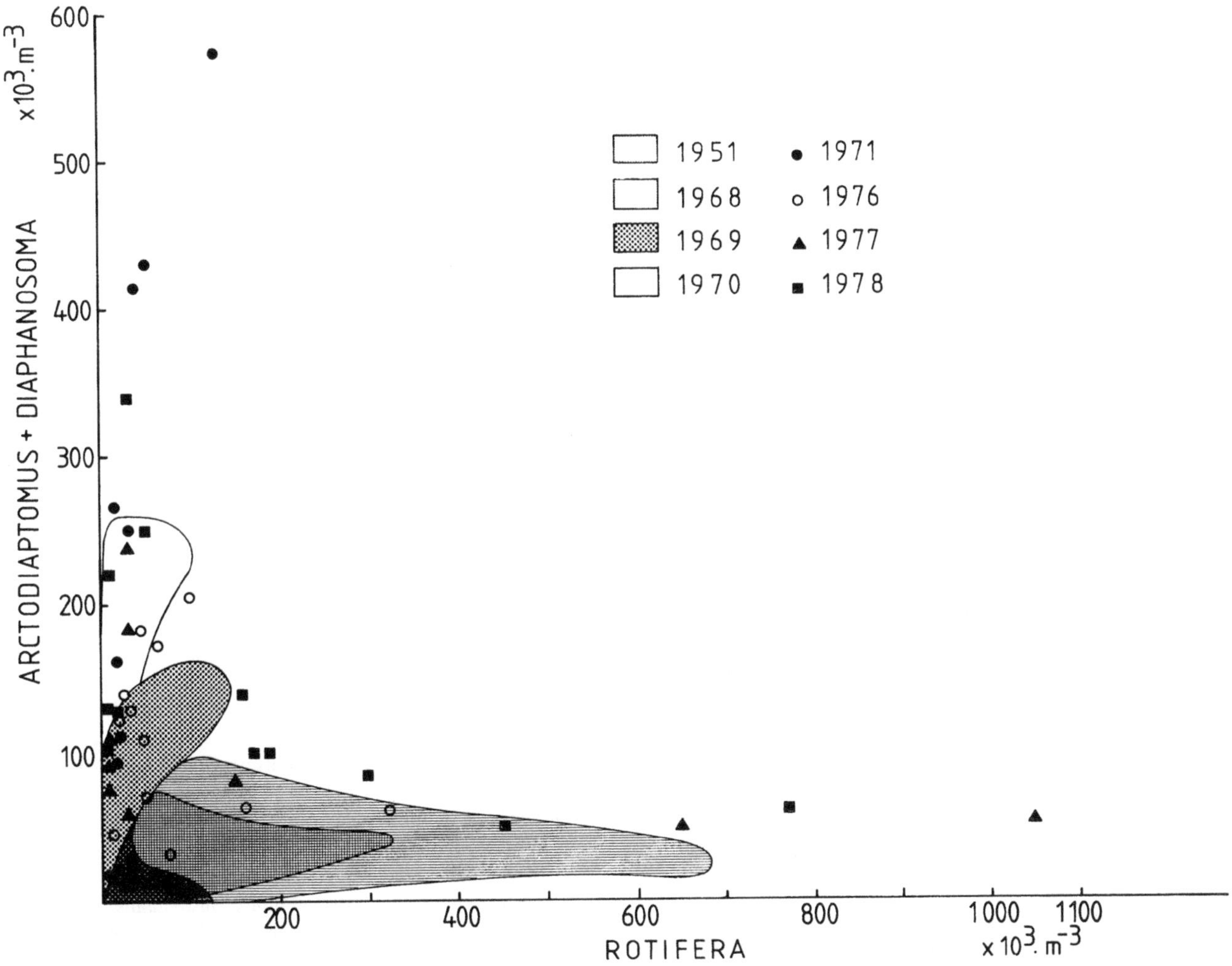

Fig. 12. Neusiedlersee: Development of Rotifera in relation to Crustacea (monthly averages).

Concluding remarks

1. In water bodies of the temperate zone climatic seasonality is reflected in the composition of the rotifer community. Long-term data show that species abundance may vary markedly from year to year, especially when the lake is disturbed, e.g. by eutrophication, but the seasonality of occurrence and growth remains constant over decades (e.g. Neusiedlersee).

2. Well-defined, short-term successions exist which can partly be explained by physical and chemical limitations, food and exploitative competition, mechanical interference competition, predation and parasitism. Most of these factors need more experimental investigation using both single factor and multi-factor analysis. Priority should be given to the study of species succession within a single genus (e.g., *Polyarthra, Synchaeta*), within one size class, or within one feeding type.

3. Intraspecific differences in adaptation phenomena are known to influence succession; much more experimental work is needed on this topic.

4. Vertical distribution and migration should be analysed over the course of the whole year, within the context of seasonality. Cold-water forms do not occur only during cold seasons but may also occur throughout the year living in the cool hypolimnion where they often use bacteria for food (e.g., Piburger See).

5. All the long-term data indicate that rotifers respond quickly to optimal or even improving conditions, whereas cladocerans and copepods need a longer time to make use of a change in trophic conditions. Frequently the crustaceans reach densities that cause an obvious suppression of the rotifer population. An inverse relationship between crustaceans and rotifer abundance may be observed which may reflect ecological interactions like competition and predation.

6. Long-term data show that every year is characterized by a few species that develop best and contribute most to the rotifer abundance; these species often are different from year to year. Can this phenomenon be explained by the common ideas about ecological interactions? Or, should we better search for other causes like quantity and quality of resting eggs, or even, long-term oscillations (10–20 years) of maximum population development which are linked to the physical features of the lake?

7. Results from Neusiedlersee indicate that the long-term development of a rotifer population (*Rhinoglena fertöensis*) is strongly dependent on the number of resting eggs produced. Therefore the question arises: How important are resting eggs for continuation of a population? How important is the number of viable resting eggs as an inoculum for the next generation? Is the maximum population size dependent on the number of resting eggs produced? This topic needs more experimentation and long-term field investigations.

8. The analysis of short-term and long-term development of rotifer populations reveals the urgent need for qualitative and quantitative observations in field and laboratory over long periods to be able to formulate meaningful testable hypotheses (cf. Likens' [1983] statement about future ecological research).

Acknowledgements

My sincere thanks are due to Dr Linda May for her constructive criticism and her help improving this manuscript. I am grateful to the Austrian Bundesministerium für Wissenschaft und Forschung and the British Council who contributed towards travel and living expenses at the Rotifer Symposium in Edingburgh.

References

Bangerter, B., E. Böhlen & H. Fankhauser, 1980. Untersuchungen über das pelagische Phyto-, Rotatorien- und Crustaceen-Plankton im Bielersee. Lizentiatsarbeit, Universität Bern, 193 Seiten,

Bernardi, de R. & E. Soldavini, 1976. Long-term fluctuations of zooplankton in Lake Mergozzo, Northern Italy. Mem. Ist. ital. Idrobiol. 33: 345–375.

Berner-Fankhauser, H., 1986. Populationsdynamik der Planktonrotatorien im Bielersee 1981/82. Unpubl. PhD. thesis, Univ. Bern, Switzerland, 88 pp.

Bogdan, K. G. & J. J. Gilbert, 1982. Seasonal patterns of feeding by natural populations of Keratella, Polyarthra, and Bosmina: Clearance rates, selectivities, and contributions to community grazing. Limnol. Oceanogr. 27: 918–934.

Bogdan, K. G., J. J. Gilbert & P. L. Starkweather, 1980. In situ clearance rates of planktonic rotifers. Hydrobiologia 73: 73–77.

Brandl, Z. & C. H. Fernando, 1978. Prey selection by the cyclopoid copepods Mesocyclops edax and Cyclops vicinus. Verh. int. Ver. Limnol. 20: 2505–2510.

Burns, C. W. & F. H. Rigler, 1967. Comparison of filtering rates of Daphnia rosea in lake water and in suspensions of yeast. Limnol. Oceanogr. 12: 492–503.

Carlin, B., 1943. Die Planktonrotatorien des Motalaström. Medd. Lunds Univ. limnol. Inst. 5: 256 pp.

Dokulil, M., 1979. Seasonal pattern of phytoplankton. In H. Löffler (ed.), Neusiedlersee. Limnology of a shallow lake in Central Europe. Dr. W. Junk bv Publ., The Hague, Boston, London, Monogr. Biol. 37: 203–231.

Dokulil, M., 1984. Assessment of components controlling phytoplankton photosynthesis and bacterioplankton production in a shallow, alkaline, turbid lake (Neusiedlersee, Austria). Int. Revue ges. Hydrobiol. 69: 679–727.

Edmondson, W. T., 1946. Factors in the dynamics of rotifer populations. Ecol. Monogr. 16: 357–372.

Edmondson, W. T., 1960. Reproductive rates of rotifers in natural populations. Mem. Inst. ital. Idrobiol. 12: 21–77.

Edmondson, W. T., 1965. Reproductive rate of planktonic rotifers as related to food and temperature in nature. Ecol. Monogr. 35: 61–111.

Einsle, U., 1983. Die Limnologie des Mindelsees. Natur- und Landschaftsschutzgebiete Bad.-Würt. 11: 161–218.

Elliott, J. I., 1977. Seasonal changes in the abundance and distribution of planktonic rotifers in Grasmere (English Lake District). Freshwat. Biol. 7: 147–166.

Eloranta, P. V., 1982. Zooplankton in the Vasikkalampi pond, a warm water effluent recipient in Central Finland. J. Plankton Res. 4: 813–837.

Frank, V., D. Bosch & J. Ringelberg, 1985. Seasonal succession and population dynamics of Keratella cochlearis (Ehrb.) and Kellicottia longispina (Kellicott) in lake Maarsseveen I (Netherlands). Arch. Hydrobiol. 103: 273–290.

Fuller, D. R., R. S. Stemberger & J. E. Gannon, 1977. Limnetic rotifers as indicators of trophic change. J. Elisha Mitchell Scient. Soc. 93: 104–113.

Gilbert, J. J., 1980. Feeding in the rotifer Asplanchna: Behavior, cannibalism, selectivity, prey defenses, and impact on rotifer communities. In W. C. Kerfoot (ed.), Evolution and Ecology of Zooplankton Communities. The University Press, New England, Hanover (N.H.): 158–172.

Gilbert, J. J. & K. G. Bogdan, 1981. Selectivity of Polyarthra and Keratella for flagellate and aflagellate cells. Verh. int. Ver. Limnol. 21: 1515–1521.

Gilbert, J. J. & K. G. Bogdan, 1984. Rotifer grazing: In situ studies on selectivity and rates. In D. G. Meyers & J. R. Strickler (eds.), Trophic interactions within aquatic ecosystems. American Association for the Advancement of Science, Selected Symposium 85: 97–133.

Gilbert, J. J. & R. S. Stemberger, 1984. Control of Keratella populations by interference competition from Daphnia. Limnol. Oceanogr. 30: 180–188.

Gilbert, J. J. & C. E. Williamson, 1978. Predator-prey behavior and its effect on rotifer survival in associations of Mesocyclops edax, Asplanchna girodi, Polyarthra vulgaris, and Keratella cochlearis. Oecologia 37: 12–22.

Herzig, A., 1979. The zooplankton of the open lake. In H. Löffler (ed.), Neusiedlersee. Limnology of a shallow lake in Central Europe. Dr. W. Junk bv Publ., The Hague, Boston, London, Monogr. Biol. 37: 281–335.

Herzig, A:, 1980. Ten years quantitative data on a population of Rhinoglena fertöensis (Branchionidae, Monogononta). Hydrobiologia 73: 161–167.

Herzig, A:, 1983. Comparative studies on the relationship between temperature and duration of embryonic development of rotifers. Hydrobiologia 104: 237–246.

Herzig, A., 1984. Temperature and life cycle strategies of Diaphanosoma brachyurum: An experimental study on development, growth, and survival. Arch. Hydrobiol. 101: 143–178.

Hofmann, W., 1983. Interactions between Asplanchna and Keratella cochlearis in the Plußsee (North Germany). Hydrobiologia 104: 363–365.

Jones, H. R., T. J. Lack & C. S. Jones, 1979. Population dynamics and production of Daphnia hyalina var. lacustris in Farmoor I, a shallow eutrophic reservoir. J. Plankton Res. 1: 45–65.

Larsson, P., 1978. The life cycle dynamics and production of zooplankton in övre Heimdalsvatn. Holarct. Ecol. 1: 162–218.

Likens, G. E., 1983. A priority for ecological research. Bull. Ecol. Soc. America 64: 234–243.

Likens, G. E., 1984. Beyond the shoreline: a watershed-ecosystem approach. Verh. int. Ver. Limnol. 22: 1–22.

May, L., 1980a. Ecology of planktonic rotifers at Loch Leven, Kinross-Shire. Unpubl. Ph.D. thesis, Paisley College of Technology, 158 pp.

May, L., 1980b. On the ecology of Notholca squamula Müller in Loch Leven, Kinross, Scotland. Hydrobiologia 73: 177–180.

May, L., 1983. Rotifer occurrence in relation to water temperature in Loch Leven, Scotland. Hydrobiologia 104: 311–315.

Miracle, M. R., 1977. Epidemiology in rotifers. Arch. Hydrobiol. Beih. 8: 138'41.

Nauwerck, A., 1963. Die Beziehung zwischen Zooplankton und Phytoplankton im See Erken. Symb. bot. upsal. 17, 5: 1–163.

Nauwerck A., 1978. Notes on the planktonic rotifers of Lake Ontario. Arch. Hydrobiol. 84: 269–301.

Pejler, B., 1961. The zooplankton of Ösbysjön, Djursholm. I. Seasonal and vertical distribution of the species. Oikos 12: 225–248.

Pejler, B., 1962. The zooplankton of Ösbysjön, Djursholm. II. Further ecological aspects . Oikos 13: 216–231.

Pejler, B., 1977. On the global distribution of the family Brachionidae (Rotatoria). Arch. Hydrobiol. Suppl. 53: 253–306.

Pourriot, R., 1977. Food and feeding habits of Rotifera. Arch. Hydrobiol. Beih. 8: 243–260.

Rott, E., 1983. Sind die Veränderungen im Phytoplanktonbild des Piburger Sees Auswirkungen der Tiefenwasserableitung? Arch. Hydrobiol. Suppl. 67, 1: 29–80.

Ruttner-Kolisko, A., 1971. Rotatorien als Indikatoren für den Chemismus von Binnensalzgewässern. Sber. Akad. Wiss., math.-nat. Kl., Abt. I, 179: 283–298.

Ruttner-Kolisko, A:, 1977a. Population dynamics of rotifers as related to climatic conditions in Lunzer Obersee und Untersee. Arch. Hydrobiol. Beih. 8: 88–93.

Ruttner-Kolisko, A., 1977b. The effect of the microsporid Plistophora asperospora on Conochilus unicornis in Lunzer Untersee. Arch. Hydrobiol. Beih. 8: 135–137.

Sacreni, A., D. Ruggio & M. Nakanishi, 1978. Phytoplankton dynamics, Chlorophyll a and Phaeophytin in Lago di Mergozzo (Northern Italy). Mem. Ist. ital. Idrobiol. 36: 215–237.

Schaber, P., 1975. Rotatorien und Crustaceen des Piburger Sees 1971–1973. Jber. Abt. Limnol. Innsbruck 1: 59–72.

Schaber, P., 1976. Rotatorien und Crustaceen des Piburger Sees 1973–1975. Jber. Abt. Limnol. Innsbruck 2: 78–94.

Schaber, P., 1977. Rotatorien und Crustaceen des Piburger Sees 1976. Jber. Abt. Limnol. Innsbruck 3: 119–132.

Snell, T. W., 1979. Intraspecific competition and population structure in rotifers. Ecology 60: 494–502.

Starkweather, P. L., 1980. Aspects of the feeding behavior and trophic ecology of suspension-feeding rotifers. Hydrobiologia 73: 63–72.

Wappis, E., 1980. Zur Populationsökologie des Zooplanktons des Ossiacher Sees. Diss. Univ. Graz, 168 Seiten.

Williamson, C. E., 1980. The predatory behavior of Mesocyclops edax: predator preferences, prey defenses, and starvation induced changes. Limnol. Oceanogr. 25: 903–909.

Williamson, C. E., 1983. Invertebrate predation on planktonic rotifers. Hydrobiologia 104: 385–396.

Williamson, C. E., 1984. Laboratory and field experiments on the feeding ecology of the cyclopoid copepod, Mesocyclops edax. Freshwat. Biol. 14: 575–585.

Williamson, C. E., & J. J. Gilbert, 1980. Variation among zooplankton predators: The potential of Asplanchna, Mesocyclops, and Cyclops to attack, capture, and eat various rotifer prey. In W. C. Kerfoot (ed.), Evolution and Ecology of Zooplankton Communities. The University Press, New England, Hanover (N.H.): 509–517.

Zankai, N. P., 1984. Predation of Cyclops vicinus (Copepoda: Cyclopoida) on small zooplankton animals in Lake Balaton (Hungary). Arch. Hydrobiol. 99: 360–378.

Zimmermann, C., 1974. Die pelagischen Rotatorien des Sempachersees, mit spezieller Berücksichtigung der Brachioniden und der Ernährungsfrage. Schweiz. Z. Hydrol. 36: 205–300.

Hydrobiologia 147: 181–188 (1987)
May, L., Wallace, R. & Herzig, A. (eds), Rotifer Symposium IV

The influence of sampling strategy on the apparent population dynamics of planktonic rotifers

Heidi Berner-Fankhauser
Zoologisches Institut der Universität, Baltzerstrasse 3, CH-3012 Bern, Switzerland

Keywords: rotifers, sampling strategy, population dynamics, *Synchaeta*, competition

Abstract

Rotifer composition and density, chlorophyll-a content, secchi depth and temperature were investigated with high sampling frequency (every third day in summer, weekly in winter) in the eutrophic lake Bielersee from April 1981 to April 1982. Long intervals between sampling, especially from spring to autumn, lead to contradictory interpretation of abundance dynamics of both total rotifer community and single species. Short intervals between sampling revealed a sequence of mass developments of the different species. As an example, alternating gradation phases are shown for *Synchaeta grandis* and *Synchaeta stylata.*

Introduction

Planktonic rotifers are known as very fast developing animals. Generation time in summer is usually less than one week. Fortnightly or even monthly sampling is therefore not sufficient to obtain a realistic picture of their population dynamics and of their importance in the ecosystem. Hillbricht-Ilkowska [1965] and Halbach [1974] demonstrated this for rotifer populations in ponds. In the present paper I have tried to show the influence of sampling frequency and of the starting date of an investigation on the apparent population dynamics of rotifers in a lake.

The lake

Bielersee is situated in the northwest of Switzerland (surface area 39.3 km^2, maximal depth 74 m, mean depth 30 m); it is eutrophic and stratifies in summer. The thermocline usually lies between 15 and 25 m. Due to the main inflow, the river Aare, the theoretical water renewal time is only about 60 days. In summer epilimnetic waters may be replaced even every 10–20 days. The lake lies along the main axis of prevailing winds (NE and SW). The high flushing rate and the winds cause undefined water movements and horizontal inhomogenities in the lake (von Orelli, 1981).

Material and methods

From April 1981 to April 1982 rotifer numbers and composition, chlorophyll-a content, temperature conditions and secchi depth were investigated, usually at midday at the deepest point of the lake. In summer samples were taken every third day, in spring and autumn every fourth to fifth day and in winter once a week. (comp. Berner-Faukhauser, 1986).

From the surface to 30 m, water was taken in steps of one metre with a 5 l-Friedinger sampler. The samples were bulked for the water layers 0–2 m, 2–4 m, 4–8 m, 8–12 m, 12–20 m and 20–30 m. One litre from each mixed sample was taken for measuring the content of chlorophyll-a and its phaeopigments (following the recommendations of Marker *et al.* [1980], with 90% ethanol).

Another aliquot of the mixed sample (one litre per metre) was taken for determining rotifer numbers and composition. To inhibit the contraction of illoricate species, rotifers were narcotized with 0.02% Procainiumchloride (compare May [1980]) for 30–60 minutes. The samples were then concentrated by filtering (30 μm mesh sieve) and preserved with formalin (5 ml formalin 38% to 100 ml of water). Depending on rotifer density the whole sample, a quarter or half of it was counted with an inverted microscope (Leitz Diavert) with about 80× magnification. Rotifer species were identified according to Koste [1978].

Results

Seasonal changes of chlorophyll-a content, secchi depth and rotifer abundance (Fig. 1)

Phytoplankton revealed a pattern of development typical of temperate lakes (compare Sommer *et al.* 1986]): a spring bloom followed by an early summer minimum (clear water phase) and oscillating high densities during summer. In September a bloom of *Mougeotia* resulted in very high chlorophyll-a values. After a further maximum in November, due to a phase of relatively warm and calm weather, chlorophyll-a content decreased to low winter levels. In March and April the new spring bloom began to develop.

Transparency in Bielersee was mainly influenced by phytoplankton density and the seasonal changes of secchi depth reflected the phytoplankton development. Exceptions to this general trend were probably caused by turbid water entering the lake from the inflow after heavy rains, or by water displacement due to strong winds.

Rotifer densities tended to follow the phytoplankton development. After a spring maximum dominated first by *Synchaeta oblonga* Ehrb., then by *Polyarthra dolichoptera* Carlin, *P. vulgaris* Carlin, *Asplanchna priodonta* Gosse and *Conochilus unicornis* Rouss. rotifer densities fell to a minimum during the clear water phase. During summer, rotifer abundance oscillated due to the succession of mass development of different species. Dominant species during this period are *Synchaeta kitina* Rouss., *S. grandis* Zach., *S. stylata* Wierz., *P. vulgaris, P. major* Burckh., *P.remata* (Skorikov), *Keratella cochlearis* (Gosse) and *Trichocerca rousseleti* (Voigt).

During the bloom of *Mougeotia* total rotifer abundance remained more or less constant, although, at the species level, densities of *Keratella, Synchaeta* and *Trichocerca* decreased while numbers of *Polyarthra* increased. As soon as the density of *Mougeotia* decreased, the abundance of total rotifers increased and was again dominated by *S. kitina.* At the end of October rotifer density decreased again but remained on a relatively high level during November due to abundant *S. tremula, S. oblonga* and *S. kitina.* In winter rotifer numbers were very low. At the end of March 1982 a community of several small *Synchaeta* species developed (mainly *S. lakowitziana* Lucks and *S. oblonga*).

Influence of sampling frequency on the apparent abundance dynamics of total rotifers

To evaluate the effect of sampling frequency on the apparent population dynamics of rotifers, nine different sampling strategies were simulated from the available data (Table 1), all starting on the same date. This resulted in quite different pictures of the abundance dynamics of total rotifers (Fig. 2). With long intervals the rate and extent of changes in rotifer density were under estimated. In series 'VII' and 'IX' the spring maximum is almost lost; rotifers seemed to attain only one maximum in summer. In series 'VIII', rotifer numbers were quite low throughout the year. The shorter the sampling interval, the more complex the diagrams became. But even series 'I' and 'II' (sampling in summer every third and every sixth day, respectively) differed considerably. This was especially true in July, where series 'II' included neither of the two peaks, nor the minimum and, consequently, the abundance seemed to remain more or less constant.

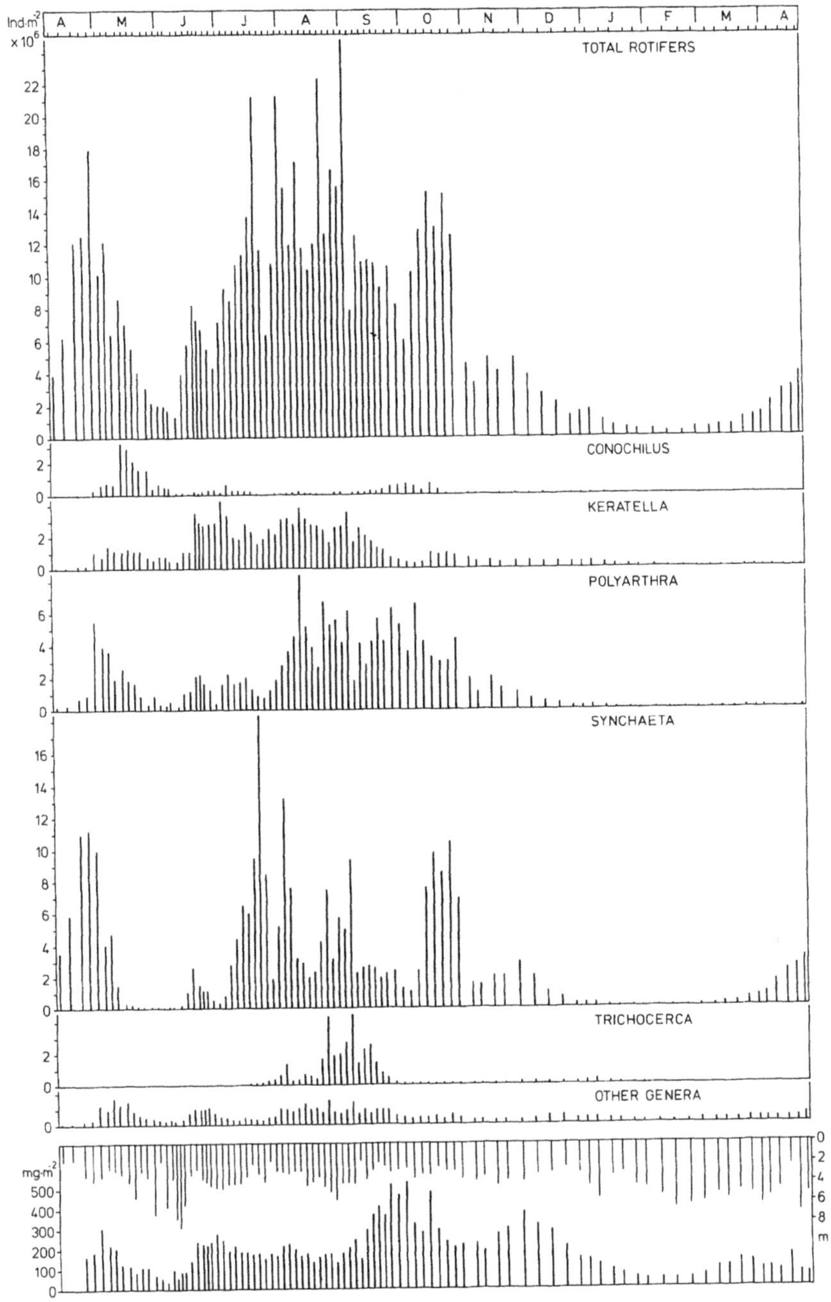

Fig. 1. Seasonal changes of rotifer abundance (0–30 m), chlorophyll-a content (0–30 m) and secchi depth.

Influence of date of starting on the apparent abundance dynamics of total rotifers

All the sampling dates were distributed consecutively among five groups. This resulted in five time-series with almost identical sampling frequency, i.e. every 15th day in summer and monthly in winter, but with different starting dates (Fig. 3). These sampling intervals are common in plankton research.

Each time series showed a different pattern of rotifer development. The maxima and minima differed in magnitude and timing, and the phases

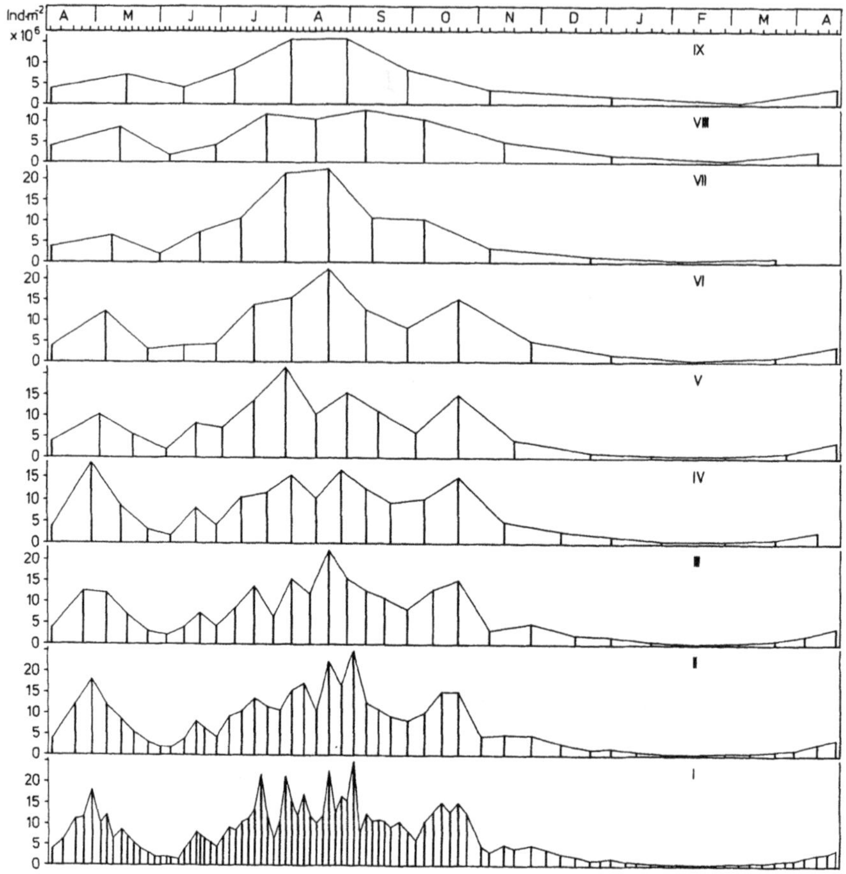

Fig. 2. Effect of sampling frequency on the apparent abundance dynamics of total rotifers.

of increase and decrease of populations were contradictory. None of the five time series reflected the development of rotifers satisfactorily, compared to short interval sampling (Fig. 2, 'I').

Table 1. Simulation of different sampling frequency.

Series	Selected sampling dates	Simulated sampling frequency in summer
I	every date	every 3rd day
II	every 2nd date	every 6th day
III	every 3rd date	every 9th day
IV	every 4th date	every 12th day
V	every 5th date	every 15th day
VI	every 6th date	every 18th day
VII	every 7th date	every 21st day
VIII	every 8th date	every 24th day
IX	every 9th date	every 27th day

Influence of sampling frequency on the apparent relationships between different species

The data of *Synchaeta grandis* and *S. stylata* were manipulated in the same way as those of total rotifers (Table 1), whereby five sampling series were constructed for the period June to November, when both species were abundant (Fig. 4). While with sampling every 15th day (graph 'V') both species seemed to develop very slowly (apart from the increase towards the begining of August) and maintain the same density over a long period, sampling every 12th day (graph 'IV') resulted in quite abrupt changes of abundance, especially of *S. stylata.* On the other hand the shorter the intervals, the better the patterns of population development became visible. Sampling every 6th day (graph 'II') and, more obviously, every 3rd day (graph 'I') revealed

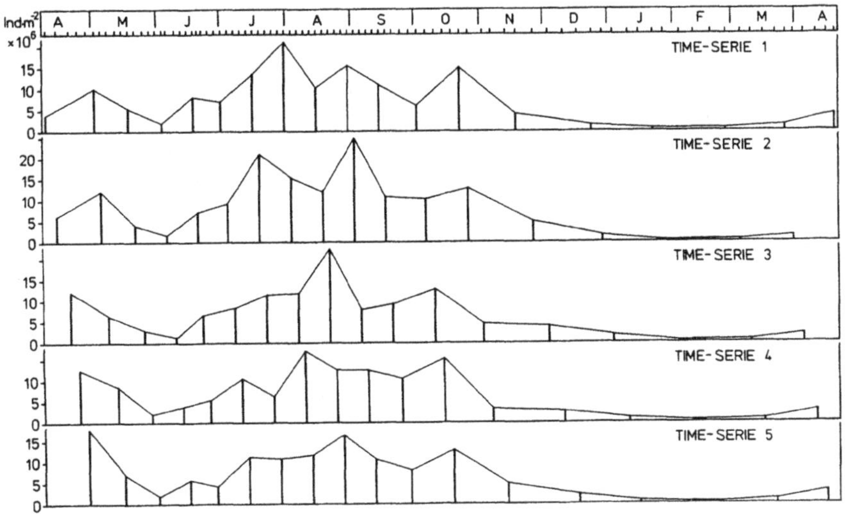

Fig. 3. Effect of starting date on the apparent abundance dynamics of total rotifers.

that both species had a sequence of gradation phases and latent phases as described by Hofmann [1972, 1977]. The fluctuations of the two species were alternating. Only at the begin of August did they reach a maximum at the same time. Between the gradation phases, the abundance of both species was very low.

Influence of starting date on the apparent relationships between different species

The above comparison (Fig. 4) demonstrates that sampling every 15th day is not sufficiently frequent to reveal the population dynamics of single species, and their inter-relationships. Consequently, varying the starting date leads to completely different pictures of the development of both species as well as their sequence (Fig. 5). In time-series 'I' *S. grandis* seems to have two or three pulses of population development, whereas *S. stylata* has only one maximum and otherwise quite low densities. In time-series '5' the case is almost constrary with two maxima of *S. stylata* and only one of *S. grandis.* In time-series '3' and '4' two to three maxima and minima of both species follow each other.

General conclusions on the relationship between the two species would be difficult to summarise and contradictory, depending on the time-series being considered.

Discussion

The total abundance of a rotifer community is the sum of the population development of the individual species (Fig. 1), which themselves are influenced by varying biotic and abiotic factors. Therefore abundance dynamics derived from sampling with short intervals (as shown in Fig. 1 and Fig. 2, 'I') is characterized by numerous maxima and minima. With longer intervals between sampling dates (Fig. 2 & 3), some of the maxima and minima and the information about the importance of the species involved are lost. With sampling intervals of 9–18 days (Fig. 2, 'III'–'VI' & Fig. 3) we may still see the general trend: Rotifer abundance has a spring maximum, followed by an early summer minimum, several maxima and minima in summer and autumn and very low values in winter. With longer intervals (Fig. 2, 'VII'–'IX') the characteristics of rotifer development are completely distorted.

For a more detailed ecological analysis, concerning influences of distinct factors on population dynamics or interactions between different plankton

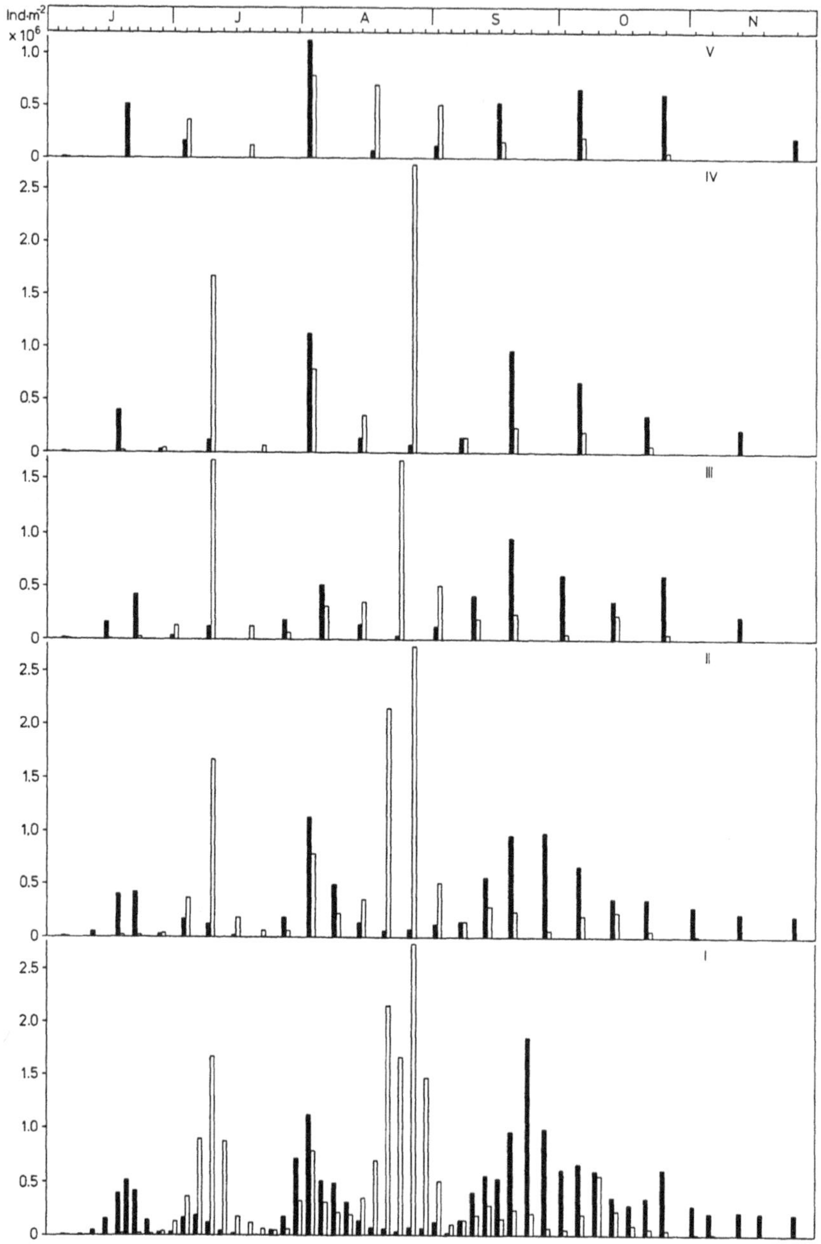

Fig. 4. Effect of sampling frequency on the apparent relationship between *Syncheata grandis* (black) and *Synchaeta stylata* (white).

components, we have to consider individual species. As the examples of *Synchaeta grandis* and *S. stylata* show, long intervals between sampling dates do not adequately reveal the relationships between different species; the interpretation of the sampling series 'IV' and 'V' (Fig. 4) and the five time-series (Fig. 5) leads to contradictory conclusions. On the other hand, high sampling frequency (Fig. 4, 'I' and 'II') reveals clearly alternating gradation phases for the two species. As both species have a similar vertical distribution (they prefer the upper 8 m) and both are at least partly carnivorous (both have been observed feeding on *S. kitina* and *P. vulgaris*) we may suggest, following Pejler [1962] and

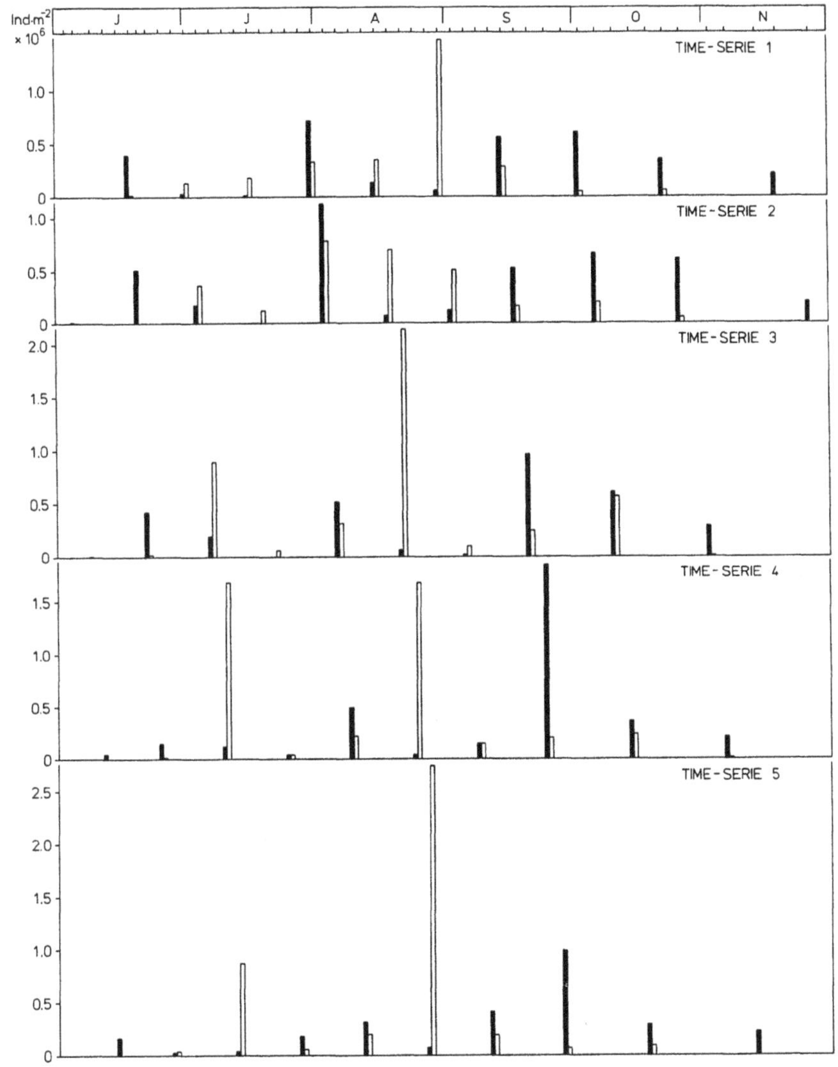

Fig. 5. Effect of starting date on the apparent relationship between *Synchaeta grandis* (black) and *Synchaeta stylata* (white).

Dumont [1977], that they obviate competition by alternating mass developments.

Conclusions

Low sampling frequency (intervals not adapted to generation time) provides, at most, only the general trend of total rotifer development and species composition in the course of a year. By no means can the population dynamics of different species be compared from this data.

High sampling frequency (intervals adapted to generation time) gives much better estimates of the population development of individual species, even though abiotic factors (e.g. high rainfall or strong winds) disturb this biological process. The absolute values are probably not as important as the pattern of development, and this remains visible in spite of sudden and short lived changes. Only reliable descriptions of population development can be used as a basis for further interpretation.

Acknowledgements

I thank Mrs. D. Zbären for improving the English. The investigation was supported by the 'Schweizerischer Nationalfonds zur Förderung der wissenschaftlichen Forschung'.

References

Berner-Fankhauser, H., 1986. Populationsdynamik der Planktonrotatorien im Bielersee 1981/82. Diss. Univ. Bern, 88 pp.

Dumont, H. J., 1977. Biotic factors in the population dynamics of rotifers. Arch. Hydrobiol. Beih. 8: 98–122.

Halbach, U., 1974. Methoden der Populationsökologie. Verh. Ges. Oekol., Erlangen: 1–24.

Hillbricht-Ilkowska, A., 1965. The effect of the frequency of sampling on the picture of the occurence and the dynamics of plankton rotifers. Ekol. pol. A 13: 101–112.

Hofmann, W., 1972. Zur Populationsökologie des Zooplanktons im Plussee. Verh. int. Ver. Limnol. 18: 410–418.

Hofmann, W., 1977. The influence of abiotic environmental factors on population dynamics in planktonic rotifers. Arch. Hydrobiol. Beih. 8: 77–83.

Koste, W., 1978. Rotatoria. Borntraeger. Berlin, 2 vol.: 673 pp, 234 plates.

Marker, A. F. H., E. A. Nusch, H. Rai & B. Riemann, 1980. The measurement of photosynthetic pigments in freshwaters and standardization of methods: Conclusions and recommendations. Arch. Hydrobiol. Beih. 14: 91–106.

May, L., 1980. On the ecology of Notholca squamula Müller in Loch Leven, Kinross, Scotland. Hydrobiologia 73: 177–180.

Orelli, J. von., 1981. Die horizontale Verteilung von Chlorophyll im Bielersee. Schweiz. Z. Hydrol. 43: 168–175.

Pejler, B., 1962. The zooplankton of Osbysjön, Djursholm. II. Further ecological aspects. Oikos 13: 216–231.

Sommer, U., Z. M. Gliwicz, W. Lampert and A. Duncan, 1986. The PEG-model of seasonal succession of planktonic events in fresh waters. Arch. Hydrobiol., 106: 432–471.

Hydrobiologia 147: 189–196 (1987)
May, L., Wallace, R. & Herzig, A. (eds), Rotifer Symposium IV

Abundance and distribution of pelagic rotifers in a cold, deep oligotrophic alpine lake (Königssee)

Ruth Laxhuber
Zoologisches Institut der Universität München Seidlstr. 25 8000 München 2 FRG

Keywords: Rotifera, oligotrophic lake, succession, abundance, distribution

Abstract

Rotifer community structure of the oligotrophic, alpine lake Königssee was investigated from October 1982 to October 1983. Twenty different species were found, including several cold-stenothermal and oligotrophic species. *Polyarthra vulgaris/dolichoptera, Kellicottia longispina* and *Keratella cochlearis* were the most abundant species throughout the year, comprising 90% of the total rotifer community (2.8×10^6 individuals m^{-2}, maximum). The remainder of the rotifer community was represented by as many as eight species which occured for a short time in summer and autumn. The rotifer community was limited to the upper 50 m (85-100%) in which most (60-80%) of the rotifers preferred the trophogenic layer (0-20 m). Species-specific depth preferences were observed. Rotifer abundance and distribution are discussed in terms of the specific environmental conditions in Königssee. The rotifer community of Königssee is compared to that of Lake Constance in its former oligotrophic state.

Introduction

Rotifers are especially sensitive to changes in water quality [Gannon & Stemberger, 1978] and are excellent indicators of eutrophication [Walz *et al.*, in press]. This paper describes the succession, abundance, and vertical distribution of rotifers in Königssee, an oligotrophic, alpine lake. The rotifer community is compared to that of Lake Constance which has recently undergone eutrophication.

Study site

Königssee is an oligotrophic, alpine lake of tectonic and glacial origin. It is situated 603 m above sea level in southeastern Bavaria, within the Alpine National Park at Berchtesgaden, and is surrounded by forests and steep rocky shores. The lake is sheltered from the wind by a range of mountains which rise to 2000 m. Königssee is 7.7 km long and 1.2 km wide, with maximum and mean depths of 190 m and 98 m, respectively. It has a surface area of 5.22 km^2. Low temperatures and temperature amplitudes, which are observed down to 30 m depth, are a feature of the temperature regime of Königssee [Siebeck, 1982; Laxhuber, 1984]. Königssee is largely free of anthropogenous influence. Nutrient input is restricted to mountain brooks which bring limited allochthonous material into the lake during the ice free period [Siebeck, 1982; Laxhuber, 1981].

Materials and methods

Plankton samples were collected biweekly from five depths (0-10 m, 10-20 m, 20-50 m, 50-75 m and 75-170 m) at two sampling stations (Eiswinkel,

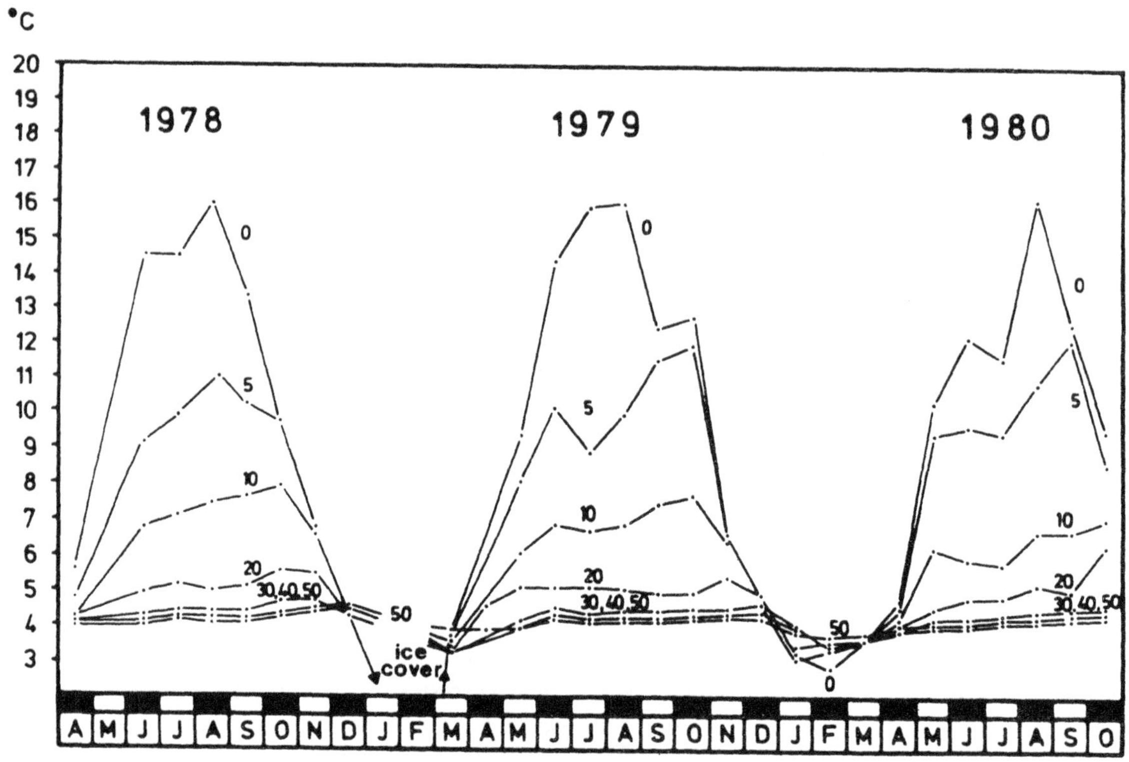

Fig. 1. Mean monthly temperatures in Königssee at different depths.

Echowand) from October 1982 to October 1983 using a closing net with a mesh size of 30 µm. Water temperature and Secchi disk transparency were measured on each sampling occasion. The animals were narcotized with 0.13% procaine hydrochloride, as described by May (1983), and preserved in 4% sugar-formalin. Samples were prepared for counting by adding a few drops of detergent to reduce surface tension and then concentrating the animals by sedimentation. Rotifers and eggs were counted using an inverted microscope. *P. vulgaris/P. dolichoptera* and *K. hiemalis/K. quadrata* were distinguishable as separate species, but for routine counts only the genus was taken into account.

Results

Temperature, species diversity, abundance, and seasonal succession

Mean monthly water temperatures are less than 8 °C at 10 m, and not above 6 °C at 20 m (Fig. 1)[Siebeck, 1982]. During summer, an epilimnion does not exist; the thermocline lies near the water surface (Fig. 2)[Müller, 1981]. At the surface, the maximum temperature may rise slightly above 20 °C in summer (e.g. 1983). Secchi disk transparency was 6–12 m during the year (18m maximum); the trophogenic layer extended down to 20 m.

In total, 20 rotifer species were found in Königs-

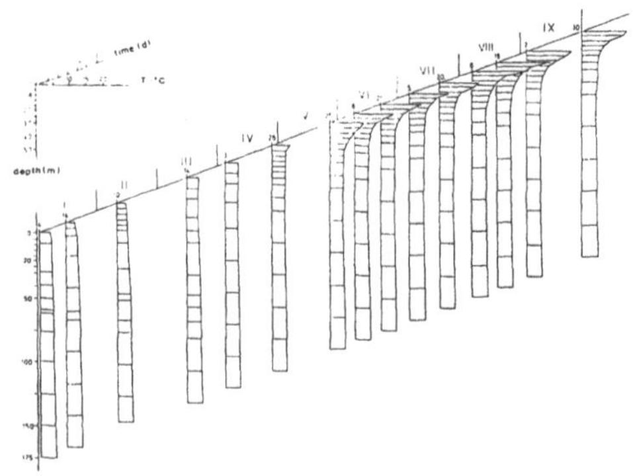

Fig. 2. Vertical temperature gradients in Königssee in 1980.

Table 1. List of the rotifer species found in Königssee.

Perennial species	
Synchaetidae:	*Polyarthra vulgaris* Carlin 1943
	Polyarthra dolichoptera Idelson 1925
Brachionidae:	*Keratella cochlearis* (Gosse 1851)
	Keratella hiemalis (Carlin 1943)
	Keratella quadrata (O.F. Müller 1786)
	Kellicottia longispina (Kellicott 1879)
Spring species	
Conochilidae:	*Conochilus unicornis* Rousselet 1892
Summer species	
Collòthecidae:	*Collotheca mutabilis* (Hudson 1885)
	Collotheca libera (Zacharias 1894)
Gastropodidae:	*Gastropus stylifer* Imhof 1891
	Ascomorpha ecaudis (Perty 1850)
	Ascomorpha saltans Bartsch 1870
	Ascomorpha ovalis (Bergendahl 1892)
Autumn species	
Filinidae:	*Filinia terminalis* (Plate 1886)
Rare species	
Synchaetidae:	*Synchaeta pectinata* Ehrenberg 1832
Brachionidae:	*Notholca* sp.
Collothecidae:	*Colletheca pelagica* (Rousselet 1893)
Lecanidae:	*Lecane* sp.
Notommatidae:	*Cephalodella* sp.
Trichocercidae:	*Trichocerca* sp.

see, including perennial species and species which occurred for only a short time during the year (Table 1). A similar seasonal succession of species has been observed in other lakes, e.g. Zimmerman [1974]; Elliott [1977]; Fuller *et al* [1977]; Fankhauser [1980]; Walz *et al.* [in press].

Figure 3 shows the percentage composition of the rotifer community, integrated over a vertical range of 0-50 m, for the two sampling stations from October 1982 to October 1983. These data include 85-100% of the total rotifer community of this very deep lake. There was no significant difference in rotifer abundance between the two sampling points (Wilcoxon, P = 0.05; matched pairs signed rank test; 95% significance level; Sachs [1984]. *Polyarthra vulgaris/dolichoptera, Kellicottia longispina* and *Keratella cochlearis* dominated the total rotifer community (90%). In autumn and early winter *Polyarthra* was the dominant rotifer genus reaching a maximum of 1.25×10^6 individuals m^{-2} or 61% of the total population (Figs. 3&4). A rapid decline in the *Polyarthra* population in January was paralleled by a gradual increase of *K. longispina*, which remained dominant until August, reaching a maximum of 75%. Its highest density in spring (1.24×10^6 individuals m^{-2} or 62%) is similar to the autumn peak exhibited by *Polyarthra* (Figs. 3&5). When *K. longispina* was dominant *Polyarthra* density remained low $\approx (7 \times 10^4$ individuals m^{-2} or 12%). However, when *Polyarthra* numbers increased in late summer, the *K. longispina* population declined to about 2.7×10^5 individuals m^{-2} (17%). On an annual basis, 35% of the rotifer community consisted of *P. vulgaris/dolichoptera* and 32% *K. longispina*, respectively. *Keratella cochlearis* comprised only 24% of the rotifer community and showed smaller fluctuations throughout the year. It reached a maximum density of 9.8×10^5 individuals m^{-2} (35%) in September (Figs. 3&6). From December to March, its density did not exceed 10^5 individuals m^{-2} (16%). *Conochilus unicornis* with its maximum in June (Fig. 7) and *Filinig terminalis* with its appearance in autumn (Fig. 8) showed a short-term seasonal occurrence. The greatest species diversity was observed in summer and autumn, however, other species contributed less than 10% to the community.

Vertical distribution

The rotifer community in Königssee was stratified with 60–80% of individuals living in the trophogenic zone (0-20 m); 85–100% were found in the upper 50 m of the water column where temperatures ranged between 3.7 °C and 4.3 °C. Species-specific depth preferences were observed. *P. vulgaris/dolichoptera* and *K. cochlearis* were found in their greatest numbers in the surface layers, 0-10 m (Figs. 4&6). *Conochilus unicornis* was also found here, although a few colonies were found between 10 m and 20 m (Fig. 7). *Kellicottia longispina* was more abundant in the deeper trophogenic zone (10–20 m) and in the tropholytic zone (20–50 m) (Fig. 5). *Filinia terminalis* was more common at depths of 10–20 m (Fig. 8). In contrast, *Keratella hiemalis/quadrata* existed exclusively in the tropholytic zone (20–75 m) (Fig. 9). Samples taken from

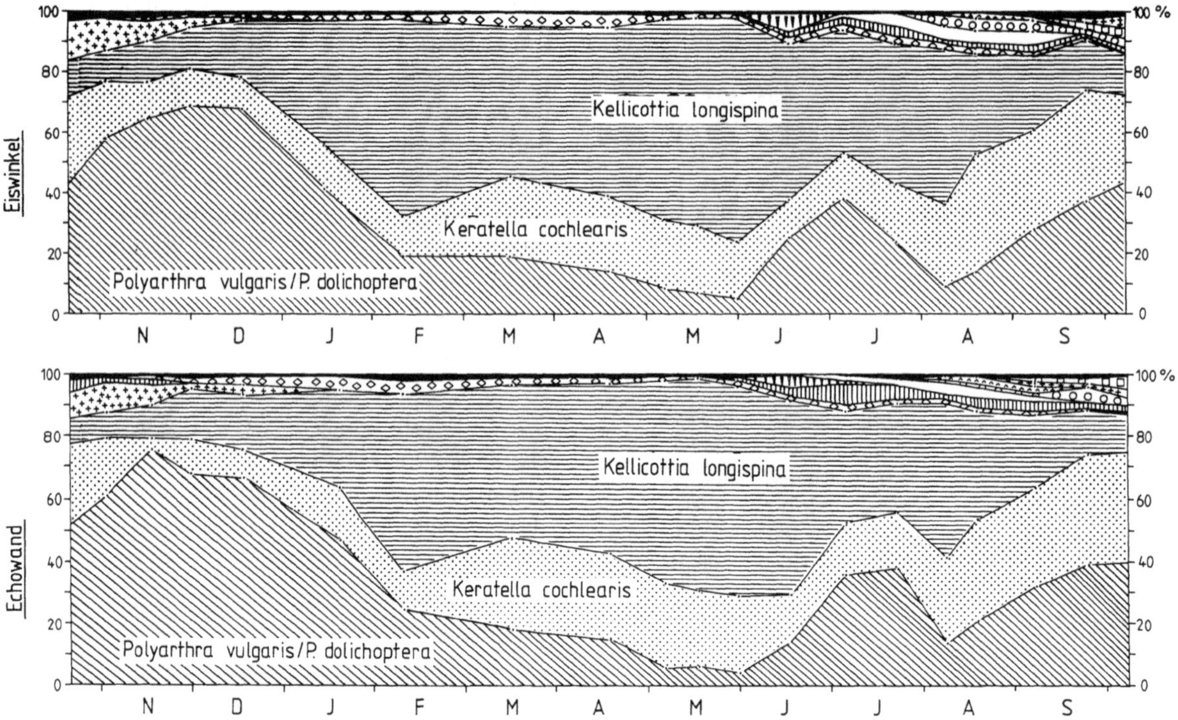

Fig. 3. Percentage distribution of rotifers at two sampling stations in Königssee:

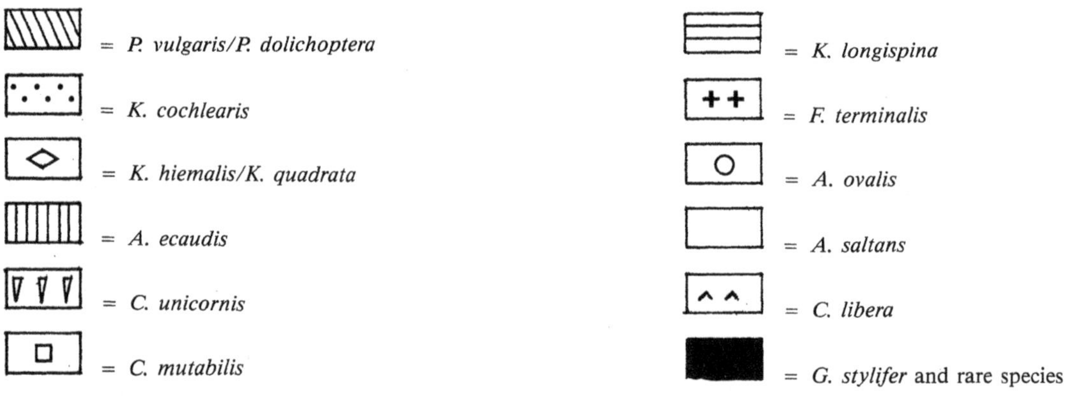

depths of 50–170 m, consisted mainly of detritus and the undamaged, empty loricas of *K. longispina* and *K. cochlearis.* [Laxhuber, 1981].

Discussion

Königssee possesses a plankton community which is typical of cold, oligotrophic lakes. Some of the rotifer species found here (e.g. *Ascomorpha ovalis, Collotheca libera*) are well known to avoid eutrophic lakes [Pejler, 1957, 1965; Ruttner-Kolisko, 1972; Gannon & Stemberger, 1978; Koste, 1978]. Additionally some cold-stenotherms, which are known to prefer oligotrophic lakes were present in Königssee (*K. longispina, P. vulgaris/dolichoptera, Conochilus unicornis, F. terminalis, K. hiemalis, Ascomorpha saltans*). Königssee shows a similarity to pre-eutrophic Lake Constance (1934/35) with 70% of the species being the same, but with densities 5 × greater [Walz *et al.*, in press]. The subse-

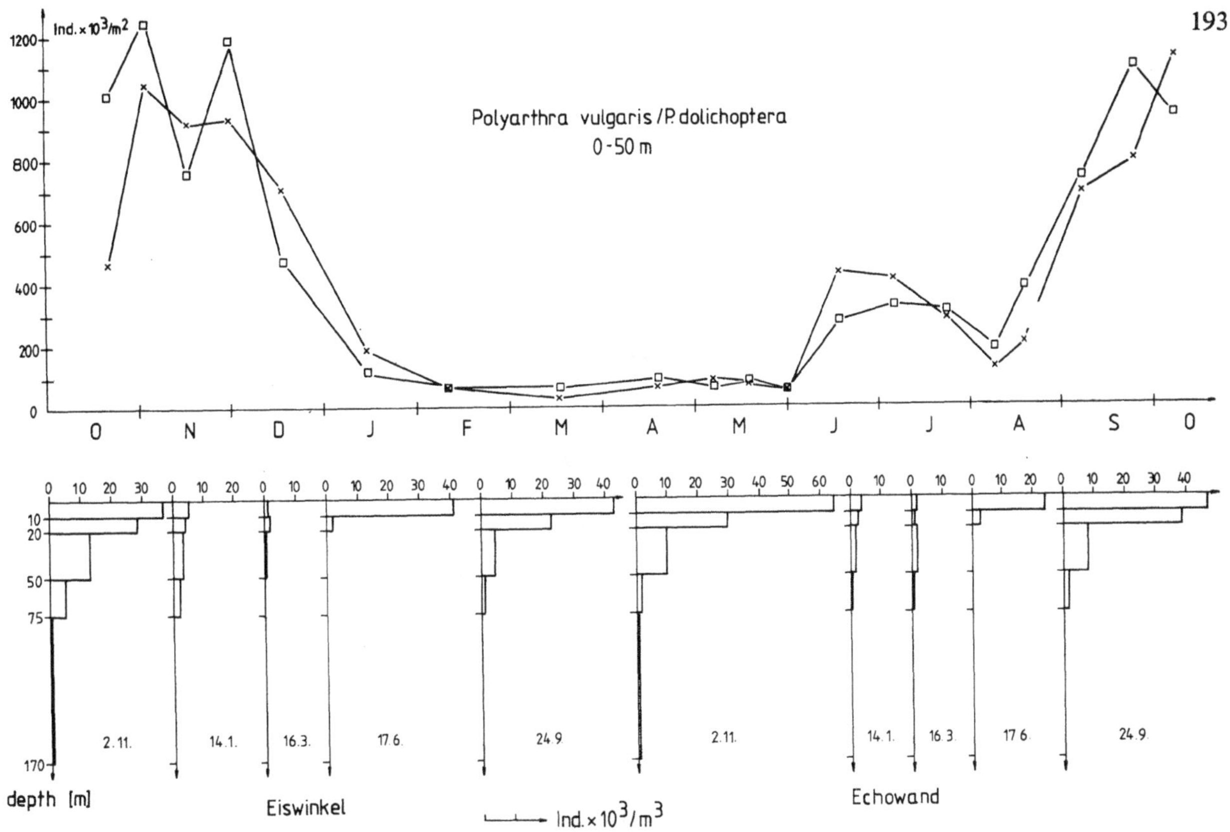

Fig. 4. Abundance (upper panel) and vertical distribution (lower panel) of *P. vulgaris/P. dolichoptera* in Königssee (× = Eiswinkel, □ = Echowand)

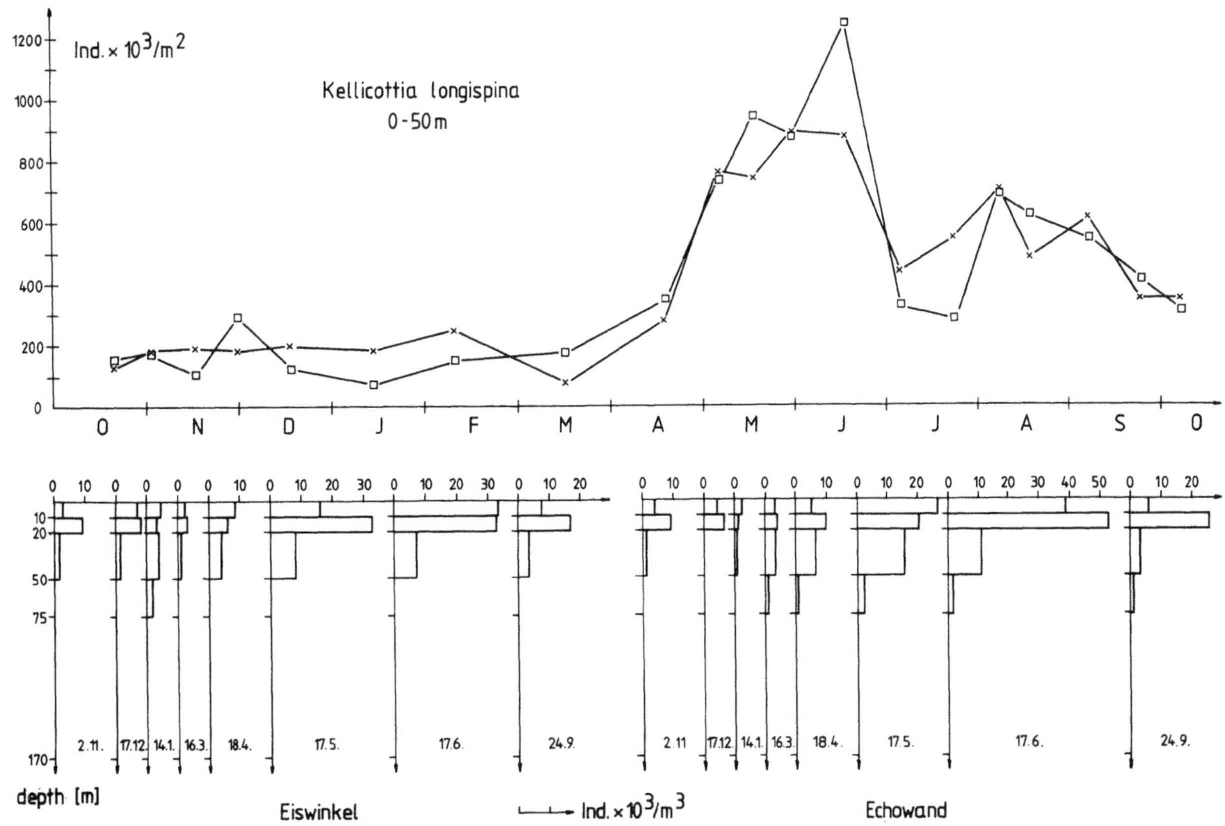

Fig. 5. Abundance (upper panel) and vertical distribution (lower panel) of *K. longispina* in Königssee (× = Eiswinkel, □ = Echowand)

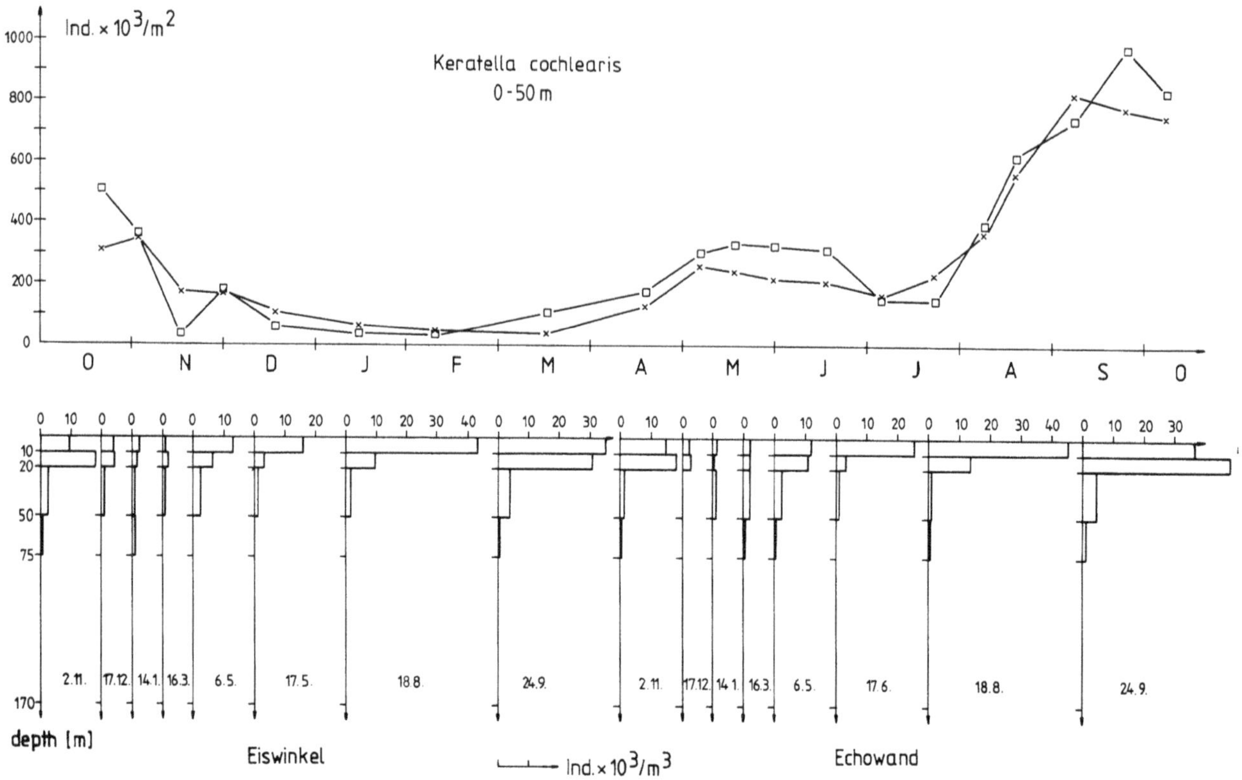

Fig. 6. Abundance (upper panel) and vertical distribution (lower panel) of *K. cochlearis in Königssee (× = Eiswinkel, □ = Echowand)*

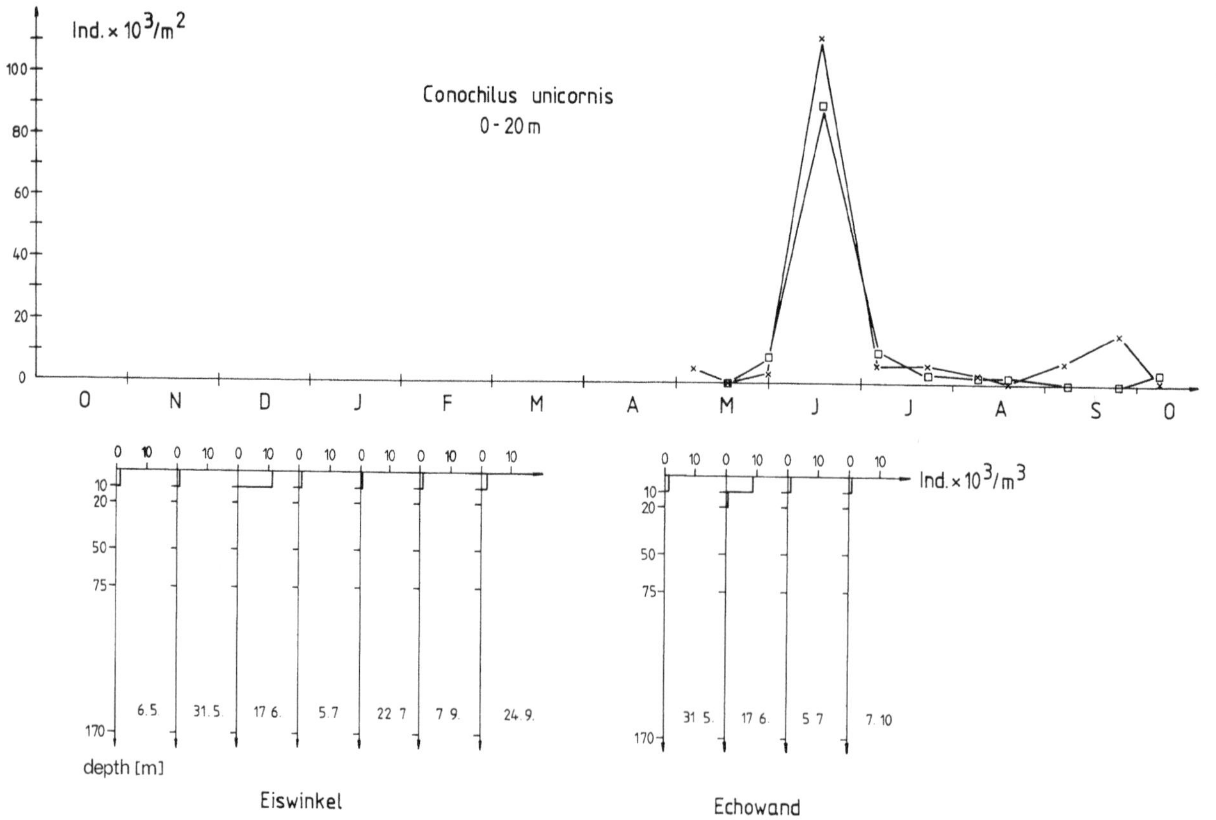

Fig. 7. Abundance (upper panel) and vertical distribution (lower panel) of *C. unicornis* in Königssee (× = Eiswinkel, □ = Echowand)

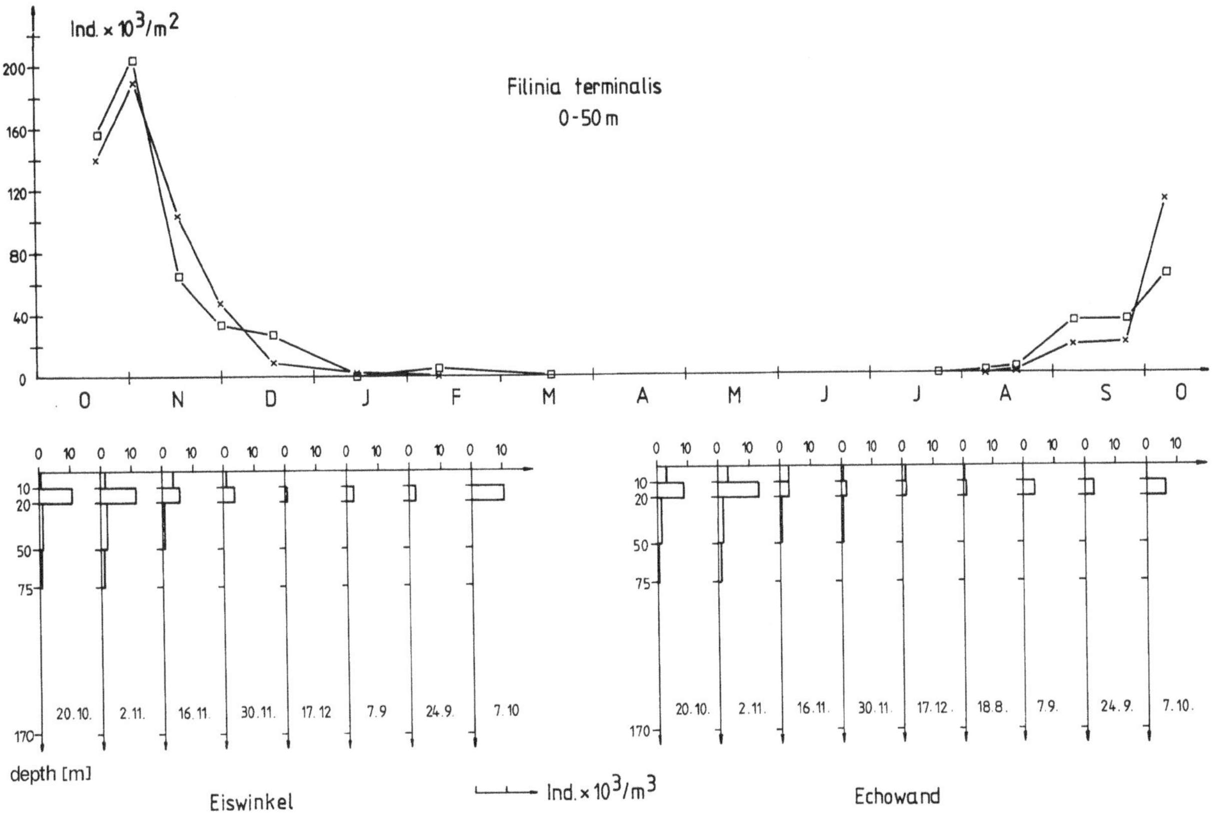

Fig. 8. Abundance (upper panel) and vertical distribution (lower panel) of *F. terminalis* in Königssee (× =Eiswinkel, □ = Echowand)

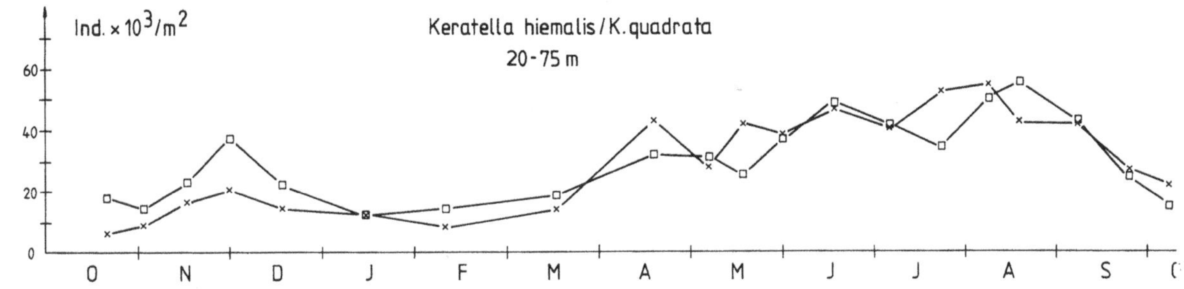

Fig. 9. Abundance of *K. hiemalis/quadrata* in 20–75 m (× = Eiswinkel, □ = Echowand)

quent shift in species composition and an increase in species number appears to be a direct consequence of the eutrophication of Lake Constance [Walz *et al.*, in press]. For example, *A. ovalis* which occurs in the late summer in Königssee is no longer present in Lake Constance and the predatory *A. priodonta* which is completely absent in Königssee has risen to high numbers in Lake Constance. In comparison with meso- and eutrophic lakes such as Lake Constance (3.6×10^7 individuals m^{-2}), such as off rotifer density in Königssee is still relatively low (2.8×10^6individuals m^{-2}) [see also Pechlaner 1970; Ruttner-Kolisko 1972; Zimmermann 1974; Frankhauser 1980; Cajander 1983]. The extreme temperature gradient, low mean temperature, great depth of the trophogenic layer, and restricted circulation probably influence rotifer abundance and distribution significantly. In pre-eutrophic Lake Constance, stratification of the three dominant rotifer species into narrow bands near the surface is comparable to our observations in Königssee [Walz *et al.*, in press]. In the course of the eutrophication of Lake Constance this stratification has become broader and has extended to greater depths [Walz *et al.*, in press]. Availability of food resources is probably an important factor in determining rotifer vertical distribution.

Acknowledgements

I thank Prof. O. Siebeck for reviewing the manuscript and Dr. N. Walz for helpful discussions and the use of some of his data on the rotifers of Lake Constance. I am grateful to Dr. W. Koste for his aid in species identification. Thanks are due especially to my husband and friends, without whose assistance the field work would not have been possible. This work is part of a research report to the administration of National Park of Berchtesgaden in conjunction with the Zoological Institute of Munich. The work was supported by the Bayerisches Staatsministerium für Landesentwicklung und Umweltfragen.

References

Cajander, V.-R., 1983. Production of planktonic Rotatoria in Ormajärvi, and eutrophicated lake in southern Finland. Hydrobiologia 104: 329–333.

Elliott, J., 1977. Seasonal changes in the abundance and distribution of planktonic rotifers in Grasmere (English Lake District). Freshwat. Biol. 7: 147–166.

Frankhauser, H., 1980. Untersuchungen über das pelagische Rotatorienplankton im Bielersee. Lizentiatsarbeit Zool. Inst. Univ. Bern, pp. 73–145.

Fuller, D. R., R. S. Stemberger & J. E. Gannon, 1977. Limnetic rotifers as indicators of trophic change. J. Elisha Mitchell Sci. Soc. 93: 104–113.

Gannon, J. E. & R. S. Stemberger, 1978. Zooplankton (especially Crustaceans and Rotifers) as indicators of water quality. Trans. am. microsc. Soc. 97: 16–35.

Koste, W., 1978. Rotatoria. Die Rädertiere Mitteleuropas (Textband, Tafelband). Gebrüder Bornträger, Berlin, Stuttgart, 673 pp, 234 Tafeln.

Laxhuber, R., 1981. Sedimentationsmessungen im Königssee – Biomasse, Sedimentation und Mineralisation des Phytoplanktons in verschiedenen Tiefenstufen eines oligotrophen Sees. Diplomarbeit, Universität München, pp. 119.

Laxhuber, R., 1984. Entfaltung, Sukzession und Dynamik der pelagischen Rotatorien (Rädertiere) des Königssees in Wechsel der Jahreszeiten. Schlussbericht zum Projekt des Bayerischen Staatsministeriums für Landesentwicklung und Umweltfragen und der Nationalparkverwaltung Berchtesgaden, pp. 72.

May, L., 1985. The use of procaine hydrochloride in the preparation of rotifer samples for counting. Verh. internat. Ver. Limnol. 22: 2987–2990.

Müller, R., 1981. Die Verteilung des Sauerstoffs im Königssee und ihre Ursachen. Diplomarbeit, Universität München, pp. 132.

Pechlaner, R., G. Bretschko, P. Gollmann, H. Pfeifer, M. Tilzer & H. P. Weissenbach, 1970. The production processes in two high-mountain lakes (Vorderer und Hinterer Finstertaler See, Kühtai, Austria). In Z. Kajak, & A. Hillbricht-Ilkowska (eds), Productivity Problems of Freshwaters. Warsaw. PWN Polish Scientific Publishers, pp. 239–269.

Pejler, B., 1957. Taxonomical and ecological studies on plankton Rotatoria from Northern Swedish Lapland. Svensk. Vetensch. Handl. 6: 1–67.

Pejler, B., 1965. Regional-ecological studies of Swedish freshwater zooplankton. Zool Bidr. Upps. 36: 407–515.

Ruttner-Kolisko, A., 1972. Rotatoria. In Die Binnengewässer, XXVI. Das Zooplankton der Binnengewässer I. Teil, Schweizerbartsche Verlagsbuchhandlung, Stuttgart, pp. 99–234.

Sachs, L., 1984. Angewandte Statistik. Springer Verlag, Berlin, Heidelberg, New York, 552 pp.

Siebeck, O., 1982. Der Königssee. Eine limnologische Projektstudie. Forschungsberichte 5, Nationalparkverwaltung Berchtesgaden, 131 pp.

Walz, N., H. J. Elster & H. Mezger, in press. The development of the rotifer community structure in Lake Constance during its eutrophication. Arch. Hydrobiol., Monogr. Beitr.

Zimmermann, C., 1974. Die pelagischen Rotatorien des Sempachersees, mit spezieller Berücksichtigung der Brachioniden und der Ernährungsfrage. Schweiz. Z. Hydrol. 36: 205–300.

Hydrobiologia 147: 197–201 (1987)
May, L., Wallace, R. & Herzig, A. (eds), Rotifer Symposium IV

Population dynamics of hypolimnetic rotifers in the Pluss-see (North Germany)

Wolfgang Hofmann
Max-Planck-Institut für Limnologie, Abt. Mikrobenökologie, Postfach 165, D-2320 Plön, F.R.G.

Keywords: rotifers, hypolimnion, population dynamics, population response

Abstract

The hypolimnetic rotifer populations of the Pluss-see (*Keratella hiemalis, Filinia terminalis, Filinia hofmanni*) show similar patterns of changing population parameters indicating that they respond to particular environmental signals in the same way.

Introduction

The planktonic rotifer community in Pluss-see has a distinct vertical structure and can be separated into an epilimnetic and a hypolimnetic assemblage [Hofmann, 1975, 1985]. The hypolimnion is inhabited by only a few species and is a more stable and uniform environment than the epilimnion. This biotope, therefore, represents a relatively simple system for ecological analysis.

In this paper, the relationship between the hypolimnetic rotifer assemblage and its environment are demonstrated by pattern analysis of population parameters.

Materials and Methods

The results presented were obtained in 1972 and 1981 from weekly vertical profiles (1 m intervals) using a 5 l water sampler. The zooplankton were concentrated with a 55 μm plankton net and counted under a stereo microscope at $\times 30$ magnification.

Results and discussion

At least temporarily high abundance and egg production in the hypolimnion was shown by four rotifer species, *Filinia terminalis* (Plate), *Filinia hofmanni* Koste, *Keratella hiemalis* Carlin, and *Keratella cochlearis* (Gosse). The first three of these species are typical of regions near the chemocline and the thermocline [Larsson, 1971; Ruttner-Kolisko, 1975, 1980; Miracle, 1977; Miracle & Vicente, 1983]. In 1972 and 1981 each of these species showed a distinct pattern of changing vertical distribution from spring to summer (Fig. 1 & Hofmann [1975]. In March and April *Keratella hiemalis* and *Filinia hofmanni* were restricted to the lowermost hypolimnion. During May and June, the both migrated upwards. In contrast, *Filinia terminalis* and *Keratella cochlearis* were concentrated in the epilimnion in early spring, and then moved downwards at the beginning of May; thus meeting the *K. hiemalis* and *F. hofmanni* populations. In 1972, this downward shift in mean depth of *K. cochlearis* was caused by a rapid decline of the population in the upper 5 m due to predation by *Asplanchna priodonta* [Hofmann, 1983]. In June/July *Filinia terminalis* showed the same upward shift as *K. hiemalis* and *F. hofmanni* whereas

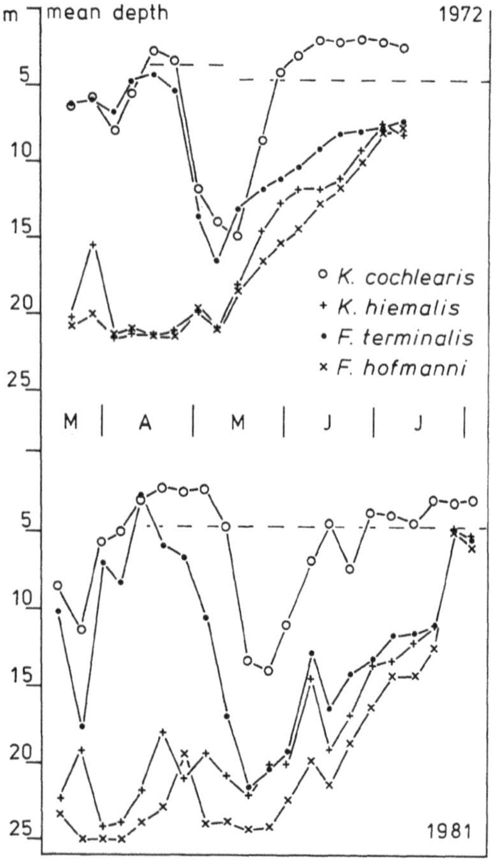

Fig. 1. Mean residence depths of *Keratella cochlearis, K. hiemalis, Filinia terminalis, F. hofmanni* in Pluss-see from March to July of 1972 and 1981. Broken line indicates the position of the thermocline.

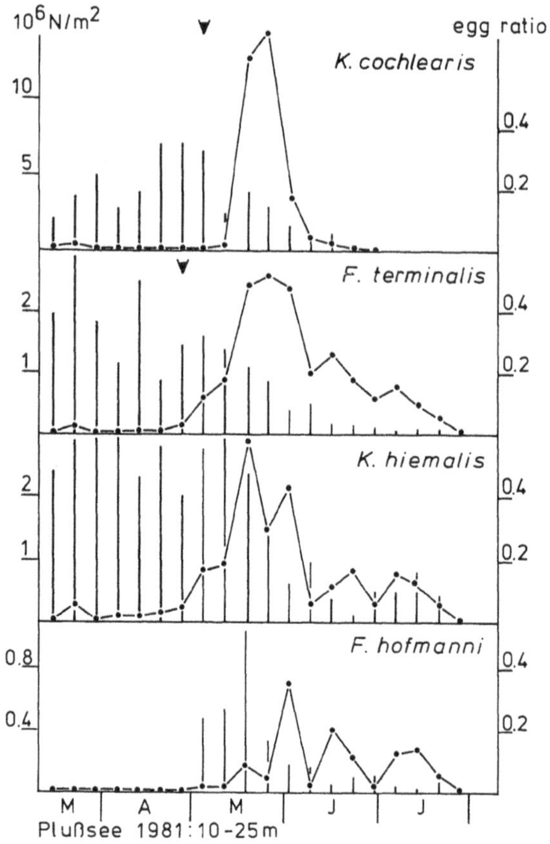

Fig. 2. Abundance (•) and egg per female ratios (vertical bars) of *Keratella cochlearis, K. hiemalis, Filinia terminalis* and *F. hofmanni* in the 10 to 25 m layer of Pluss-see from March to July 1981; arrows indicate maximum abundance in the epilimnion.

K. cochlearis disappeared from the hypolimnion, and subsequently re-established a population in the epilimnion. *K. cochlearis* occurred in the hypolimnion for a short time only, while *F. terminalis* showed a distinct shift, from an epilimnetic to a hypolimnetic distribution.

K. hiemalis, F. terminalis and *F. hofmanni* remained abundant in the hypolimnion until the end of July, although they no longer occurred in the deeper water, below 10 m. By August they had disappeared from the pelagic zone altogether. Maximum abundance of the four species occurred just before the upward migration, when *Keratella cochlearis* and *Filinia terminalis* had reached their deepest point of downward migration and there was considerable spatial overlap with the other two species.

Abundance and egg per female ratios (egg ratios) of *K. hiemalis, F. hofmanni* and the hypolimnetic populations of *F. terminalis* and *K. cochlearis* changed almost simultaneously (Fig. 2). During March and April egg ratios were high and abundance low. In May abundance increased rapidly, followed by a decrease in egg ratio and a subsequent decline in abundance. This pattern indicates a close relationship between fluctuations in abundance and fecundity, i.e. birth rates, in all three species. Since egg ratio is, itself, closely related to food availability these results also suggest that the three populations were feeding on very similar food.

In 1972 and 1981, the temporal and spatial distribution of rotifer abundances in excess of 500 ind l^{-1} were very similar for each of the four species (Figs. 4 & 6). *Filinia* spp. and *K. hiemalis* estab-

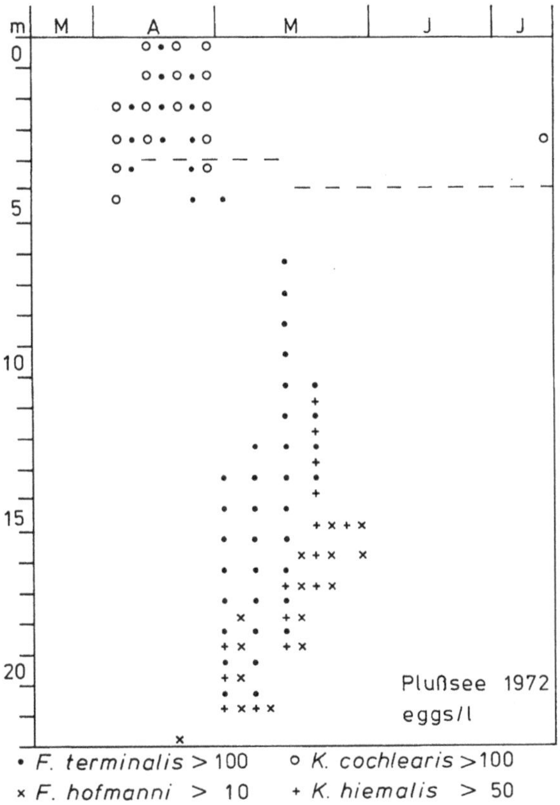

Fig. 3. Temporal and spatial distribution of amictic egg densities of *Keratella cochlearis, K. hiemalis, Filinia terminalis, F. hofmanni* in Pluss-see, March–July 1972. Broken line indicates the position of the thermocline.

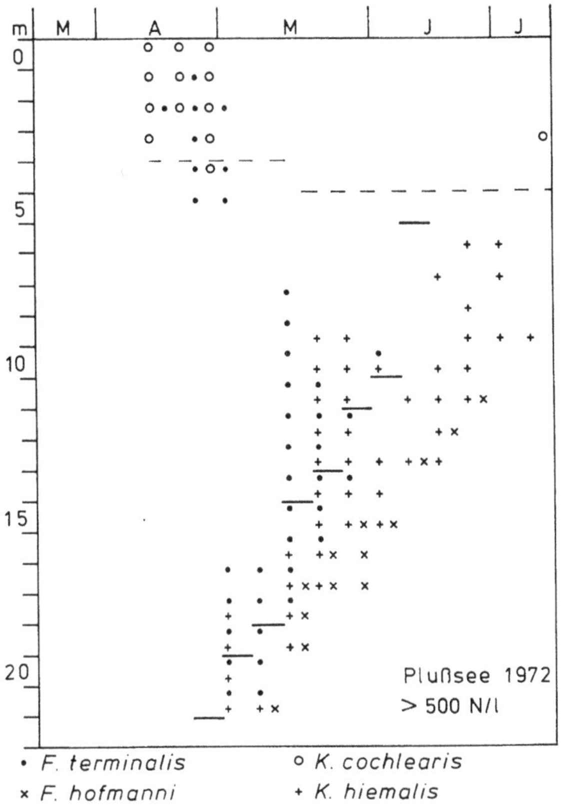

Fig. 4. Temporal and spatial distribution of rotifer population densities exceeding 500 ind l^{-1} in Pluss-see, March–July 1972; horizontal lines indicate the position of the 1 mg 0_2 l^{-1} line. Broken line indicates the position of the thermocline.

lished a joint centre of high abundances in the lower hypolimnion at the beginning or middle of May. Subsequently, these dense patches shifted upward, together, with a high degree of spatial overlap. Figures 3 and 5 indicate that these temporal and spatial abundance patterns developed from the centres of egg production, as indicated by the areas of maximum egg numbers per litre.

In *F. terminalis* the epilimnetic and hypolimnetic egg abundance maxima were separated, whereas for *K. cochlearis* (in 1981) they were connected. Egg numbers first increased in the epilimnion and about two weeks later in the hypolimnion. Hence, displacement of mean residence depth from the epilimnion to the hypolimnion evolved by different mechanisms in *F. terminalis* and *K. cochelaris.* In *F. terminalis* the shift was due to disappearance of the animals from the epilimnion with the establishment of a separate centre of high production in the lower hypolimnion. In contrast, in *K. cochlearis* the high hypolimnetic abundance of 1981 originated from an epilimnetic abundance peak from which animals shifted downwards, maintaining a high birth rate.

The upward movement of *K. hiemalis* and the two *Filinia*-species was closely related to increasing oxygen depletion in the lower hypolimnion, as indicated by the position of the 1 mg 0_2 l^{-1} line, and the production of H_2S. Decrease in oxygen concentration began about three weeks later in 1981 than in 1972, and the upward migration was correspondingly delayed. During this shift, most representatives of these populations were located below the 1 mg 0_2 l^{-1} boundary, which shows that all three species are able to tolerate very low oxygen concentration for long periods of time.

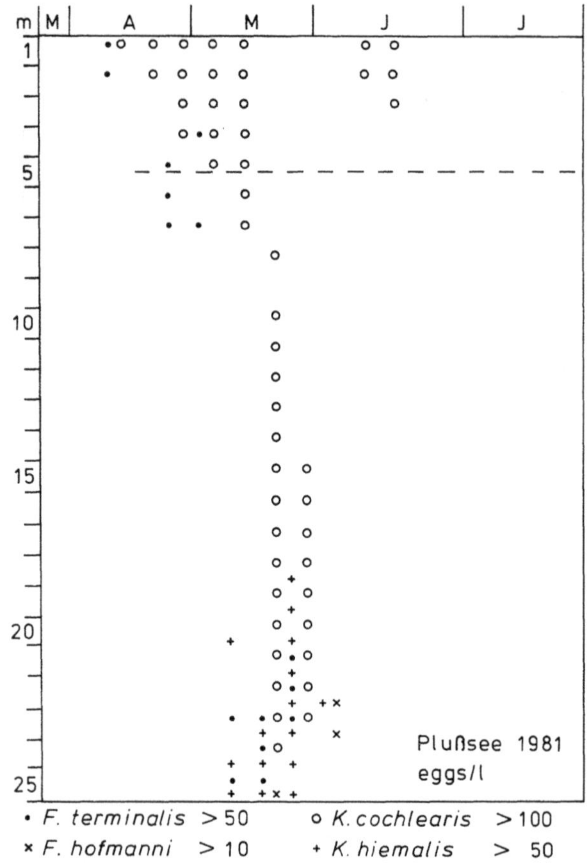

Fig. 5. Temporal and spatial distribution of high egg abundances of *Keratella cochlearis. K. hiemalis, Filinia terminalis, F. hofmanni* in Pluss-see, March–July 1981. Broken line indicates the position of the thermocline.

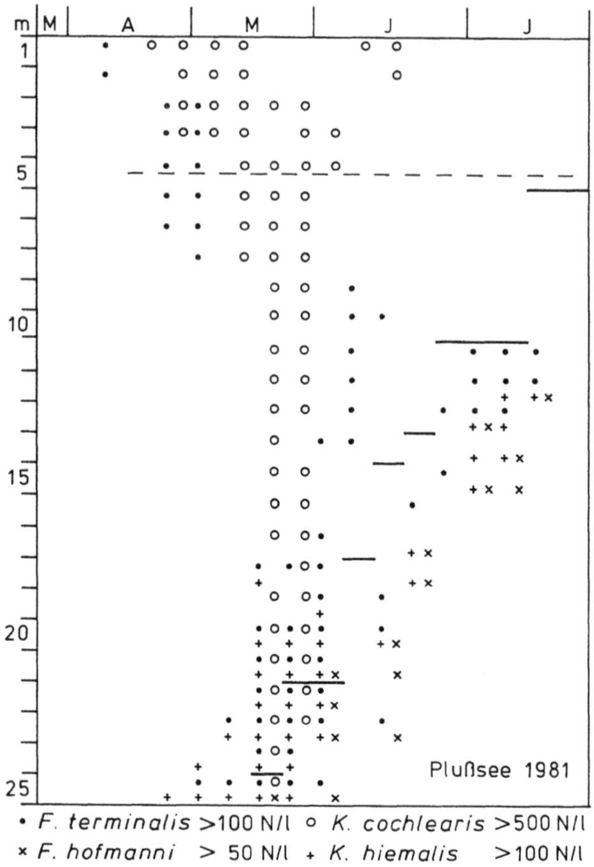

Fig. 6. Temporal and spatial distribution of high abundances of *Keratella cochlearis, K. hiemalis, Filinia terminalis , F. hofmanni* in Pluss-see, March–July 1981. Broken line indicates the position of the thermocline. Horizontal line indicates the position of the 1 mg O_2 l^{-1} line.

Oxygen conditions were most extreme in *F. hofmanni* which was generally the deepest of the three species in the vertical profiles. More than 80% of *F. hofmanni* experienced environmental oxygen conditions of 0–1.7 mg O_2 l^{-1}, from May to July (temperature: 4–4.6 °C). The corresponding ranges for the other species were: *K. hiemalis:* 0–9 mg O_2 l^{-1}, 3.4–8.0 °C; *F. terminalis:* 0–13.9 mg O_2 l^{-1}, 3.6–8.6 °C; *K. cochlearis:* 0.6–14.7 mg O_2 l^{-1}, 3.5–19.2 °C (March–July 1981).

In autumn, the thermocline and the oxygen boundary again shifted downwards. However, no hypolimnetic rotifer assemblage was established during this period. This may be due to unfavorable food supply, or the fact that the boundary is displaced downwards from the epilimnion.

In summary, the results suggest that *K. hiemalis* and the two *Filinia* species have some common characteristics. They have similar food requirements (detritus, bacteria), prefer low temperatures, and are able to tolerate low oxygen concentrations. As these factors changed from spring to summer the three species responded in a similar way. Thus, a similar pattern of population dynamics occurred in all three species, with a high degree of temporal and spatial overlap.

Makarewicz & Likens [1975] found that the response surfaces (areas of high production) of epilimnetic zooplankton species with similar food requirements were separated, thus reducing competition and favouring species coexistence. However, in the hypolimnion of Pluss-see, almost complete overlap of response surfaces (areas of maximum egg abundance) of the rotifer species occurred. This suggests that competition in the hypolimnion is already reduced by various factors, such as variable conditions, importance of density independent factors (oxygen, temperature), the temporal character of an unsaturated environment and prevailing r-selection [Hutchinson, 1961; Pianka, 1972; Nichols, *et al.* 1976].

References

Hofmann, W., 1975. The influence of spring circulation on zooplankton dynamics in the Pluss-see. Verh. int. Ver. Limnol. 19: 1241–1250.

Hofmann, W., 1983. Interactions between Asplanchna and Keratella cochlearis in the Pluss-see (north Germany). Hydrobiologia 104: 363–365.

Hofmann, W., 1985. Dynamics of vertical zooplankton community structure in the Pluss-see: Cluster analysis. Verh. int. Ver. Limnol. 22: 2983–2986.

Hutchinson, G. E., 1961. The paradox of the plankton. Amer. Nat. 95: 137–146.

Larsson, P., 1971. Vertical distribution of planktonic rotifers in a meromictic lake; Blankvatn near Oslo, Norway. Norw. J. Zool. 19: 47–75.

Makarewicz, J. C. & G. E. Likens, 1975. Niche analysis of a zooplankton community. Science 190: 1000–1003.

Miracle, M. R., 1977. Migration, patchiness and distribution in time and space of planktonic rotifers. Arch. Hydrobiol. Beih. 8: 19–37.

Miracle, M. R. & E. Vicente, 1983. Vertical distribution and rotifer concentrations in the chemocline of meromictic lakes. Hydrobiologia 104: 259–267.

Nichols, J. D., W. Conley, B. Batt, & A. R. Tipton, 1976. Temporally dynamic reproductive strategies and the concept of r- and K-selection. Am. Nat. 110: 995–1005.

Pianka, E., 1972. r and K seleaction or b and d selection? Am. Nat. 106: 581–588.

Ruttner-Kolisko, A., 1975. The vertical distribution of plankton rotifers in a small alpine lake with sharp oxygen depletion. Verh. int. Ver. Limnol. 19: 1286–1294.

Ruttner-Kolisko, A., 1980. The abundance and distribution of Filinia terminalis in various types of lakes as related to temperature, oxygen and food. Hydrobiologia 73: 169–175.

Hydrobiologia 147: 203–208 (1987)
May, L., Wallace, R. & Herzig, A. (eds), Rotifer Symposium IV

Environmental factors influencing the vertical migration of planktonic rotifers in a hypereutrophic tarn

L. J. Stewart[1] & D. G. George[2]
[1]*Freshwater Biological Association, The Ferry House, Ambleside, Cumbria LA22 0LP, U.K. and Dept. of Plant Biology and Genetics, The University of Hull, Hull HU6 7RX, U.K.*
[2]*Freshwater Biological Association, The Ferry House, Ambleside, Cumbria LA22 0LP, U.K.*

Keywords: rotifers, vertical migration, environmental variables

Abstract

The diel vertical migration of planktonic rotifers in a small, hypereutrophic tarn was investigated on four occasions in 1983. When the tarn was isothermal the rotifers were distributed throughout the water column. After stratification, the rotifers were confined to the top 1–2 m of oxygenated water. On all four dates the rotifers were aggregated at specific depths in the water column. On some occasions, the pattern of aggregation changed as the animals performed distinct diurnal migrations. *Keratella cochlearis, K. quadrata* and *Polyarthra vulgaris* usually followed the reverse migrations of the phytoplankton. In contrast, the movements of *Anuraeopsis fissa* were less pronounced and were associated with variations in the depth of the oxycline.

Introduction

There are numerous studies of diel migration in planktonic crustacea [see reviews by Cushing, 1951 and Longhurst, 1976] but relatively few accounts of diel movement in rotifers. Most accounts of rotifer migration describe the general pattern of movement, without relating these movements to changes in the animals' environment. This paper describes the diel movements of planktonic rotifers in a small, intensely stratified tarn and relates these movements to changes in the animals' physical, chemical and biotic environment.

Description of site

Priest Pot is a highly eutrophic tarn in the English Lake District (54°22′31″N, 2°58′55″W). It is one hectare in area, on average 2.3 m deep and is physically and chemically stratified throughout the summer. Few planktonic crustacea are present and the zooplankton is dominated by rotifers, at densities often exceeding 40000 individuals per litre. Further details of this tarn are given by Goulder [1972, 1980], Finlay and Berninger [1984] & Hewitt and George [1987].

Methods

The diel vertical movements of planktonic rotifers in Priest Pot were investigated over four 24 h periods in April, May, July and September. In April samples were collected every three hours but on the other dates sampling was more frequent at sunrise and sunset. In April, May and September the sampling depth interval was 0.4 m, but in July the interval was changed to 0.2 m in the epilimnion with additional samples taken at 1.4, 1.8 and 2.8 m in the hypolimnion.

Samples were collected using a 1.5 l min^{-1} capacity peristaltic pump (Glen Creston Ltd., Stanmore, Middlesex). The inlet pipe was a 5 m length of reinforced plastic tubing fitted with a laminar flow head [Cunningham & Davison, 1980] which allowed discrete water samples to be collected at 5 cm intervals. Temperature and oxygen profiles were recorded with a temperature compensated dissolved oxygen probe (Yellow Springs Instrument Co., Yellow Springs, Ohio). Surface light intensities and under water light attenuation were measured using a Lambda Flat-plate sensor (Lambda Instrument Co., Lincoln, Nebraska) connected to a Fluke multimeter (Fluke (Nederlander) B.V., Tilburg, The Netherlands).

Sub-samples of water for chlorophyll extraction and chemical analysis were processed within 12–24 hours of collection. The concentration of chlorophyll *a* was determined by filtering 100–250 ml of the water sample through a Whatman GF/C glass fibre filter and extracting the pigment by boiling in 90% methanol for 1–2 minutes. The total chlorophyll *a* concentration (uncorrected for degradation products) was calculated using the abbreviated equation of Talling [1974].

Processing of all rotifer samples was begun within two hours of collection. In April, when rotifer numbers were low, 2-litre sub-samples were concentrated by reverse filtration through a 35 μm filter before sedimentation [Dodson & Thomas, 1964]. Otherwise, 1-litre samples were concentrated directly by sedimentation. All samples were killed and fixed in non-acetic Lugol's iodine. Sub-samples of the concentrate were counted in a glass Sedgewick-Rafter cell using a modified microcomputer as a tally counter [Cunningham & Purewal, 1983].

Animal densities at each depth were expressed as percentages and mean depth of the population ($\bar{D}$) was calculated using the weighted expression:

$$\bar{D} = \Sigma_i D_i N_i / \Sigma_i N_i$$

where D_i is the depth of the ith sample and N_i the concentration of animals at that depth.

Results

Patterns of vertical migration in Priest Pot

Figure 1 shows the diel change in the vertical distribution of the four most abundant rotifer species in Priest Pot on the 13–14 April, 1983. Sunset was at 19.00 hours and sunrise at 05.00 hours GMT and 80–90% of incident light was attenuated in the top 0.5 m (Fig. 1a). The weather was wet with winds of up to 8 m s^{-1}, and the tarn was isothermal at 7–8 °C. The phytoplankton was dominated by

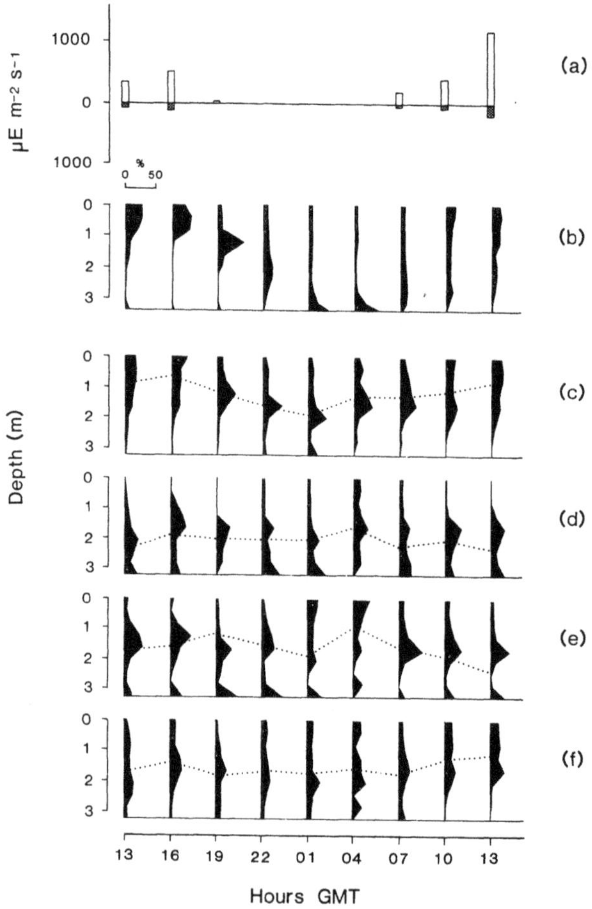

Fig. 1. Physical, chemical and biological parameters measured during a study of vertical migration of rotifers in Priest Post, (13/14 April 1983). (a) Light intensity at the surface (open blocks) and at 0.5 m depth (shaded blocks) (μE m^{-2} s^{-1}). (b) depth profiles of chlorophyll *a*. (c–f) depth profiles of rotifers with mean depths at each sampling time joined by a dotted line. Species shown are *K. cochlearis* (c), *K. quadrata* (d), *F. terminalis* (e) and *P. vulgaris* (f).

Synura sp. (Chrysophyceae) which moved away from the surface at night and returned the following day (Fig. 1b). *Keratella cochlearis* (Gosse) (Fig. 1c) followed the *Synura*, with the mean depth decreasing during the day and increasing at night. The diel movements of *Keratella quadrata* (Müller) (Fig. 1d) were less pronounced, but this species also moved into deeper water at night. In contrast, *Filinia terminalis* (Plate) (Fig. 1e) performed a nocturnal migration with most of the animals moving towards the surface at night. *Polyarthra vulgaris* (Carlin) (Fig. 1f) showed no clear pattern of movement but tended to aggregate at depth throughout the sampling period.

Figure 2 shows the diel movements of four

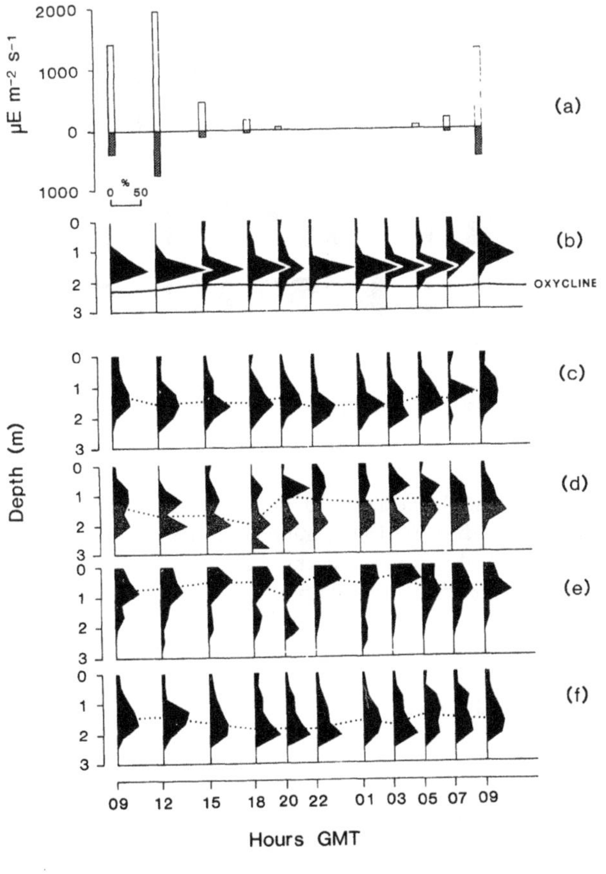

Fig. 2. Physical, chemical and biological parameters measured during a study of vertical migration of rotifers in Priest Pot, (23/24 May 1983). (a) light intensity at the surface (open blocks) and at 0.5 m depth (shaded blocks) (μE m^{-2} s^{-1}). (b) depth profiles of chlorophyll *a* with oxycline marked. (c–f) depth profiles of rotifers with mean depth joined by a dotted line. Species as for Fig. 1.

rotifer species on 23–24 May 1983. Sunset was at 20.15 hours and sunrise at 04.00 hours GMT and 60–80% of incident light was attenuated in the top 0.5 m (Fig. 2a). The weather was cool with winds of up to 5 m s^{-1}. The tarn was thermally stratified and the temperature in the epilimnion ranged from 11–16 °C. A pronounced oxycline had formed between 2.0 and 2.5 m (Fig. 2b) and the concentration of oxygen in the epilimnion ranged from 10 to 20 mg l^{-1}.

The phytoplankton was dominated by *Peridinium lomnickii* (Woloszynska) which was concentrated at a depth of 1.5–1.6 m on the afternoon of 23 May and moved towards the surface on the morning of the 24th (Fig. 2b). Throughout the 24 hour period, the rotifers were confined to the top 2.25 m of oxygenated water. The rotifer migrations were not especially pronounced but *K. cochlearis* (Fig. 2c) and *P. vulgaris* (Fig. 2f) moved deeper at night and *F. terminalis* (Fig. 2e) performed a weak nocturnal migration. The movements of *K. quadrata* (Fig. 2d) were irregular and the depth distributions frequently bimodal.

Figure 3 shows the diel movements of four rotifer species on 18–19 July 1983. Sunset was at 20.30 hours and sunrise at 04.00 hours GMT and 60–80% of incident light was attenuated in the top 0.5 m (Fig. 3a). The weather was calm and sunny, temperatures in the epilimnion ranged from 20–24 °C and a very steep oxycline was present between 1.1 and 1.3 m (Fig. 3b). The epilimnetic phytoplankton was dominated by the Chrysophyte *Synura* but a dense population of photosynthetic bacteria was present in the hypolimnion (Fig. 3b).

Four species of rotifers were abundant in the tarn: *K. cochlearis, K. quadrata, P. vulgaris* and *Anuraeopsis fissa* (Gosse). There was again no clear pattern of migration, but all four species dispersed upwards during the night and most reaggregated at depth the following day (Fig. 3c–f). Throughout the 24 hour period, *A. fissa* (Fig. 3f) tended to remain deeper in the water column than the other three species and at times appreciable numbers were recovered from hypoxic water.

Figure 4 shows the diel movements of five rotifer species on 5–6 September 1983. Sunset was at 18.30 hours and sunrise at 05.00 hours GMT and

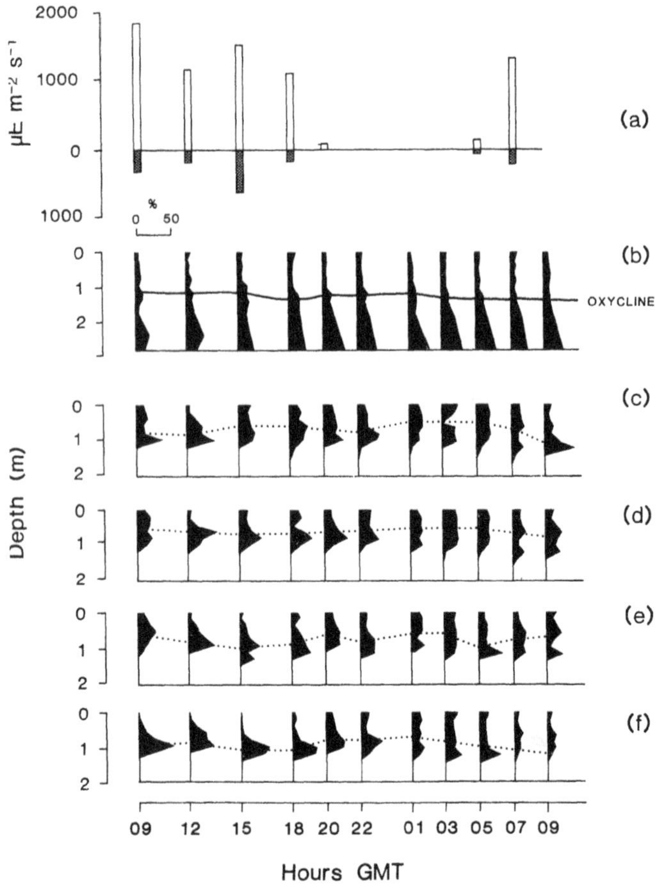

Fig. 3. Physical, chemical and biological parameters measured during a study of vertical migration of rotifers in Priest Pot, (18/19 July 1983). (a) light intensity at the surface (open blocks) and at 0.5 m depth (shaded blocks) (μE m^{-2} s^{-1}). (b) depth profiles of chlorophyll *a* with oxycline marked. (c–f) depth profiles of rotifers with mean depths joined by a dotted line. Species shown are *K. cochlearis* (c), *K. quadrata* (d), *P. vulgaris* (e) and *A. fissa* (f).

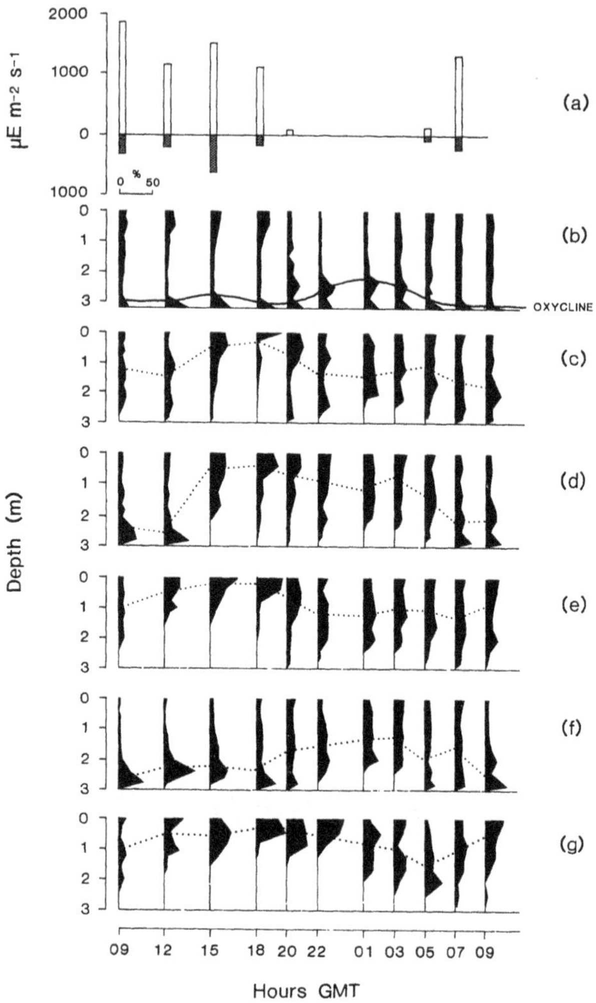

Fig. 4. Physical, chemical and biological parameters measured during a study of vertical migration of rotifers in Priest Pot. (5 September 1983). (a) Light intensity at the surface (open blocks) and at 0.5 m depth (shaded blocks) (μE m^{-2} s^{-1}). (b) depth profiles of chlorophyll *a* with oxycline marked. (c–g) depth profiles of rotifers with mean depths joined by a dotted line. Species shown are *K. cochlearis* (c), *K. quadrata* (d), *P. vulgaris* (e), *A. fissa* (f) and *F. brachiata* (g).

80–100% of incident light was attenuated in the top 0.5 m (Fig. 4a). The weather was dry with winds of 6 m s^{-1} gusting to 12 m s^{-1}. The water column was relatively well mixed but there was still some deoxygenation above the sediment (Fig. 4b). The phytoplankton was again dominated by *Synura* which performed a very definite reverse migration in the top 2.0–2.5 m of water (Fig. 4b). The movements of *K. cochlearis* (Fig. 4b), *K. quadrata* (Fig. 4d), *P. vulgaris* (Fig. 4e) were similar to those of the *Synura* but the three rotifers did not form such pronounced deep water maxima. *Filinia brachiata* (Rousselet) (Fig. 4g) did become concentrated above the oxycline at night, but the timing of their movement was very different to that of the *Synura*. The movements of *A. fissa* (Fig. 4f) were quite unlike those of other rotifers. This species was initially concentrated above the oxycline and only moved upwards when the depth of the oxycline changed.

Table 1. Summary of the results from four twenty-four hour studies of diel migration in rotifers.

Species	Date	Type of Migration	Amplitude of Migration (m)	Maximum Speed of Ascent cm s^{-1}	Maximum Speed of Descent cm s^{-1}
K. cochlearis	April	R	1.36	0.006	0.005
	May	R	0.48	0.003	0.003
	July	N	0.59	–	–
	Sept	MR	1.39	0.008	0.009
K. quadrata	April	R	1.11	0.004	0.006
	May	I	0.89	0.008	0.003
	July	N	0.23	–	–
	Sept	MR	2.09	0.018	0.012
P. vulgaris	April	I	0.73	0.005	0.004
	May	R	0.40	–	–
	July	N	0.46	–	–
	Sept	MR	1.07	0.005	0.007
F. terminalis	April	N	1.63	0.009	0.006
	May	N	0.59	0.005	0.004
F. brachiata	Sept	MR	1.28	0.007	0.008
A. fissa	July	N	0.48	–	–
	Sept	N	1.43	0.012	0.009

R : reverse migration
MR: modified reverse migration
N : nocturnal movements
I : irregular movements

Discussion

The various patterns of rotifer migration observed in Priest Pot in 1983 are summarized in Table 1. The individual species did not follow any consistent pattern of movement, but the most common behaviour was some form of reverse migration. Reverse migration is usually thought of as abnormal behaviour in planktonic crustacea [Hutchinson, 1967] but has often been reported in planktonic rotifers [Pennak, 1944; George & Fernando, 1970; Dumont, 1972; Cruz-Pizarro, 1978].

The migrations with the highest amplitudes were recorded in April and September when the rotifers were able to move through the whole water column. In summer, hypolimnetic anoxia confined the rotifers to the top 1–2 m and the amplitudes of migration were correspondingly reduced. The maximum ascending and descending velocities recorded in Table 1 are nevertheless comparable to those reported by George & Fernando (1970) in the much deeper Sunfish Lake, Ontario.

Light is generally regarded as the most important factor influencing the vertical migration of plankton. Diel changes in light intensity can act both as a stimulus for initiating vertical migration and as a means of controlling the speed and amplitude of movement [Blaxter, 1975]. Most of the rotifers in Priest Pot appear to use light as an initiating stimulus rather than a controlling factor. The two most common species (Fig. 5) certainly make no attempt to regulate their depth within a 'preferred' range of light intensities. This behaviour pattern is very different to that commonly reported in Crustacea where the animals remain within a relatively narrow range of light intensities and move towards the sur-

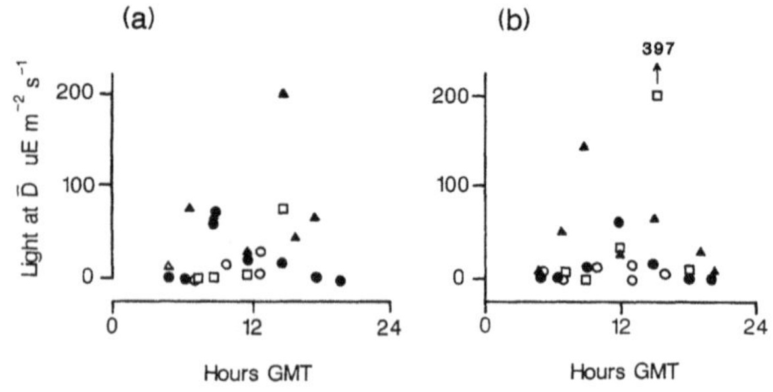

Fig. 5. Diel variations in the light intensity at the mean depth of (a) *K. cochlearis*, (b) *P. vulgaris*. The different symbols represent sampling dates.

face as the light fades [see Harris & Wolfe, 1956; Boden & Kampa, 1967].

In Priest Pot the most pronounced reverse migrations were associated with similar diel movements of the phytoplankton. It is not at all clear why the rotifers follow the migrating phytoplankton. Many investigators have suggested that food conditions can influence vertical migration but the motile phytoplankton in Priest Pot are generally too large to ingest. Perhaps the spatial association between the rotifers and the phytoplankton simply reflects a common response to the diel light cycle. The only rotifer species which persistently moved independently of the phytoplankton was *A. fissa*. This species is known to feed almost exclusively on detritus and bacteria [Pourriot, 1977], and was invariably associated with the oxycline where bacterial numbers were high.

Acknowledgements

We wish to thank Diane Hewitt and Nicholas White for assisting with field work. We are grateful to Dr R. Goulder for helpful comments on the manuscript and Trevor Furnass for drawing the figures.

This research was supported by the Natural Environment Research Council. One of us (LJS) was in receipt of a NERC(CASE) research studentship.

References

Blaxter, J. H. S., 1975. The role of light in the vertical migration of fish – a review. Symp. Br. ecol. Soc. 16: 189–210.

Boden, B. P. & E. M. Kampa, 1967. The influence of natural light on the vertical migration of an animal community in the sea. Symp. zool. Soc. Lond. 19: 15–26.

Cruz-Pizarro, L., 1978. Comparative vertical zonation and diurnal migration among crustacea and rotifera in the small, high mountain lake, La Caldera (Granada, Spain). Verh. int. Ver. Limnol. 20: 1026–1032.

Cunningham, C. R. & W. Davison, 1980. An opto-electronic sediment detector and its use in the chemical profiling of lakes. Freshwat. Biol. 10: 413–418.

Cunningham, C. R. & J. S. Purewal, 1983. A microcomputer based aid for counting plankton samples. J. Plankton Res. 5: 783–786.

Cushing, D. H., 1951. The vertical migration of planktonic crustacea. Biol. Rev. 26: 158–192.

Dodson, A. N. & W. H. Thomas, 1964. Concentrating plankton in a gentle fashion. Limnol. Oceanogr. 9: 455–456.

Dumont, H. J., 1972. A competition based approach of the reverse migration in zooplankton and its implication, chiefly based on a study of the interactions of the rotifer Asplanchna priodonta (Gosse) with several crustacean Entomostraca. Int. Revue ges. Hydrobiol. 57: 1–38.

Finlay, B. J. & U.-G. Berninger, 1984. Co-existence of congeneric ciliates (Karyorelictida: Loxodes) in relation to food resources in two freshwater lakes. J. anim. Ecol. 53: 929–943.

George, M. G. & C. H. Fernando, 1970. Diurnal migration in three species of rotifers in Sunfish Lake, Ontario. Limnol. Oceanogr. 15: 218–223.

Goulder, R., 1972. The vertical distribution of some ciliated protozoa in the plankton of a eutrophic pond during summer stratification. Freshwat. Biol. 2: 163–176.

Goulder, R., 1980. The ecology of two species of primitive ciliated protozoa commonly found in standing freshwaters (Loxodes magnus Stokes and L. striatus Penard). Hydrobiologia 72: 131–158.

Harris, J. E. & U. K. Wolfe, 1956. A laboratory study of vertical migration. Proc. Roy. Soc. B 144: 329–354.

Hewitt, D. P. & D. G. George, 1987. The population dynamics of Keratella cochlearis in a hypereutrophic tarn and the possible impact of predation by young roach. Hydrobiologia, this volume.

Hutchinson, G. E., 1967. A treatise on limnology. Vol. 2. Introduction to lake biology and the limnoplankton. John Wiley & Sons, N.Y., 1115 pp.

Longhurst, A. R., 1976. Vertical migration. In: D. H. Cushing & J. J. Walsh (eds) The Ecology of the Seas. Blackwell, Oxford: 116–137.

Pennak, R. W., 1944. Diurnal movements of zooplankton organisms in some Colorado mountain lakes. Ecology 25: 387–403.

Pourriot, R., 1977. Food and feeding habits of Rotifera. Arch. Hydrobiol. Beih. 8: 243–260.

Talling, J. F., 1974. In Standing Waters. In: R. A. Vollenweider (ed.) A Manual on Methods for Measuring Primary Production in Aquatic Environments. IBP Handbook no. 12, 2nd edn., Blackwell, Oxford: 119–123.

Hydrobiologia 147: 209–213 (1987)
May, L., Wallace, R. & Herzig, A. (eds), Rotifer Symposium IV

Comparative population dynamics of the rotifers *Brachionus angularis* and *Keratella cochlearis*

Norbert Walz
Zoologisches Institut der Universität München, Seidlstr. 25, D-8000 München 2, FRG

Keywords: Rotifera, life table analysis, r- and K- strategy, *Brachionus angularis, Keratella cochlearis*

Abstract

The results indicate that the two rotifer species *Brachionus angularis* and *Keratella cochlearis* have different life history strategies. *Brachionus angularis* appeared to be an r-strategist, showing higher reproductive, mortality and population growth rates than *K. cochlearis.* Additional evidence is offered suggesting that *K. cochlearis* is a K-strategist. Although *K. cochlearis* had a lower growth rate than *B. angularis*, this was compensated by a lower mortality rate, especially in pre-reproductive individuals. In addition, *K. cochlearis* allocated energy to reproduction more efficiently than *B. angularis.*

Introduction

The population growth characteristics of two pelagic rotifer species, *Keratella cochlearis* and *Brachionus angularis* were described by Walz (1983b). *Brachionus angularis* appeared to be an r-strategist because of its large clutch size and high population growth rate. In contrast, *K. cochlearis* achieved low birth rates only and seemed to be a K-strategist, allocating energy to reproduction more efficiently than *B. angularis.* This paper considers how *K. cochlearis* compensates for the disadvantage of having a lower growth rate than *B. angularis.* The demographic factors that determine the life strategy of *B. angularis* are discussed.

Methods

Life table analysis was carried out for *Brachionus angularis* (Gosse) (hereafter *Brachionus*) and *Keratella cochlearis* (Gosse) (hereafter *Keratella*) at 5 °C, 10 °C, 15 °C, 20 °C and 25 °C. The animals were cultured individually in glass spot plates. The green alga *Stichococcus bacillaris* was supplied as food at a concentration of 5 μg C ml^{-1}. Individuals were transferred daily to new spot plates with fresh food suspension. Further details of the culture conditions are given by Walz [1983a].

Individuals were observed up to four times per day, depending on the incubation temperature (*Brachionus*: 5 °C, 10 °C, 15 °C – 2 ×; 20 °C, 25 °C – 4 ×; *Keratella*: 5 °C, 10 °C, 15 °C – 1 ×; 20 °C, 25 °C – 2 ×). The times of egg laying, egg hatching and moralities were noted and life tables were constructed. The growth rate of the populations were calculated with Lotka's equation [Birch, 1948], and the population birth rates were calculated by the egg ratio method of Paloheimo [1974].

Results

Figure 1 shows the individual cumulative production of the two species over a range of incubation temperatures. The production of *Brachionus* increased quickly to values of up to 0.15 μg C ind^{-1}, depending on incubation temperature. *Keratella*

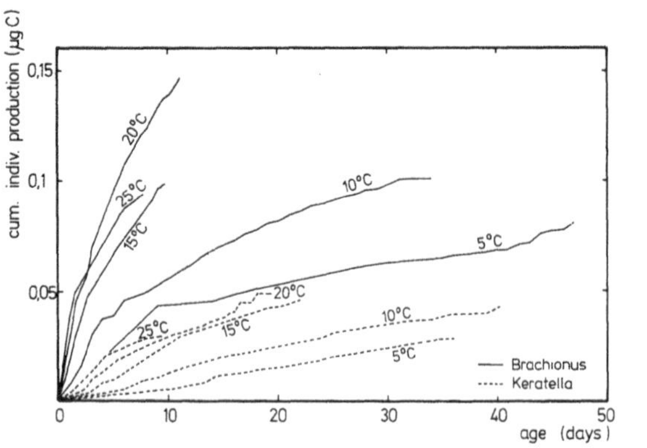

Fig. 1. Cumulative individual production (somatic and reproductive) of *Brachionus angularis* (solid lines) and *Keratella cochlearis* (broken lines) as a function of age of individual (days).

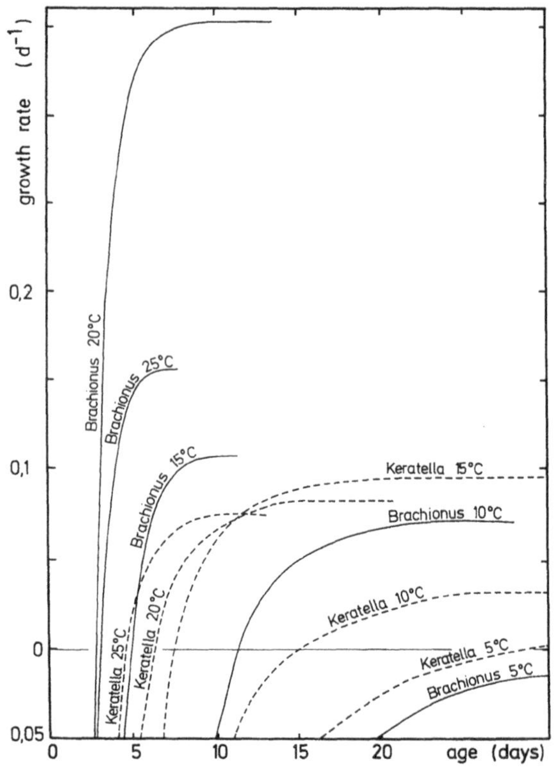

Fig. 2. Population growth rate (d^{-1}) of *Brachionus angularis* (solid lines) and *Keratella cochlearis* (broken lines) as a function of the maximum age of individuals (days).

production increased more slowly, reaching maximum values of only 0.05 μg C ind^{-1}; production of *Keratella* at 20 °C was still lower than *Brachionus* at 5 °C. However, taking into consideration that *Brachionus* has three times the body weight of *Keratella* [Walz, 1983b], the production per unit body weight was similar for both species at all of the incubation temperatures (but not production rate!).

Figure 2 shows the growth rate of the populations of *Keratella* and *Brachionus*, in relation to the maximum age of the individuals, over a range of incubation temperatures and at a food concentration of 5 μg C ml^{-1}. Growth rates were calculated by reducing the age interval in Lotka's equation [Birch, 1948] in a stepwise manner. A positive growth rate was achieved only after the populations had attained a threshold age which will be referred to as 'age of zero growth'. Population growth rate then increased until a plateau was reached, indicating that older individuals were no longer contributing to population growth. At temperatures above 15 °C, *Brachionus* attained positive growth rates during the first few days of development, when individuals were still very young. In contrast, *Keratella* achieved positive growth rates much later, mostly due to the longer embryonic and juvenile development times and increased egg laying interval (Table 1) that are characteristic of this species. The life length of the two species differed significantly at every incubation temperature (t- and F-tests, $P < 0.05$). In general, *Brachionus* could grow faster than *Keratella* under the prevailing food conditions.

Table 2 shows that (1) the rapid population growth rate of *Brachionus* can be attributed to its high gross reproductive rate and (2) its high birth rate can be attributed to its low 'age of zero growth'. The growth rate of *Keratella* was more closely related to its net reproductive rate (R_0). Net reproductive rate is related to the number of surviving offspring per female and this element of 'survival' seems to be very important for *Keratella.* As the number of offspring is low, a much higher juvenile survival rate is necessary to maintain a population. (The data for these correlations are shown in Table 3.)

Mortality rates of age cohorts were higher in *Brachionus* than *Keratella* at all incubation temperatures (Fig. 3). In particular, survival of juveniles was significantly lower in *Brachionus* than *Keratel-*

Table 1. Duration of embryonic and juvenile development stages (days) of *Brachionus angularis* and *Keratella cochlearis* as a function of incubation temperature (±95% confidence interval).

Life history features	Temperature (°C)				
	5	10	15	20	25
Embryonic development					
B. angularis	4.57 ± 0.19	2.92 ± 0.08	1.06 ± 0.03	0.73 ± 0.02	0.63 ± 0.02
K. cochlearis	4.49 ± 0.12	2.51 ± 0.10	2.02 ± 0.10	1.25 ± 0.05	0.88 ± 0.06
Juvenile development					
B. angularis	9.71 ± 0.40	6.06 ± 0.20	2.50 ± 0.12	1.25 ± 0.07	1.26 ± 0.05
K. cochlearis	12.90 ± 0.92	7.82 ± 0.46	4.62 ± 0.42	3.02 ± 0.22	2.00 ± 0.20
Egg laying interval					
B. angularis	2.52 ± 0.42	1.64 ± 0.18	0.80 ± 0.13	0.49 ± 0.05	0.50 ± 0.06
K. cochlearis	3.18 ± 0.53	2.90 ± 0.28	2.08 ± 0.20	1.15 ± 0.16	0.53 ± 0.11

Table 2. Relationships of growth and birth rates of *Brachionus angularis* and *Keratella cochlearis* to some demographic factors as shown by partial correlation coefficients. Correlations were calculated for constant temperature: * indicates $P < 0.05$.

Demographic factor	Gross reproductive rate	Net reproductive rate	Age of zero growth
B. angularis			
growth rate	0.979*	0.896	−0.095
birth rate	0.803	0.537	−0.932*
K. cochlearis			
growth rate	0.837	0.960*	−0.830
birth rate	−0.715	0.337	0.029

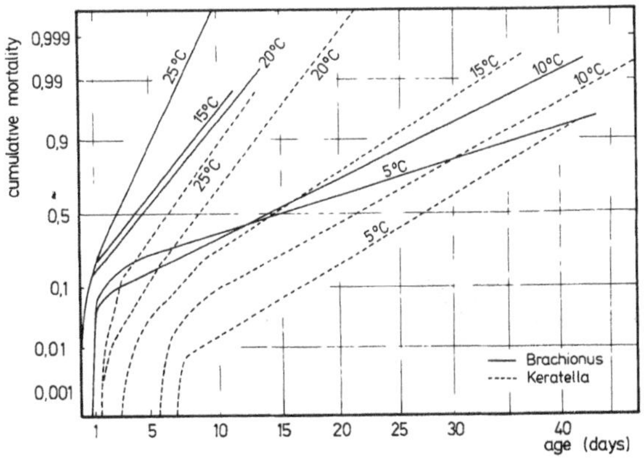

Fig. 3. Cumulative relative mortality of age-cohorts of *Brachionus angularis* (solid lines) and *Keratella cochlearis* (broken lines) as a function of time after hatching (days).

Table 3. Gross and net reproductive rates, growth and birth rates, and "age of zero growth" of *Brachionus angularis* and *Keratella cochlearis* (in numbers, days and d^{-1}).

temperature (°C)	*Brachionus angularis*					*Keratella cochlearis*				
	5	10	15	20	25	5	10	15	20	25
sum of m_x	3.70	6.19	5.31	10.37	5.20	1.11	3.01	4.20	4.39	1.83
sum of $l_x m_x$	0.86	2.47	1.60	3.61	1.51	1.12	1.56	2.15	1.52	1.29
growth rate	−0.005	0.073	0.103	0.352	0.154	0.006	0.034	0.095	0.082	0.075
birth rate	0.083	0.151	0.348	0.704	0.631	0.058	0.094	0.172	0.203	0.300
age of zero growth	–	11.42	4.96	2.77	6.35	28.50	15.12	7.50	6.35	4.50

Table 4. Length of median life time (LT 50), until 50% of the individuals were dead (days) (±95% confidence limits).

Species	Temperature (°C)				
	5	10	15	20	25
E. angularis	19.4±1.7	14.9±1.1	4.1±0.5	4.5±0.5	2.8±0.2
K. cochlearis	26.8±1.5	21.4±1.0	14.7±1.0	8.8±0.5	6.2±0.5

la ($P < 0.001$; Mann-Whitney U-test). In *Keratella*, 97% of individuals reached maturity; in *Brachionus* this value was only 74%. After this initial period, mortality rates followed a log normal distribution in both species. On approaching generation time, the mean survival of *Keratella* (92%) and that of *Brachionus* (63%) were very significantly different ($P < 0.001$). The median life lengths of *Keratella* were significantly higher than those of *Brachionus* (t-test, $P < 0.0001$) over the range of incubation temperatures (Table 4).

Discussion

Two life history parameters most influenced the growth of *Brachionus* populations, number of offspring (generally high) and embryonic development time (relatively short). In contrast, the growth rate of *Keratella* responded to variation in fecundity, only. According to Snell (1978), the influence of fecundity on population growth is dependent on the level of fecundity itself, and tends to be higher at low fecundities. The influence of development time on growth rate is lower at lower fecundity levels.

The comparative advantage of *Brachionus* seemed to be in building a greater individual body size. In contrast, *Keratella* allocated relatively more ingested material to reproduction than *Brachionus*, forming its eggs more efficiently and needing less material for the growth of somatic tissues. As a result the eggs of *Keratella* had relatively more yolk material and an egg volume of 65.4% of adult volume, compared to 17.0% in *Brachionus* [Walz, 1983b].

Although the mean number of eggs produced by each female during her life-time was not significantly lower in *Keratella* (2.3) than in *Brachionus* (2.6), reproduction at an early stage in *Brachionus* diminished further survival, as demonstrated by Snell and King [1977]. *Keratella* reproduced at a low rate over many age classes. In spite of these different life strategies, the overall fitness (success of reproduction expressed as the number of young surviving at least until the hatching of their first egg) was not significantly different between *Keratella* (1.34) and *Brachionus* (1.41) (Mann-Whitney U-test, $P<0.05$). Both species could maximize their fitness, but in different ways.

In general, the hypothesis of different life strategies of *Keratella* (K-strategy) and *Brachionus* (r-strategy) was confirmed. However, the relationship was more complex at 5 °C and 10 °C, as the population dynamics of *Brachionus* resembled that of *Keratella* at these temperatures. In chemostats, *Brachionus* was able to switch between these strategies, to a certain degree. At high food levels, its clutch size was large but the eggs were small. At low food levels, *Brachionus* produced fewer eggs of larger size indicating a temporally dynamic strategy [Nichols *et al.* 1976]. This could not be shown in individual cultures as no more than three eggs per female appeared under the prevailing food concentration. So, variation in fitness among the offspring could not be tested. *Keratella*, like *Asplanchna girodi* [Robertson & Salt, 1981], did not change the size of its offspring when exposed to different food levels.

The results of these experiments show that it is an over-simplification to classify all rotifers as zooplankton with the highest growth rates [Allan, 1976]. This is true only for some species, e.g., *Euchlanis* [King, 1967], *Asplanchna* [Robertson & Salt, 1981], *Synchaeta* spp. [Walz, 1987; Stemberger & Gilbert, 1985]) and other *Brachionus* spp. [Halbach, 1970]. In the field, cladocerans often reach growth rates as high as, or higher than, rotifers [Allan & Goulden, 1978; Lynch, 1980]. However, the r-strategic rotifers seem to have a much shorter lag time than cladocerans in reacting to changing food conditions. This gives them a decisive time advantage in exploiting new sources of food. *Daphnia* populations are unable to achieve a positive growth rate in less than 12 days [Jacobs, 1978]. K-strategists, on the other hand, seem to be able to utilize much lower food levels than are accessible to r-strategic rotifers. Living longer, K-strategists do not need to balance a high mortality by a high growth rate.

Acknowledgements

This study was supported by the Deutsche Forschungsgemeinschaft (DFG).

References

Allan, J. D., 1976. Life history patterns in zooplankton. Am. Nat. 110: 165–180.

Allan, J. D. & C. E. Goulden, 1980. Some aspects of reproductive variation among freshwater zooplankton. In W. C. Kerfoot (ed.), Evolution and Ecology of Zooplankton Communities. The University Press of New England, Hanover (N.H.): 388–410.

Birch, L. C., 1948. The intrinsic rate of natural increase of an insect population. J. anim. Ecol. 17: 15–26.

Halbach, U., 1970. Einfluss der Temperatur auf die Populationsdynamik des planktischen Rädertiers Brachionus calyciflorus Pallas. Oecologia 4: 176–207.

Jacobs, J., 1978. Coexistence of similar zooplankton species by differential adaptation to reproduction and escape in an environment with fluctuating food and enemy densities. III. Laboratory experiments. Oecologia 35: 35–54.

King, C. E., 1967. Food, age, and the dynamics of laboratory populations of rotifers. Ecology 48: 111–128.

Lynch, M., 1980. The evolution of Cladoceran life histories. Quart. Rev. Biol. 55: 23–42.

Nichols, J. D., W. Conley, B. Batt & A. R. Tipton, 1976. Temporally dynamic reproductive strategies and the concept of r- and K-selection. Am. Nat. 110: 995–1005.

Paloheimo, J. E., 1974. Calculation of instant birth rates. Limnol. Oceanogr. 19: 692–694.

Robertson, J. R. & G. W. Salt, 1981. Responses in growth, mortality, and reproduction to variable food levels by the rotifer, Asplanchna girodi. Ecology 62: 1585–1596.

Snell, T. W., 1978. Fecundity, development time, and population growth rate. Oecologia 32: 119–125.

Snell, T. W. & C. E. King, 1977. Lifespan and fecundity patterns in rotifers: the cost of reproduction. Evolution 31: 882–890.

Stemberger, R. S. & J. J. Gilbert, 1985. Body size, food concentration, and population growth in planktonic rotifers. Ecology 66: 1151–1159.

Walz, N., 1983a. Individual culture and experimental population dynamics of Keratella cochlearis (Rotatoria). Hydrobiologia 107: 35–45.

Walz, N., 1983b. Continuous culture of the pelagic rotifers Keratella cochlearis and Brachionus angularis. Arch. Hydrobiol. 98: 70–92.

Walz, N., 1987. Wie werden Rotatorienpopulationen reguliert? Beziehungen zwischen Rotatorien. Crustaceen, und Phytoplankton. Laufener Seminarbeitr. (in press.)

Hydrobiologia 147: 215–219 (1987)
May, L., Wallace, R. & Herzig, A. (eds), Rotifer Symposium IV

Changes in the population dynamics of *Keratella cochlearis* (Gosse), *Kellicottia longispina* (Gosse) and *Polyarthra vulgaris* Carlin in a fertilized enclosure

Reidar Grundström
Institute of Limnology, Box 557, S-751 22 Upsala, Sweden

Keywords: Feeding, predation, competition, food availability, Cladocera, Copepoda, Rotifera, *Coccomyxa*

Abstract

The population dynamics of phytoplankton and zooplankton in a fertilized enclosure were studied from April 1977 to June 1978. During the first spring period, the rotifers *Keratella cochlearis, Kellicottia longispina,* and *Polyarthra vulgaris* were scarce. During the following spring, all three species were abundant. An attempt is made to explain these differences. Food resource competition from cladocerans, and the low food quality and possible inhibitory effects of *Coccomyxa* sp. were found to be the most likely factors limiting the growth of these rotifers during the first spring period.

Introduction

Food availability, competition and predation are important factors in the regulation of zooplankton communities. Several recent studies have investigated the effect of these three factors on the population dynamics of rotifers [e.g. Hofmann, 1983; Walz, 1983; Gilbert & Bogdan, 1985; Gilbert & Stemberger, 1985]. This paper evaluates the importance of food and competition in regulating populations of *Keratella cochlearis* (Gosse), *Kellicottia longispina* (Gosse), and *Polyarthra vulgaris* Carlin in a semi-natural ecosystem of the limnocorral type.

Materials and methods

In situ *enclosure*

A non-transparent plastic cylinder, 10 m diameter, open at both ends, was anchored in water 10 m deep in Lake Siggeforasjön, Sweden. The lower end of the cylinder was pressed into the sediment and the upper end was supported by a wooden frame kept 50 cm above the water surface by large styrofoam blocks.

The enclosure was kept in the lake from October 1976 until mid-April 1977, at which time fresh lake water was pumped into the cylinder and the experiment was begun. The enclosure was fertilized several times during the ice-free season with phosphoric acid and ammonium chloride to maintain high nitrogen and phosphores concentrations. Fertilization resulted in total-N and total-P concentrations of 0.9–2.2 mg l^{-1} and 65–150 $\mu g\ l^{-1}$, respectively, throughout the study period.

The water temperatures recorded within the enclosure were similar in 1977 and 1978, although May 1977 was slightly warmer than May 1978 (8.4 °C on 8 May 1977; 6.9 °C on 8 May 1978) and the beginning of June was slightly colder in 1977 than 1978 (15.2 °C on 4 June 1977; 20.1 °C on 4 June 1978).

Sampling and plankton counting

Plankton was sampled in the enclosures from 8 May 1977 to 4 June 1978. Samples were collected every third day from 8 May to 28 July, 1977, and from 5 May to 4 June, 1978. Longer interval sampling was carried out during the remainder of the study period. The sampler consisted of a 2 m long (2 l volume) plexiglass tube with a closing mechanism [Ramberg, 1979]. Composite samples, representing epilimnion (surface to thermocline) and meta-plus hypolimnion (thermocline to eight metres depth), were prepared by mixing 1–3 water columns from the respective stratum.

The samples were subsampled with a Folson sample splitter [McEwan *et al.,* 1954] and the phytoplankton and zooplankton were counted using an inverted microscope [Utermöhl, 1958]. Further details of the methods are given by Blomqvist *et al.,* [1981]. All individuals were classified to species and developmental stage, except cladocerans, rotifers and nauplii. Calanoid copepods were divided into two groups, (1) animals smaller than development stage four and (2) development stages four and five.

Phytoplankton biovolumes

Phytoplankton biovolumes were calculated using the mean cell volumes found in the lake [Blomqvist *et al.,* 1981] or in the enclosure. Phytoplankton was classified in two different ways, firstly excluding and secondly including *Coccomyxa* sp. In the first case, ambient food availability was estimated by classifying the algae into 'edible' and inedible' groups according to cell volume. Cells between 14 μm^3 and 1000 μm^3 were considered to be edible, while cells outside this range e.g. *Coccomyxa* sp. (3.6 μm^3), large *Cryptomonas* (1766 μm^3), *Synedra* sp. and *Oscillatoria argardhii*, were considered to be inedible.

Although *Coccomyxa* sp. was considered to be inedible it is possible that it may still have been ingested. In order to estimate the extent to which *Coccomyxa* sp. was ingested, the efficient food concentration (C_{eff}) was calculated for all cells up to 1000 μm^3. Biovolumes of these cells were multiplied by a weighting factor, which decreased with decreasing cell size, and then summed to give C_{eff}:

$$C_{eff} = \sum_{n}^{max} I.L.C._{max}/I.L.C._{n} * C_n$$

where I.L.C. is the Incipient Limiting Concentration calculated according to Duncan & Gulati [1983: 123, eq. 2], 'max' refers to the size of the largest 'edible' cells (900 μm^3) and 'n' to the size of smaller cells. $I.L.C._{max}/I.L.C._n$ is the weighting factor and C_n the biovolume of cells of size 'n'. The percentage contribution of *Coccomyxa* sp. to the total efficient food concentration was also calculated.

Zooplankton dry weights

Cladoceran body length were measured, excluding the mucro, at three to six day intervals. Between 30 and 300 individuals of each common species were measured from each sample. Mean body lengths for each date were calculated. Copepod body length was assumed to be constant for each developmental stage. Dry weights of rotifers, copepods and cladocerans were estimated from body lengths and previously published data [Bottrell *et al.,* 1976; Johansson *et al.,* 1976; Latja & Salonen, 1978; Rosen, 1981]. These were used in conjunction with abundance estimates to calculate total dry weights per litre.

Results

1. Zooplankton

Variation in the total dry weight of zooplankton within the enclosure is shown in Fig. 1. From early to mid June 1977 *Asplanchna priodonta* Gosse and *Synchaeta pectinata* Ehbg. dominated the rotifer community. Other species of rotifer were scarce throughout this period. The population maximum of *Keratella cochlearis* in 1977 was only 9 ind l^{-1}. Similarly, *Kellicottia longispina* and

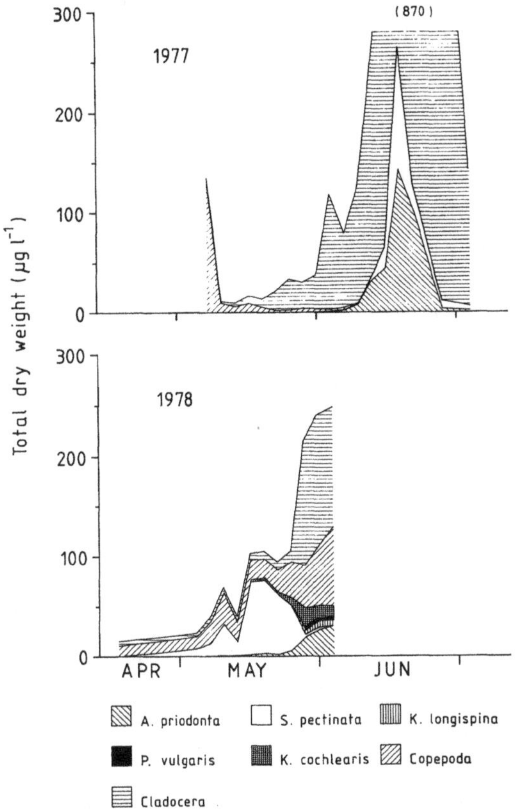

Fig. 1. Total dry weight ($\mu g\ l^{-1}$) of zooplankton populations during spring 1977 (upper panel) and 1978 (lower panel).

Polyarthra vulgaris attained only 8 ind l^{-1} and 1 ind l^{-1}, respectively.

In 1978, a more diverse rotifer community became established. The maximum densities of the most abundant species (indicated as ind l^{-1}, in parentheses) were recorded as follows: *K. cochlearis* (377), *Keratella hiemalis* Carlin (97), *K. longispina* (162) and *P. vulgaris* (55). These species developed large populations alongside *A. priodonta* (76) and *S. pectinata* (268).

In summary, the main differences between the rotifer populations of spring 1977 and spring 1978 were that (1) the rotifers developed later in 1977 than 1978 and (2) *K. cochlearis, K. longispina* and *P. vulgaris* did not develop noticeable populations in 1977, but were abundant in 1978. The biomass values of *A. priodonta* and *S. pectinata* were higher in 1977 than 1978.

Copepods were rare in the enclosure. The Cladocera of the enclosure was dominated by *Bosmina longirostris* (Müll.) and *Daphnia cristata* Sars in spring 1977. *Chydorus sphaericus* Müll. and *Bosmina coregoni* s.l. Baird were most abundant during summer and autumn 1977.

2. Phytoplankton

Temporal variation in algal biovolume within the enclosure is shown in Fig. 2. The 'edible' fraction consisted mainly of flagellated chrysophyceans and small cryptophyceans. The 'inedible' fraction was dominated by *Coccomyxa* sp. and a large *Cryptomonas* before mid June 1977, after which *Oscillatoria* dominated. In 1978, the 'inedible' fraction was dominated by *Synedra* sp. and a large *Cryptomonas* sp. There were only minor quantitative differences in the edible fraction between the two years, although that fraction peaked nine days earlier in 1977 than 1978.

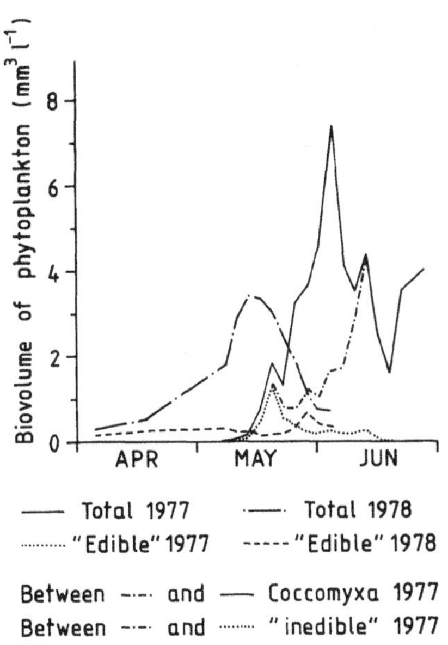

Fig. 2. Biovolume ($mm^3\ l^{-1}$) of phytoplankton in relation to time 'Edible' cells have a volume of 14–1000 μm^3, and are mainly *Chrysophyceae* and *Cryptophyceae*. 'Inedible' cells, falling outside this size range, are mostly *Coccomyxa* sp., large *Cryptomonas* (1766 μm^3), *Synedra* sp and *Oscillatoria agardhii. Coccomyxa* sp. is shown separately for 1977, although they belong to the range of 'inedible' algae.

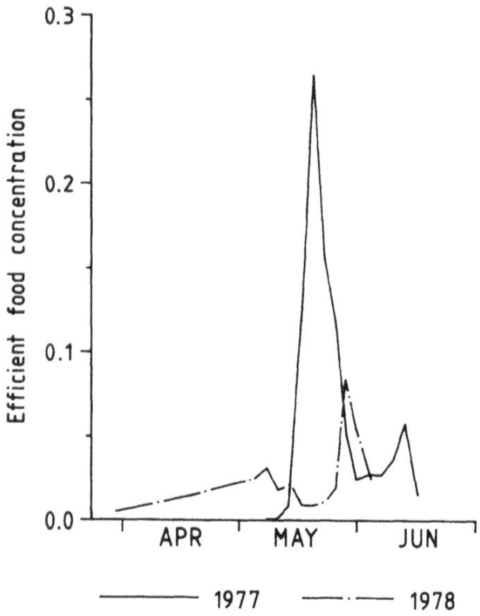

Fig. 3. Efficient food concentration (C_{eff}) in relation to time.

$$C_{eff} = \sum_{n}^{max} I.L.C._{max}/I.L.C._{n} \cdot C_n$$

where I.L.C. is the Incipient Limiting Concentration calculated according to Duncan & Gulati [1983: 123, eq. 2], 'max' refers to the size of the largest 'edible' cells (900 μm^3) and 'n' to the size of smaller cells. $I.L.C._{max}/I.L.C._{n}$ is the weighting factor and C_n the biovolume of cells of size 'n'.

The efficient food concentration varied as shown in Fig. 3. *Coccomyxa* sp. never contributed more than 7% to this value. After 17 May 1978, C_{eff} rarely exceeded the values recorded in 1977.

Discussion

The possible reasons for the failure of *K. cochlearis, K. longispina* and *P. vulgaris* to develop large populations in 1977 are predation, toxins, resource competition, mechanical interference and food availability. These factors are discussed below.

Predation seems unlikely to have been an important factor in limiting rotifer densities in 1977. Fish predation could not have affected the rotifer populations as fish were excluded from the enclosures. The main invertebrate predators, cyclopoid copepods, also appear to have had little effect on the growth of the rotifer community, since both predator and prey were more abundant in 1978 than 1977. Finally, none of the smaller rotifers were ever observed in the stomach of *A. priodonta,* suggesting that the impact of this potential predator was also negligible.

The growth of rotifer populations in 1977 may have been limited by toxins released into the environment by *Coccomyxa* sp. which was more abundant in 1977 than 1978. The author could find no evidence in the literature to suggest that this alga is toxic to zooplankton; indeed, if it were, one might expect the other zooplankton species to have been similarly depleted in 1977 and they were not. Nevertheless, this factor cannot be completely disregarded as it is possible that the toxin was only effective after digestion and that this alga was only ingested by te three rotifer species.

Cladocerans may compete with rotifers for food, but this competition does not necessarily affect the population growth of rotifers, as long as food resources are not limiting. Both the edible food concentration (Fig. 2) and C_{eff} (Fig. 3) were higher in 1977 than in 1978, suggesting that food concentration was not the crucial limiting factor. However, cladoceran biomass indicates more intense competition in 1977 than 1978. Thus, food resource competition cannot be completely excluded as one of the possible factors regulating rotifer populations in the enclosure.

Another factor of potential importance is mechanical interference between cladocerans and rotifers, as demonstrated for *Daphnia* and *K. cochlearis* [Gilbert & Stemberger, 1985]. The *Daphnia* of the present study (*D. cristata*) were smaller than those used by Gilbert & Stemberger [1985] and the feeding chamber may have been too small to catch *K. cochlearis.* Furthermore, the number of *Daphnia* per litre in the present study was usually lower than in the cited experiment. Nevertheless, *D. cristata* was less abundant in 1978 when *K. cochlearis* grew well, which may support the conjecture that mechanical interference can affect the development of rotifer populations.

The role of the small alga *Coccomyxa* sp. (very abundant in 1977 and absent in 1978) as a food resource is of special interest. The results of Chalk [1981] and Duncan & Gulati [1983] indicate that small cells are consumed much less efficiently than large cells by *D. magna* and *Brachionus* spp. Similarly, Gilbert & Bogdan [1985] showed that *K. cochlearis* and *Kellicottia bostoniensis* did not selectively feed on very small food particles, such as bacteria (1.5 μm^3). *K. cochlearis* and *K. bostoniensis* preferred larger cells (e.g., yeast – 35 μm^3), while *Polyarthra* sp. selected even larger cells, preferring *Cryptomonas* sp. (460 μm^3) to *Chlamydomonas* (127 μm^3). Cells of *Coccomyxa* sp. are very small and, although ingested, are probably unsuitable as food for *K. cochlearis* and *K. longispina*, and even less suitable for *P. vulgaris*. This could have inhibited the growth of these rotifers during 1977.

The scarcity of *K. cochlearis, K. longispina* and *P. vulgaris* during 1977, and their subsequent abundance during 1978, indicates that certain environmental factors were limiting the growth of these populations in 1977. Although competition from cladocerans was more intense in 1977 than 1978, it was probably not the only factor contributing to the enormous variations in rotifer population development recorded. The low food quality and possible inhibitory effects of *Coccomyxa* sp. may also have inhibited the population growth of these rotifers.

Acknowledgements

I thank Professor B. Pejler and Dr. G. Persson for their advice on on preparing this manuscript. I also thank Dr. P. Blomqvist for allowing me to use his phytoplankton data, and to him and Dr. U Heyman for their help in collecting and analysing samples.

References

Blomqvist, P., U. Heyman & R. Grundström, 1981. The structure of the pelagic escosystem in lake Siggeforasjön. Scripta Limnol. Upsal. A 522.

Bottrell, H. H., A. Duncan, Z. M. Gliwicz, E. Grygierek, A Herzig, A. Hillbricht-Ilkowska, H. Kurasawa, P. Larsson & T. Weglenska, 1976. A review of some problems in zooplankton production studies. Norw. J. Zool. 24: 419–456.

Chalk, E. A., 1981. Cladoceran filter feeding in a Thames Valley Reservoir. Ph.D. thesis. CNAA (Central London Polytechnic/Thames Water Authority), 266 pp.

Duncan, A. & R. D. Gulati, 1983. Feeding studies with natural food particles on tropical species of planktonic rotifers. In F. Schiemer (ed.). Limnology of Parakrama Samudra-Sri Lanka, Developments in Hydrobiology, Dr. W. Junk Publishers, The Hague.

Edmondson, W. T., 1965. Reproductive rate of planktonic rotifers as related to food and temperature in nature. Ecol. Monogr. 35: 61–111.

Gilbert, J. J. & K. G. Bogdan, 1985. Rotifer grazing: in situ studies on selectivity and rates. In D. G. Meyers & J. R. Strickler (eds.), Trophic interactions within aquatic ecosystems. Amer. Ass. Sci. 97–133.

Gilbert, J. J. & R. S. Stemberger, 1985. Control of Keratella populations by interference competition from Daphnia. Limnol Oceanogr. 30: 180–188.

Hofmann, W., 1983. Interactions between Asplanchna and Keratella cochlearis in the Plussee (north Germany). Hydrobiologia 104: 363–365.

Johansson, J.-Å., H. Olofsson & L. Ramberg, 1976. Studier av zooplanktons konsumtion i Botjärn. Klotenprojektet rapport nr 5. Uppsala Universitet, Limnologiska inst. 27 pp. (In Swedish).

Latja, R. & K. Salonen, 1978. Carbon analysis for the determination of individual biomasses of planktonic animals. Verh. Int. Verein. Limnol. 20: 2556–2560.

McEwen, G. F., M. W. Johnson & T. R. Folsom, 1954. A statistical analysis of the performance of the Folsom plankton sample splitter, bassed upon test observations. Arch. Meteor. Geophys. Bioklim. A7: 502–527.

Ramberg, L., 1979. Relations between phytoplankton and light climate in two Swedish forest lakes. Int. Revue ges. Hydrobiol. 64: 749–782.

Rosen, R. A., 1981. Length – dry weight relationships of some freshwater zooplankton. J. Freshwat. Ecol. 1: 225–229.

Utermöhl, H., 1958. Zur vervollkommung der quantitativen Phytoplankton methodik. Mitt. Int. Ver. Limnol. 9: 1–39.

Walz, N., 1983. Individual culture and experimental population dynamics of Keratella cochlearis (Rotatoria). Hydrobiologia. 107: 35–45.

Hydrobiologia 147: 221–227 (1987)
May, L., Wallace, R. & Herzig, A. (eds), Rotifer Symposium IV

The population dynamics of *Keratella cochlearis* in a hypereutrophic tarn and the possible impact of predation by young roach

D. P. Hewitt & D. G. George
Freshwater Biological Association, The Ferry House, Ambleside, Cumbria, England LA22 OLP

Abstract

Priest Pot is a shallow hypereutrophic tarn that supports large numbers of rotifers and young roach (*Rutilus rutilus* (L.).). In this paper we describe the planktonic rotifer community of the tarn and show that the young roach feed extensively on *Keratella cochlearis*. A two year study of the population dynamics of *K. cochlearis*, nevertheless, failed to detect any obvious demographic effects of roach predation. Some general comparisons of fish biomass and rotifer production suggest that the fish are unlikely to influence the development of their rotifer prey in such a productive system.

Introduction

A number of studies have shown that fish predation can have a pronounced effect on the population dynamics of planktonic crustacea [Galbraith, 1967; Jacobs, 1977]. The young of some fish feed extensively on rotifers, and may, under certain conditions, have an important regulating effect on their prey [Duncan & Gulati, 1983].

Priest Pot is a hypereutrophic tarn in the English Lake District (54°22′N, 2°58′W), approximately 1 hectare in area and on average 2.3 m deep. The tarn is unusual in that it supports large numbers of rotifers (>10000) individuals l^{-1}), but very few planktonic crustacea. The fish community is dominated by roach (*Rutilus rutilus* (L.)) with very small numbers of perch (*Perca fluviatilis* L.) and pike (*Esox lucius* L.). During the summer the tarn becomes thermally stratified and the rotifers and fish are then confined to the well oxygenated epilimnion.

In this paper we examine the possible effects of roach predation on the dynamics of one rotifer species – *Keratella cochlearis* Gosse. This species is usually abundant in the tarn when the young roach start to feed, and forms an important component of their diet.

Methods

Rotifer samples were collected at weekly intervals throughout 1980 and 1981 by lowering a 2.5 litre Friedinger sampler to successive depths in the water column. On each occasion additional horizontal samples were collected at ten randomly selected locations by lowering the sampler to the mid-point of the epilimnion (ca. 0.5 m). Temperature and dissolved oxygen were measured at 0.25 m intervals with a Yellow Springs Instruments oxygen electrode (Yellow Springs, Ohio).

The rotifers in each sample were concentrated by reverse filtration through a 35 μm filter [Dodson & Thomas, 1964]. The concentrate was sedimented in Lugols' iodine and decanted before sub-sampling with a wide mouth pippette. Sub-samples of the sedimented concentrate were placed in a Sedgewick-Rafter cell and counted at a magnification of 40×. All attached eggs were counted, but no attempt made to count any loose eggs. Unknown species were identified under a high-power microscope using the keys in Ruttner-Kolisko [1974].

No quantitative samples of fish were collected in 1980 and 1981, but observations during regular rotifer sampling suggested that young roach were present in the tarn over the same time period as previous years [Hewitt, unpublished data]. In these earlier studies the roach were sampled quantitatively over a period of 5 years, using the buoyant nets described by Bagenal [1974] and tested by Hewitt [1979]. Where samples were required for diet analysis, the nets were set every two hours over a 24 hour period. The catch was anaesthetised using tricainmethansulfonate and preserved in the field to prevent regurgitation or further digestion of the gut contents. Most of the material in the gut could be identified to species but some undifferentiated organic remains were classed as 'phytoplankton'. For each fish the abundance of each food item was expressed as a percentage of the total volume eaten [see Windell, 1971]. Individual results were then summed to give the average composition of the diet on a particular day.

Results

The rotifer populations of Priest Pot

Sixteen species of planktonic rotifers were collected from Priest Pot in 1980 and 1981 but only nine species were ever common or abundant. The mean and maximum densities of the more common species and genera are given in Table 1, which lists the rotifers in order of their frequency of occurrence. In many temperate lakes there is a seasonal succession of species related to the change in temperature [Elliott, 1977; May, 1983]. In Priest Pot the three most common species appear to tolerate a wide range of temperatures. The two *Keratella* species are generally regarded as eurythermal forms [Herzig, 1983; May, 1983] but *Polyarthra vulgaris* is usually considered a late summer form [Elliott, 1977].

The most striking feature of the Priest Pot plankton is the very high concentration of rotifers in the epilimnion. Orcutt and Pace [1984] present a summary of the rotifer densities found in a variety of lakes and record a maximum density of 17700 individuals l^{-1} from an effluent pond in Finland [Eloranta, 1982]. The concentration of rotifers in Priest Pot frequently exceeded this total. Densities of over 20000 individuals l^{-1} were fairly common in summer and densities of over 40000 individuals l^{-1} were occasionally recorded.

The food of young roach in Priest Pot

In Priest Pot the roach spawn among littoral vegetation in May and the young fish start to feed in mid June. Several feeding studies on young roach have shown that, as the young fish grow, their food changes from phytoplankton, to rotifers and crustacea and finally to larger invertebrates [Grigorash *et al.*, 1972; Hammer, 1985]. Figure 1 shows the average composition of the young roach diet on three occasions in 1975. On 25-26 June (Fig. 1a) the average size of the young roach was 18 mm (range 16–20 mm). *K. cochlearis* was the dominant food item, but the guts also contained the remains of some small crustacea and dipteran larvae. On 30-31 July (Fig. 1b) the average size of

Table 1. The mean and maximum population densities of the most common rotifers in Priest Pot between June 1979 and December 1981. The '% occurrence' is the proportion of occasions on which a particular species accounted for at least 1% of the total number.

Species	Mean density (individuals l^{-1})	Maximum density (individuals l^{-1})	% Occurrence (n = 122)
Polyarthra vulgaris Carlin	1962	15402	83
Keratella cochlearis (Gosse)	828	10960	76
Keratella quadrata (O.F. Müller)	1317	23448	70
Brachionus calyciflorus Pallas	492	5238	66
Synchaeta sp.	218	2660	61
Filinia terminalis Plate	33	650	30
Trichocerca pusilla Jennings	68	1078	25
Hexarthra mira Hudson	82	3256	11
Asplanchna priodonta Gosse	76	3104	10

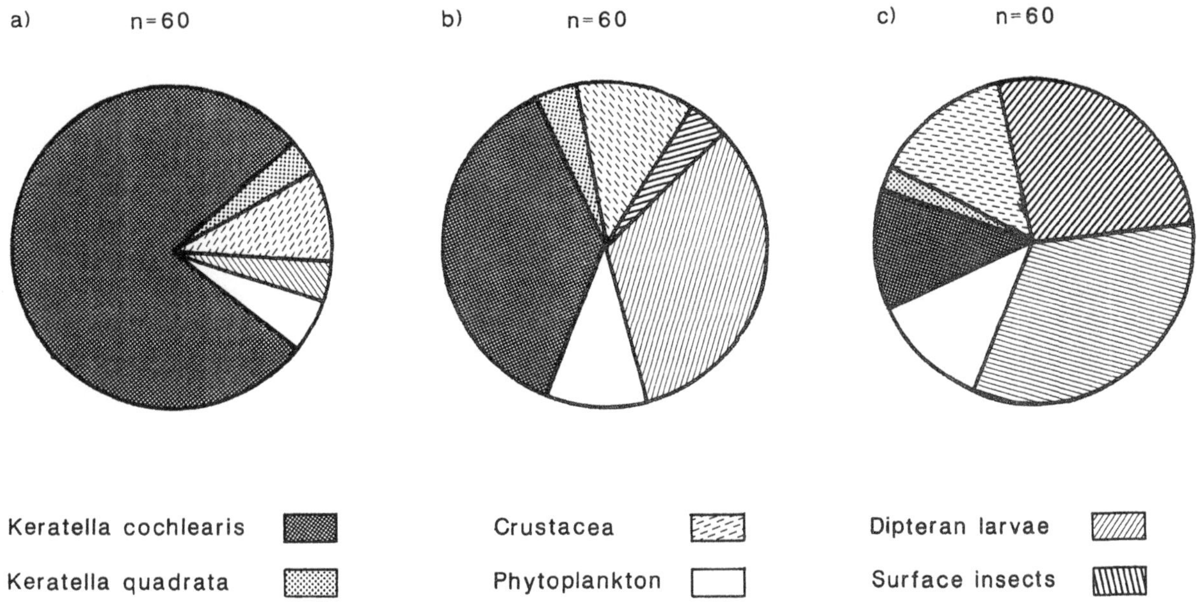

Fig. 1. The percentage composition of roach gut contents on three occasions in 1975. (a) 25-26 June. (b) 30-31 July (c) 27-28 August.

the young roach was 34 mm (range 32 – 37 mm). The fish were still feeding on *K. cochlearis* but microcrustacea and dipteran larvae now formed a larger proportion of their diet. By 27-28 August (Fig. 1c) the young roach had grown to 38 mm (range 35 – 43 mm). They were now feeding on surface insects as well as dipteran larvae but quite large numbers of *K. cochlearis* were still found in their guts.

No rotifer samples were collected in 1975 so we cannot relate the diet of the fish to the specific composition of the plankton. Long-term studies in the tarn, nevertheless, suggest that the young roach feed selectively on *K. cochlearis*. In 1973, for example, samples of young roach were collected at weekly intervals between June and December and their gut contents examined. *K. cochlearis* was usually the only rotifer found in the guts even though *K. quadrata* and *P. vulgaris* were also abundant in the tarn.

The population dynamics of K. cochlearis

K. cochlearis was present in the tarn throughout the year but only became abundant when the tarn was thermally stratified (Fig. 2a). In 1980 a maximum population density of nearly 9000 individuals l^{-1} was reached in early June, whereas in 1981 the population only reached a maximum of 3000 individuals l^{-1} in mid August. The instantaneous birth rate (b) of *K. cochlearis* was calculated according to the familiar equation of Paloheimo [1974]. The development time of the eggs was taken from Edmondsons data presented in the graph in Downing and Rigler [1984]. The temperature used in the estimation was the average for the proportion of the water column occupied by the rotifers. Some *K. cochlearis* populations exhibit a diel pattern of egg laying which can lead to erroneous estimates of the birth rate [Bosch & Ringelberg, 1985]. None of our diel studies showed any cyclical change in the rate of egg laying, so 'daytime' estimates of the birth rate have been taken as representative. The seasonal fluctuations in the birth rate (Fig. 2b) show that the birth rate per individual was higher in 1981, when the population density was lower than in 1980 when the population density was high. The average birth rate for the May to September period was 0.12 individuals day^{-1} in 1980 and 0.16 individuals day^{-1} in 1981. These mean values are

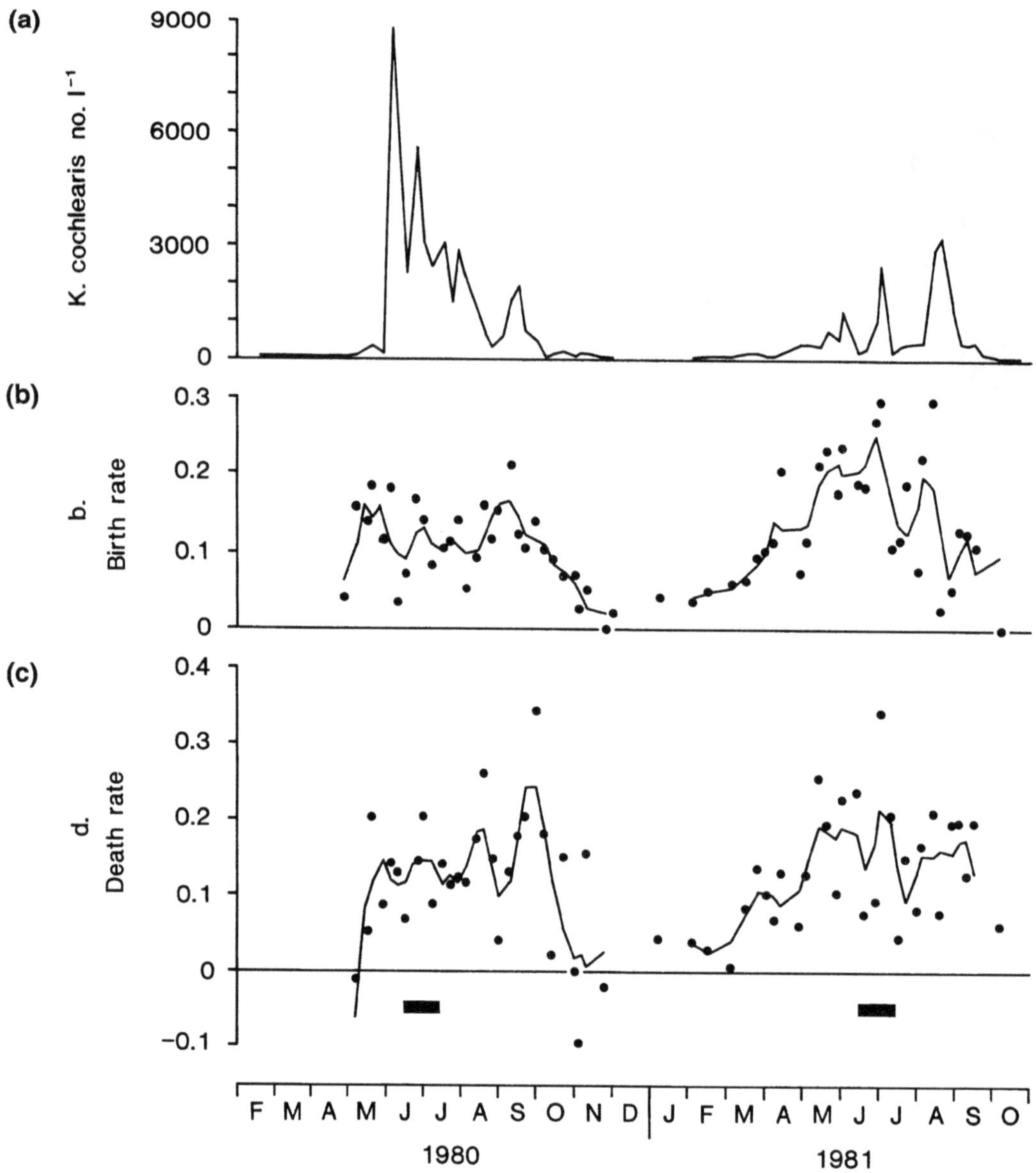

Fig. 2. The population dynamics of *K. cochlearis* in Priest Pot during 1980 and 1981. (a) – population density; (b) – birth rate; (c) – death rate). Solid lines in middle and lower panels show running means of birth and death rates. The shaded bars in the lower panel show when the young roach are likely to be feeding on rotifers.

very similar to the spring and early summer averages of 0.14 and 0.16 individuals day^{-1} quoted for this species by Hillbricht-Ilkowska [1967].

Estimates of the instantaneous death rate (d) were derived by subtracting interpolated values of the instantaneous rate of increase (r) from the calculated instantaneous birth rates (b). The death rate is the most unreliable statistic associated with the egg ratio method since it depends on the difference between two quantities already estimated with error. Figure 2c shows the seasonal fluctuations in the death rate of *K. cochlearis* during 1980 and 1981. The potential rate of increase (b) usually exceeded the observed rate of increase (r) but there were a few occasions when low population densities gave rise to negative values of d. There was no consistent seasonal trend in the death rate, and no indication that the death rate was any higher in early summer when shoals of young roach were present in the tarn.

Biomass and production

The biomass of *K. cochlearis* was estimated from the weight and number of individuals present on each sampling date. The average weight of *K. cochlearis* was taken to be 0.90 μg ww individual^{-1} [the average of two published dry weight values, Dumont *et al.*, 1975 and Makarewicz & Likens, 1979] assuming dry weight to be 10% of wet weight. The daily production of *K. cochlearis* was obtained by converting the instantaneous death rates (d) to finite death rates (D) and assuming that the turnover time for numbers was equivalent to 1/D [see George & Edwards, 1974].

The mean biomass of *K. cochlearis* for the May to September period was 1.97 g ww m^{-3} in 1980 and 0.76 g ww m^{-3} in 1981. The corresponding production estimates were 25.64 g ww m^{-3} in 1980 and 12.38 g ww m^{-3} in 1981. These biomass estimates are exceptionally high for a temperate lake, and are almost two orders of magnitude higher than those reported for the species in Lake Esrom [Bosselmann, 1979a]. The ratio of production to biomass (P/B) is an estimate of the turnover time of a population. Several workers have noted a tendency for zooplankton P/B coefficients to increase with the productivity of the system [Patalas, 1970]. The turnover time of *K. cochlearis* from May to October in Priest Pot (P/B = 22) was, however, practically identical to the comparable turnover of the same species in Lake Esrom [P/B = 21, Bosselmann, 1979a].

The potential impact of roach predation on *K. cochlearis*

It is sometimes possible to demonstrate a predator prey interaction by relating the death rate of the prey to the abundance of the predator [Dodson, 1972; Duncan & Gulati, 1984]. The roach in Priest Pot were not sampled systematically in 1980 and 1981, but casual observations showed that the young fish were abundant in June and July as in previous years. The average instantaneous death rate for *K. cochlearis* during this period was 0.128 ± 0.057 in 1980 and 0.136 ± 0.105 in 1981. These death rates are not particularly high [cf. Bosselmann, 1979b] and are generally lower than those recorded later in the year.

Some simple calculations of fish biomass and rotifer production show why the young roach have so little effect on the population dynamics of their prey. Previous studies on the roach populations of Priest Pot [Hewitt, unpublished data] suggest that the biomass of young roach averages 1 500 mg ww m^{-2} in June and July. If we assume that these fish consume aroung 15% of their body weight per day [see Grigorash *et al.*, 1972] then their total food requirement would be 225 mg ww m^{-2}. Our feeding studies (Fig. 1) show that *K. cochlearis* can account for 80% of this total, giving a possible biomass of *K. cochlearis* consumed of 180 mg ww m^{-2}.

In 1980 and 1981 the daily production of *K. cochlearis* in June and July ranged from 500–1 000 mg ww m^{-2}. The estimated maximum demand of the roach can therefore account for only 18–36% of the observed turnover rate.

In Priest Pot the rotifers dominate the plankton both numerically and in terms of biomass throughout the year. A similar preponderance of rotifers has been noted in other eutrophic systems [Pace & Orcutt, 1981] and may be related to the size distribution of the available algal food [cf. Finlay & Berninger, 1984].

In lakes where crustacea and small invertebrates are abundant young roach progress from feeding on rotifers to larger prey. We show here that when crustacea are scarce the fish are quite capable of feeding on rotifers for an extended period of time. Mikheev [1984] has shown that the minimum size of prey ingested by young roach is determined by the distance between the gill rakers. In his study, roach larger than 18 mm seldom ingested rotifers, whereas in Priest Pot fish larger than 30 mm still took large numbers of *K. cochlearis*. There is strong circumstantial evidence to suggest that the young roach feed preferentially on *K. cochlearis*. *K. cochlearis* was the only rotifer found in quantity in the fish guts even though *P. vulgaris* and *K. quadrata* were often more common in the tarn. Gilbert and Williamson [1978] have shown that *P. vulgaris* has a very effective escape mechanism when approached by a predator, but it is not clear why the roach take so few *K. quadrata*.

There are very few water bodies where it is possible to relate the mortality rate of the zooplankton to the standing crop of planktivorous fish. In this study we were unable to make a direct comparison, but previous estimates of the fish stock allowed us to draw some general conclusions. Our calculations suggest that the young roach in Priest Pot have little effect on the dynamics of their rotifer prey. The biomass of roach in the tarn is comparable to that recorded in other productive systems [e.g. Mathews, 1971], but the rotifer densities are quite exceptional. In Parakrama Samudra reservoir (Sri Lanka) Duncan and Gulati [1983] showed that the daily death rate of the rotifers was primarily influenced by fish predation. The mean density of rotifers in this reservoir were comparable to those recorded in Priest Pot, but the biomass of fish was four or five times greater [Newrkla & Duncan, 1984].

Acknowledgements

We wish to thank Dr T. B. Bagenal for initiating the roach study, Dr J. M. Thompson for advice on the rotifer sampling and Janet Pollard and Janice Fletcher for field assistance. We also wish to thank Trevor Furnass for drawing the diagrams and Joyce Hawksford for typing the manuscript.

References

Bagenal, T. B., 1974. A buoyant net designed to catch freshwater fish larvae quantitatively. Freshwat. Biol. 4: 107–109.

Bosch, F. v. d. & J. Ringelberg, 1985. Seasonal succession and population dynamics of Keratella cochlearis (Ehrb.) and Kellicottia longispina (Kellicott) in Lake Maarsseveren I (Netherlands). Arch. Hydrobiol. 103: 273–290.

Bosselmann, S., 1979a. Population dynamics of Keratella cochlearis in Lake Esrom. Arch. Hydrobiol. 87: 152–165.

Bosselmann, S., 1979b. Production of Keratella cochlearis in Lake Esrom. Arch. Hydrobiol. 87: 304–313.

Dodson, S. I., 1972. Mortality in a population of Daphnia rosea. Ecology 53: 1011–1023.

Dodson, A. N. & W. H. Thomas, 1964. Concentrating plankton in a gentle fashion. Limnol. Oceanog. 9: 455–456.

Downing, J. A. & F. H. Rigler (eds.), 1984. A manual on methods for the assessment of secondary productivity in freshwaters. 2nd edition. IBP Handbook No. 17. Blackwell Scientific Publications, Oxford, 525 pp.

Dumont, H. J., I. Van de Velde & S. Dumont, 1975. The dry weight estimate of biomass in a selection of Cladocera, Copepoda and Rotifera from the plankton, periphyton and benthos of continental waters. Oecologia 19: 75–97.

Duncan, A. & R. D. Gulati, 1983. A diurnal study of the planktonic rotifer populations in Parakrama Samudra Reservoir, Sri Lanka. In F. Schiemer (ed.), Limnology of Parakrama Samudra – Sri Lanka. Junk, The Hague, 236 pp.

Elliott, J. I., 1977. Seasonal changes in the abundance and distribution of planktonic rotifers in Grasmere (English Lake District). Freshwat. Biol. 7: 147–166.

Eloranta, P. V., 1982. Zooplankton in the Vassikkalampi pond, a warm water effluent recipient in central Finland. J. Plankton Res. 4: 813–837.

Finlay, B. J. & U.-G. Berninger, 1984. Coexistance of congeneric ciliates (Karyorelictida: Loxodes) in relation to food resources in two freshwater lakes. J. anim. Ecol. 53: 919–943.

Galbraith, M. G., 1967. Size-selective predation on Daphnia by rainbow trout and yellow perch. Trans. am. Fish. Soc.1: 1–10.

George, D. G. & R. W. Edwards, 1974. Population dynamics and production of Daphnia hyalina in a eutrophic reservoir. Freshwat. Biol. 4: 445–465.

Gilbert, J. J. & C. E. Williamson, 1978. Predator-prey behaviour and its effect on rotifer survival in associations of Mesocyclops edax, Asplanchna girodi, Polyarthra vulgaris and Keratella cochlearis. Oecologia 37: 13–22.

Grigorash, V. A., V. D. Spanovskaya & T. V. Lebdeva, 1972. The availability of food to underyearling roach [Rutilus rutilus (L.)] in Mozhysk reservoir. J. Ichthyol. 12: 457–465.

Hammer, C., 1985. Feeding behaviour of roach (Rutilus rutilus) larvae and the fry of perch (Perca fluviatilis) in Lake Lankau. Arch. Hydrobiol. 103: 61–74.

Herzig, A., 1983. Comparative studies on the relationship between temperature and duration of embryonic development of rotifers. Hydrobiologia 104: 237–246.

Hewitt, D. P., 1979. Tests to confirm quantitative sampling of young fish by the Bagenal buoyant net. Freshwat. Biol. 9: 339–341.

Hillbricht-Ilkowska, A., 1967. Attempt at evaluation of the production and turnover of plankton rotifers on the example of Keratella cochlearis (Gosse). Bull. Acad. pol. Sci. Cl. II Ser. Sci. Biol. 15: 35–40.

Jacobs, J., 1977. Coexistence of similar zooplankton species by differential adaption to reproduction and escape in an environment with fluctuating food and enemy densities. II. Field data analysis of Daphnia. Oecologia 30: 313–329.

Makarewicz, J. C. & G. E. Likens, 1979. Structure and function of the zooplankton community of Mirror Lake, New Hampshire. Ecol. Monogr. 49: 109–127.

Mathews, C. P., 1971. Contribution of young fish to total production of fish in the River Thames near Reading. J. Fish Biol. 3: 157–180.

May, L., 1983. Rotifer occurrence in relation to water temperature in Loch Leven, Scotland. Hydrobiologia 104: 311–315.

Mikheev, V. M., 1984. Prey size and food selectivity in young fishes. J. Ichthyol. 24: 66–76.

Newrkla, P. & A. Duncan, 1984. The biology and density of Ehirava fluviatilis (clupeoiod) in Parakrama Samudra, Sri Lanka. Verh. internat. Ver. Limnol. 22: 1572–1578.

Orcutt, J. D. & M. L. Pace, 1984. Seasonal dynamics of rotifer and crustacean zooplankton populations in a eutrophic monomictic lake with a note on rotifer sampling techniques. Hydrobiologia 119: 73–80.

Pace, M. L. & J. D. Orcutt, 1981. The relative importance of protozoans, rotifers, and crustaceans in a freshwater zooplankton community. Limnol. Oceanogr. 26: 822–830.

Paloheimo, J. E., 1974. Calculation of instantaneous birth rate. Limnol. Oceanogr. 19: 692–694.

Patalas, K., 1970. Primary and secondary production in a lake heated by thermal power plant. Proc. 16th annual tech. Meeting of Inst. of env. Sciences. Mt. Prospect, Ill. 1267–1271.

Ruttner-Kolisko, A., 1974. Plankton rotifers: biology and taxonomy. Binnenge wasser 26 (1 suppl) 146 pp.

Windell, J. T., 1971. Food analysis and rate of digestion. In W. E. Ricker (ed.), Methods for assessment of fish production in freshwaters. I.B.P. Handbook 3. 2nd edition. Blackwell Scientific Publications, Oxford, 348 pp.

Hydrobiologia 147: 229–233 (1987)
May, L., Wallace, R. & Herzig, A. (eds), Rotifer Symposium IV

Post-encounter vulnerability of some rotifer prey types to predation by the copepod *Acanthocyclops robustus*

Kennedy F. Roche
Laboratoria voor Ecologie, Rijksuniversiteit Gent, K. L. Ledeganckstraat 35, B-9000 Gent, Belgium

Keywords: Rotifera, cyclopoid predation, selectivity, prey features

Abstract

Prey characteristics affecting vulnerability of a number of rotifer species to the predator *Acanthocyclops robustus* were investigated. Body volume was the main feature affecting the probability of attack after encounter and, with the exception of *Polyarthra major,* capture after attack. The lorica provided protection against predation in some species and a skipping escape response reduced the likelihood of capture in *Polyarthra major.*

Introduction

Invertebrate predation can have an important influence on the population dynamics and community structure of freshwater zooplankton, and on the behaviour and morphology of individual species [Kerfoot, 1977; Lane, 1979]. The characteristics which determine the susceptibility of zooplankton to invertebrate predation can be divided into pre-encounter and post-encounter stages. The former includes prey swimming speed [Gerritsen & Strickler, 1977] and spatial distribution [Melville & Maly, 1981], while the latter consists, primarily, of escape response abilities and morphological features such as size and shape [Williamson, 1983a, b; Li & Li, 1979; Stemberger, 1985]. This paper investigates the post-encounter vulnerability of rotifers to predation by the cyclopoid copepod *Acanthocyclops robustus.* The relative importance of various prey characteristics in predator avoidance are discussed.

Materials and methods

Experimental rotifers (see Table 1), apart from *Filinia terminalis* and *Synchaeta pectinata,* and the predatory copepod *Acanthocyclops robustus* were collected from Lake Donk, Belgium. Lake Donk is a shallow, eutrophic water-body situated about 20 km east of Gent. *F. terminalis* and *S. pectinata* were obtained from nearby ponds. Prey body volumes were determined from modifications of the formulae of Ruttner-Kolisko [1977].

The predation experiments were carried out in glass, round-bottomed vials containing about 0.5 ml of glass-fibre filtered lake water. In general, the experimental procedure was similar to that described by Gilbert & Williamson [1978]. One to six rotifers were placed in each vial with a single, adult, female *A. robustus.* The predator had been starved for 2.5 h to 9 h and acclimated to the experimental temperature. The experimental animals were observed for 1 min to 60 mins (average 15 mins) and all encounters, attacks, captures and ingestions, as defined below, were recorded.

Encounter – any physical contact between prey and predator, excluding contact with the furcal rami of the copepod; attack reactions by the predator; any case when the prey was brushed aside by

Table 1. Body size and vulnerability to predation of rotifers when preyed upon by *Acanthocyclops robustus*, where
P[E→A] = Probability of attack after encounter
P[A→C] = Probability of capture after attack
P[C→I] = Probability of ingestion after capture
P[E→I] = Probability of ingestion after encounter

Type of prey	Volume ($\times 10^5$ μm^3)	No. of predators tested	No. of encounters recorded	P[E→A]	P[A→C]	P[C→I]	P[E→I]
Brachionus calyciflorus	122.97	21	591	0.48	0.06	0.00	0.00
Brachionus calyciflorus	64.52	9	126	0.49	0.29	0.54	0.05
Brachionus calyciflorus	45.15	8	87	0.54	0.85	1.00	0.45
Brachionus calyciflorus	35.08	10	77	0.36	0.93	1.00	0.34
Brachionus diversicornis	30.72	16	244	0.56	1.00	0.00	0.00
Notholca acuminata	8.59	15	–	–	0.96	0.24	–
Keratella quadrata	11.59	16	774	0.09	1.00	0.31	0.03
Keratella cochlearis	0.97	19	584	0.03	1.00	0.00	0.00
Keratella cochlearis	0.53	35	1318	0.02	0.96	0.39	0.01
Asplanchna girodi	?>300	17	–	–	0.01	1.00	–
Asplanchna priodonta	113.19	9	111	0.57	0.25	1.00	0.09
Synchaeta pectinata	29.92	12	–	–	0.94	1.00	–
Synchaeta pectinata		17	198	0.39	0.87	1.00	0.35
Polyarthra major	15.84	27	647	0.14	0.55	1.00	0.07
Pompholyx sulcata	4.07	11	361	0.04	1.00	0.13	0.01
Pompholyx sulcata		9	–	–	1.00	0.07	–
Filinia terminalis	3.45	23	753	0.06	0.96	0.97	0.06

the displaced water in front of the moving copepod; any occasion when the prey used its escape response to leave the path of an approaching predator.

Attack– any leaping or pouncing movement of the predator towards the prey.

Capture– prey held by the mouthparts of the predator.

Ingestion– more than 50% of the prey ingested by the predator.

The probability of each step in the predation cycle occurring was calculated as the mean of the values recorded for each individual copepod. Standard errors of the means were also determined. The following abbreviations are used in the text:

P[E→A] = Probability of attack after encounter
P[A→C] = Probability of capture after attack
P[C→I) = Probability of ingestion after capture
P[E→I) = Probability of ingestion after encounter

Results

The rotifer prey varied in their response to the approaching predator. *Polyarthra major* avoided the predator by making one or two skipping movements away from the stimulus as described by Gilbert [1985]. *Filinia terminalis* simply raised its two anterior spines above its corona. In contrast, the other rotifer species tended to show a 'dead-man' response [Kerfoot, 1977, 1978; Jamieson, 1980). On sensing the predator, they retracted any protruding appendages and stopped swimming.

The probability of attack after encounter

The probability of attack after encounter (P[E→A]) varied from 0.02 for *Keratella cochlearis* (small morph) to 0.57 for *Asplanchna priodonta* (Table 1). Prey body volume had a major influence on

P[E→A] (Fig. 1a). P[E→A] increased with prey volume until a threshold size of about 30×10^5 μm^3 was reached; P[E→A] then levelled off at about 0.5, despite increases in prey size. An inverse relationship between predator activity and P[E→A] was also recorded; 14% of encounters involving moving predators resulted in attacks, whereas 73% of encounters involving motionless predators resulted in attacks.

Probability of capture after attack

The probability of capture after attack (P[A→C]) ranged from 0.01 for *Asplanchna girodi* to 1.00 for *Brachionus diversicornis, Keratella quadrata, Keratella cochlearis* and *Pompholyx sulcata* (Table 1). Smaller rotifers were more easily captured than larger animals (Fig. 1b). *Polyarthra major,* however, was the exception. P[A→C] for this species was approximately half the value expected for its body size (Fig. 1b). This is probably due to the effectiveness of its escape response, described above.

Probability of ingestion after capture

The main feature affecting the probability of ingestion after capture in the species studied was the nature of the lorica. Loricate rotifers, apart from juvenile *Brachionus calyciflorus,* had a low P[C→I] while illoricate rotifers were almost always ingested if captured. In *Brachionus calyciflorus* P[C→I] was inversely related to body size.

Probability of ingestion after encounter

The probability of ingestion after encounter (P[E→I]) was low (<0.1) for most species, apart from *Synchaeta pectinata* (P[E→I]=0.35) and juvenile *Brachionus calyciflorus* (P[E→I]=0.34 to 0.45) (Table 1, Fig. 1c).

Discussion

Probability of attack after encounter

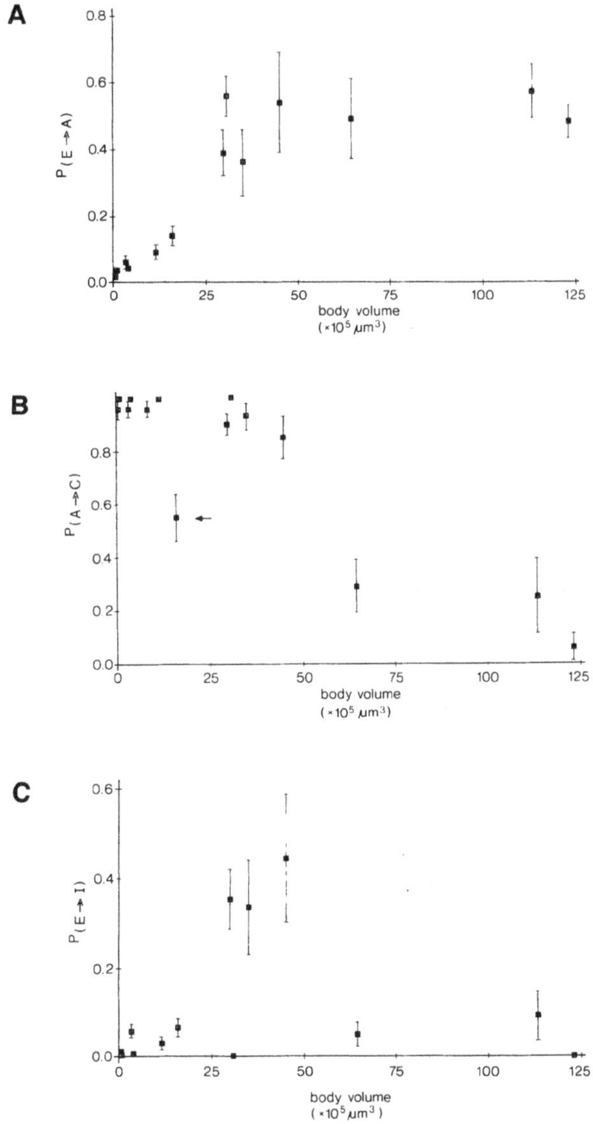

Fig. 1. The relationship between prey body volume and
a) Probability of attack after encounter (P[E→A] ± s.e.)
b) Probability of capture after attack (P[A→E] ± s.e.)
c) Probability of ingestion after encounter (P[E→I] ± s.e.)
for rotifers preyed upon by the copepod *Acanthocyclops robustus. Polyarthra major* is indicated by an arrow in (b).

The probability of attack after encounter was found to be closely related to prey body volume, up to a threshold body volume of about 30×10^5 μm^3. This is probably explained by the fact that copepods mainly detect their prey through mechanoreceptors which sense hydrodynamic disturbances [Strickler, 1975] and touch contacts. A larger prey size would increase the amount of hydrodynamic disturbance and the probability of detection by the predator [Gerritsen & Strickler, 1977; Gilbert & Williamson, 1978; Kerfoot, 1978; Wong, 1981; Williamson, 1983a; Cooper, 1983].

The inverse relationship detected between copepod activity and P[E→A] was probably due to differences in the sensitivity of the copepods receptors due to changes in water pressure around the predator [Gerritsen & Strickler, 1977]. More active copepods generate more pressure and shear disturbances than less active animals. For this reason, most attacks are thought to occur during the sinking phase of 'hop-and-sink' swimming style of copepods [Jamieson, 1980; Kerfoot *et al.* 1980]. Predator swimming speed is affected by starvation level, prey density and other environmental factors and these, therefore, have an indirect effect on P[E→A].

Probability of capture after attack

Among animals without specialised escape responses, it is clear that P[A→C] was closely related to body size; animals smaller than 50×10^5 μm^3 were more easily captured than larger individuals. *Polyarthra major* was much less likely to be captured than other animals of similar size, due to its skipping escape response. This is in contrast to the results described for *P. major* and the predator *Diacycops thomasi* by Stemberger [cit. Williamson, 1983b and Stemberger, 1985].

Probability of ingestion after capture

It is generally thought that the lorica of rotifers provides protection against predators [Jamieson, 1980; Williamson, 1983b; Stemberger, 1985]. The present study confirms this. Loricate rotifers were, generally, less likely to be ingested than illoricate forms. The spines of *Keratella cochlearis, K. quadrata* and *Brachionus diversicornis* may also have provided additional protection, but this is not confirmed as spineless forms were not tested. The 'dead-man' responses did not seem to affect P[C→I], although they must have protected the prey from damage during handling by the predator.

Probability of ingestion after encounter

Most of the prey types tested had a similar overall vulnerability to predation (P[E→I], but the reasons for this varied from species to species. For small prey the main factor reducing vulnerability to predation was P[E→A]; for large prey the main limiting factor was a low value of P[A→C]. Figure 1c resembles the electivity curve for invertebrate predators shown in Zaret [1980] whereby maximum electivity is shown for prey of intermediate size; see also Cooper [1983]. Upon this was superimposed the effects of prey body armour and escape responses. The importance of these effects depends on body size.

Acknowledgements

I thank H. J. Dumont and K. Martens for reading and improving the manuscript, P. Herman for help in analysing the data, and M. Bruyneel for drawing the figures. The author is financially supported by a traveling studentship from the National University of Ireland.

References

Cooper, S. D., 1983. Selective predation on cladocerans by common pond insects. Can. J. Zool. 61: 879–886.

Gerritsen, J. & J. R. Strickler, 1977. Encounter probabilities and community structure in zooplankton: a mathematical model. J. Fish. Res. Bd Can. 34: 73–82.

Gilbert, J. J., 1985. Escape responses of the rotifer Polyarthra: a high-speed cinematographic analysis. Oecologia 66: 322–331.

Gilbert, J. J. & C. E. Williamson, 1978. Predator-prey behaviour and its effect on rotifer survival in associations of Mesocyclops edax, Asplanchna girodi, Polyarthra vulgaris and Keratella cochlearis. Oecologia 37: 13–22.

Jamieson, C. D., 1980. The predatory feeding of copepodid stages III to adult Mesocyclops leuckarti (Claus). In W. C. Kerfoot (ed.), Evolution and Ecology of Zooplankton Communities. The University Press of New England, Hanover (N.H.); Lond.: 518–537.

Kerfoot, W. C., 1977. Implications of copepod predation. Limnol. Oceanogr. 22: 316–325.

Kerfoot, W. C., 1978. Combat between predatory copepods and their prey: Cyclops, Epischura and Bosmina. Limnol. Oceanogr. 23: 1089–1102.

Kerfoot, W. C., D. L. Kellogg, Jr., & J. R. Strickler, 1980. Visual observations of live zooplankton: Evasion, escape, and chemical defenses. In W. C. Kerfoot (ed.), Evolution and Ecology of Zooplankton Communities. The University Press of New England, Hanover (N.H.); Lond.: 10–27.

Lane, P. A., 1979. Vertebrate and invertebrate predation intensity on freshwater zooplankton communities. Nature 280: 391–393.

Li, J. L. & H. W. Li, 1979. Species-specific factors affecting predator-prey interactions of the copepod Acanthocyclops vernalis with its natural prey. Limnol. Oceanogr. 24: 613–626.

Melville, G. E. & E. J. Maly, 1981. Vertical distribution and zooplankton predation in a small temperate pond. Can. J. Zool. 59: 1720–1725.

Ruttner-Kolisko, A., 1977. Suggestions for biomass calculation of plankton rotifers. Arch. Hydrobiol. 8: 71–76.

Stemberger, R. S., 1985. Prey selection by the copepod Diacyclops thomasi. Oecologia 65: 492–497.

Strickler, J. R., 1975. Intra- and interspecific information flow among planktonic copepods: Receptors. Verh. int. Ver. Limnol. 19: 2951–2958.

Williamson, C. E., 1983a. Behavioral interactions between a cyclopoid copepod predator and its prey. J. Plankton. Res. 5: 701–712.

Williamson, C. E., 1983b. Invertebrate predation on planktonic rotifers. Hydrobiologia 104: 385–396.

Wong, C. K., 1981. Predatory feeding behaviour of Epischura lacustris (Copepoda, Calanoida) and prey defense. Can. J. Fish. aquat. Sci. 38: 275–279.

Zaret, T. M., 1980. Predation and freshwater communities. Yale University Press, New Haven, London, 187 pp.

Hydrobiologia 147: 235–238 (1987)
May, L., Wallace, R. & Herzig, A. (eds), Rotifer Symposium IV

The *Polyarthra* escape from response: Defense against interference from *Daphnia*

John J. Gilbert
Department of Biological Sciences, Dartmouth College, Hanover, New Hampshire 03755, U.S.A.

Keywords: competition, *Daphnia*, defense, interference, *Polyarthra*

Abstract

Direct observations show that almost all (97%) *Polyarthra remata* caught in the inhalant current of *Daphnia pulex* avoid inhalation by initiating escape responses. Microvideographic analysis shows that these responses are initiated, on average, 0.24 mm (about 2.5 body lengths) from the inhalant opening of the branchial chamber, when the *Polyarthra* are moving at a velocity of about 1.4 mm s^{-1} (about 6 times their normal swimming speed – 0.24 mm s^{-1}). The stimulus for the escape response could be rapid acceleration or, more likely, shear. The average distance traveled during an escape response is at least 1.25 mm (about 12 body lengths).

The *Polyarthra* escape response can provide a very effective defense against interference from *Daphnia* and may greatly increase the ability of *Polyarthra* spp. to coexist with *Daphnia*.

Introduction

The escape response of *Polyarthra*, recently described and analyzed using high-speed microcinematography [Gilbert, 1985], can be a very effective defense against some invertebrate predators, such as *Asplanchna* [Gilbert & Williamson, 1978; Gilbert, 1980], *Diacyclops* [Stemberger, 1985], and first instar *Chaoborus* larvae [Moore and Gilbert, 1987]. The most extensive observations on *Polyarthra*'s escape behavior have been made with *Asplanchna*. In these interactions the escape response is triggered when *Polyarthra* is physically contacted by the *Asplanchna* [Gilbert, 1985].

In the present study I test the hypothesis that the *Polyarthra* escape response also provides an effective defense against interference from *Daphnia*. Rotifers drawn into the branchial chamber of a *Daphnia* are susceptible to several types of damage. They may be ingested, or they may be killed, injured, or lose attached eggs when rejected by the postabdomen. These interactions, discovered by Gilbert & Stemberger [1985] and further analyzed by Burns & Gilbert [1986a, b], may result in very high rotifer mortality rates. *Polyarthra* would be protected against such interference if they could escape from the inhalant current of a feeding *Daphnia* before being taken into the branchial chamber.

Materials and methods

Polyarthra remata was collected from a pond in Thetford, Vermont. *Daphnia pulex* was collected from a pond in Lyme, New Hampshire. Both species were cultured in filtered (Whatman 934-AH) lake water (Post Pond, Lyme, New Hampshire) with the alga *Cryptomonas* sp. (isolated from Pleasant Lake, New London, New Hampshire) at 20 °C in a photocycle (LD 16:8). The *Cryptomonas* was cultured in an artificial medium [Stemberger, 1981]. The mean (± S.D.) body length of ovigerous *P. remata* was 109 ± 7 μm. The body lengths (head to base of tail spine) of the adult *D. pulex* used were 1.7–2.0 mm.

Interactions between *P. remata* and actively feeding *D. pulex* at 21 °C were observed on several occasions by placing 50–75 *P. remata* with *Cryptomonas* in a 1 ml observation chamber with a single *D. pulex* attached by one of its lateral surfaces to the bottom of the chamber with silicone grease. Visual observations were made with a Wild M-5 stereomicroscope. Microvideographic observations were made with a Zeiss IV/b stereomicroscope fitted with a Zeiss 63 mm/0.1 Luminar objective lens, a monocular tube, and a 4× eyepiece. Videotapes were made using a Dage-MTI 68 series camera, a Sony VO-5800H recorder with Sony KCA-60K videotape, a Panasonic WV-5410 videomonitor, and a Panasonic WJ-810 time date generator having a stop-watch display accurate to 0.01 s. Videotapes were analyzed by slow motion playback of sequences and by tracing the positions of objects onto a thin plastic film (Saran Wrap) applied to the surface of the screen. Magnification on the screen was 43X, one millimetre on the screen being equivalent to 20.6 μm actual distance.

Visual observations were used to determine the probability that *P. remata* entrained in the inhalant current of a *D. pulex* would avoid entering the branchial chamber by initiating an escape response. Microvideography was used to determine: 1) the distances of *P. remata* from the inhalant opening of the branchial chamber of *D. pulex* at the start of their escape responses, 2) the distances traveled by these escaping *P. remata*, 3) the swimming speed of *P. remata*, and 4) the velocities of the inhalant current at different distances from the inhalant opening, as determined from the movements of inert objects entering the branchial chamber (distances were from the inhalant opening to the mid-points of the 4–6 sections of path traveled by each object).

All observations and analyses of interactions between *P. remata* and *D. pulex* were made during periods when the second antennae of the *D. pulex* were inactive. Distances of *P. remata* from the inhalant opening of *D. pulex* were measured only when the *D. pulex* were lying perfectly flat, with the margins of both carapace valves superimposed.

Results and discussion

Visual observations showed that 97% of the 257 *P. remata* caught in the inhalant currents of five *D. pulex* (1.7–2.0 mm body length) initiated escape responses and thereby avoided entry into the branchial chamber. Thus, the escape response of *P. remata* should provide a very effective defense against interference from *D. pulex*.

Other small rotifers without pronounced escape responses, such as *Keratella cochlearis* and *Synchaeta oblonga*, almost invariably enter the branchial chambers of *D. pulex* when entrained in their inhalant currents [(Gilbert, unpublished]; then they are likely to be damaged on rejection or ingested [Gilbert & Stemberger, 1985; Burns & Gilbert, 1986a, b; Gilbert, unpublished].

On average, *P. remata* initiated escape responses 0.24 mm (about 2.5 body lengths) from the inhalant openings of the branchial chambers of *D. pulex* (Table 1). The distances traveled by the *P. remata* during escape responses could be determined for only about one-half of the responses, because many of the escaping rotifers moved out of the field of view. The mean (± S.D.) distance traveled during 26 complete escape responses was 1.25 ± 0.54 mm, the shortest and longest distances traveled being 0.25 and 2.3 mm. These mean and maximum distances certainly are underestimates, since most of the longest jumps terminated beyond the field of view.

The distances traveled during the escape responses of *P. remata* in this study are similar to those of *P. vulgaris* in a previous study [Gilbert, 1985]. The mean and greatest distances traveled by *P. vulgaris* were 1.95 and 3.07 mm.

Elapsed times during the escape responses of *P. remata* could not be resolved accurately enough to provide good estimates of body velocities. Three estimates, made from relatively long responses, were about 13, 15, and 30 mm s^{-1}. These values are lower than the accurate, cinematographically-determined values for *P. vulgaris*, of which the mean was 35.7 mm s^{-1} [Gilbert, 1985].

The velocity of the inhalant current of one *D. pulex* (2.0 mm body length) was estimated from the movements of three different inert objects (about

100 μm long) as they approached the inhalant opening of the branchial chamber (Fig. 1). The velocity increased exponentially with decreasing distance from the opening. For example, velocities 1.0 and 0.2 mm from the opening were about 0.2 and 1.5 mm s^{-1}. For the *Daphnia* individual studied, the velocity (V) at any given distance from the opening (D) can be estimated from the linear regression formula: $V = \text{antilog}_{10}\ (0.407 - 1.122\ D)$ ($r^2 = 0.89$).

The inhalant current velocities shown in Fig. 1 undoubtedly are a great oversimplification and should be viewed cautiously. The three-dimensional flow field around the inhalant opening of the branchial chamber probably is complex. The inhalant opening along the ventral margin of the carapace is quite large, and current velocities into this opening may vary considerably along the antero-posterior axis. Also, there must be a velocity gradient along the plane normal to the dorso-ventral axis, at any given distance from the ventral surface, with velocities being greater near the midline. The velocities in Fig. 1 probably are close to maximal inhalant velocities. Furthermore, the actual velocity of the inhalant current entering the branchial chamber may vary greatly with the species, size, and physiological state of the *Daphnia*, with the temperature, and with the concentration of food available to the *Daphnia*. Finally, attachment of the *Daphnia* onto the bottom of a small vessel in this study probably caused some unnatural currents.

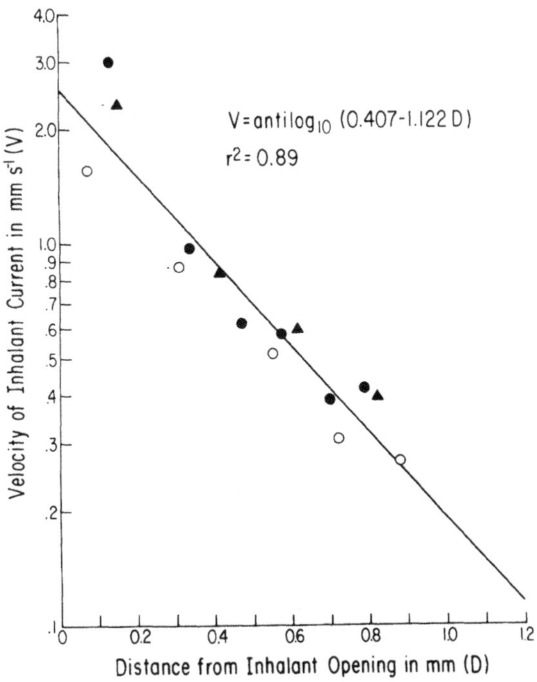

Fig. 1. Velocity of the inhalant current of a *Daphnia pulex* (2.0 mm body length) as a function of distance from the inhalant opening of the branchial chamber. The three different symbols represent the velocities of three different, inert objects caught in the inhalant current.

The velocity pattern in Fig. 1 suggests that small rotifers, which seem to have swimming speeds of 0.2–0.5 mm s^{-1} [Gilbert, 1985; Stemberger, 1985], would be unable to swim against or resist the inhalant current when closer than several body lengths away from the inhalant opening. This hypothesis is confirmed by observations of *Keratella cochlearis* and *Synchaeta oblonga* [Gilbert, unpublished].

Only because of its escape response can *Polyarthra* regularly get out of *Daphnia*'s inhalant current. The mean (± S.D.) swimming speed of the *P. remata* in this study, determined from the movements of 5 individuals for periods of 7.05–12.71 s, was 0.24 ± 0.04 mm s^{-1}. The inhalant current velocity of the *D. pulex* observed was about 5 times this value (1.2 mm s^{-1}) 0.3 mm (or about 3 *P. remata* body lengths) from the inhalant opening (Fig. 1).

The stimulus eliciting the escape response of *P. remata* caught in the inhalant current of *D. pulex* is not known. It is clear that this response is not dependent upon physical contact with another object. Only 2 of 58 responses (Table 1) were initiated close enough to a *Daphnia* to have been stimulated by contact. As a *P. remata* is caught in the inhalant current, its body will experience a rapid acceleration (Fig. 1). This acceleration may provide the stimulus. The *P. remata*, on average, initiated es-

Table 1. Distances of *Polyarthra remata* from the inhalant opening of *Daphnia pulex* at the start of their escape responses (determined from microvideographic analysis).

Daphnia		Distance from inhalant opening (mm)			
Specimen	Size (mm)	Mean	N	S.D.	Range
1	2.0	0.28	12	0.14	0.08–0.54
2	1.8	0.22	46	0.13	0 –0.57
1–2		0.24	58	0.13	0 –0.57

cape responses 0.24 mm away from the inhalant opening (Table 1). At this distance, they would be moving at a velocity of about 1.5 mm s^{-1} (Fig. 1), more than 5 times their normal swimming speed. Alternatively, a *P. remata* might enter a velocity gradient and experience shear when it is caught in the inhalant current. This shear is, perhaps, most likely to provide the stimulus.

General ecological considerations

This study shows that the escape response of *Polyarthra remata* provides an extremely effective defense against inhalation by, and hence interference from, immobilized but actively feeding *Daphnia pulex*. This conclusion leads to three predictions or hypotheses. First, the escape response of *P. remata* should be very effective against interference from naturally-swimming *D. pulex*. Second, *P. remata* should experience much less mortality from *D. pulex* interference than some rotifers that cannot escape inhalation. Third, other species of *Polyarthra* also should be defended against interference competition from cladocerans. The various *Polyarthra* species, however, may not be equally well defended. For example, *P. vulgaris* was much better able to escape predation by the copepod *Diacyclops thomasi* than either *P. major* or *P. remata* [Stemberger, 1985].

The first two hypotheses are nicely supported by some laboratory experiments comparing the death rates of *P. remata* and *Keratella cochlearis tecta* cultured together with *D. pulex* [Gilbert, unpublished]. The instantaneous, *Daphnia*-induced death rates of the *Keratella* were 11–22 times higher than those of the *P. remata*.

Some indirect evidence from field studies suggests that populations of *Polyarthra* are less inhibited by *Daphnia* than those of some other rotifers. Adalsteinsson [1979] presents data from Lake Mývatn, Iceland indicating that *P. dolichoptera* is less suppressed by *D. longispina* than either *Keratella cochlearis* or *Synchaeta* cf *lakowitziana*. Stemberger & Evans [1984] and Evans [unpublished] show that the offshore Lake Michigan population of *P. major* and *P. remata* was most dense between August and October, when *Daphnia* spp. were most abundant and when *Keratella cochlearis* death rates were at a maximum (0.11–0.15 day^{-1}). Invertebrate predators could not account for the *K. cochlearis* mortality.

In conclusion, it seems quite certain that susceptibility to interference by *Daphnia*, and probably other cladocerans, often may be an important factor determining the species structure of natural rotifer communities. Because of its escape response, *Polyarthra* may largely avoid such interference and thus be more likely to coexist with *Daphnia* than some other rotifers.

Acknowledgements

This study was supported by National Science Foundation research grant BSR 8415024. I thank Carolyn Burns, Kevin Kirk, and Richard Stemberger for improving the manuscript.

References

Adalsteinsson, H., 1979. Zooplankton and its relation to available food in Lake Mývatn. Oikos 32: 162–194.

Burns, C. W. & J. J. Gilbert, 1986a. Effects of daphnid size and density on interference between Daphnia and Keratella cochlearis. Limnol. Oceanogr. 31: 848–858.

Burns, C. W. & J. J. Gilbert, 1986b. Direct observations of the mechanisms of interference between Daphnia and Keratella cochlearis. Limnol. Oceanogr. 31: 859–866.

Gilbert, J. J., 1980. Feeding in the rotifer Asplanchna: behavior, cannibalism, selectivity, prey defenses, and impact on rotifer communities. In W. C. Kerfoot (ed.), Evolution and Ecology of Zooplankton Communities. The University Press of New England, Hanover (N.H.): 158–172.

Gilbert, J. J., 1985. Escape response of the rotifer Polyarthra: a high-speed cinematographic analysis. Oecologia 66: 322–331.

Gilbert, J. J. & R. S. Stemberger, 1985. Control of Keratella populations by interference competition from Daphnia. Limnol. Oceanogr. 30: 180–188.

Gilbert, J. J. & C. E. Williamson, 1978. Predator-prey behavior and its effect on rotifer survival in associations of Mesocyclops edax, Asplanchna girodi, Polyarthra vulgaris, and Keratella cochlearis. Oecologia 37: 13–22.

Moore, M. V. & J. J. Gilbert, 1987. Age-specific Chaoborus predation on rotifer prey. Freshwat. Biol., in press.

Stemberger, R. S., 1981. A general approach to the culture of planktonic rotifers. Can. J. Fish. aquat. Sci. 38: 721–724.

Stemberger, R. S., 1985. Prey selection by the copepod Diacyclops thomasi. Oecologia 65: 492–497.

Stemberger, R. S. & M. S. Evans, 1984. Rotifer seasonal succession and copepod predation in Lake Michigan. J. Great Lakes Res. 10: 417–428.

Hydrobiologia 147: 239–243 (1987)
May, L., Wallace, R. & Herzig, A. (eds), Rotifer Symposium IV

Rotifers, cladocerans and planktivorous fish: What are the major interactions?

Stephen T. Threlkeld & Elizabeth M. Choinski
Biological Station and Department of Zoology, University of Oklahoma, Kingston, OK 73439 U.S.A.

Keywords: Rotifera, Cladocera, fish, interactions, correlation, multiple regression analyses

Abstract

Results of ten outdoor tank experiments on the effects of two kinds of planktivorous fish (*Dorosoma cepedianum* and *Menidia beryllina*) on phytoplankton and zooplankton were reviewed for evidence of rotifer-cladoceran interactions. Correlation and multiple regression analyses of the responses of populations of six rotifer species in ten experiments were examined for evidence of interference competition with daphnids, resource exploitation, or invertebrate predation by crustaceans associated with daphnid-dominated communities. A separate enclosure experiment was conducted to assess the potential for short-term effects of dense daphnid populations on four rotifer populations. Indirect resource-mediated effects on rotifer populations associated with manipulations of planktivorous fish were more numerous and consistently expressed than direct negative effects of cladocerans or copepods on rotifer populations.

Introduction

Interaction of planktonic rotifers and cladocerans is usually studied in the context of experimentally controlled or natural variation in the density of planktivorous fish. Three specific mechanisms that have received close attention in explaining the inverse relationship between rotifers and cladocerans are (1) differential resource exploitation [Neill, 1984 and references therein], (2) mechanical interference competition [Gilbert & Stemberger, 1985], and (3) predation on rotifers by zooplankton that are associated with cladoceran-dominated communities (e.g., copepods) [see review by Williamson, 1983]. This paper examines results of ten outdoor tank experiments in which the responses of rotifers, cladocerans, copepods, and phytoplankton to two kinds of planktivorous fish were investigated. Differences between the food webs generated by the fish were used to elucidate rotifer-cladoceran interactions.

Methods and materials

Ten experiments were conducted in 18 large (7000 L), outdoor fiberglass tanks at the University of Oklahoma Biological Station (Kingston, Oklahoma, U.S.A.) on the effects of two kinds of planktivorous fish (*Dorosoma cepedianum* and *Menidia beryllina*) on zooplankton and phytoplankton community structure. *Dorosoma* is a pump filter-feeding clupeid and is known to capture rotifers, slowly moving cladocerans, and large algal cells [Drenner *et al.*, 1982, 1984]. *Menidia*, a particulate feeding atherinid, feeds selectively on large-bodied zooplankton [McComas & Drenner, 1982]. Four experiments conducted in 1982–83 and involving both *Dorosoma* and *Menidia* will be reported in greater detail elsewhere [Drenner *et al.*, 1986]; two experiments with both fish conducted in 1984 were described by Threlkeld [1986]; four additional experiments involving *Menidia* only will be reported elsewhere [Threlkeld & Soballe, unpubl. obs.].

All tank experiments were initiated by filling the tanks with water from Lake Texoma, a large southwestern U.S. reservoir. Fish were then stocked and measurements of phytoplankton and zooplankton communities usually made at 5 day intervals for periods of 30–45 days [Drenner *et al.*, 1986; Threlkeld, 1986]. Tank-specific parameter means for several experiments were subjected to correlation and multiple regression analyses. Correlation analysis was used to show the magnitude and kind of association between each pair of variables in the food web, even though it only suggests causal relationships among variables if they are themselves not correlated to other variables. Multiple regression analysis was used to identify partial correlation coefficients between functional groups within each planktonic food web when other predictor variables were kept constant at their means. Standardized regression coefficients that express the relationship between two variables in standard deviation units were also calculated. This standardization allows all food web components to be discussed in terms of their own variability, and thus their real ecological weight appreciated fully [Sokal & Rohlf, 1981]. For analysis of *Dorosoma*-dominated food web interactions, only experiments in which *Dorosoma* was manipulated are included (along with corresponding fishless controls). Data for 42 tanks from 6 experiments were used in the analysis of *Dorosoma* effects. In a similar fashion, 80 tanks from 10 experiments provided data for examining *Menidia* effects (Table 1).

For correlation and regression analyses, some individual food web components were pooled into functional groups. Small cladocerans included *Bosmina, Ceriodaphnia*, and *Diaphanosoma*; daphnids included *Daphnia parvula, Daphnia galeata*, and *Daphnia pulicaria*; copepods include only the adult and copepodid stages of *Eurytemora affinis, Diaptomus siciloides*, and all cyclopoids. Fish mortality refers to biomass of fish introduced to the tanks that was not recovered at the end of an experiment, and is assumed to have decomposed [Threlkeld, 1986]. Rotifers were identified using the key of Stemberger [1979].

One enclosure experiment was conducted during tank experiment 9 (see Table 1), in which nine densities of *Menidia* were stocked in replicate tanks. On day 30 of this experiment, 40 combinations of rotifers and cladocerans from the 18 tanks were incubated in 12-liter plastic containers in the tanks. Each container received half its water from a tank then dominated by rotifers or cladocerans, and the other half from another tank of similar or dissimilar composition. Water from each tank was involved in 4 or 5 such pairings, resulting in a broad range of combinations of rotifer and daphnid densities. After five days the remaining zooplankton were collected, and examined microscopically.

Table 1. Experiment dates, temperature ranges and number of experimental tanks used in *Dorosoma* or *Menidia* treatments and fishless controls.

Experiment	Dates	Temp. range (°C)	Number of tanks used for		
			Dorosoma	*Menidia*	Controls
1	5 Nov – 28 Dec 82	2.9 – 17.1	6	2	2
2	11 Mar – 26 Apr 83	9.5 – 16.7	6	3	3
3	15 Jun – 30 Jul 83	24.5 – 30.0	6	3	3
4	23 Sep – 12 Nov 83	10.4 – 23.3	3	3	3
5	24 May – 16 Jun 84	22.0 – 27.0	3	3	3
6	26 Jul – 25 Aug 84	25.0 – 27.2	0	3	3
7	31 Oct – 28 Nov 84	7.2 – 19.1	2	2	2
8	18 Jan – 17 Feb 85	0.0 – 7.0	0	3	3
9	6 Mar – 15 Apr 85	12.1 – 20.3	0	16	2
10	8 May – 2 Jun 85	22.0 – 27.0	0	15	3

Results

The correlation matrices for food web components show that *Dorosoma* and *Menidia* treatments were strongly correlated with a nuisance variable in the experiments, fish mortality, which was also correlated with the occurrence of bluegreen algae and algal chlorophyll fluorescence (Table 2). Strongly positive and negative rotifer-rotifer correlations were both present, which is indicative of strong seasonal associations or differential responses to food regimes.

Multiple regression analysis clearly showed that both fish have strong negative effects on cladoceran and copepod populations (Fig. 1). There were few significant negative coefficients between rotifer and crustacean taxa. Some copepod and daphnid effects on rotifers were positive. Only *Polyarthra vulgaris* was negatively associated with daphnids. Rotifers were independent of small cladocerans. Most significant connections to rotifer populations were with algal variables (bluegreen filaments and algal fluorescence).

One experiment was conducted in the spring of 1985 (Experiment 9, Fig. 2) when rotifers and cladocerans are most abundant in Lake Texoma [Threlkeld, unpubl. obs.]. During that experiment special attention was paid to the possibility of

Table 2. Product-moment correlations among food web components in 42 tanks involving *Dorosoma* treatments or fishless controls (upper panel) or 80 tanks with *Menidia* or controls (lower panel). Critical values for individual coefficients are shown for each food web. Abbreviated column headings, from left to right, are in the same order as food web components listed on the left.

	Fili	Noth	Quad	Poly	Brac	Coch	Daph	Smal	Cope	*Dor/Men*	Loss	Alga	BGfl
Dorosoma cepedianum (P = 0.05; ± 0.304)													
Filinia	1.00												
Notholca	0.80	1.00											
K. quadrata	0.96	0.76	1.00										
Polyarthra	−0.14	−0.11	−0.16	1.00									
Brachionus	−0.23	−0.21	−0.27	−0.25	1.00								
K. cochlearis	−0.15	−0.17	−0.19	−0.07	0.46	1.00							
Daphnids	0.30	0.29	0.27	−0.19	−0.18	−0.06	1.00						
Small cladocerans	−0.17	−0.15	−0.11	−0.04	−0.11	−0.05	0.17	1.00					
Copepods	0.50	0.53	0.50	0.10	−0.47	−0.36	0.50	−0.05	1.00				
Dorosoma	0.01	0.11	0.02	0.07	0.05	−0.17	−0.52	−0.35	−0.28	1.00			
Fish loss	−0.15	−0.08	−0.17	−0.05	0.11	−0.10	−0.18	−0.25	−0.18	0.53	1.00		
Algal Fluorescence	−0.21	−0.13	−0.22	0.19	−0.11	−0.17	−0.29	−0.21	−0.23	0.78	0.73	1.00	
Bluegreen filaments	−0.15	−0.13	−0.16	−0.09	−0.09	−0.10	−0.18	−0.14	−0.13	0.04	−0.08	−0.06	1.00
Menidia beryllina (P = 0.05; ± 0.220)													
Filinia	1.00												
Notholca	0.70	1.00											
K. quadrata	0.79	0.77	1.00										
Polyarthra	−0.03	0.27	−0.07	1.00									
Brachionus	−0.11	−0.13	−0.11	−0.02	1.00								
K. cochlearis	−0.06	−0.09	−0.15	0.03	0.51	1.00							
Daphnids	0.02	−0.14	−0.03	−0.13	−0.20	−0.24	1.00						
Small cladocerans	−0.07	−0.13	0.01	−0.04	−0.15	−0.17	0.18	1.00					
Copepods	0.29	0.10	0.19	−0.03	−0.06	−0.05	0.03	0.12	1.00				
Menidia	−0.10	−0.07	−0.08	−0.06	0.15	0.05	−0.25	−0.20	−0.19	1.00			
Fish loss	0.00	0.06	−0.07	−0.05	0.04	0.40	−0.30	−0.27	−0.24	0.37	1.00		
Algal Fluorescence	0.06	0.17	0.01	−0.05	0.22	0.36	−0.46	−0.02	0.14	0.23	0.37	1.00	
Bluegreen filaments	−0.08	−0.08	−0.13	−0.06	−0.06	−0.06	−0.16	−0.13	0.02	0.42	0.30	0.04	1.00

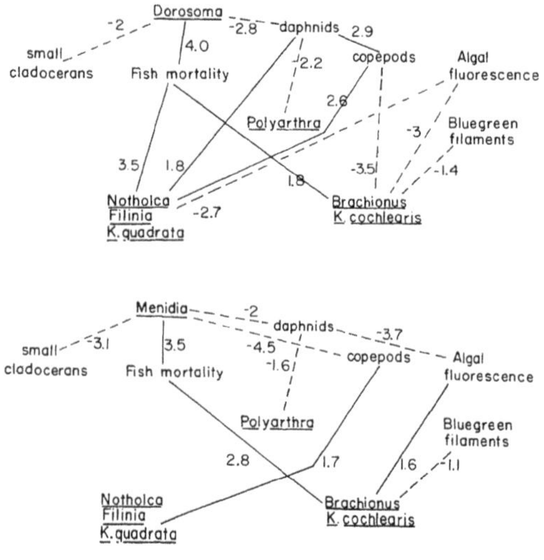

Fig. 1. Functional relationships among food web components in experiments with *Dorosoma* (upper panel) of *Menidia* (lower panel). Negative relationships (shown by dashed lines) often imply predatory or inhibitory effects, while positive effects (solid lines) suggest common linkage to other, unspecified food web components (e.g., daphnids and *Notholca acuminata, Filinia longiseta*, and *K. quadrata* in the *Dorosoma* food web may benefit from the same resource base or occur at the same temperatures). Coefficients greater than ± 2 indicate probabilities of random correlation < 0.05; other coefficients less than 2 are indicated when the relationship in one food web is consistent with one shown in the other food web or where the joint probability of the two pathways < 0.05. Where food web components are combined (*K. cochlearis* and *Brachionus*) in the figures, individual regression coefficients to other components did not differ by greater than 10%, and the mean coefficient was shown.

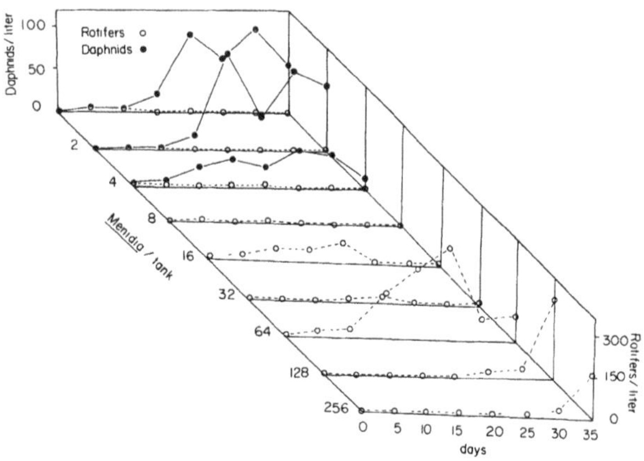

Fig. 2. Temporal development of rotifer and daphnid populations in experiment 9 under nine logarithmically-ordered densities of *Menidia beryllina.* Note the range of fish densities (8–32 fish/tank) where neither rotifers nor daphnids are abundant.

cladoceran-rotifer interference competition [Gilbert & Stemberger, 1985]. Experiment 9 examined the responses of phytoplankton and zooplankton to nine densities of *Menidia* over a period of 35 days. The divergence of rotifers and daphnids was gradual (10–30 days) and left unresolved the relative importance of resource exploitation or interference. Rotifers did not respond positively to the rapid elimination of daphnids in tanks containing 8–32 fish/tank. Some fish mortality was associated with the higher (≥ 64) fish densities, and bacterial growth associated with decomposing fish may have been a key feature of the rotifer response. This may be similar to the effects of nutrient manipulations [Neill, 1984]. It was not possible to evaluate fish effects related to sediment stirring or disruption of littoral vegetation in ponds [Hillbricht-Ilkowska, 1964] in the fiberglass tanks.

The 5-day enclosure experiment conducted during experiment 9 was used to examine if rotifer exclusion by daphnids would occur as rapidly (3–5 days) as had been shown by Gilbert & Stemberger [1985]. The exponential rate of change in density of each rotifer population in the 40 mixtures of daphnids and rotifers was regressed against daphnid density. None of the rotifer species were significantly affected by daphnid density: *Filinia longiseta* ($P = 0.14$), *Brachionus calyciflorus* ($P = 0.94$), *Keratella cochlearis* ($P = 0.37$), and *Keratella quadrata* ($P = 0.85$). In *Brachionus* and *K. cochlearis* (two abundant rotifers in these experiments), there were no significant changes (all $P > 0.2$) observed in egg ratios (eggs/female) and in the frequency of occurrence of empty loricas, considered previously as a possible index of a fatal encounter with predators [Threlkeld, 1983].

Discussion

Three approaches were used to examine the relationships among rotifers, cladocerans, and copepods in the context of experimental manipulations of two kinds of planktivorous fish. By correlation and multiple regression analyses, it was shown that direct negative interactions of rotifers and crustaceans were minimal compared to those involving indirect connections to food resources. Interactions between cladocerans or copepods and rotifers were occasionally positive, indicating mutual dependency on other unidentified predictor variables (e.g., temperature). Only in the case of *Polyarthra* was there a consistent negative interaction with daphnids. In spite of *Polyarthra*'s well developed escape response to daphnids [Gilbert, this symposium] its lack of a hard lorica (as in *Keratella, Brachionus*, or *Notholca*) or well developed spines (as in *Filinia*) may account for its net susceptibility to damage by daphnids. In contrast to the findings of Gilbert & Stemberger [1985], daphnids did not negatively affect *K. cochlearis*, except through food web components related to the resource base. This suggests that resource exploitation may be a more widespread (seasonally invariant) mechanism of rotifer-cladoceran antagonism than interference competition associated with rotifers being swept into daphnid branchial chambers. Because fish mortality and decomposition are often associated with manipulations of planktivorous fish [see Threlkeld, 1986 for a review], enhancement of rotifers associated with fish predation also may be due to phytoplankton enhancement by decomposing fish. The importance to rotifer population growth of pulsed nutrient additions similar to fish decomposition has been demonstrated by Neill [1984]. A single enclosure experiment completed in association with tank experiment 9 showed no short-term effect of dense daphnid populations (up to 64/liter) on four species of planktonic rotifers. Absence of a clear attribute of interference competition makes it difficult to evaluate the relative importance of interference competition under field conditions, even though it may be readily observed in resource-rich conditions in the laboratory [Gilbert & Stemberger, 1985; Gilbert, unpubl. obs.]. In the broad context of the tank experiments presented here, it is unclear how important interference competition is where resource limitation, invertebrate predation, and predation by filter-feeding fish may also occur.

Acknowledgements

The fine experimental work of Drs. John Gilbert, Rich Stemberger, and Bill Neill provided the stimulus for this examination of rotifer-cladoceran interactions. Comments of Dr. Gilbert prior to the IV*th* International Rotifer Symposium were also very helpful. Dr. Ray Drenner provided an endless supply of ideas and comments on the tank experiments; his unselfish collaboration is gratefully acknowledged. The tank experiments were supported by NSF grants BSR 8206850, BSR 8206865, and BSR 8315127.

References

Drenner, R. W., F. deNoyelles Jr. & D. Kettle, 1982. Selective impact of filter-feeding gizzard shad on zooplankton community structure. Limnol. Oceanogr. 27: 956–968.

Drenner, R. W., J. R. Mummert, F. DeNoyelles Jr. & D. Kettle, 1984. Selective particle ingestion by a filter-feeding fish and its impact on phytoplankton community structure. Limnol. Oceanogr. 29: 941–948.

Drenner, R. W., S. T. Threlkeld & M. D. McCracken, 1986. Experimental analysis of the direct and indirect effects of an omnivorous filter-feeding clupeid on plankton community structure. Can. J. Fish. Aquatic Sci. 43: 1935–1945.

Gilbert, J. J. & R. S. Stemberger, 1985. Control of Keratella populations by interference competition from Daphnia. Limnol. Oceanogr. 30: 180–188.

Hillbricht-Ilkowska, A., 1964. The influence of the fish population on the biocenosis of a pond, using Rotifera fauna as an illustration. Ekol. Pol., Ser. A. 12: 454–503.

McComas, S. R. & R. W. Drenner, 1982. Species replacement in a reservoir community: silverside feeding mechanisms and competition. Can. J. Fish. Aquat. Sci. 38: 815–821.

Neill, W. E., 1984. Regulation of rotifer densities by crustacean zooplankton in an oligotrophic montane lake in British Columbia. Oecologia 61: 175–181.

Sokal, R. R. & F. J. Rohlf, 1981. Biometry. 2nd edition, W. H. Freeman and Co., 859 pp.

Stemberger, R. S., 1979. A guide to rotifers of the Laurentian Great Lakes. EPA-600/4–79–021. 186 pp.

Threlkeld, S. T., 1983. Empty loricas and the dynamics of Kellicottia longispina in a subalpine, oligotrophic lake. Hydrobiologia 104: 367–372.

Threlkeld, S. T., 1986. Experimental evaluation of trophic-cascade and nutrient-mediated effects of planktivorous fish on plankton community structure. In: W. C. Kerfoot and A. Sih (eds). Predation: Direct and indirect impacts on aquatic communities. The University Press of New England. pp. 161–173.

Williamson, C. E., 1983. Invertebrate predation on planktonic rotifers. Hydrobiologia 104: 385–396.

Hydrobiologia 147: 245–255 (1987)
May, L., Wallace, R. & Herzig, A. (eds), Rotifer Symposium IV

Raising rotifers for use in aquaculture

E. Lubzens
Israel Oceanographic & Limnological Research Tel-Shikmona, P.O.B. 8030, Haifa 31080, Israel

Keywords: *Brachionus plicatilis*, mass culture

Abstract

Of the three Brachionus species used in aquaculture, *Brachionus rubens, B. calyciflorus* and *B. plicatilis*, the latter is most widely used in raising marine fish and shrimp larvae due to its tolerance to the marine environment. In freshwater aquaculture the use of *B. rubens* and *B. calyciflorus* is limited, probably because inert food products are readily available as feed for freshwater larvae.

The rotifer *Brachionus plicatilis* is used in large numbers as the first food organism in intensive cultures of marine fish and shrimp larvae. An adequate supply of these rotifers relies on mass cultures. The reproductive rate of rotifers in these cultures depends on food quality and quantity, salinity, temperature and pH of the medium. Removal of waste products from culture tanks leads to higher and more efficient production of rotifers over extended periods of time. Rotifers have to be enriched with polyunsaturated fatty acids, which are essential for proper development and survival of marine fish and shrimp larvae.

The future use of preserved rotifers and their resting eggs may help to overcome unforeseen failures of live cultures and may lead to more efficient use of these organisms in raising freshwater and marine fish and shrimp larvae.

Introduction

Rotifers are considered to be an excellent food source for newly hatched fish larvae [Howell, 1973], even better than copepods [Theilacker & Kimball, 1984]. In freshwater aquaculture, *Brachionus calyciflorus* and *Brachionus rubens* have been proposed as food for fish larvae [Galkovskaya, 1963; Groeneweg & Schluter, 1981; Schluter & Groeneweg, 1981], and their limited use is probably due to convenient inert food being available to feed freshwater fish larvae.

In the marine environment inert food is unavailable and the euryhaline rotifer, *Brachionus plicatilis*, is extensively used as a food organism for a large variety of finfish and crustacean larvae [for references see Hirata, 1980; Lubzens, 1981; James *et al.*, 1983; Kafuku & Ikenoue, 1983]. This rotifer, once considered a pest in Japanese eel ponds [Hirata, 1980], is now indispensible in raising marine fish larvae because of its size (130–320 μm), relatively slow motility, caloric value, and the possibility of artificially manipulating its nutritional qualities [Watanabe *et al.*, 1983; Theilacker & Kimball, 1984].

Rotifers must be cultured at relatively high densities to provide an adequate supply of food for mariculture hatcheries producing vast numbers of fingerlings [Kafuku & Ikenoue, 1983]. Fish larvae production in Japan is probably limited by the number of rotifers raised in these hatcheries [Hirata, 1980]. It has been estimated that 40000 to 100000 rotifers are required to feed one fish larva from the day it hatches until it can use another source of food [Okauchi *et al.*, 1980; Kafuku & Ikenoue, 1983]. Clearly, the production of millions of fish larvae, which takes place in many countries today, [Gordin, 1980; Hirata, 1980; James *et al.*,

1983; Kafuku & Ikenoue, 1983] involves the production of vast numbers of rotifers. The rotifer is looked upon as a living food capsule which transmits adequate supplies of macro- and micronutrients, vitamins, or even antibiotics to the fish larvae [Gatesoupe, 1982]. Providing rotifers in adequate numbers during crucial periods is the main problem of most marine hatcheries.

Many hatcheries practice mass cultivation of rotifers (Table 1). In initiating mass cultivation of rotifers, the following problems must be considered:

1. A rotifer strain must be selected which will be an appropriate size for the larvae and suitable to the culture conditions.
2. The food quality and quantity must be maintained at an adequate level.
3. Water quality parameters in culture tanks (i.e. salinity, temperature, pH) must be controlled and waste products must be removed.

The main aim is to obtain large numbers of rotifers at predictable qualities and production rates, although the economic cost of achieving these aims must be taken into consideration.

Rotifer culture

1. Rotifer strain

In choosing the rotifer strain, selection of body size is very important. Prey selection by fish larvae is size dependent, and preferred size increases as the larvae grow [Hunter, 1980]. For example, newly hatched larvae of *Sparus aurata* preferred small rotifers (191 μm × 138 μm), while larvae more than 6 days old selected larger (258 μm × 171 μm) rotifers [Helps, 1982]. Small rotifers must also be supplied in greater numbers than larger ones, to meet the metabolic demands of the developing larvae [Hunter, 1980]. In Japan, cyclic changes in rotifer size in outdoor tanks were reported by Fukusho & Iwamoto [1980]. During July, when temperatures were high, S-type rotifers (lorica length 205–304 μm) prevailed, while during February, when low temperatures were recorded, L-type rotifers (lorica length 306–347 μm) were most common. Fukusho & Iwamoto [1980] attributed these differences in rotifer size to cyclomorphosis, while Snell & Carrillo [1984] suggested that they may have been due to clonal replacement. Data presented recently by Okauchi & Fukusho [1985] support the assumption that L and S type rotifers are genetically different.

Cyclomorphosis was also disclosed in a study made by Serra & Miracle [1983] on *Brachionus plicatilis* collected from Spanish lagoons. Using allometric coefficients, a wider dispersion in size was revealed in field collected rotifers when compared to laboratory cultured ones. The analysis of results suggested that allometric coefficients were influenced by environmental factors, mainly temperature and salinity. As in Japan, the largest rotifer forms appeared in winter and the smallest in summer, with intermediate sizes in autumn and spring. Also in this study, it was not known whether the diversity in size was associated with phenotypic variation of parthenogenic lineages or with clonal displacement.

Alloenzyme variation in six gene systems confirmed that clones showing variation in size were genetically distinguishable [Serra & Miracle, 1985]. Genetic variation was also resolved among 17 strains of *Brachionus plicatilis* originating from different geographical locations [Snell & Winkler, 1984], and cultured under identical laboratory conditions. Isozyme analysis of five rotifer enzymes showed that geographic proximity of strains was not always reflected in electrophoretic patterns. Among five strains collected in Tampa Bay, Florida, U.S.A., three were 93% similar to strains collected in the Soviet Union and Austria. One strain (WP) appeared genetically distinct from all the other tested strains.

Following the work of Yufera [1982] and Snell & Carrillo [1984], which showed clearly that lorica length in *B. plicatilis* is genetically determined, it is now possible to culture a large variety of size specific strains of *B. plicatilis* to suit the various demands of fish larvae.

Only small modifications in size (up to 15%) were possible by changing diet or salinity, but it was possible to encourage growth of large or small rotifers, which differed in their specific growth rate, by manipulation of food concentration [Endo & Mochizuki, 1979]. Also, the effect of temperature on rotifer size was found to be strain dependent [Snell & Carrillo, 1984].

In choosing the appropriate rotifer strain for mass culture, its reproductive rate at the available environmental conditions should be carefully considered. Rotifer strains used for the above-mentioned genetic studies vary considerably in their reproductive potentials, with r values from 0.55 to 1.15 [Snell & Carrillo, 1984]. Larger stock cultures must be kept of rotifers with low reproductive rates than of those with fast reproduction rates. It is therefore essential that a rotifer strain is chosen which will reproduce rapidly under the available environmental conditions.

2. *Food quality and quantity*

Brachionus plicatilis is a filter feeder and may be fed on a variety of food types, including algae, yeast, bacteria or inert food such as microcapsules. In filter feeding rotifers, the size of ingested particles is thought to be closely related to that of the mouth opening [Pourriot, 1977], and a positive correlation between body size and the maximum size of particles ingested has been shown [Hino & Hirano, 1980]. Similarly, a low correlation between body size and the minimum size of these ingested particles has also been reported [Hino & Hirano, 1984]. These results agree with those obtained on several other zooplankton species [Bogdan & Gilbert, 1984].

a. Algae. Chotiyaputta & Hirayama [1978] showed that, of two algal species, *B. plicatilis* preferred *Chlamydomonas* sp. over *Olisthodiscus* sp., and algae in their exponential phase of growth over senescent ones. While some studies have suggested that algal diet has little effect on reproductive rates [Ito, 1960; Theilacker & McMaster, 1971; Scott & Baynes, 1978], others [Hirayama *et al.*, 1979; Okauchi & Fukusho, 1984] reported that different species of algae resulted in substantially different reproductive rates. Of the eight species examined by Hirayama *et al.* (1979), two were evaluated as excellent food (*Synechococcus* sp. and *Chlorella* sp.), four were average (*Chlamydomonas* sp., *Monochrysis lutheri, Dunaliella tertiolecta* and *Cyclotella cryptica*), and two were found inadequate (*Eutreptiella* sp. and *Nitzschia clostelium*). *Tetraselmis tetrathele* was recently reported to yield higher reproductive rates than *Chlorella* sp. [Okauchi & Fukusho, 1984]. These authors also showed a delay in reproduction following a transfer of rotifers fed on *Tetraselmis tetrathele* to *Chlorella*. This delay was not observed if rotifers were first fed on *Chlorella* and later on *T. tetrathele*.

Mixed diets containing two different algae (*Chlorella* and the blue-green *Schizothrix*) were found to increase the reproductive rate of *B. plicatilis* substantially (by an average of 2.7 times) compared to those obtained by feeding rotifers on either alga alone [Snell *et al.*, 1983a]. The blue-green alga was found to contain a heat labile substance whose effect was mediated through the medium; ingestion was not required.

The reproductive rate and survival of *B. plicatilis* depends on the concentration of food in the culture medium [Hirayama *et al.*, 1973, 1979; Endo & Mochizuki, 1979; Lubzens, 1981; Snell *et al.*, 1983a; Yamasaki *et al.*, 1984]. Although maximal conversion rates in rotifers fed on *Chlorella saccharophila* var. *saccharophila* were observed in cultures containing 10^6 cells ml^{-1}, maximal population growth was obtained in cultures containing at least 3×10^6 cells ml^{-1} [Yamasaki *et al.*, 1984]. Rotifers cultured on *Chlorella stigmatophora* exhibited maximal reproductive rates in algal concentrations from 2 to 8×10^6 cells ml^{-1}, depending on the salinity of the culture [Lubzens, 1981].

The amount of food that should be provided to rotifer cultures depends on rotifer density, salinity and algal species. Hirayama & Ogawa [1972] showed that filtration rates of *B. plicatilis* change with temperature, chlorinity and food concentration. Satiated rotifers maintained under optimal conditions (algal food concentrations 2.13×10^6 cells ml^{-1}, temperature 22 °C, chlorinity 7.8‰, pH 8.0) ingested 200 cells min^{-1} $individual^{-1}$. Although rotifer density was found to have little effect on ingestion rate at this food concentration, almost all the algae are consumed in less than one hour in cultures maintained at a density of 200 rotifers ml^{-1}. Food must either be supplied initially at much higher concentrations or several times a day, preferably by continuous feeding, in order to support dense rotifer populations.

In evaluating the conditions prevailing in rotifer mass cultures, Yufera & Pascual [1985] showed that filtration and ingestion rates were dependent on algal concentration and varied among four algal species. With *Nannochloris oculata, Nannochloris maculata* and *Nannochloropsis oculata*, filtration rates decreased with increasing cell concentrations, reaching a constant level at concentrations exceeding 60 – 80 $\mu g\ ml^{-1}$. Bimodal ingestion rates were observed by increasing food concentrations. These rates increased with food density, reaching a plateau at 15 – 30 $\mu g\ ml^{-1}$ and increased again after the food concentration exceeded 50 – 100 $\mu g\ ml^{-1}$. Rotifers fed on *Nannochloropsis gaditana* showed a decrease in filtration and ingestion rates at food concentrations above 40 $\mu g\ ml^{-1}$. Food consumption and utilization were also found to depend on the rotifer strain. At moderate algal concentrations, the small rotifer strain (Bs) was found to consume about 0.16 ng $ind^{-1}\ min^{-1}$, while the larger size strain (S-1) consumed 0.30 ng $ind^{-1}\ min^{-1}$. Bs strain rotifers ingested each day about three times their own dry weight, while S-1 rotifers consumed only one and a half times their own dry weight and yielded more biomass for the same food quantities [Yufera *et al.*, 1983].

To meet food requirements of rotifers, the volumes of algal cultures required for feeding the rotifers are 5 – 10 times greater than the volumes of rotifer cultures, depending on the algal concentrations obtained [Hirata, 1980]. Using algae alone as food for rotifers would have made mass production of rotifers prohibitively expensive. Furthermore, rotifer cultures would have to depend on another live organism with all its associated culture problems.

b. Yeast. A major breakthrough in rotifer culture was made by Hirata & Mori [1967], who found that rotifers could be raised on baker's yeast (*Saccharomyces cerevisiae*) which is inexpensive and commercially available in most countries.

Rotifers cultured exclusively on baker's yeast were found to have lower reproductive rates than those cultured on algae and could not be maintained for long periods [Hirata, 1979, 1980; Yufera & Pascual, 1980]. The yeast used by Japanese workers lacked vitamin B12 and cystine which are essential for rotifer growth [Scott, 1981; Hirayama & Funamoto, 1983]. Providing yeast and *Chlorella*, in a ratio of 1:1, resulted in rotifer reproductive rates similar to those obtained by feeding *Chlorella* alone. This showed that yeast could partially replace the more expensive *Chlorella* in rotifer cultures. Following this work, several species of marine yeast were isolated and found to be an adequate food for rotifers [Furukawa & Hidaka, 1973; Kawano *et al.*, 1976]. Matsuda *et al.*, [1980] isolated a marine yeast which, when fed to rotifers, resulted in population growth rates two to three times higher than those achieved with *Chlorella*. However, the use of these marine yeasts is rather limited since they require special culture techniques, making them relatively expensive.

Using baker's yeast as food for rotifers, which are later fed to marine larvae, requires special attention. Rotifers fed exclusively on baker's yeast have insufficient amounts of highly unsaturated fatty acids, mainly eicosapentaenoic acid (20:5 ω3) and docosahexaenoic acid (22:6 ω3), which are essential for survival and development of marine larvae [(see review by Watanabe *et al.*, 1983]. Although these acids can be synthesized in yeast fed rotifers either *de novo* or through elongation and desaturation of precursor molecules, this process is rather slow and rotifers do not accumulate sufficient amounts [Lubzens *et al.*, 1984a, 1985a]. Enrichment of rotifers with these essential acids can take place by feeding them for 8 – 24 hours on algae, ω-yeast or a suspension of egg yolk and fish oil containing these acids [Watanabe *et al.*, 1983].

c. Bacteria. The bacteria present in the culture tanks of *Chlorella* fed rotifers were also consumed by the animals and this improved their reproduction [Ushiro *et al.*, 1980; Yamasaki & Hirata, 1984]. Yasuda & Taga [1980] found that at least two strains of marine bacteria (*Pseudomonas* P-1 and P-7) supported rotifer growth. Recently, the possible use of inexpensive supplies of bacteria produced in wastewater treatment plants or from alcohol fermentation slops has been examined with promising results [Hino *et al.*, 1981; Fukuhara *et al.*, 1982]. The nutritional quality of rotifers raised on these bacteria awaits analysis.

d. Inert food. The search for inexpensive convenient and reliable food sources for rotifer cultures led several workers to examine powdered algae or synthetic food sources. Spray dried *Chlorella* fed to rotifers was less effective than live algae, but better than caked yeast [Hirayama & Nakamura, 1976]. Similarly, the use of freeze dried *Spirulina* or *Platymonas suecica* resulted in low density rotifer cultures [Person-le Ruyet, 1975]. Gatesoupe & Robin [1981] showed that successful use of commercially available *Spirulina, Chlorella* and methanol grown yeast depended on the salinity of the rotifer culture and the amount of food provided. Low salinity (18‰), and adjustment of food level to rate of increase in rotifer number, resulted in a daily production of 35–55 rotifers ml^{-1} day^{-1}, with poor food conversion ratios. Higher food conversion ratios were achieved in formulated diets containing vitamins and other additives in addition to yeast and algae.

Although microencapsulated diets did not support rotifer cultures over extended periods of time [Teshima *et al.*, 1981], this feeding method may permit the enrichment of rotifers with micronutrients required by fish larvae.

3. *Salinity*

Rotifer reproductive rates are strongly influenced by the salinity of the medium [Ito, 1960; Ruttner-Kolisko, 1972; Pascual & Yufera, 1983; Lubzens *et al.*, 1985b]. *Brachionus plicatilis* tolerates a wide range of salinities from 1–97 g l^{-1} [Walker, 1981]; its reproductive rate is most optimal at 4–35‰, depending on strain and culture conditions. Filtration rates are reduced at high salinity [Hirayama & Ogawa, 1972], and this may account, at least in part, for the lower reproductive rates. Thus, the amount of food provided to rotifers has to be adjusted according to the salinity in which they are cultured.

In choosing the salinity for culturing rotifers, the salinity of the predator must be taken into account. Transferring rotifers from low to high salinity or vice versa may result in death of rotifers. Transfer should be gradual to allow rotifers to adapt to new salinity [Gatesoupe & Luquet, 1981; Pascual & Yufera, 1983]. By culturing rotifers at a salinity of 20‰, relatively high reproductive rates are obtained. Rotifers from this culture may then be transferred to fish larvae maintained either in fresh water [Lubzens *et al.*, 1984b] or in 40‰ salinity sea water. In the first case rotifers have to be transferred for about one day to 10‰ salinity sea water, while in the second case they must first be transferred to 30‰ salinity sea water.

Raising rotifers at salinities lower than 35‰ results in mixis in certain strains [Lubzens *et al.*, 1985b]. This results in lower yields and the possible collapse of the culture. However, this can be avoided by the isolation of clones which do not respond to salinity changes in this way [Lubzens *et al.*, 1984a].

4. *Temperature*

Optimal culture temperature for rotifers is dependent on the strain being cultured [Snell & Carrillo, 1984]. Hirayama & Kusano [1972] recommended culturing rotifers at the optimum temperature for reproduction, 25 °C. Using a thermophilic small-sized rotifer strain, Pascual & Yufera [1983] observed maximal densities and greatest yield in rotifers raised at 30 to 35 °C.

Temperature also affects the biochemical composition of rotifers [Scott & Baynes, 1978]. At 18 °C, rotifers consumed their food slowly and maintained relatively high levels of lipids and carbohydrates for a long period. However, at 28 °C, food was consumed rapidly, and the biochemical composition of the rotifers changed quickly when the food was gone.

5. *Water quality*

Accumulation of excretory products produced by

rotifers (feces, urea, ammonia and phosphates) and excess uneaten food contribute to the deterioration in water quality in tanks of rotifers cultured for extended periods.

Hirata & Nagata [1982] showed that *B. plicatilis* fed on *Chlorella* excretes ammonia, urea and phosphates ($1.41 \pm 0.87 \times 10^{-4}$ μg NH_4-N, 1.17 ± 1.31 μg urea$-N$ and $0.27 \pm 0.29 \times 10^{-4}$ μg PO_4-P, respectively, hour^{-1} individual^{-1}). While these excretory rates increased with the amount of food available and the number of eggs produced, they decreased rapidly in rotifer cultures containing more than 100 individuals ml^{-1}. These results were obtained in 1 litre beakers where temperature ($20 \pm 0.2\,°C$), salinity (20‰) and illumination (16:8 photoperiod) were kept constant.

Little information is available on the toxicity of the nitrogeneous compounds (i.e. ammonia, urea, nitrite and nitrate) that may accumulate in culture tanks. Lincoln *et al.*, [1983] reported that free ammonia levels of 17 mg l^{-1} were found to be toxic to rotifers. Furukawa & Hidaka [1973] observed relatively high levels of ammonia (40–50 mg l^{-1}) in their culture tanks. Slow accumulation of NO_3-N, NO_2-N and NH_4-N levels in rotifer tanks was observed in cultures maintained for 70 days, reaching levels of 170, 716 and 2310 μg l^{-1}, respectively [Hirata *et al.*, 1983]. Preliminary observations have shown that while ammonia levels of 9–12 mg l^{-1} NH_4-N (at pH 7.6–7.8) did not affect rotifer reproduction, nitrite (NO_2-N) levels of 90–140 μg l^{-1} resulted in collapse of some cultures [Lubzens, Holtkamp & Gordin, unpublished]. Groeneweg & Schluter [1981] showed that growth of *B. rubens* was not affected by high levels of nitrites (10–20 mg l^{-1}), probably as a result of adaptation. The interrelationship among the nitrogeneous excretory products in culture tanks requires further detailed attention.

6. Rotifer culture methods

A large array of culture methods, some adapted to specific local requirements, is found in the literature (Table 1). These can be sorted into three basic methods: (1) batch cultures, (2) semicontinuous cultures, and (3) 'feedback' culture systems.

a. Batch cultures. In using this method, rotifers are introduced at low density into dense algal cultures ('green water') and harvested once all the algae are consumed. The duration of these cultures and the rotifer densities obtained depends on the initial concentration of algae. Using this method, rotifers are either cultured in large outdoor tanks where volumes may be 10000 litres and rotifer densities are rather low [Kawano *et al.*, 1978; Rothbard, 1979; Watanabe, 1982; James *et al.*, 1983], or in indoor plastic bags (50 litres) where rotifer densities are much higher [Trotta, 1980, 1981]. An early version of this method called for daily harvest of 1.5 ton tanks, where 10–30% of the rotifers were removed to be used as food, and the remaining rotifers served to inoculate new algal culture tanks [Hirata, 1979, 1980].

b. Semicontinuous cultures. This method is an extension of the previous one. Rotifers are inoculated at low density into dense algal cultures, but once algae are consumed, rotifers are either fed on yeast or on an external supply of algae until densities exceed 100 rotifers ml^{-1}. Every 1–3 days, a certain volume of the culture is removed and replaced by sea water containing algae or pure sea water. In the latter case, rotifers are fed on yeast. About 10–30% of the water is removed from the culture tanks, depending on the rotifer reproductive rate. Using this method, rotifers are maintained at relatively high densities. After several harvests, rotifer reproductive rates decrease, probably as a result of accumulated excretory products and uneaten food. At that time, all the rotifers are harvested and used as food or to inoculate new cultures. Attaching a biological filter to these culture tanks extends the harvest period for 2–3 months [Fujita, 1979; Witt *et al.*, 1981; Watanabe, 1982; James *et al.*, 1983]. The culture volumes used in this method vary greatly from a few hundred litres to 200000 litre cultures [Fujita, 1979; Hirata, 1980; Watanabe, 1982]. Relatively higher rotifer densities are obtained in smaller volume cultures.

Table 1. Daily production of *Brachionus plicatilis* in mass culture systems at various volumes and salinities.

Reference	Production rotifers l^{-1} day^{-1}	Maximum rotifer density ml^{-1}	Salinity ‰	Volume litre	Food organism
Theilacker & McMaster 1971	5387	N.S.*	33	464	*Dunaliella* sp.
Furukawa & Hidaka 1973	16300	1000	N.S.	1000	Marine yeast
Cruz & Millares 1974	24719	360	diluted sea water	14	*Nannochloris* sp.
Devauchelle & Girin 1974	24250	150	N.S.	250	*Tetraselmis* sp.
Alessio 1974	40000	55	N.S.	600	A mix of: *Dunalliella salina* *Platymonas suecica* *Chroomonas fragarioides* *Phaeodactylum tricornutum* *Chlorella ovalis* *Glenodinium* sp.
Amat 1975	42571	300	33	60	*Tetraselmis* sp.
Millares & Gonzalez 1975	31000	420	35	20 – 100	*Platymonas* sp.
Person-le Ruyet 1975	55000	340	N.S.	100	*Tetraselmis suecica*
Hirayama & Nakamura 1976	28500	434	22	12	*Chlorella* sp.
Kawano *et al.* 1978	59633	1100	N.S.	1090	*Chlorella* dried bread yeast
	7143	90	N.S.	10000 or 20000	*Chlorella* marine yeast
Hirata 1979	57000	530	N.S.	550	*Chlorella* sp.
Fontaine & Revera 1980	34000	583	N.S.	500	Turulose yeast
Hirata 1980	80000	514	N.S.	500	*Chlorella*, yeast
Trotta 1980	18860 – 34780	± 300	N.S.	30	*Chlorella* sp.
Yufera & Pascual 1980	63733	540	36	70	*Nannochloris* sp. *Tetraselmis suecica*
Gatesoupe & Luquet 1981	122000	790	33	60	*Tetraselmis suecica*
Witt *et al.* 1981	47143	219	15	1400	*Nannochloris*
	22955	100		2200	*Nannochloris*
Reguera *et al.* 1982	51750	786	36	750	Bread yeast
Ortega *et al.* 1983	77000	230	N.S.	6000	*Tetraselmis* sp., yeast
Trotta 1983	57000	428	36	50	*Tetraselmis suecica*
Meragelman, Lubzens & Minkoff (unpublished)	70000	500	40	200	Baker's yeast

* N.S. – Not Specified.

c. 'Feedback' culture systems. Culturing rotifers at high densities results in the accumulation of high levels of particulate matter and excretory products. Hirata [1974] described a simple method for removal of particulate matter, by passing the rotifer culture through a gravel bed. A more efficient method was achieved by using a stream unit with partitions through which the culture was circulated 20 times each day [Hirata, 1979]. Feces and accumulated particulate matter were removed every few days from the stream unit into a decomposer tank, where bacterial decomposition took place. The decomposed matter was used as fertilizer for algal cultures, which were then used to feed the rotifer cultures.

Rotifers in this system are cultured in 500 litre tanks attached to a 150 litre stream unit. Rotifers are inoculated at low density, and harvesting begins once rotifer density has reached 100–150 rotifers ml^{-1} and may continue to increase to 500 ml^{-1}. At this density a daily harvest of 80.5×10^6 rotifers was recorded for at least 20 days [Hirata, 1979]. By recirculation of waste products accumulated in rotifer tanks, the apparent conversion rate, calculated as the amount of food supplied to the weight of harvested rotifers, was 89.2% in wet weight and 29.0% in caloric value.

By using these culture methods, production rates ranging from 5387 to 122800 rotifers l^{-1} day^{-1} were recorded from various locations (Table 1). In general, higher rotifer densities in cultures lead to higher yields per volume. Thus, relatively low yields were obtained in large outdoor tanks (10–20 tons), where rotifer concentration was relatively low [Kawano *et al.*, 1978].

7. *Use of preserved rotifers*

Unpredictable events occasionally result in failure of rotifer cultures, leaving the fish culturist without an available food source. In extreme cases this may lead to a loss of one year's production. The use of preserved rotifers may help to overcome the consequences of such disastrous events. Four methods for preserving rotifers can be considered: (1) freezing, (2) cryopreservation, (3) refrigeration, and (4) production of resting eggs.

Frozen rotifers have been successfully used as food for shrimp larvae [Yamasaki & Hirata, 1982], although their nutritional quality probably deteriorates upon thawing as a result of leaching [Grabner *et al.*, 1981]. No record is available on the successful use of frozen rotifers as food for marine fish larvae, although this method is routinely used in the fish hatchery at the Kagoshima Mariculture Center, Tarumizu City, Japan (Fujita, personal communication).

Attempts to preserve rotifer using cryopreservation have met with limited success [King *et al.*, 1983; Euteneuer *et al.*, 1984]. This method, which is relatively expensive, may prove valuable in the preservation of important isolated clones over extended periods of time.

Live rotifers were found to survive refrigeration (4 °C) for at least 22 days, at densities of 2000 individuals ml^{-1}. During this period rotifers were fed *Isochrysis* sp. and the water was changed every four days [Lubzens, Kolodny & Perry, unpubl. obs.]. This method would permit short term storage of surplus rotifers or the advance preparation of rotifers in anticipation of high demands in the hatchery.

Long-term preservation of live rotifers can be achieved by using their resting eggs, which are available from natural deposits [Snell *et al.*, 1983b] or from laboratory cultures [Lubzens, 1981]; they may be hatched under controlled conditions [Blanchot & Pourriot, 1982; Minkoff *et al.*, 1983]. Today the production efficiency of these eggs is rather low [Lubzens, 1981], making them unattractive for commercial production and widespread use. Research into factors controlling resting egg production will permit their use, if not directly as food for larvae, at least as a reserve stock for initiating large scale rotifer cultures.

Conclusions

The extensive use of *Brachionus plicatilis* as food for marine fish and shrimp larvae led to the development of several methods for its mass cultivation, the most efficient and reliable one being the 'feedback' system of Hirata [1979]. In the search for conditions leading to high production of rotifers, the effects of environmental factors such as food quality and quantity, temperature and salinity on the rotifer's physiology have been studied extensively. In order to overcome unforeseen failures of live cultures, the use of cryopreservation, refrigeration and resting egg production has been examined. Economic considerations will dictate the future use of rotifers as food for marine larvae or their replacement by inert particulate foods, as used today in freshwater fish cultures.

References

Alessio, G., 1974. Riproduzione artificiale di orata, Sparus aurata (L.) (Osteichtyes, Sparidae). 3. Produzione su vasta scala di fito e zooplancton per l'alimentazione delle larve e degli avannoti. Boll. Pesca Piscic. Idrobiol. 29: 133–147.

Amat, F., 1975. Cultivo masivo del rotífero mixohalino Brachionus plicatilis O. F. Müller. Publ. técn. junt. Est. Pesca 11: 387–400.

Blanchot, J. & R. Pourriot, 1982. Influence de trois facteurs de l'environnement, lumière, temperature et salinitè, sur l'éclosion des oeufs de durée d'un clone de Brachionus plicatilis (O. F. Müller) Rotifére. C.R. Acad. Sc. Paris, Ser. III, 295: 243–246.

Bogdan, K. G. & J. J. Gilbert, 1984. Body size and food size in freshwater zooplankton. Proc. natn. Acad. Sci. U.S.A. 81: 6427–6431.

Chotiyaputta, C. & K. Hirayama, 1978. Food selectivity of the rotifer Brachionus plicatilis feeding on phytoplankton. Mar. Biol. 45: 105–111.

Cruz, A. & D. Millares, 1974. Metodo de cultivo masivo de Brachionus plicatilis (Rotifera) a escala experimental. Ciencias, Ser. 8, Inv. Mar. 11: 60–62.

Devauchelle, B. & M. Girin, 1974. Production du rotifer Brachionus plicatilis O. F. Müller en elevage mixte avec le copepode Tisbe furcata (Baird). Publ. Cent. Nat. Explor. Oceans (CNEXO) 1: 87–99.

Endo, K. & T. Mochizuki, 1979. Relation between specific growth rate of rotifer Brachionus plicatilis, and concentration of food, marine species of Chlorella. J. Ferment. Technol. 57: 372–374.

Euteneuer, S., E. Lubzens & R. Fishler, 1984. A preliminary report on cold preservation of the rotifer Brachionus plicatilis (O. F. Müller). Spec. Publ. Europ. Maricult. Soc. 8: 211–228.

Fontaine, C. & D. Revera, 1980. The mass culture of the rotifer Brachionus plicatilis for use as foodstuff in aquaculture. World Maric. Soc. Proc. 11: 211–218.

Fujita, S., 1979. Culture of red sea bream Pagrus major and its food. Spec. Publ. Europ. Maricult. Soc. 4: 183–197.

Fukuhara, O., H. Yukawa & T. Nishimura, 1982. Culture of rotifer, Brachionus plicatilis (Müller) in an air lift aquarium using alcohol fermentation slops. Bull. Nansei Fish. Res. Lab. 14: 1–11.

Fukusho, K. & H. Iwamoto, 1980. Cyclomorphosis in size of the cultured rotifer Brachionus plicatilis. Bull. Natl. Res. Inst. Aquacult. 1: 29–37. (In Japanese, English summary).

Furukawa, I. & K. Hidaka, 1973. Technical problems encountered in the mass culture of the rotifer using marine yeast as food organisms. Bull. Plankton Soc. Japan 20: 61–71.

Galkovskaya, F. A., 1963. On the utilization of food for growth and conditions of highest yield in rotatoria, Brachionus calyciflorus, Pallas. Zool. Zh. 42: 505–512.

Gatesoupe, F. J., 1982. Nutritional and antibacterial treatments of live food organisms: the influence on survival, growth rate and weening success of turbot (Scophthalmus maximus). Ann. Zootech. 31: 353–368.

Gatesoupe, F. J. & P. Luquet, 1981. Practical diet for mass culture of the rotifer Brachionus plicatilis: application to larval rearing of sea bass Dicentrarchus labrax. Aquaculture 22: 149–163.

Gatesoupe, F. J. & J. H. Robin, 1981. Commercial single-cell proteins either as sole food source or in formulated diets for intensive and continuous production of rotifers (Brachionus plicatilis). Aquaculture 25: 1–15.

Gordin, H., 1980. Aquaculture: potential development. In P. G. Brewer (ed.), Oceanography. Springer-Verlag, New York: 347–361.

Grabner, H., Wieser & R. Lackner, 1981. The suitability of frozen and freeze-dried zooplankton as food for fish larvae: a biochemical test program. Aquaculture 26: 85–94.

Groeneweg, J. & M. Schluter, 1981. Mass production of freshwater rotifers on liquid wastes. II. Mass production of Brachionus rubens Ehrenberg 1838 in the effluent of high-rate algal ponds used for the treatment of piggery waste. Aquaculture 25: 25–33.

Helps, S. K., 1982. An examination of prey size selection and its subsequent effect on survival and growth of gilthead sea bream (Sparus aurata). M.Sc. Thesis, Plymouth Polytechnic, U.K.

Hino, A. & R. Hirano, 1980. Relationship between body size of the rotifer Brachionus plicatilis and the maximum size of particles ingested. Bull. Jpn. Soc. Sci. Fish. 46: 1217–1222.

Hino, A. & R. Hirano, 1984. Relationship between body size of the rotifer Brachionus plicatilis and the minimum size of particles ingested. Bull. Jpn. Soc. Sci. Fish. 50: 1139–1144.

Hino, A., Y. Nogami & R. Hirano, 1981. Fundamental studies on the mass culture techniques of the rotifer Brachionus plicatilis making use of the secondary products of waste water treatment. I. Rotifer culture provided with activated sludge. Aquiculture 28: 174–176. (In Japanese).

Hirata, H., 1974. An attempt to apply an experimental microcosm for the mass culture of marine rotifer Brachionus plicatilis. Mem. Fac. Fish. Kagoshima Univ. 23: 163–172.

Hirata, H., 1979. Rotifer culture in Japan. Spec. Publ. Europ. Maricult. Soc. 4: 361–388.

Hirata, H., 1980. Culture methods of the marine rotifer Brachionus plicatilis. Min. Rev. Data File Fish. Res. 1: 27–46.

Hirata, H. & Y. Mori, 1967. Mass culture of the rotifer fed baker's yeast. Saibai Gyogyo 5: 36–40.

Hirata, H. & W. Nagata, 1982. Excretion rates and excreted compounds of the rotifer Brachionus plicatilis O. F. Müller in culture. Mem. Fac. Fish. Kagoshima Univ. 31: 161–174.

Hirata, H., S. Yamasaki, T. Kawaguchi & M. Ogawa, 1983. Continuous culture of the rotifer Brachionus plicatilis fed recycled algal diets. Hydrobiologia 104: 71–75.

Hirayama, K. & H. Funamoto, 1983. Supplementary effect of several nutrients on nutritive deficiency of baker's yeast for population growth of the rotifer Brachionus plicatilis. Bull. Jpn. Soc. Sci. Fish. 49: 505–510.

Hirayama, K. & T. Kusano, 1972. Fundamental studies on physiology of rotifer for its mass culture. II. Influence of water temperature on population growth of rotifer. Bull. Jpn. Soc. Sci. Fish. 38: 1357–1363.

Hirayama, K. & K. Nakamura, 1976. Fundamental studies on the physiology of rotifers in mass culture. V. Dry Chlorella powder as food for rotifers. Aquaculture 8: 301–307.

Hirayama, K. & S. Ogawa, 1972. Fundamental studies on physiology of rotifer for its mass culture. I. Filter feeding of rotifer. Bull. Jpn. Soc. Sci. Fish. 38: 1207–1214.

Hirayama, K., K. Watanabe & T. Kusano, 1973. Fundamental studies on physiology of rotifer for its mass culture. III. Influence of phytoplankton density on population growth. Bull. Jpn. Soc. Sci. Fish. 39: 1123–1127.

Hirayama, K., K. Takagi & H. Kimura, 1979. Nutritional effect of eight species of marine phytoplankton on population growth of the rotifer Brachionus plicatilis. Bull. Jpn. Soc. Sci. Fish. 45: 11–16.

Howell, B. R., 1973. Marine fish culture in Britain. VIII. A marine rotifer, Brachionus plicatilis Müller, and the larvae of the mussel, Mytilus edulis L. as foods for larval flatfish. J. Cons. Int. Explor. Mer 35: 1–6.

Hunter, J. R., 1980. The feeding behaviour and ecology of marine fish larvae. In J. E. Bardach, J. J. Magnuson, R. C. May & J. M. Reinhart (eds), Fish Behaviour and Its Use in the Capture and Culture of Fishes. ICLARM Conf. Proc. 5, Manila, Philippines: 287–330.

Ito, T., 1960. On the culture of the mixohaline rotifer Brachionus plicatilis O. F. Müller in sea water. Rep. Fac. Fish. Pref. Univ. Mie 3: 708–740. (in Japanese, English summary).

James, C. M., B. Ben-Abbas, A. M. Al-Khars, S. Al-Hinty & A. E. Salman, 1983. Production of the rotifer Brachionus plicatilis for aquaculture in Kuwait. Hydrobiologia 104: 77–84.

Kafuku, T. & H. Ikenoue, 1983. Modern methods of aquaculture in Japan. Developments in Aquaculture and Fisheries Science, 11. Kodansha Ltd., Tokyo, and Elsevier, Amsterdam, 216 pp.

Kawano, T., H. Kojima, H. Ohosawa & K. Morinaga, 1976. Feed and method of aquianimals cultivation. U.S. patent 3,939,279: 1–16.

Kawano, T., A. M. Al-Khars, S. Al-Mattar & S. M. Kamel, 1978. Marine microorganism culture and production under Kuwait conditions. Kuwait Inst. Sci. Res., Ann. Res. Rep.: 63–67.

King, C. E., H. B. Bayne, T. K. Cannon & A. E. King, 1983. Cryopreservation of monogonont rotifers. Hydrobiologia 104: 85–88.

Lincoln, E. P., T. W. Hall & B. Koopman, 1983. Zooplankton control in mass algal cultures. Aquaculture 32: 331–337.

Lubzens, E., 1981. Rotifer resting eggs and their application to marine aquaculture. Spec. Publ. Europ. Maricult. Soc. 6: 163–179.

Lubzens, E., A. Marko & A. Tietz, 1984a. Lipid synthesis in the rotifer Brachionus plicatilis. Spec. Publ. Europ. Maricult. Soc. 8: 201–210.

Lubzens, E., G. Sagie, G. Minkoff, E. Meragelman & A. Schneller, 1984b. Rotifers (Brachionus plicatilis) improve growth rate of carp (Cyprinus carpio) larvae. Bamidgeh 36: 41–46.

Lubzens, E., A. Marko & A. Tietz, 1985a. De novo synthesis of fatty acids in the rotifer Brachionus plicatilis. Aquaculture 47: 27–37.

Lubzens, E., G. Minkoff & S. Marom, 1985b. Salinity dependence of sexual and asexual reproduction in the rotifer Brachionus plicatilis. Mar. Biol. 85: 123–126.

Matsuda, M., T. Makita & S. Kadonaki, 1980. Feeding examination of yeast cultured in seawater as food for the rotifer Brachionus plicatilis. Min. Rev. Data File Fish. Res. 1: 150.

Millares, N. & E. Gonzalez, 1975. Cultivo del rotifero Brachionus plicatilis. Res. Invest. 2: 60–62.

Minkoff, G., E. Lubzens & D. Kahan, 1983. Environmental factors affecting hatching of rotifer (Brachionus plicatilis) resting eggs. Hydrobiologia 104: 61–69.

Okauchi, M. & K. Fukusho, 1984. Food value of a minute alga, Tetraselmis tetrathele, for the rotifer Brachionus plicatilis culture. I. Population growth with batch culture. Bull. Natl. Res. Inst. Aquacult. 5: 13–18. (iIn Japanese, English summary).

Okauchi, M., T. Oshiro, S. Kitamura, A. Tsujigado & K. Fukusho, 1980. Number of rotifers, Brachionus plicatilis, consumed daily by a larva and juvenile of porgy, Acanthopagrus schlegeli. Bull. Natl. Res. Inst. Aquacult. 1: 39–45.

Okauchi, M. & K. Fukusho, 1985. Different modes in carrying resting eggs between two wild types of the rotifer Brachionus plicatilis from Matsusaka. Bull. Jpn. Soc. Sci. Fish. 51: 1907.

Ortega, A., E. Santaella, A. Garcia, M. Olmedo & J. B. Peleteiro, 1983. Cultivo de dorata Sparus aurata L. en el centro costero del mar Menor durante la temporada 1978–1979. Inf. Téc. Inst. Esp. Oceanogr. 5: 1–29.

Pascual, E. & M. Yúfera, 1983. Crecimento en cultivo de una cepa de Brachionus plicatilis O. F. Müller en funcion de la temperatura y la salinidad. Inv. Pesq. 47: 151–159.

Person-le Ruyét, 1975. Techniques d'élevage en masse d'un Rotifére (Brachionus plicatilis Müller) et d'un Crustacé Branchiopode (Artemia saline L.). 10th Europ. Symp. Mar. Biol., Ostend, Belgium, Sept. 17–23, 1975, 1: 331–343.

Pourriot, R., 1977. Food and feeding habits of Rotifera. Arch. Hydrobiol. Beih. Ergebn. Limnol. 8: 243–260.

Reguera, B., C. Mosquera & C. Fernandez, 1982. Consideraciones acerca de la produccion masiva del rotifero Brachionus plicatilis O. F. Müller alimentado con levadura de panificacion. Inf. Tec. Inst. Esp. Oceanogr. 2: 1–19.

Rothbard, S., 1979. Practical mass culture of the rotifer Brachionus plicatilis Müller. Eifac Tech. Pap. No. 35, Suppl. 1: 194–202.

Ruttner-Kolisko, A., 1972. Der Einfluss von Temperatur und Salzgehalt des Mediums auf Stoffwechsel und Vermehrungs intensitat von Brachionus plicatilis (Rotatoria). Verh. dt. zool. Ges. 65: 89–95.

Schluter, M. & J. Groeneweg, 1981. Mass production of freshwater rotifers on liquid wastes. I. The influence of some environmental factors on population growth of Brachionus rubens. Aquaculture 25: 17–24.

Scott, J. M., 1981. The vitamin B12 requirement of the marine rotifer Brachionus plicatilis. J. mar. biol. Ass. U.K. 61: 983–994.

Scott, A. P. & S. M. Baynes, 1978. Effect of algal diet and temperature on the biochemical composition of the rotifer Brachionus plicatilis. Aquaculture 14: 247–260.

Serra, M. & M. R. Miracle, 1983. Biometric analysis of Brachionus plicatilis ecotypes from Spanish lagoons. Hydrobiologia 104: 279–291.

Serra, M. & M. R. Miracle, 1985. Enzyme polymorphism in Brachionus plicatilis populations from several Spanish lagoons. Verh. Int. Verein. Limnol. 22: 2991–2996.

Snell, T. W. & K. Carrillo, 1984. Body size variation among strains of the rotifer Brachionus plicatilis. Aquaculture 37: 359–367.

Snell, T. W. & B. C. Winkler, 1984. Isozyme analysis of rotifer proteins. Biochem. Sys. Ecol. 12: 199–202.

Snell, T. W., C. J. Bieberich & R. Fuerst, 1983a. The effects of green and blue-green algal diets on the reproductive rate of the rotifer Brachionus plicatilis. Aquaculture 31: 21–30.

Snell, T. W., B. E. Burke & S. D. Messur, 1983b. Size and distribution of resting eggs in a natural population of the rotifer Brachionus plicatilis. Gulf Res. Rep. 7: 285–287.

Teshima, S., A. Kanazawa & M. Sakamoto, 1981. Attempts to culture the rotifers with microencapsulated diets. Bull. Jpn. Soc. Sci. Fish. 47: 1575–1578.

Theilacker, G. H. & A. S. Kimball, 1984. Comparative quality of rotifers and copepods as foods for larval fishes. CalCOFI Rep. 25: 80–86.

Theilacker, G. H. & M. F. McMaster, 1971. Mass cultivation of the rotifer Brachionus plicatilis and its evaluation as a food for larval anchovies. Mar. Biol. 10: 183–188.

Trotta, P., 1980. A simple and inexpensive system for continuous monoxenic culture of Brachionus plicatilis Müller as a basis for mass production. In G. Shelef & C. J. Soeder (eds), Algae Biomass. Elsevier/North Holland Biomedical Press: 307–313.

Trotta, P., 1981. A simple and inexpensive system for continuous monoxenic mass culture of marine microalgae. Aquaculture 22: 383–387.

Trotta, P., 1983. An indoor solution for mass production of the marine rotifer Brachionus plicatilis Müller fed on the marine microalga Tetraselmis suecica Butcher. Aquacult. Eng. 2: 93–100.

Ushiro, M., S. Yamasaki & H. Hirata, 1980. Examination of bacteria as food for Brachionus plicatilis, in culture. Min. Rev. Data File Fish. Res. 1: 96–106.

Walker, K. F., 1981. A synopsis of ecological information on the saline lake rotifer Brachionus plicatilis Müller 1786. Hydrobiologia 81: 159–167.

Watanabe, T., 1982. The production of food organisms with particular emphasis on rotifers. Report of training course on sea bass spawning and larval rearing. FAO UNDP Training Course, Songkhla, Thailand, June 1–20, 1982.

Watanabe, T., C. Kitajima & S. Fujita, 1983. Nutritional values of live organisms used in Japan for mass propagation of fish. A review. Aquaculture 34: 115–143.

Witt, U., P. H. Koske, D. Kuhlmann, J. Lenz & W. Nellen, 1981. Production of Nannochloris spec. (Chlorophyceae) in large scale outdoor tanks and its use as a food organism in marine aquaculture. Aquaculture 23: 171–181.

Yamasaki, S. & H. Hirata, 1982. Rearing of the prawn, Penaeus japonicus, fed on frozen and living rotifers. Min. Rev. Data File Fish. Res. 2: 87–89.

Yamasaki, S. & H. Hirata, 1984. An observation of bacterial consumption by rotifer, Brachionus plicatilis. Min. Rev. Data File Fish. Res. 3: 55–58.

Yamasaki, S., T. Nishihara & H. Hirata, 1984. Influence of marine Chlorella density on food consumption and growth rate of rotifer, Brachionus plicatilis. Mem. Fac. Fish. Kagoshima Univ. 33: 57–61. (iIn Japanese, English summary).

Yasuda, K. & N. Taga, 1980. Culture of Brachionus plicatilis Müller using bacteria as food. Bull. Jpn. Soc. Sci. Fish. 46: 933–939.

Yúfera, M., 1982. Morphometric characterization of a small sized strain of Brachionus plicatilis in culture. Aquaculture 27: 55–61.

Yúfera, M. & E. Pascual, 1980. Estudio del rendimiento de cultivos del rotifero Brachionus plicatilis O. F. Müller alimentados con levadura de panificacion. Inv. Pesq. 44: 361–368.

Yúfera, M. & E. Pascual, 1985. Effects of algal food concentration on feeding and ingestion rates of Brachionus plicatilis in mass culture. Hydrobiologia 122: 181–187.

Yúfera, M., L. M. Lubián & E. Pascual, 1983. Efecto de cuatro algas marinas sobre el crecimiento poblacional de dos cepas de Brachionus plicatilis (Rotifera: Brachionidae) en cultivo. Inv. Pesq. 47: 325–337.

Hydrobiologia 147: 257–261 (1987)
May, L., Wallace, R. & Herzig, A. (eds), Rotifer Symposium IV

Production and nutritional quality of the rotifer *Brachionus plicatilis* fed marine *Chlorella* sp. at different cell densities

T. A. Rezeq & C. M. James
Kuwait Institute for Scientific Research, Mariculture and Fisheries Department, P.O. Box 1638, Salmiya, Kuwait

Keywords: *Brachionus plicatilis,* production, *Chlorella,* nutrition

Abstract

The effect of different cell densities of marine *Chlorella* sp. on the growth rate, doubling time and production of the rotifer *Brachionus plicatilis* was investigated. A significant increase in rotifer production was achieved at a density of 50×10^6 *Chlorella* cells ml^{-1}. The nutritional quality of rotifers grown at different concentrations of *Chlorella* is discussed.

Introduction

The rotifer *Brachionus plicatilis* is indispensable for aquaculture since it is the first food of the initial larval stages of many finfish and crustacean larvae. In recent years, many attempts have been made to improve rotifer culture conditions and the nutritional quality of the rotifers produced for aquaculture. In addition to inert foods, several species of algae have been tried as food for these rotifers, and the effect of phytoplankton quality on rotifer population growth has been described [Ito, 1955, 1960; Hirayama *et al.,* 1973, 1979; Hirayama & Nakamura, 1976; Endo & Mochizuki, 1979; Preisser & Spittler, 1979; Gatesoupe & Robin, 1981; James *et al.,* 1983; Snell *et al.,* 1983; Yúfera *et al.,* 1983]. Some investigators have discussed the use of different cell densities of *Chlorella* for feeding rotifers [Hirayama & Ogawa, 1972; Hirayama *et al.,* 1973; Halbach & Halbach-Keup, 1974; Pilarska 1977a, b; Endo & Mochizuki, 1979], but the effects of high *Chlorella* densities ($>25 \times 10^6$ cells ml^{-1}) on production rates and the nutritional quality of the rotifers produced is not well documented. This study examines the production dynamics and nutritional quality of *B. plicatilis* cultured on high densities of marine *Chlorella* sp.

Materials and methods

The experiments were carried out using a local strain of marine *Chlorella* sp. from an algal chemostat maintained at pH 6, temperature 25 °C and salinity 30‰ (ambient salinity of seawater 38‰). *Chlorella* in the exponential growth phase, with a cell density of more than 60×10^6 cells ml^{-1}, was diluted to 12.5, 25.0, 37.5 and 50.0×10^6 cells ml^{-1} using 30‰ diluted seawater. The experiments were conducted in 5 l beakers kept in a water bath at 25 °C. Point aeration was provided at a rate of about 400 ml min^{-1}. Four replicates were set up for each treatment.

Amictic rotifers, acclimatised to *Chlorella* as a food source, were inoculated into the cultures at an initial density of about 50 rotifers ml^{-1}. Rotifer population density, pH and DO were monitored daily. From the third day of the experimental period 50% of the culture volume was replaced with *Chlorella* suspension, at the appropriate cell density, at 48 h intervals. The experiment was terminated after 15 days.

Doubling times were calculated by dividing $\log_e 2$ by the instantaneous growth rate (K) from the expression:

$$K = \frac{\ln N_t - \ln N_o}{t} + D'$$

where N_o = initial number of rotifers
N_t = number of rotifers after t days
D' = dilution rate (ml d^{-1})

To study their nutritional quality, rotifers were lyophilized at −100 °C and analysed for the presence of fatty acids and amino acids. To quantify the highly unsaturated fatty acids (ω3 HUFA), the extracted fat (using a chloroform methanol mixture) from lyophilized samples was esterified and analysed by gas chromatography (Perkin-Elmer Sigma 3 B) using high-resolution fused silica capillary columns. Amino acids were analysed by a Beckman amino acid analyser (model 119 BL). A Schimadzu (C-R 1A) Chromatopac recorder and integrator was used to quantify the data.

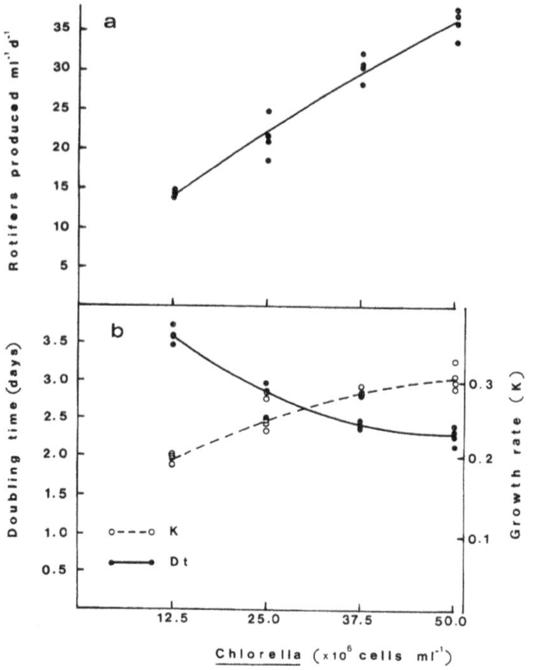

Fig. 1 a. Rotifers produced in relation to different cell densities of marine *Chlorella* sp. Each point represents the individual value of each replicate. (Rotifers produced ml^{-1} d^{-1} = $4.91 + 0.768C - 0.003C^2$; $r^2 = 0.965$ and n = 16); *Fig. 1 b.* Instantaneous growth rate (K) and doubling time (Dt) of rotifers in relation to different cell densities of marine *Chlorella* sp. ($K = 0.124 + 0.006C - 0.00005C^2$; $r^2 = 0.93$ and n = 16. $Dt = 4.61 - 0.094C + 0.0009C^2$; $r^2 = 0.947$ and n = 16).

Results

Rotifer production increased with increasing *Chlorella* densities (Fig. 1a). A maximum production of 36.33 ± 1.53 rotifers ml^{-1} d^{-1} was recorded at 50×10^6 cells ml^{-1} of *Chlorella*. Population doubling time decreased from 3.59 ± 0.09 days to 2.29 ± 0.1 days as the food density increased from 12.5×10^6 to 50×10^6 cells ml^{-1} (Fig. 1b). The doubling times recorded at 37.5×10^6 and 50×10^6 cells ml^{-1} *Chlorella* were not significantly different ($P > 0.05$).

Rotifer population growth rate showed a curvilinear relationship with increasing cell densities of *Chlorella* (Fig. 1b). The growth rate ranged from 0.193 at 12.5×10^6 cells ml^{-1} to 0.325 at 50×10^6 cells ml^{-1} *Chlorella*.

Rotifer population density fluctuated from the third day of the experiment due to the harvests made at two day intervals. The mean rotifer densities recorded were 203.5, 157.7, 118.25 and 112.5 rotifers ml^{-1} at *Chlorella* cell densities of 50×10^6, 37.5×10^6, 25×10^6, 12.5×10^6 cells, respectively. The pH ranged from 7.87 to 8.26 and DO from 4.7 to 5.5 ppm; no significant fluctuation in pH or DO was recorded between treatments.

The fatty acid composition of the rotifers (Table 1) shows that the total ω3 highly unsaturated fatty acids (HUFA) increased with increasing *Chlorella* cell densities, except in rotifers fed with 50×10^6 cells ml^{-1} *Chlorella*. The essential fatty acids (EFA) methyl linolenate (linolenic acid 18:3 ω3) and eicosapentaenoic acid (20:5 ω3) showed a similar pattern. In contrast, docosapentaenoic acid (22:5 ω3) and docosahexaenoic acid (22:6 ω3) were more abundant in rotifers fed with 50×10^6 *Chlorella* cells ml^{-1}.

The amino acid content (Table 2) was higher in

Table 1. Fatty acid composition (area %) of rotifers fed on different cell densities of *Chlorella*.

Fatty acid	*Chlorella* density (× 10^6 cells ml^{-1})			
	12.5	25	37.5	50
14:0	4.30	5.13	4.10	5.10
14:1ω9	0.30	0.20	0.17	0.13
15:0	0.75	0.60	0.50	0.75
16:0	15.71	16.42	13.07	19.04
16:1ω9	10.00	10.51	10.00	10.81
17:0	0.75	0.84	1.20	1.10
18:0	3.24	3.02	3.94	7.23
18:1ω9	7.56	5.70	6.10	6.91
18:2ω6	4.40	4.60	3.80	6.83
18:3ω6	0.28	0.32	0.60	0.21
18:3ω3	0.90	1.00	1.25	0.67
19:0	0.29	0.38	0.36	0.21
20:0	2.20	2.04	1.50	3.53
20:1ω9	0.67	0.66	0.67	0.91
20:3ω3	2.96	3.42	4.05	3.54
20:5ω3	9.72	11.16	13.41	6.96
22:1ω9	1.00	1.17	1.10	0.84
22:5ω3	4.46	3.90	4.01	4.70
22:6ω3	0.17	0.19	0.13	0.80
Σ ω3 HUFA	18.21	19.67	22.85	16.67
Total fat %	23.62	22.84	36.98	24.82

Table 2. Amino acid composition (μg 10 mg^{-1}) in rotifers fed with different cell densities of *Chlorella*.

Amino acid	*Chlorella* density (10^6 cells ml^{-1})			
	12.5	25	37.5	50
Aspartic acid	233.539	245.235	294.850	410.401
Threonine	88.007	90.653	113.994	81.379
Serine	120.666	134.769	150.391	129.390
Glutamic acid	287.206	365.350	373.610	320.016
Proline	128.733	190.699	198.453	66.378
Cystine	13.147	18.770	21.712	10.374
Glycine	83.295	119.830	112.284	87.488
Alanine	99.871	141.148	131.429	105.594
Valine	82.965	120.804	104.989	66.655
Methionine	25.583	30.119	31.861	27.376
Isoleucine	62.068	99.376	82.587	51.582
Leucine	156.037	228.904	196.972	154.054
Tyrosine	142.217	214.690	173.319	172.109
Phenylalanine	105.241	165.696	113.661	107.647
Histidine	33.706	55.964	27.617	31.461
Lysine	135.096	203.082	115.286	140.410
Arginine	189.981	184.527	88.739	111.287
Total %	19.87	26.10	23.32	20.74

rotifers fed with 25 × 10^6 and 37.5 × 10^6 *Chlorella* cells ml^{-1}. Basic amino acids such as histidine and lysine were more abundant in rotifers fed with 25 × 10^6 *Chlorella* cells ml^{-1}, and arginine in rotifers fed with 12.5 × 10^6 *Chlorella* cells ml^{-1}. The essential amino acids, such as proline and methionine, were more abundant in rotifers fed with 37.5×10^6 *Chlorella* cells ml^{-1}.

Discussion

The present investigation shows that increased population density and production in *Brachionus plicatilis* is associated with increasing initial concentrations of the alga *Chlorella.* However, the doubling time and instantaneous growth rate show that beyond 37.5 × 10^6 *Chlorella* cells ml^{-1}, the rotifer population may reach a steady state. This is in contrast to the observations of Hirayama *et al.* [1973] who suggested that the most suitable *Chlorella* cell density for optimum population growth and net reproduction rate is 1.5 × 10^6 *Chlorella* cells ml^{-1}. Pilarska [1977a, b] observed the highest daily growth rate of *Brachionus rubens* at *Chlorella vulgaris* concentrations ranging from 0.4 × 10^6 to 1.0 × 10^6 cells ml^{-1}.

The results of the present investigation are in accordance with the observations of Endo & Mochizuki [1979] who concluded that it would be possible to produce the large (satiated) type of rotifer in a practical mass culture only at concentrations of more than 2.6 × 10^6 cells ml^{-1} of marine *Chlorella.* Hirayama & Ogawa [1972] also observed that rotifers cannot reach a satiated condition until the *Chlorella*-density increases to 2.3 × 10^6 cells ml^{-1}.

The presence of highly unsaturated ω-3 fatty acids in rotifers, which is indispensible for rearing marine fish larva [Watanabe *et al.,* 1983], shows that up to a food density of 37.5 × 10^6 cells ml^{-1} *Chlorella* there is no inhibition in the biosynthesis of these essential fatty acids and the high initial algal density enhances the fatty acid content. Although there is a decline of 18:3 ω3 and 20:5 ω3 acids at 50 × 10^6 *Chlorella* cells ml^{-1}, the long-chain 22:6 ω3 is more abundant in these rotifers than in other treatments. Watanabe *et al.* [1978,

1983] observed no marked difference in the amino acid content of rotifers fed with various species of algae and yeast. During the present investigation, the amino acid content was higher in rotifers fed with 25×10^6 and 37.5×10^6 *Chlorella* cells ml^{-1} compared to other treatments.

Watanabe *et al.* [1983] observed that when marine *Chlorella* was used to culture rotifers, the HUFA increased in proportion to the length of the culture period due to incorporation of 20:5 ω3 from the marine *Chlorella*. The fatty acid and amino content in different species of *Chlorella* has also been discussed. There is, however, no information available on the effect of different cell densities of *Chlorella* on the fatty acid and amino acid content in rotifers. Furthermore, a concentration of 12% of 20:5 ω3 in rotifers has been normally used for feeding marine fish larvae at the Nagasaki Prefectural Institute of Fisheries, Japan [Watanabe *et al.*, 1983]. During the present investigation a concentration of 13.41% 20:5 ω3 has been observed in rotifers fed with 37.5×10^6 *Chlorella* cells ml^{-1}. This is a unique investigation in which the fatty acid and amino acid content of rotifers in relation to different cell densities of marine *Chlorella* sp. has been studied in detail. The above observations show that the nutritional quality of rotifers could vary with different cell densities of *Chlorella* utilized in the culture system.

Acknowledgement

Our sincere thanks are due to Dr Ziad H. Shehadeh, Kuwait Institute for Scientific Research, for his valuable suggestions and encouragement.

References

Endo, E. & T. Mochizuki, 1979. Relation between specific growth rate of rotifer, Brachionus plicatilis, and concentration of food, marine species of Chlorella. J. Ferment. Technol. Osaka 57: 372–374.

Gatesoupe, F. J. & J. Robin, 1981. Commercial single-cell proteins either as sole food source or in formulated diets for intensive and continuous production of rotifers (Brachionus plicatilis). Aquaculture 25: 1–15.

Halbach, U. & G. Halbach-Keup, 1974. Quantitative Beziehungen zwischer Phytoplankton und der Populationsdynamik des Rotators Brachionus calyciflorus Pallas. Befunde aus Laboratorium experimenten und freilanduntersuchungen. Arch. Hydrobiol. 73: 273–309.

Hirayama, K. & S. Ogawa, 1972. Fundamental studies on physiology of rotifer for its mass culture – I. Filter feeding of rotifer. Bull. jap. Soc. scient. Fish. 38: 1207–1214.

Hirayama, K., K. Watanabe & K. Kusano, 1973. Fundamental studies on physiology of rotifer for its mass culture – III. Influence of phytoplankton density on population growth. Bull. jap. Soc. scient. Fish. 39: 1123–1127.

Hirayama, K. & K. Nakamura, 1976. Fundamental studies on the physiology of rotifers in mass culture. 5. Dry Chlorella powder as a food for rotifers. Aquaculture 8: 301–308.

Hirayama, K., K. Takagi & H. Kimura, 1979. Nutritional effect of eight species of marine phytoplankton on population growth of the rotifer, Brachionus plicatilis. Bull. jap. Soc. scient. Fish. 45: 11–16.

Ito, T., 1955. Studies on the 'Mizukawari' in eel culture ponds. I. The feeding activity of Brachionus plicatilis on phytonanoplankton. Rep. Fac. Fish., Prefect. Univ. Mie 2: 162–167.

Ito, T., 1960. On the culture of the mixohaline rotifer Brachionus plicatilis O. F. Müller, in seawater. Rep. Fac. Fish., Prefect. Univ. Mie 3: 708–740.

James, C. M., M. Bou-Abbas, A. M. Al-Khars, S. Al-Hinty & A. E. Salman, 1983. Production of the rotifer Brachionus plicatilis for aquaculture in Kuwait. Hydrobiologia 104: 77–84.

Leimeroth, N., 1980. Untersuchungen der Respiration und Energiebilanzen verschiedener Stadien des Rotators Brachionus calyciflorus. Thesis, University of Frankfurt.

Pilarska, J., 1977a. Ecophysiological studies on Brachionus rubens Ehrbg (Rotatoria). I. Food selectivity and feeding rate. Pol. Arch. Hydrobiol. 24: 319–328.

Pilarska, J., 1977b. Ecophysiological studies on Brachionus rubens Ehrbg (Rotatoria). II. Production and respiration. Pol. Arch. Hydrobiol. 24: 329–341.

Preisser, B. & P. Spittler, 1979. The mass culture of rotifers and a phyllopod on a laboratory scale. European Mariculture Society 4: 389–401.

Rothhaupt, K. O., 1985. A model approach to the population dynamics of the rotifer Brachionus rubens in two-stage chemostat culture. Oecologia 65: 252–259.

Scot, A. P. & S. M. Baynes, 1978. Effect of algal diet and temperature on the biochemical composition of the rotifer Brachionus plicatilis. Aquaculture 14: 247–260.

Snell, T. W., C. J. Bieberich & R. Fuerst, 1983. The effects of green and blue-green algal diets on the reproductive rate of the rotifer Brachionus plicatilis. Aquaculture 31: 21–30.

Theilacker, G. H. & M. F. McMaster, 1971. Mass cultivation of the rotifer Brachionus plicatilis and its evaluation as a food for larval anchovies. Mar. Biol. 10: 183–188.

Vuckovic, M., 1981. Einfluß der Temperatur und Nahrungsquantität auf die Populationsdynamik des planktischen

Rädertieres Brachionus rubens Ehrbg. Thesis, University of Frankfurt.
Watanabe, T., T. Arakawa, C. Kitajima & S. Fujita, 1978. Nutritional evaluation of proteins of living feeds used in seed production of fish. Bull. jap. Soc. scient. Fish. 44: 985–988.
Watanabe, T., C. Katajima & S. Fujita, 1983. Nutritional values of live organisms used in Japan for mass propagation of fish: a review. Aquaculture 34: 115–143.
Yüfera, M., L. M. Lubian & E. Pascual, 1983. Effecto de cuatro algas marinas sobre el crecimiento poblacional de dos cepas de Brachionus plicatilis (Rotifera: Brachionidae) en cultivo. Inv. Pesq. 47: 325–337.

Hydrobiologia 147: 263–268 (1987)
May, L., Wallace, R. & Herzig, A. (eds), Rotifer Symposium IV

The use of marine yeast (*Candida* sp.) and bakers' yeast (*Saccharomyces cerevisiae*) in combination with *Chlorella* sp. for mass culture of the rotifer *Brachionus plicatilis*

Charles M. James, P. Dias & Assad E. Salman
Kuwait Institute for Scientific Research, Mariculture and Fisheries Department, P.O. Box: 1638, Salmiya, Kuwait

Keywords: *Brachionus plicatilis,* production, feed, yeast, nutrition

Abstract

Use of marine yeast and bakers' yeast in combination with *Chlorella* sp. for the large-scale production of the rotifer *Brachionus plicatilis* was investigated. The culture density of marine yeast fed rotifers was significantly higher than rotifers fed bakers' yeast. Rotifer production was significantly higher and the doubling time was lower for marine yeast fed rotifers than for bakers' yeast fed rotifers. It appears that the addition of marine yeast to the feed enhances the birth rate and overall production of rotifers in the culture system. The nutritional quality of rotifers is discussed.

Introduction

Mass cultures of the rotifer *Brachionus plicatilis* are an indispensible source of live food for many finfish and crustacean larvae grown in commercial hatcheries. *Chlorella* is an excellent source of food for these rotifers but it is a space and time consuming task to produce this alga in sufficient quantities to feed the rotifer cultures. Hirayama and Watanabe [1973] suggested the use of caked yeast (*Rhodotorula* sp.) as an alternative food for rotifers and many hatcheries now use bakers' yeast instead of *Chlorella* as the sole diet for their rotifers [Hirata & Mori, 1967; Mori & Ishii, 1970; Yamagata, 1972; Furukawa & Hidaka, 1973; Fushimi, 1975; Fukusho *et al.,* 1976; Al-Mattar & Kawano, 1978]. This culture method owes its success to nutritive supplements released by decomposing yeast cells, and the phytoplankton and bacteria which thrive on the by-products of this decomposition [Hirayama & Watanabe, 1973].

It is generally known that mixed diets of algae and yeast produce higher population growth rates in *B. plicatilis* cultures [Hirayama & Funamoto, 1983; James *et al.,* 1983]. The present investigation examines the use of the marine yeast *Candida* sp. and the bakers' yeast *Saccharomyces cerevisiae,* in combination with a marine species of *Chlorella,* to improve the production and nutritional quality of the rotifer *Brachionus plicatilis* for aquaculture.

Materials and methods

The marine yeast *Candida* sp. was cultured in 600 l capacity fermenters using a synthetic culture medium containing 0.15% ω-feed oil (fish oil) [Al-Hinty & James, 1983; James *et al.,* 1983]. Bakers' yeast was available as a commercial product (Fermipan, Gist-Brocades, Holland). *Chlorella* sp. was cultured in 2 m^3 capacity fibreglass tanks maintained under outdoor culture condition [Al-Khars & James, 1983; James *et al.,* 1983].

Rotifers were cultured in six 15 m^3 capacity concrete tanks, heated in winter (November to March) to 25–27 °C. Aeration was provided at the bottom of each tank using 12 mm diameter PVC pipes, perforated with 1 mm diameter holes at 50 cm intervals. The cultures were inoculated with amictic females and fed with either *Chlorella* + marine yeast + bakers' yeast (marine yeast fed) or *Chlorella* + bakers' yeast (bakers' yeast fed). Yeast cells in suspension (30‰ salinity) were allowed to drip into the culture tank for 20 h d^{-1}. At the end of the experiment some marine yeast fed rotifers were additionally fed with *Chlorella* for 24 h.

Rotifer harvesting/thinning was carried out periodically using a 69 μm mesh plankton net. After each harvest, the culture volume was adjusted by the addition of 30‰ diluted sea water (ambient salinity 36–42‰). Chemical and biological parameters were monitored daily.

Population growth rates were calculated from initial and final aliquot samples. Doubling times were calculated by dividing $\log_e 2$ by the instantaneous growth rate (K) from the expression:

$$K = \frac{\ln N_t - \ln N_o}{t} + D'$$

where N_o = initial number of rotifers
N_t = number of rotifers after t days
D' = dilution rate ($m^3\ d^{-1}$)

The instanteneous birth rate of rotifers was calculated from the expression:

$$b = \frac{\ln (E + 1)}{D}$$

where E = eggs ml^{-1}/rotifers ml^{-1}
D = duration of embryo development (0.5 d)

Rotifer dry weights were determined from freeze-dried and oven-dried material and averaged 0.331 μg per rotifer. Feeding efficiency was calculated as the ratio of the total biomass of food provided per day and the biomass of rotifers produced each day.

Nutritional quality was studied by lyophilizing rotifers, yeast and *Chlorella* at −100 °C and analysing for the presence of fatty acids and amino acids.

To quantify the highly unsaturated fatty acids (ω3 HUFA), the extracted fat (using a chloroform methanol mixture) from lyophilized samples was esterified and analysed by gas chromatography (Perkin-Elmer Sigma 3 B) using high-resolution fused silica capillary columns. Amino acids were analysed by a Beckman amino acid analyser (model 119 BL). A Schimadzu (C–R 1A) Chromatopac recorder and integrator was used to quantify the data.

Results

The culture density recorded at the end of the experiment was significantly higher ($P < 0.01$) in marine yeast fed rotifers (mean 454 ind. ml^{-1}) than bakers' yeast fed rotifers (mean 277 ind. ml^{-1}). Doubling time was significantly lower ($P < 0.05$) for rotifers fed marine yeast than for rotifers fed bakers' yeast. Rotifer production was significantly higher ($P < 0.05$) with marine yeast (range 33.82–157.65, mean 65.32 ind. $ml^{-1}\ d^{-1}$) than with bakers' yeast (range 17.55–46.27, mean 27.43 ind. $ml^{-1}\ d^{-1}$). No significant difference ($P > 0.05$) in feeding efficiency was observed between the two types of food. Feeding efficiency of rotifers fed marine yeast varied from 1.88–2.85 with a mean of 2.04, while that of bakers' yeast fed animals ranged from 1.52–3.68 with a mean of 2.74.

The birth rate of rotifers declined gradually during the experimental period under both feeding regimes (Fig. 1). The rate of decline was significantly higher ($P < 0.05$) for bakers' yeast fed rotifers than marine yeast fed rotifers.

The total content of ω3 highly unsaturated fatty acids (HUFA) was higher in marine yeast fed rotifers (8.41%) than bakers' yeast fed rotifers (7.66%) (Table 1). Although marine yeast fed rotifers contained less linolenic acid (18:3 ω3) and docosahexaenoic acid (22:6 ω3) than bakers' yeast fed rotifers, they contained more of the essential fatty acid (EFA) eicosapentaenoic acid (20:5 ω3). The total fat content was higher in marine yeast fed rotifers (34.87%) than bakers' yeast fed rotifers (26.12%). Furthermore, rotifers produced using

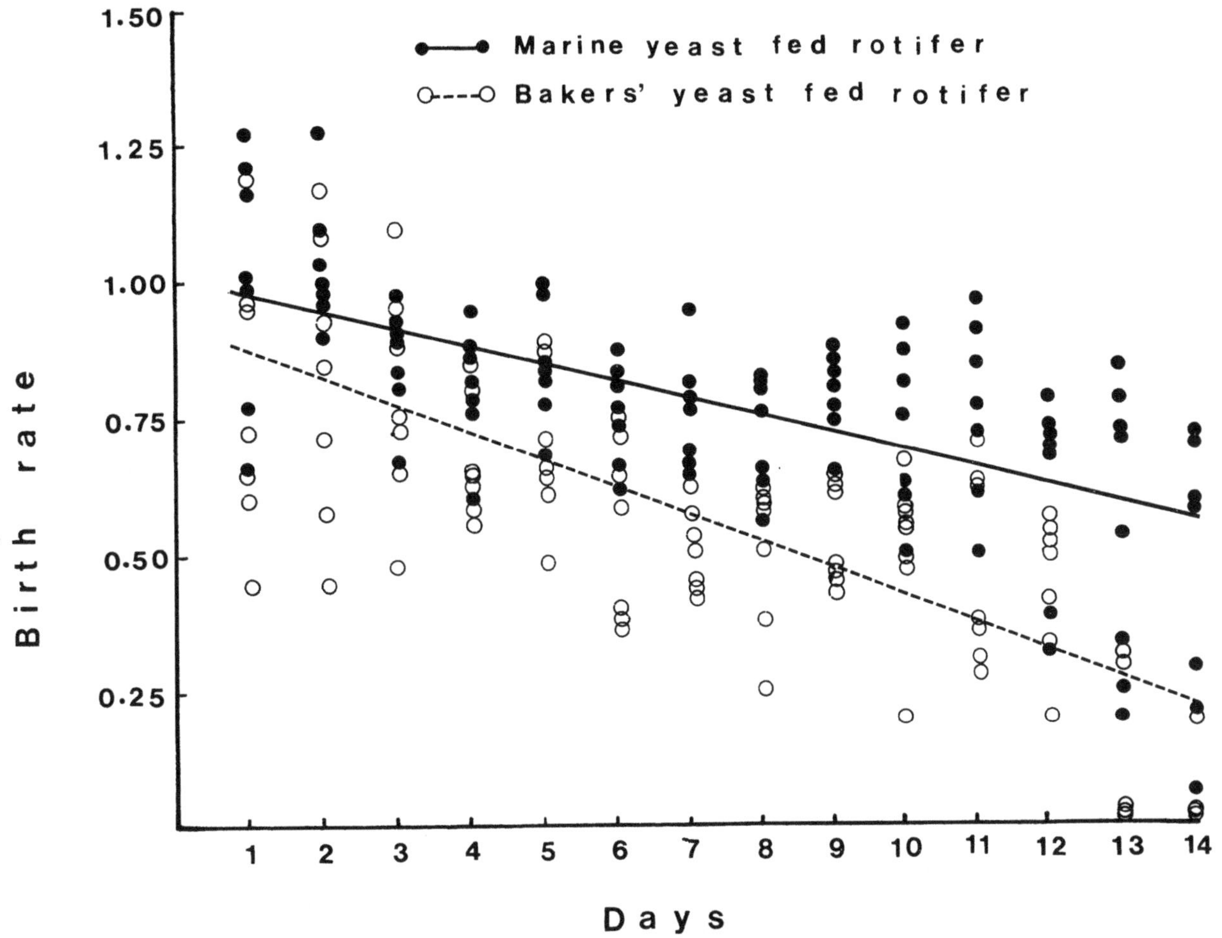

Fig. 1. The relationship between rotifer birth rate and time for marine yeast fed rotifers ($y = 1.02 - 0.03\times$; $r^2 = 0.40$; $n = 98$) and bakers' yeast fed rotifers ($y = 0.93 - 0.05$; $r^2 = 0.56$; $n = 98$). Each point represents the individual value of each replicate.

marine yeast and treated with *Chlorella* sp. for 24 h contained more ω3 HUFA (13.06%), particularly EFA eicosapentaenoic acid (7.99%), than marine yeast fed rotifer (ω3 HUFA 8.41% and 20:5 ω3 = 3.61%).

Total amino acid content was higher in marine yeast fed rotifers (32.51%) than animals fed with bakers' yeast (22.60%) (Table 2). The amino acid profile also shows that all the amino acids, especially the essential amino acids such as proline, alanine and methionine, and the basic amino acids such as histidine, lysine and arginine were present in greater quantities in marine yeast fed rotifers than in bakers' yeast fed rotifers (Table 2).

Discussion

The present investigation shows a highly significant increase ($P < 0.01$) in rotifer population density and yield when marine yeast is introduced into the culture system. Rotifer production of up to 157.65 rotifers ml^{-1} d^{-1} were achieved by this method which is considerably higher than that recorded by Furukawa and Hidaka [1973]. These authors obtained only 18.6–27.63 rotifers ml^{-1} d^{-1} while using marine yeast. Furthermore, the present yield is higher than that of Gatesoupe and Robin [1981] who obtained 35–55 rotifers ml^{-1} d^{-1} while using commercial single cell proteins, such as *Spirulina,*

Table 1. Fatty acid composition (area %).

Fatty acid	Rotifers		Yeast		*Chlorella* density (25×10^6 cells ml^{-1})	24 h *Chlorella* treated rotifers
	fed M.Y.	fed B.Y.	*Candida* sp.	*Saccharomyces cerevisiae*		
14:0	1.40	4.16	1.34	0.80	4.89	2.33
14:1 ω9	0.73	0.31	0.10	0.10	–	1.26
15:0	0.79	0.51	0.24	0.31	0.25	0.72
16:0	8.98	9.12	10.57	23.78	15.03	11.35
16:1 ω9	14.31	13.51	7.80	11.83	18.69	15.79
17:0	1.35	0.67	0.75	0.71	0.15	1.14
18:0	6.65	27.00	53.20	43.40	0.30	6.27
18:1 ω9	24.70				1.50	18.64
18:2 ω6	7.45	7.27	9.51	6.70	2.05	5.10
18:3 ω6	–	0.07	–	–	0.82	–
18:3 ω3	1.10	1.33	1.44	0.92	0.15	1.02
19:0	0.25	0.07	–	0.22	–	0.26
20:0	0.36	7.52	2.52	3.47	–	0.23
20:1 ω9	4.19	–	–	0.14	–	3.02
20:2 ω6	1.00	–	–	–	–	0.46
20:3 ω3	2.20	0.72	–	–	5.12	2.21
20:5 ω3	3.61	3.21	0.79	–	34.53	7.99
22:1 ω9	1.21	7.31	2.30	0.83	–	0.86
22:5 ω3	0.82	0.52	–	–	–	1.33
22:6 ω3	0.68	1.88	0.74	0.18	0.04	0.51
Σ ω3 HUFA	8.41	7.66	2.97	1.10	39.84	13.06
Total fat %	34.87	26.12	9.2	3.29	34.72	27.77

M.Y. = Marine yeast; B.Y. = Baker's yeast.

Table 2. Amino acid composition (μg 10 mg^{-1}) in rotifers fed with the marine yeast *Candida* sp. and the bakers' yeast *Saccharomyces cerevisiae*.

Amino acid	Marine yeast fed rotifers	Bakers' yeast fed rotifers
Aspartic acid	406.821	293.110
Threonine	145.802	91.924
Serine	198.589	173.534
Glutamic acid	442.172	379.555
Proline	275.763	99.229
Cystine	29.544	12.574
Glycine	123.492	93.823
Alanine	163.642	140.252
Valine	138.929	106.754
Methionine	21.585	12.893
Isoleucine	119.853	76.459
Leucine	266.659	184.974
Tyrosine	219.903	152.988
Phenylalanine	186.746	114.704
Histidine	59.226	37.109
Lysine	239.800	166.296
Arginine	212.337	123.780
Total %	32.51	22.60

Chlorella and methanol grown yeast, for feeding rotifers.

The present rotifer culture density (up to 671 rotifers ml^{-1}) is higher than that of James *et al.* [1983] who reported a maximum density of about 500 rotifers ml^{-1} while using marine yeast in the culture system. In contrast, Furukawa and Hidaka [1973] reported a culture density of about 1000 rotifers ml^{-1} while using marine yeast; rotifer production, however, was much lower than in the present study. The culture density of < 1000 rotifers ml^{-1} observed during the present investigation could be due to frequent harvests made during the observation period.

In a practical batch culture of rotifers with bakers' yeast, the food concentration is kept as low as possible because excess feeding causes ciliate contamination and deterioration of the culture medium [Kureha & Amagaya, 1978]. In the present investigation, the rotifer birth rates decline most rapidly in bakers' yeast fed cultures, indicating more rapid deterioration of this culture system than of marine yeast fed cultures.

The present investigation suggests that the nutritional quality of marine yeast is better than that of bakers' yeast. The higher amino acid content of marine yeast compared to bakers' yeast, especially of essential amino acids such as cystine, shows that it could improve the feeding efficiency of rotifers. Bearing in mind that the nutritional quality of yeast could be improved by supplementing it with several nutrients [Hirayama & Funamoto, 1983], it is possible that the nutritional quality of marine yeast could be further enhanced, perhaps even to the level of the marine *Chlorella* usually used in the hatcheries.

In fish hatcheries ω-yeast (yeast supplemented with fish oil) is used to improve the nutritional quality of rotifers for feeding fish larvae rather than promoting population growth in rotifers [Imada *et al.,* 1979; Fukusho *et al.,* 1980; Arakawa *et al.,* 1979; Kitajima *et al.,* 1980a, b]. Nevertheless, Hirayama and Funamoto [1983] observed that the ω-feed oil (fish oil) also supplements the nutritional deficiency of bakers' yeast and improves the population growth of the rotifer. The marine yeast utilized during the present investigation was produced with 0.15% ω-feed oil. However, since the HUFA content in the marine yeast and in the marine yeast-fed rotifers is lower than that in the marine *Chlorella* and in the rotifers treated for 24 h with *Chlorella,* it is recommended that the ω-feed oil content be increased during the production of marine yeast. This could further enhance rotifer production. The level of the essential fatty acid 20:5 ω3 in rotifers fed with marine yeast or marine yeast fed rotifers treated with *Chlorella* for 24 h almost corresponds to that reported for rotifers cultured with marine *Chlorella* and bakers' yeast [Watanabe *et al.,* 1978, 1983], but is lower than that of rotifers fed with *Chlorella* as a sole diet.

Acknowledgement

Our sincere thanks are due to Dr Ziad H. Shehadeh, Kuwait Institute for Scientific Research, for his valuable suggestions and encouragement.

References

Al-Hinty, S. & C. M. James, 1983. Production and nutritional evaluation of the marine yeast Candida sp. for aquaculture in Kuwait. Poster paper presented at the 14th Annual Meeting of the World Mariculture Society, January 9–13, Washington, D.C., 19 pp.

Al-Khars, A. M. & C. M. James, 1983. Production of microalgae Chlorella sp. under Kuwaiti environmental conditions. Presented at the Arab Aquaculture Conference, Kuwait, 28–30 May (Arabic).

Al-Mattar, S. M. & T. Kawano, 1978. First experiment in continuous method for mass culture of rotifer Brachionus plicatilis using forzen bread yeast. Kuwait Institute for Scientific Research, KISR-tech. Rep. Microorganisms Cult. Proj. 014/78, 8 pp.

Arakawa, T.; T. Yogata & T. Watanabe, 1979. Rearing of larvae of puffer (Fugurubripes rubripes) by rotifer (Brachionus plicatilis) cultured with various feeds. Nagasaki-ken Suisan Shikenjo Hokoku 5: 5–8 (in Japanese).

Fukusho, K., O. Hara & J. Yoshio, 1976. Mass production of rotifers in large volumes with Chlorella and yeast. Aquiculture 24: 96–101.

Fukusho, K., T. Arakawa & T. Watanabe, 1980. Food value of a copepod, Tigriopus japonicus, cultured with ω-yeast for larvae and juveniles of mud dab, Limanda yokochamae. Bull. jap. Soc. scient. Fish 46: 499–503.

Furukawa, K. & K. Hidaka, 1973. Technical problems encountered in the mass culture of the rotifer using marine yeast and food organisms. Bull. Plankton Soc. Japan 20: 61–71.

Fushimi, T., 1975. Feeding and growth of larval fish (ed. by Japan Soc. Sci. Fish.), Koseisha Koseikaku, Tokyo, pp. 67–73.

Gatesoupe, F. J. & J. Robin, 1981. Commercial single-cell proteins either as sole food source or in formulated diets for intensive and continuous production of rotifers (Brachionus plicatilis). Aquaculture 25: 1–15.

Hirata, H. & Y. Mori, 1967. Mass culture of marine rotifer (Brachionus plicatilis) fed on the bread yeast. Saibai-Gyogyo 5: 36–40.

Hirayama, K. & K. Watanabe, 1973. Fundamental studies on physiology of rotifer for its mass culture 4. Nutritional effect of yeast on population growth of rotifer. Bull. jap. Soc. scient. Fish. 39: 1129–1133.

Hirayama, K. & H. Funamoto, 1983. Supplementary effect of several nutrients on nutritive deficiency of bakers' yeast for population growth of the rotifer Brachionus plicatilis. Bull. jap. Soc. scient. Fish. 49: 505–510.

Imada, O., Y. Kageyama, T. Watanabe, C. Kitajima, S. Fujita & Y. Yone, 1979. Development of a new yeast as a culture medium for living feeds used in the production of fish. Bull. jap. Soc. scient. Fish. 45: 955–959.

James, C. M., M. Bou-Abbas, A. M. Al-Khars, S. Al-Hinty & A. E. Salman, 1983. Production of the rotifer Brachionus plicatilis for aquaculture in Kuwait. Hydrobiologia 104: 77–84.

Kitajima, C., T. Arakawa, S. Fujita, O. Imada, T. Watanabe & Y. Yone, 1980a. Dietary value for red sea bream larvae of rotifer Brachionus plicatilis cultured with a new type of yeast. Bull. jap. Soc. scient. Fish. 46: 43–46.

Kitajima, C., M. Yoshida & T. Watanabe, 1980b. Dietary value for ayu Plecoglossus altivelis of rotifer Brachionus plicatilis cultured with bakers' yeast Saccharomyces cerevisiae supplemented with cuttle fish liver oil. Bull. jap. Soc. scient. Fish. 46: 47–50.

Kureha, H. & A. Amagaya, 1978. Mass culture of rotifers. The Aquiculture 26: 88 (in japanese).

Mori, S. & S. Ishii, 1970. Studies on the culture of food organisms – III. Mass culture of the rotifer (Brachionus plicatilis). Rep. Gifu Project. Fish. exp. stat. 15: 79–81.

Watanabe, T., C. Kitajima, T. Arakawa, K. Fukusho & S. Fujita, 1978. Nutritional quality of rotifer Brachionus plicatilis, as a living feed from the viewpoint of essential fatty acids. Bull. jap. Soc. scient. Fish. 44: 1109–1114 (in japanese).

Watanabe, T., C. Kitajima & S. Fujita, 1983. Nutritional values of live organisms used in Japan for mass propagation of fish: a review. Aquaculture 34: 115–143.

Yamagata, Y., 1972. Culture of the rotifer by feeding yeast-1. Culture temperature and ration. Mie-Ken Naisuimen Suisan Shikenjo Nenpo 1972: 7–9.

Yüfera, M., E. Pascual, 1980. Estudio del rendimiento de cultivos del rotifero Brachionus plicatilis O. F. Müller alimentados con levadura de panificacion. Inv. Pesq. 44: 361–368.

Hydrobiologia 147: 269–270 (1987)
May, L., Wallace, R. & Herzig, A. (eds), Rotifer Symposium IV

A consideration of why mass culture of the rotifer *Brachionus plicatilis* with baker's yeast is unstable

K. Hirayama
Faculty of Fisheries, Nagasaki University, Nagasaki 852, Japan

Keywords: Rotifera, mass culture, baker's yeast

Abstract of Oral Communication

Since the introduction of the rotifer *Brachionus plicatilis* as a food source for larval fish [Ito, 1960], fingerling production in an average Japanese commercial marine hatchery has increased to more than one million fingerlings of red sea bream, *Pagrus major*, per year. However, to make the release of artificially produced fingerlings more effective as a coastal management project, it is necessary to increase our current production of *B. plicatilis*.

In order to do this, it is necessary not only to construct more tanks, but also to improve mass culture techniques. *Chlorella* is the best food for *B. plicatilis* in mass culture, but *Chlorella* is an unreliable source of food as *Chlorella* cultures are sometimes unstable. Baker's yeast is now used as a substitute food, but there are still many problems with its use, including the lack of highly unsaturated fatty acids in rotifers fed baker's yeast alone [Watanabe *et al.*, 1983]. In addition, mass cultures of *B. plicatilis* are sometimes subject to a sudden unexpected decrease or suppressed growth of the population, especially when fed baker's yeast. This phenomenon now forms a great obstruction to fingerling production through the uncertain supply of the rotifer food. Without a solution to this phenomenon, it may be difficult to obtain the goal of the fingerling release project.

Hirayama and Funamoto [1983] tested the dietary value of baker's yeast, and the supplementary effect of vitamin B_{12}, on rotifers grown under bacteria free conditions. Rotifers fed with the yeast alone could not grow, and their eggs were not viable. However, when supplemented with vitamin B_{12}, the rotifers grew well and their eggs hatched successfully. These results supported the findings of Scott [1981] that vitamin B_{12} is an essential vitamin for *B. plicatilis*. Nevertheless, the dietary value of baker's yeast supplemented with vitamin B_{12} was still much lower than that of *Chlorella*.

There are, however, many cases in which *B. plicatilis* cultures fed baker's yeast alone have been successful. Experiments have shown that the dietary value of marine yeast is improved when decomposing yeast remains in the culture vessels. The addition of marine *Chlorella* at extremely low densities similarly increases the nutritive value of the yeast [Hirayama & Watanabe, 1973]. These result indicate that the success of mass cultures of rotifers fed baker's yeast may be improved by the by-products of decomposition, or by the growth of phytoplankton or bacteria in the culture tanks. These provide a supplementary source of the nutrients lacking in yeast, such as vitamin B_{12}.

Imada [1984] measured the vitamin B_{12} content of filtrates of culture media, several kinds of food and the body tissues of rotifers cultured on those foods. Some types of food, such as baker's yeast, contained little vitamin B_{12}, although rotifers grown successfully on such diets contained high levels of vitamin B_{12}. These results also suggested that a supply of vitamin B_{12} as a metabolic product of micro-organisms may be essential for rotifer growth.

Imada [1984] observed an inverse relationship

between the oxygen content of mass culture tanks and the growth rate of rotifers fed with ω-yeast (yeast strengthened with essential fatty acids), alone. This appears to indicate that *B. plicatilis* grows better at low oxygen levels. However, this was not the case when *Chlorella* was provided as food. It seems likely that this relationship occurred because the bacteria which produce vitamin B_{12} in yeast fed cultures are facultative anaerobic bacteria.

Excess food and animal excretion products tend to pollute culture media, increasing ammonia levels. Un-ionized ammonia is usually considered to be one of the pollutants responsible for limiting the population growth of aquatic animals. Yu and Hirayama [1986] investigated the harmful effect of ammonia on rotifers. They found that un-ionized ammonia in mass culture tanks sometimes reached harmful concentrations and, as a result, limited the rate of increase of the rotifers. Since rotifers have a high selectivity for food particles [Funamoto & Hirayama, 1982], we should pay more attention to pollution when using an unfavourable food source. Food rejected by the rotifer pollutes the water, increasing ammonia levels and reducing growth rates.

A sudden decrease in the rotifer population or suppressed growth dùring mass culture has been discussed in relation to nutritional requirement, especially vitamin B_{12}, and pollution. However, in many cases, a sudden decrease or suppressed growth of rotifers seems to have no connection with the causes described above. An abrupt change of water temperature, especially a fall of 4 °C, may directly or indirectly, affect the growth of rotifer cultures [Kitajima, 1983]. Aquaculturists in hatcheries sometimes experience suppressed growth at the period of shift of the dominant strains. The formation of large numbers of dormant eggs also results in a sudden decrease of the rotifer population. We often observe that the suppressed growth of the rotifer is accompanied by an increase in diatoms or protozoa in the cultures [Reguera, 1984]. We must consider many factors, including those mentioned above, to systematically and scientifically improve the techniques of mass production of rotifers.

References

Funamoto, H. & K. Hirayama, 1982. Food selectivity of the rotifer Brachionus plicatilis expressed by using the frequency of mastax movement as an index. Suisanzoshoku 29: 246–250 (in Japanese).

Hirayama, K. & H. Funamoto, 1983. Supplementary effect of several nutrients on nutritive deficiency of baker's yeast for population growth of the rotifer Brachionus plicatilis. Bull. Japan. Soc. Sci. Fish. 49: 505–510.

Hirayama, K. & K. Watanabe, 1973. Fundamental studies on physiology of rotifer for its mass culture IV. Nutritional effect of yeast on population growth of rotifer. Bull. Japan. Soc. Sci. Fish. 39: 1129–1133.

Imada, O., 1984. Hakko boeki no shigenka nitsuite. Hakko to kogyo 42: 103–113 (in Japanese).

Ito, T., 1960. On the culture of maxoharine rotifer Brachionus plicatilis O. F. Muller in the sea water. Rep. Fac. Fish. Mie Pre. Univ. 3: 708–740 (in Japanese with English abstract).

Kitajima, C., 1983. Mass culture – Actual examples. In Japan. Soc. Sci. Fish. (ed.), The rotifer Brachionus plicatilis – Biology and mass culture. Koseisha Koseikaku, Tokyo: 102–128 (in Japanese).

Reguera, B., 1984. The effect of ciliate contamination in mass cultures of the rotifer, Brachionus plicatilis O. F. Muller. Aquaculture 40: 103–108.

Scott, J. M., 1981. The vitamin B_{12} requirement of the marine rotifer Brachionus plicatilis. J. mar. Bio. Ass. U.K. 61: 983–994.

Watanabe, T., C. Kitajima & S. Fujita, 1983. Nutritional value of live organisms used in Japan for mass culture of fish: A review. Aquaculture 34: 115–143.

Yu, J. P. & K. Hirayama, 1986. The effect of un-ionized ammonia on the population growth of the rotifer in mass culture. Bull. Japan. Soc. Sci. Fish. 52: 1509–1513.

Hydrobiologia 147: 271–281 (1987)
May, L., Wallace, R. & Herzig, A. (eds), Rotifer Symposium IV

The components of feeding behavior in rotifers

George W. Salt
Department of Zoology, University of California, Davis CA 95616, USA

Keywords: Rotifera, feeding, behavior, *Asplanchna*, *Brachionus*

Abstract

Feeding behavior of a rotifer can be broken into two classes of activities: the rate of successful search and the handling process. The former consists of the following components: Perceptual field (in planktonic rotifers the area of the corona), swimming rate, and attack rate. The second class consists of capture rate, handling time, rejection rate, ingestion, digestion, and assimilation. All evidence indicates that the perceptual field cannot be varied by the rotifer. Swimming rate is variable and under rotifer control, but does not appear to vary with degree of starvation. Attack rate is also under control of the rotifer, at least in the genus *Asplanchna*. Capture rates vary with the species of food item from zero to 100%. Handling times are longer than one would expect, as are rejection times. Digestion and assimilation appear to vary inversely with rate of ingestion. There is some suggestion in the literature that feeding on very small particles differs from that on larger ones.

Introduction

A study of the feeding of rotifers may be approached from several points of view. A proportional catalogue of the foods consumed tells us the rotifer's place in the aquatic food web. Measurements of clearance rates provide estimates of the impact of the rotifer on the dynamics of the food populations. Studies of the rate of population increase and the degree of mixis relative to food quality and quantity provide insights into rotifer populations in the wild.

However, the fundamental purpose of food consumption is to provide assimilable energy and nutrients for the rotifer's maintenance, growth, and reproduction. It is this aspect of feeding which will be emphasized in this review. More specifically an attempt will be made to identify the environmental and physiological factors which control feeding rate in rotifers. Most of the information available is derived from studies on *Brachionus* sp. or *Asplanchna* sp.

Encountering the food items

Current models which consider collisions between an organism and its food items [e.g., Stanley, 1932; Gerritsen & Strickler, 1977; Cox, 1983] contain five major variables: (1) the size of the perceptual field of the organism, (2) the organism's speed of movement, (3) the size of the food items, (4) the density of the food items, and (5) the rate of movement of the food.

An organism's perceptual field is determined by the distance at which its sensory organs can detect a food item. *Asplanchna brightwelli* was found by Wurdak *et al.* [1983] to have seven receptors in its anterior end. These workers suggested functions of chemoreception for the ciliated pit located between the corona and the lateral horns, tactile recognition for the ventrolateral sensory bristles and the pseudotrochus, and mechanoreception for the ventrolateral sensory bristles. Gilbert [1977] demonstrated chemoreception in campanulate morphs of *Asplancha intermedia* and *A. sieboldi* and concluded that prey recognition and the initiation of feeding behavior is mediated via chemoreception in these species. Gilbert (1976, 1980a: Table 2) found that campanulate females of *A. intermedia* can distinguish individuals of their own species from that of another, *A. sieboldi*, presumably via chemoreception. Male *Brachionus* are also known to possess chemoreceptors for females [Gilbert, 1963].

After extended electronmicroscopic studies of the sensory organs related to feeding in *Notommata copeus, Philodina roseola, Brachionus calyciflorus, Asplanchna brightwelli, Trichocera rattus,* and *Rhinoglena frontalis,* Clément *et al.* [1983] concluded that tactile bristles are the most widespread rotifer sensory receptors and that some are developed to the point of showing tactile coordination at the level of perception itself. To the chemoreceptors, which they tentatively identified, these workers assigned functions of distance perception and contact recognition.

The question of distance perception is an important one, for were such an ability present, it would enlarge the perceptual field of the rotifer and hence its food gathering efficiency. On behavioral grounds, Dumont [1972] suggested that *Asplanchna priodonta* possessed the ability to detect vibrations in the water, but no evidence has been produced to support this speculation.

The sessile predatory rotifer, *Cupelopagis vorax,* is able to detect prey by vibrations and orientate its corona on the potential prey, but the item must come close to *Cupelopagis* before an attack may be attempted [Dr. R. L. Wallace, pers. commun.]. The majority position is exemplified by Gilbert & Stemberger's [1985a] statement, '*A. girodi* responded to prey only when prey contacted their coronae.' The perceptual field regulating the encounter rate of a rotifer with food items then, is not body size but the size of the corona.

Rotifers have the ability to withdraw the corona into the body cavity. In theory then, they should be ably to vary the diameter of the corona and hence the perceptual field. If such a capability exists and is used, no one has reported it. *Brachionus calyciflorus* has pseudotrochal cirri which close over the mouth opening to form a screen which prevents the ingestion of food particles while still permitting the animal to swim [Gilbert & Starkweather, 1977, 1978). The pseudotrochal screen is formed when the food items are large (*Euglena*) but not when they are small (*Rhodotorula*). The screen serves the same purpose as a reduction in the size of the corona. This mechanism is the only observation of variation in the perceptual field or its analogue reported to this time.

Swimming speed is the second component of the encounter process. Recorded values for mobile forms range up to a maximum of slightly less than 3 mm s^{-1}, recorded in larval *Ptygura beauchampi* [Wallace, 1975, 1980]. Epp & Lewis [1984] measured maximum speeds in *A. sieboldi* of about 1.3 mm s^{-1} and in *Brachionus plicatilis* at 0.8 mm s^{-1}, a value nearly the same as that of 0.75 mm s^{-1} reported by Luciani *et al.* [1983] for mature individuals in a medium with calcium. Swimming speed varies with age and size [Epp & Lewis, loc. cit.], larger individuals swimming faster. A large campanulate *A. sieboldi* swims about 1.5 to 2.3 mm s^{-1}, while a saccate form about half as long swims at about 0.6 to 1.3 mm s^{-1} (Salt pers. obs.). Epp & Lewis (loc. cit.) concluded from their measurements on the swimming speeds of different sized *A. sieboldi* that although larger individuals swam faster, the increase with increase in body length was not linear. '...absolute speed increases with increasing body size, but at a rate less than the rate of increase in size. Relative speed therefore decreases with increasing size within clone 10C6 of *Asplancha.*' They report a similar relationship in *B. plicatilis.*

However, there are some other considerations worth noting in this context. Suppose one assumes that the shape of a rotifer closely resembles that of a cylinder or a cone in which diameter is one-third the length. Then, a doubling of the length results in an eight-fold increase in volume, a better measure of size than length. The circumference of the base doubles with a two-fold increase in length. More importantly, the area of the base per unit volume decreases by one-half with a two-fold increase in length. In other words, if the body proportions remain the same, an individual which is twice as long as another, has only half the food finding area per unit body volume of the smaller individual. This general morphological relationship may underlie the fact that larger campanulate morphs of trimorphic species of *Asplanchna* have proportionately larger coronae and are thought to be restricted to food-rich habitats [Stemberger & Gilbert, 1984a].

Luciani *et al.* [1983] found that swimming speed in *B. plicatilis* varied with age and presence or absence of calcium in the medium. Young and old individuals swam slower than mature adults. If the medium lacked calcium, swimming speeds were slower. Swimming speed is also temperature dependent. There is a regular increase in swimming rate from 14 °C to 26 °C after which the rate becomes somewhat less [Epp & Lewis loc. cit.]. The authors point out that mean absolute speeds in the range 20 °C to 32 °C are statistically indistinguishable.

This lack of statistical significance results from the high degree of variation in measurements at a given set of conditions, a characteristic of swimming speeds with which anyone who has measured them is familiar. This variability was dramatically documented by Coulon *et al.* [1983]. During 40 minutes of continuous recording under constant conditions, the swimming speed of a single female *A. brightwelli* varied from a low value of 0.81 mm s^{-1} to a high value of 1.04 mm s^{-1}, 28% higher than the low speed. If this kind of variation is a general one, then convincing conclusions about variation in swimming speed in response to changes in either internal or external conditions will require a very large number of replications.

For sessile rotifers, the rate at which ciliary action sweeps water past the buccal field is the functional equivalent of swimming speed. This is difficult to measure. Clearance rate can be substituted for it only if one assumes that the collection of food items is complete or if one has some independent estimate of the percentage of food items collected of those passing the corona.

Swimming speed in *A. brightwelli* and *B. calyciflorus* increases with light intensity and wave length [Clément, 1977; Clément *et al.*, 1983; Cornillac, 1982, cit. Clément *et al.*, 1983]. These findings agree with field observations that, in highly turbid water, *Asplanchna* sp. densities are highest in the water near the surface where their swimming, and hence food gathering, would be the fastest.

All other things being equal, the probability of a rotifer encountering a large food item is greater than that of colliding with a small one. If food items are assumed to be spherical, each doubling of the diameter increases the cross-sectional area with which the rotifers collide by four times. Consequently, if large and small food items were present in the water in equal densities, the rotifer would encounter a higher proportion of large particles than of small. Such an equal density is not common in natural situations, so a verification of this hypothesis will have to take place in the laboratory.

Density of food particles is one of the major factors determining the rate at which rotifers encounter prey. The various models mentioned earlier describe encounter rate as a direct function of food density. In theory, if the density of the food items is high enough, the rotifer would be in continuous contact with a food particle. Food intake does not show this undiminished rate of increase with food density in most cases, but the phenomenon is apparently approached at the densities used in Starkweather *et al.*'s [1970] measurements of *B. calyciflorus* feeding on *Enterobacter* (= *Aerobacter*) *aerogenes*. Reasons for this lack of direct response to food density at higher concentrations will be considered later.

Movement by prey is the fifth component usually included in encounter models. If movements of the rotifer and its target organism are random with respect to one another, then the principal effect of movement by the food item is to cancel the loss in efficiency which the seeker generates by re-crossing its previous path. If, on the other hand, the food item (generally animals) actively avoid the rotifer, then the rate of encounter is reduced.

C. I. Haydock (pers. comm.) observed that individuals of *Paramecium aurelia* derived from laboratory cultures showed no response to the presence of *A. sieboldi* and remained dispersed randomly throughout the medium. When individuals of *P. aurelia*, *P. bursaria*, or *P. multimicronucleatum*, taken from a pond in which *A. sieboldi* was abundant, were placed alone in a drop of medium, they too remained dispersed randomly. When, however, an *A. sieboldi* was placed in the drop of fluid, the *Paramecium* immediately swam to the periphery of the drop of fluid and remained there. Similar behavior was observed in large test tubes where *Asplanchna* swam in the upper levels and *Paramecium* remained in the bottom of the tube. This kind of avoidance behavior may be more common than is suspected, as investigators rarely use stocks of food species derived from the same environment as the rotifers. Obviously such a response would appear only in food species which posses directional swimming abilities.

Inasmuch as most rotifers have no sensory organs which detect distant objects, every particle with which they collide must be identified by touch or by chemoreception. If the object is a conspecific or an organism that may represent a threat, or even an inert particle of debris, the corona is withdrawn into the body and the rotifer stops swimming momentarily. This halt has the effect of decreasing the swimming rate and hence the rate at which food particles are encountered. There is, to my knowledge, no data to support the idea, but it seems likely that the feeding rate of a rotifer could be significantly depressed in natural habitats by the occurrence of bits of decaying vegetation (e.g., in marshes where emergent vegetation provides a dense suspension of decaying leaf particles). One would expect to see the same effect in waters where cladocera and copepods were present in high numbers. It is possible that to some degree the inverse correlation of rotifer densities with cladoceran numbers may be due to this form of interference competition rather than to competition for food items [see also Gilbert & Stemberger, 1985b].

The decision to attack

Having encountered a food item, the rotifer must either respond or not. This choice can be considered a decision in a mechanistic sense, although not in a cerebral one. By far the most important factor determining the decision is the species of food item encountered. This subject has been extensively reviewed [Pourriot, 1965, 1977; Dumont, 1977; Starkweather, 1980; Wallace, 1980] and will not be considered here. However, factors other than species identity of the food item also play a part in this decision.

Rotifers show definite size preferences in their choice of food items. By analogy, one would expect to find intraspecific size differences between food items too. There are, however, few data on this subject in which other factors such as different encounter probabilities, different morphologies, or clonal differences are not confounded with size differences. Gilbert [1980b] recorded the feeding of *A. girodi* on large and small forms of *Keratella cochlearis*. One must assume that these two size forms do not differ in the chemical stimuli they emit in order to regard his data as reflecting size selectivity. *A. girodi* which were starved 2–5 h showed a higher percentage of attacks after an encounter with the large form than with the small one – 61.3% versus 43.1%. In *B. plicatilis*, feeding on sludge particles, the maximum size ingested is correlated with lorica length, but the smallest size ingested is correlated with the size of the trophi [Hino & Hirano, 1980].

Degree of satiation or starvation may also affect attack rate. Most investigators studying feeding performance of rotifers assume this to be true and starve the rotifers used in their experiments before testing them. There is, however, little data to support this general idea. Gilbert [1980a] stated that satiated *A. sieboldi* and *A. intermedia* do not exhibit feeding behavior. *Asplanchna girodi* offered small *K. cochlearis* after 2–5 h starvation initiated an attack in 43% of the encounters and 92.3% of the encounters when starved only 1.5–2.5 h [Gilbert, 1980a]. The subject of hunger will be considered further in a later section.

Capture rate

The degree of success of an attack depends first of all on the species of food attacked. Most of our information is derived from studies on *A. girodi* [Gilbert 1980b; Gilbert & Williamson, 1978]. *Paramecium aurelia*, which are small and without defensive structures, are captured in 100% of the attacks on them, as are *Peridinium*. By contrast, *Kellicottia bostoniensis,* which has long spines, is only successfully captured in 5.6% of the attacks on it, while *Keratella c. cochlearis* is intermediate with a capture success of 24% to 52%. *Brachionus calyciflorus* is captured in 18% to 26% of the attacks made on them.

In part, the differences in the success of attacks by *A. girodi* are explained by the method of capture described by Gilbert & Stemberger [1985a]. Prey are captured by aspiration resulting from a sudden dilation of the mastax brought about by changes in the shape of the peripharynx. Food items are swept into the mastax by the water current thus generated. The time required is about 50 ms. Trophi play no part in the capture of food items, at least of small prey such as *Polyarthra vulgaris* or *Synchaeta oblongata.*

Given that the capture mechanism is a short rush of fluid into the mastax, it is apparent that any obstacle to the free movement of the food item will hinder its passage into the mastax. This may account for the low capture percentage of *Kellicottia* for example. Gilbert & Williamson [1978] report that it is difficult for *A. girodi* to ingest *Keratella* if the posterior spine is oriented toward the esophagus. Similar obstruction may take place in the capture process. Presumably this protection from capture is the selective force behind the evolution of spines in *Brachionus* and *Keratella* [Gilbert, 1980a; Stemberger & Gilbert, 1984b].

Structural features of the food item may enhance capture. *Polyarthra vulgaris* and *P. dolichoptera* both ingest twice as many flagellated *Chlamydomonas reinhartii* as they do non-flagellated ones [Gilbert & Bogdan, 1981]. In some of the tests the flagella were removed by sonification, so there could have been no difference in the chemosensory character of the cells, nor could their nutritive qualities have been different. One must conclude that the flagella aided in the capture of the algal cells.

A problem arises in classifying avoidance by the food item after an encounter and before an attack can be initiated. Gilbert & Williamson [1978] scored 346 encounters between *A. girodi* and *P. vulgaris* as resulting in no attacks. Yet Gilbert & Stemberger [1985a] report on the feeding behavior of *A. girodi* on *P. vulgaris.* The implication of zero attacks in 346 encounters is that *A. girodi* is unresponsive to *P. vulgaris,* which is obviously not the case. Rather, it would appear that the escape response of *P. vulgaris* is so fast that a true attack cannot usually take place. These incidents should more properly be considered escapes by the prey item.

Ingestion and rejection

Considering the time and energy spent in affecting a capture, one would expect that all captured food items would be ingested. However, except for the small and morphologically simple food items, this is not the case. Some food items may actively escape, although there are no observations to verify this statement, and some may be rejected. Gosse (1856] provides one of the earliest descriptions. A *Brachionus* had captured an algal cell too large to be manipulated by the unci. It was rejected by the seemingly 'indignant' *Brachionus* by the use of the trophi. By contrast, Chotiyaputta & Hirama [1978] report that *B. plicatilis* rejected cells of *Olisthodiscus* sp. from the oral tube by ciliary action.

Asplanchna rarely capture *Bosmina* and frequently reject them from the mastax. Both *A. girodi* and *A. brightwelli* reject *Cephalodella* from the mastax, and *A. girodi* rejects *Proalides tentaculus* [De Beauchamp, 1951].

In most instances, however, rejection is not inevitable for a given species of food item. In one set of observations, *A. girodi* ingested all the *P. vulgaris* which could be captured whereas they ingested only 73% of the *K. cochlearis* captured [Williamson & Gilbert, 1980]. Short-spined *B. calyciflorus* and *P. aurelia* are nearly always ingested after capture by *A. girodi,* while the proportion of captures of *K. cochlearis* that result in ingestions is highly variable, ranging from 100% to a low of 17% [Gilbert, 1980b; Gilbert & Williamson, 1978]. *Kellicottia bostoniensis,* the ciliate peritrich *Rhabdostyla,* and *Peridinium* were never ingested after capture in Gilbert's [1980b] study. It seems likely, from the information above, that a large proportion of rejections result from mechanical difficulties in transferring the food item into the stomach rather than because the food item is inherently unsuitable.

Gilbert & Starkweather [1977] observed that *B. calyciflorus* exhibited two methods for rejection of *Euglena gracilis*. (1) Some *Euglena* which had been transported through the buccal field and down the oral canal were expelled through the ventral cleft. (2) *Euglena* accepted into the distensible portion of the oral canal could be forced back into the buccal funnel by the trophi and rejected. Masses of *Rhodotorula glutinosus* might be rejected in this same fashion.

The times required for ingestion or rejection of *K. cochlearis* by *A. girodi* are surprisingly long. Gilbert & Williamson [1978] report mean values, from similar but not identical tests, of 40.4 to 95.0 s for ingestion times and even longer ones for rejection – 52.3 to 151.7 s.

Assimilation rate

Once in the digestive system of the rotifer, all, or more commonly, a portion of the food items may be hydrolyzed and assimilated by extra- and intracellular processes [Wurdack, 1986]. In nearly all organisms the fraction of the food that is assimilated is a function of the food supply. When food is abundant and the ingestion rate is high, the fraction of food assimilated is low or moderate. As food supplies become scarce, the fraction assimilated increases. Assimilation rates in rotifers are measured either directly through the use of radio-tracers or indirectly. In the latter case, the sum of the energy used in respiration and that which appears as production is considered to be assimilation. The results of indirect calculations have been compared with values resulting from the use of tracers and been found to be in substantial agreement with them [Doohan, 1973; Pilarska, 1977].

An inverse relationship between the density of food and the assimilation efficiency in *Brachionus rubens* ranges from 60% when food concentration (*Chlorella vulgaris*) is low to 15% at high food densities [Pilarska, 1977]. *Brachionus calyciflorus* shows a similar response. Its assimilation efficiency, calories assimilated per calories consumed, when feeding on *Kirchneriella lunaris* at a concentration of 5×10^6 cell ml^{-1} is 20%. The efficiency increases as food concentration decreases to a value of 49% at a food concentration of 4×10^5 cells ml^{-1} [Liemeroth, 1980]. Galkovskaya [1963] found a similar relationship in this species.

By contrast, Borass [1983] studied *B. calyciflorus* feeding on *Chlorella pyrenoidosa* and concluded that assimilation did not decrease with increasing food concentration. His data do show, however, that ingestion rate does not decline in parallel with yield, which suggests that some process in the system is declining in efficiency as food density increases. Doohan [1973] concluded that the variability in assimilation efficiency in *B. plicatilis* feeding on different densities of *Dunaliella salina* was so high that one could not support the inverse efficiency hypothesis with these two species. She does point out, however, that the densities of food were at the high end of the scale for *Brachionus*, so that the degree of variation in food density may not have been great enough to show a response. This suggestion is supported somewhat by the low efficiency she reports, 19.4%.

Decrease in assimilation efficiency is apparently the result of rapid passage of food through the gut and a resulting failure in the digestion of a proportion of the food items. Both King [1967] in his studies on *Euchlanis dilatata* and Doohan [1973] in her investigations on *B. plicatilis* observed undigested cells in the feces of rotifers feeding at high food concentrations.

Droop & Scott [1978] concluded from their studies on *B. plicatilis* that assimilation efficiency did not show any obvious dependence on population growth rate.

Assimilation efficiency varies with age in *B. rubens* [Pilarska, 1977]. Young individuals are less efficient than mature ones, especially females with eggs. Overall, young individuals are about 70 to 80% as efficient as females with eggs.

There is evidence that the amount of assimilation of ingested food varies from one clone to another. Snell [1979] recorded the rates of increase of two strains of *A. brightwelli* when individuals of each clone were allowed a fixed number of *Paramecium tetraurelia* per day. The ration at which the rate of increase was zero is a valid comparison point between the two clones. For one clone (RBH) this point was at a ration of about 67 *Paramecium* per rotifer per 24 h. For the other clone (9Al), zero rate of increase was at a ration of 44 Paramecium per rotifer per 24 h. These figures suggest that the second clone more efficiently digested the prey than the first, assuming that the body sizes of individuals in both clones were the same.

Using this same argument, the data of Stemberger & Gilbert [1984a] can be used to infer relative assimilation efficiencies in species of *Asplanchna.* The ration to maintain population growth at zero, the threshold ration, varies with body size for species in this genus. Calculations of the threshold ration per μg body weight normalizes for body size. On this latter basis, the most efficient form is the campanulate morph of *A. silvestrii.* The other species investigated, in declining order of efficiency, were: *A. girodi,* saccate *A. silvestrii, A. priodonta, A. brightwelli.*

The digestive tract of a rotifer is a finite space. No new food items can be ingested until there is room for them. Therefore, the time that food items remain in the digestive tract affects the feeding rate and the assimilation efficiency. Further, different foods have different residence times in the digestive tracts of a given species. Starkweather & Gilbert [1977], using radio-tracer techniques, determined that the gut passage time for *E. gracilis* eaten by *B. calyciflorus* is 15 minutes. Wallace and Starkweather [1983] found the gut passage time of the yeast *Rhodotorula glutinis* eaten by either *Ptygura cristallina* or *P. pilula* to be greater than 30 minutes. When *A. priodonta* feeds on *Ceratium hirundinella,* the mean time the food items remain in the stomach is about 14 minutes [Ejsmont-Karabin, 1974]. By contrast, Sawada & Enesco [1984] found that *A. brightwelli* required 12 hours to completely digest and assimilate a stomach full of *Paramecium caudatum.*

Feeding and clearance rates

In many studies, no attention is paid to the steps between the availability of food items and their appearance in the stomach of the rotifer. The number of food items ingested per unit time is compared to the density of the food items in the surrounding medium and the results are presented as the feeding rate. Frequently the converse of feeding rate is calculated, i.e., the clearance rate – the volume of water which, at the prevailing food density, would contain the number of food items consumed per unit time.

Most animals respond to an increase in food density by increasing the rate of food intake asymptotically until a food density is reached beyond which no increase in feeding rate takes place. The model most commonly fitted to these data is Holling's [1959] 'disc equation', so called because it was derived from observations of a blind-folded technician searching for and picking up discs of sandpaper. The numerator contains two terms, food density and the rate of successful search, which includes all the components considered above up to and including the attack. The denominator contains both those terms plus handling time, which includes the capture, rejection, ingestion, and egestion times. The value of the handling time sets the position of the asymptote or the density of food items at which the plateau in feeding rate occurs.

With two notable exceptions, the model appears to fit the feeding behavior of rotifers fairly well. Some examples are: *B. plicatilis* [Chotiyaputta & Hirama, 1978], *B. calyciflorus* [Erman, 1962; Starkweather, 1981; Starkweather & Gilbert, 1977] and *B. rubens* [Pilarska, 1977]. Despite many studies made on feeding rates in rotifers, it is difficult to find clear cut tests of the fit of the model because investigators frequently do not include measurements of feeding rate over a wide enough range of food densities.

Halbach & Halbach-Keup [1974] found that *B. calyciflorus* feeding on *Chlorella pyrenoidosa* followed the expected pattern of a hyperbolic increase in ingestion with increasing density of food, up to a point. As food density increased beyond that point, ingestion declined. This kind of performance can only be considered maladaptive, and it is difficult to see how it could evolve. The explanation for the phenomenon must lie in the performance of the food organism, as suggested by the authors. Starkweather [1980] considered a number of other explanations for this behavior, some of which would require a failure in the performance of the regulatory mechanisms associated with feeding.

Another lack of fit with the model is displayed by *B. calyciflorus* feeding on small food items – *E. aerogenes* or *R. glutinis* [Starkweather & Gilbert, 1977; Starkweather *et al.,* 1979]. In both these circumstances, no critical food concentration was recorded. That is, no plateau in ingestion occurred even at high food densities. The only explanation for this lack of response consistent with the ideas presented here, is that the handling time for these small particles is very short and its role in regulating the feeding rate was not involved.

One can use morphological information, swimming speed, and clearance rate to derive an estimate of the efficiency of all facets of the feeding behavior combined. For example, Doohan [1973] gives the clearance rate of *B. calyciflorus* as 1 μl individual^{-1} h^{-1} when the food concentration is greater than 0.5×10^6 cell ml^{-1}. An individual *Brachionus* with a corona diameter of 0.19 mm, swimming at a speed of 0.75 mm s^{-1}, will search a volume of 77 μl h^{-1}. Consequently, one can conclude that the rotifer ingested only one out of every 77 food items it encountered, a collection efficiency of 1.3%.

The role of hunger

The phenomenon of hunger is not well understood in higher vertebrates, much less in rotifers. Nonetheless, it plays a role in regulating feeding. It appears to determine the attack rate when food is abundant. Two direct series of observation by C. I. Haydock (pers. comm.) illustrate this response. The first records the cumulative consumption of *P. tetraurelia* by *A. sieboldi* and *A. brightwelli* through time. (Figs. 1, 2). It will be noted that in the presence of abundant food, *Paramecium* were consumed at a regular rate which increased as the *Asplanchna* increased in age and hence size. Given the density of prey, the rate of consumption was far lower than the encounter rate. Consequently, one can infer that only at periodic intervals did the *Asplanchna* respond to contact with the *Paramecium*, presumably when its hunger level had built up to a critical level.

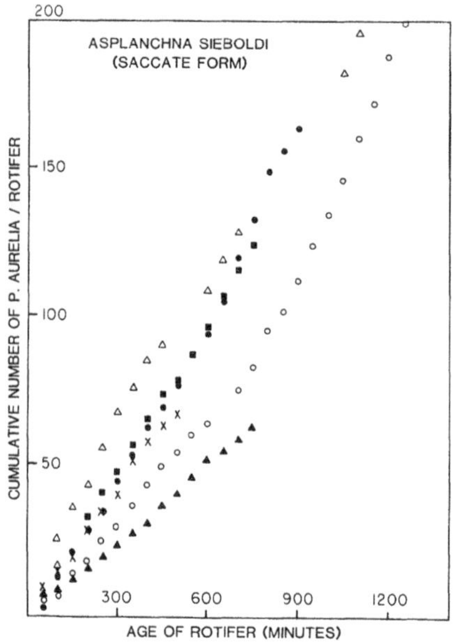

Fig. 1. Cumulative consumption of *Paramecium tetraurelia* from time of birth by six different females of the saccate form of *Asplanchna sieboldi.*

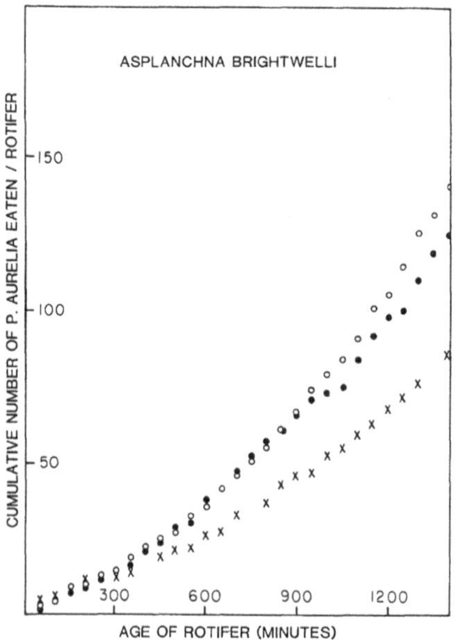

Fig. 2. Cumulative consumption of *Paramecium tetraurelia* from time of birth by three females of *Asplanchna brightwelli.*

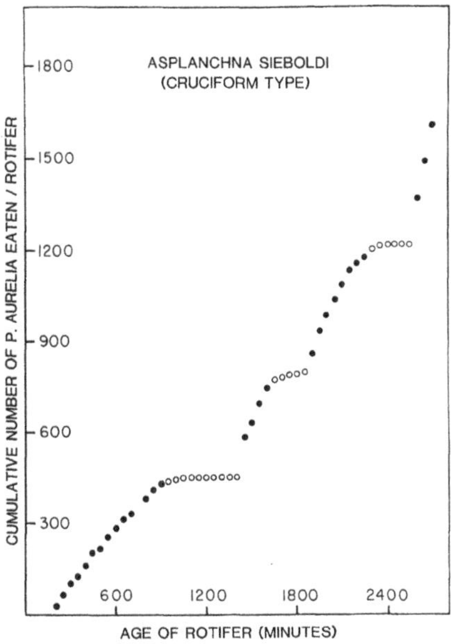

Fig. 3. Cumulative consumption of *Paramecium tetraurelia* by a female *Asplanchna sieboldi* with intermittent periods of starvation. Period without food are shown by open circles.

This interpretation is strengthened by a second series of observations on the same system. Following a period of starvation, the rate at which *Asplanchna* consumed *Paramecium* increased dramatically until it had returned to the previous well fed state, whereupon it resumed its original rate of food consumption (Fig. 3). Despite this evidence, there are no data to demonstrate that varying degrees of hunger produce proportional changes in swimming rate (Salt pers. obs.) or in any other facet of feeding behavior. Until more studies are completed, we shall have to conclude provisionally that hunger acts in an all or nothing fashion to turn feeding behavior on and off. Further, although we can be fairly sure that hunger plays a role in feeding, we have no information of how this condition is sensed by the rotifer nor how it is integrated by the neural system.

Conclusion

One can conclude from this review that, to a large degree, the feeding rate is set by conditions external to the rotifer. Temperature (May 1980), prey density, chemical composition of the water and other factors all have a strong influence on the rate of capture of food items. Among components of capture rate, only swimming and attack rate are potentially variable by the rotifer. Swimming rate certainly varies, but it does not seem to change in response to success or lack of it in capturing food. The rate of attack does change, at least in *Asplanchna.* It does not seem to in *Brachionus* where the screening of the mouth by the pseudotrochal cirri performs that function.

Within the handling process, there is little direct evidence that any of the components are varied by the rotifer in response to feeding success. Assimilation rate varies with ingestion rate, but as this response is an inverse one, it does little more than compensate for the lack of control in the ingestion rate.

In general, one can conclude that rotifers have relatively little control over their feeding rate to use to counteract variations in food availability. Maybe they have evolved the ability to manipulate their reproductive activities with variations in the food supply to compensate for their lack of regulatory ability in their feeding [King, 1967; Pourriot, 1983; Pourriot & Rougier, 1979; Robertson & Salt, 1981; Walz, 1983].

There are also hints throughout the literature that the processing of very small particles is not the same as that for large ones. Perhaps the trophi play no part in the handling of small particles. Maybe there is reduced control over the ingestion rate of small particles. Only further investigation aimed directly at this problem will answer these questions.

References

Borass, M. E., 1983. Population dynamics of food-limited rotifers in two-stage chemostat culture. Limnol. Oceanogr. 28: 546–563.

Chotiyaputta, C. & K. Hirama, 1978. Food selectivity of the rotifer Brachionus plicatilis feeding on phytoplankton. Mar. Biol. 45: 105–111.

Clément, P., 1977. Phototaxis in rotifers (action spectra). Arch. Hydrobiol. Beih. 8: 47–49.

Clément, P., E. Wurdak & J. Amsellem, 1983. Behavior and ultrastructure of sensory organs in rotifers. Hydrobiologia 104: 89–130.

Cornillac, A. M., 1982. Yeux cérébraux et réponses motrices à la lumière chez Brachionus calyciflorus et Asplanchna brightwelli (Rotifères). These Doct. spec. 1212, Univ. Lyon 1, 80 p.

Coulon, P. Y., J. P. Charras, J. L. Chassé, P. Clément, A. Cornillac, A. Luciani & E. Wurdak, 1983. An experimental system for the automatic tracking and analysis of rotifer swimming behavior. Hydrobiologia 104: 197–202.

Cox, P. A., 1983. Search theory, random motion, and the convergent evolution of pollen and spore morphology in aquatic plants. Amer. Nat. 121: 9–31.

De Beauchamp, P., 1951. Sur la variabilité spécific dans le genre Asplanchna (Rotifères). Bull. biol. Fr. Belg. 85: 137–175.

Doohan, M., 1973. An energy budget for adult Brachionus plicatilis Müller (Rotatoria). Oecologia 13: 351–362.

Droop, M. R. & J. M. Scott, 1978. Steady-state energetics of a planktonic herbivore. J. mar. biol. Ass. U.K. 58: 749–772.

Dumont, H. J., 1972. A competition-based approach to the reverse vertical migration in zooplankton and its implications, chiefly based on a study of the interactions of the rotifer Asplanchna priodonta (Gosse) with several crustacea entomostraca. Int. Revue. ges. Hydrobiol. 57: 1–38.

Dumont, H. J., 1977. Biotic factors in the population dynamics of rotifers. Arch. Hydrobiol. Beih. 8: 98–122.

Ejsmont-Karabin, J., 1974. Studies on the feeding of planktonic polyphage Asplanchna priodonta Gosse (Rotatoria). Ekol. pol. 22: 311–317.

Epp, R. W. & W. M. Lewis, Jr., 1984. Cost and speed of locomotion for rotifers. Oecologia 61: 289–292.

Erman, L. A., 1962. On the quantitative aspect of feeding and selectivity of food in the plankton rotifer Brachionus calyciflorus Pall. Zool. Zh. Moskva 41: 34–48. (In Russian with English summary.)

Galkovskaya, G. A., 1963. On the utilization of food for growth and conditions of the highest production yield in rotatoria, Brachionus calyciflorus Pallas. Zool. Zh. 42: 506–512. (In Russian with English summary.)

Gerritsen, J. & J. R. Strickler, 1977. Encounter probabilities and community structure in zooplankton: a mathematical model. J. Fish Res. Bd Can. 34: 73–82.

Gilbert, J. J., 1963. Contact chemoreception, mating behavior, and sexual isolation in the rotifer genus Brachionus. J. exp. Biol. 40: 625–641.

Gilbert, J. J., 1976. Selective cannibalism in the rotifer Asplanchna sieboldi: Contact recognition of morphotype and clone. Proc. natn. Acad. Sci. U.S.A. 73: 3233–3237.

Gilbert, J. J., 1977. Control of feeding behavior and selective cannibalism in the rotifer Asplanchna. Freshwat. Biol. 7: 337–341.

Gilbert, J. J., 1980a. Feeding in the rotifer Asplanchna: Behavior, cannibalism, selectivity, prey defenses, and impact on rotifer communities. In W. C. Kerfoot (ed.), Evolution and Ecology of Zooplankton Communities. The University Press of New England, Hanover, N.H.: 158–172.

Gilbert, J. J., 1980b. Observations on the susceptibility of some protists and rotifers to predation by Asplanchna girodi. Hydrobiologia 73: 87–91.

Gilbert, J. J. & K. G. Bogdan, 1981. Selectivity of Polyarthra and Keratella for flagellate and aflagellate cells. Verh. int. Ver. Limnol. 21: 1515–1521.

Gilbert, J. J. & P. L. Starkweather, 1977. Feeding in the rotifer Brachionus calyciflorus. I. Regulatory mechanisms. Oecologia 28: 125–131.

Gilbert, J. J. & P. L. Starkweather, 1978. Feeding in the rotifer Brachionus calyciflorus. III. Direct observations on the effects of food type, food density, change in food type, and starvation on the incidence of pseudotrochal screening. Verh. int. Ver. Limnol. 20: 2382–2388.

Gilbert, J. J. & R. S. Stemberger, 1985a. Prey capture in the rotifer Asplanchna girodi. Verh. int. Ver. Limnol. 22: 2997–3000.

Gilbert, J. J. & R. S. Stemberger, 1985b. Control of Keratella populations by interference competition from Daphnia. Limnol. Oceanogr. 30: 180–188.

Gilbert, J. J. & C. E. Williamson, 1978. Predator-prey behavior and its effect on rotifer survival in associations of Mesocyclops edax, Asplanchna girodi, Polyarthra vulgaris, and Keratella cochlearis. Oecologia 37: 13–22.

Gosse, P. H., 1856. On the structure, functions, and homologies of the manducatory organs in the class Rotifera. Philos. Trans. Linn. Soc. pp. 419–452.

Halbach, U. & G. Halbach-Keup, 1974. Quantitative Beziehungen zwischen Phytoplankton und der Populationsdynamik des Rotators Brachionus calyciflorus Pallas. Befunde aus Laboratoriumsexperimenten und Freilanduntersuchungen. Arch. Hydrobiol. 73: 272–309.

Hino, A. & R. Hirano, 1980. Relationship between body size of the rotifer Brachionus plicatilis and the maximum size of particle ingested. Bull. Jap. Soc. Sci. Fish. 46: 1217–1222.

Holling, C. S., 1959. Some characteristics of simple types of predation and parasitism. Can. Ent. 91: 385–398.

King, C. E., 1967. Food, age, and the dynamics of a laboratory population of rotifers. Ecology 48: 111–128.

Liemeroth, N., 1980. Respiration of different stages and energy budgets of juvenile Brachionus calyciflorus. Hydrobiologia 73: 195–197.

Luciani, A., J-L. Chassé & P Clément, 1983. Aging in Brachionus plicatilis: The evolution of swimming as a function of age at two different calcium concentrations. Hydrobiologia 104: 141–146.

May, L., 1980. Studies on the grazing rate of Notholca squamula Müller on Asterionella formosa Hass. at different temperatures. Hydrobiologia 73: 79–81.

Pilarska, J., 1977. Eco-physiological studies on Brachionus rubens Ehrbg. (Rotatoria). I. Food selectivity and feeding rate. II. Production and respiration. III. Energy balances. Pol. Arch. Hydrobiol. 24: 319–354.

Pourriot, R., 1965. Recherches sur l'écologie des rotifères. Vie et Milieu, Suppl. 21: 1–224.

Pourriot, R., 1977. Food and feeding habits of rotifers. Arch. Hydrobiol. Beih. 8: 243–260.

Pourriot, R., 1983. Stratégiès de reproduction chez les Rotifères. C.r. Acad. Sci., Paris, Ser. III. 296: 1109–1111.

Pourriot, R. & C. Rougier, 1979. Influences conjuguées du groupement et de la qualité de la nourriture sur la reproduction de Brachionus plicatilis O. F. Müller (Rotifère). Neth. J. Zool. 29: 242–264.

Robertson, J. R. & G. W. Salt, 1981. Responses in growth, mortality, and reproduction to variable food levels by the rotifer Asplanchna girodi. Ecology 62: 1585–1596.

Sawada, M. & H. E. Enesco, 1984. A study of dietary restriction and life span in the rotifer Asplanchna brightwelli monitored by chronic neutral red exposure. Exp. Geront. 19: 329–334.

Snell, T. W., 1979. Intraspecific competition and population structure in rotifers. Ecology 60: 494–502.

Stanley, J., 1932. A mathematical theory of the growth of populations of the flour beetle, Tribolium confusum Duv. Can. J. Res. 6: 632–671.

Starkweather, P. L., 1980. Aspects of the feeding behavior and trophic ecology of suspension feeding rotifers. Hydrobiologia 73: 63–72.

Starkweather, P. L., 1981. Trophic relationships between the rotifer Brachionus calyciflorus and the blue-green alga Anabaena flos-aqua. Verh. int. Ver. Limnol. 21: 1507–1514.

Starkweather, P. L. & J. J. Gilbert, 1977. Feeding in the rotifer Brachionus calyciflorus. II. Effect of food density on feeding rates using Euglena gracilis and Rhodotorula glutinus. Oecologia 28: 133–139.

Starkweather, P. L., J. J. Gilbert & T. M. Frost, 1979. Bacterial feeding by the rotifer Brachionus calyciflorus: clearance and ingestion rates, behavior and population dynamics. Oecologia 44: 26–30.

Stemberger, R. S. & J. J. Gilbert, 1984a. Body size, ration level, and population growth in Asplanchna. Oecologia 64: 355–359.

Stemberger, R. S. & J. J. Gilbert, 1984b. Spine development in the rotifer Keratella cochlearis: Induction by cyclopoid copepods and Asplanchna priodonta. Freshwat. Biol. 14: 639–648.

Wallace, R. L., 1975. Larval behavior of the sessile rotifer Ptygura beauchampi (Edmondson). Verh. int. Ver. Limnol. 19: 2811–2815.

Wallace, R. L., 1980. Ecology of sessile rotifers. Hydrobiologia 73: 181–193.

Wallace, R. L. & P. L. Starkweather, 1983. Clearance rates of sessile rotifers: In situ determinations. Hydrobiologia 104: 379–383.

Walz, N., 1983. Continuous culture of the pelagic rotifers Keratella cochlearis and Brachionus angularis. Arch. Hydrobiol. 98: 70–92.

Williamson, C. E. & J. J. Gilbert, 1980. Variation among zooplankton predators: The potential of Asplanchna, Mesocyclops, and Cyclops to attack, capture, and eat various rotifer prey. In W. C. Kerfoot (ed.), Evolution and Ecology of Zooplankton Communities. The University Press of New England, Hanover, N.H.: 509–517.

Wurdak, E., 1986. Ultrastructure and histochemistry of the stomach of Asplanchna sieboldi. Hydrobiologia (this volume).

Wurdak, E., P. Clément & J. Amsellem, 1983. Sensory receptors involved in the feeding behavior of the rotifer Asplanchna brightwelli. Hydrobiologia 104: 203–212.

Hydrobiologia 147: 283–288 (1987)
May, L., Wallace, R. & Herzig, A. (eds), Rotifer Symposium IV

Effect of feeding on the respiration rate of the rotifer *Brachionus plicatilis*

H. Hirata & S. Yamasaki
Faculty of Fisheries, Kagoshima University, Kagoshima, Japan

Keywords: Rotifera, oxygen consumption, feeding regime, *Brachionus plicatilis*

Abstract

The effect of feeding on the respiration rate of *Brachionus plicatilis* was studied. Oxygen consumption was determined under two feeding regimes, duplicate feeding and constant feeding. Oxygen consumption rate increased during feeding. The oxygen consumption profile is discussed in relation to the following processes:
1. filtration, mastication, and locomotion during feeding
2. specific dynamic action (SDA)
3. egg formation and routine metabolism.

Introduction

The proper management of water quality is essential in maintaining a homeostatic condition within any culture system. Dissolved oxygen content is one of the factors determining this water quality. In mass cultures of the rotifer *Brachionus plicatilis* Müller dissolved oxygen content has been observed to fluctuate from near saturation to less than 1 mg l^{-1} during feeding [Imada, 1982]. This suggests that oxygen balance within the culture system is easily disturbed during a feeding period through the increased oxygen consumption of the rotifers.

Rotifer oxygen demand includes that involved in feeding, locomotion, and specific dynamic action (SDA), as has been observed in fish [Brett & Groves, 1979; Jobling & Davies, 1980; Machida, 1981]. However, the oxygen consumption of *B. plicatilis* has, so far, only been determined in the absence of food [Ito, 1971; Doohan, 1973; Epp & Winston, 1978; Epp & Lewis, 1980; Yamasaki & Hirata, 1983]. This paper examines the relationship between oxygen consumption and food availability in *B. plicatilis* in order to categorise the physiological processes contributing to the oxygen demand of this species.

Materials and methods

The strains of *Brachionus plicatilis* and its food alga *Chlorella saccharophila* Kruger used in this study originated from the Yashima Station of the Seto Inland Sea Fish Farming Association [SISFFA, 1964]. Rotifers were grown under continuous culture conditions using two different feeding regimes, duplicate feeding and constant feeding. The duplicate feeding method was modified from Hirata *et al.* [1983], where food was provided twice daily at high levels. In the constant feeding designs, food was provided at thirty minute intervals to maintain a constant food level of 3×10^6 cells ml^{-1}.

In the duplicate feeding method, continuous respiration measurements were made at 4 day intervals over an experimental period of 50 days, using the apparatus described below. In the constant feeding design, rotifers were acclimated to a constant food level for 4 days, before respiration was measured for a continuous 6 hour period. The *Chlorella* cells were then removed by filtration and the respiration rates of the rotifers alone were monitored for a further 20 hours.

The equipment used for measuring rotifer oxygen consumption is shown schematically in Fig. 1.

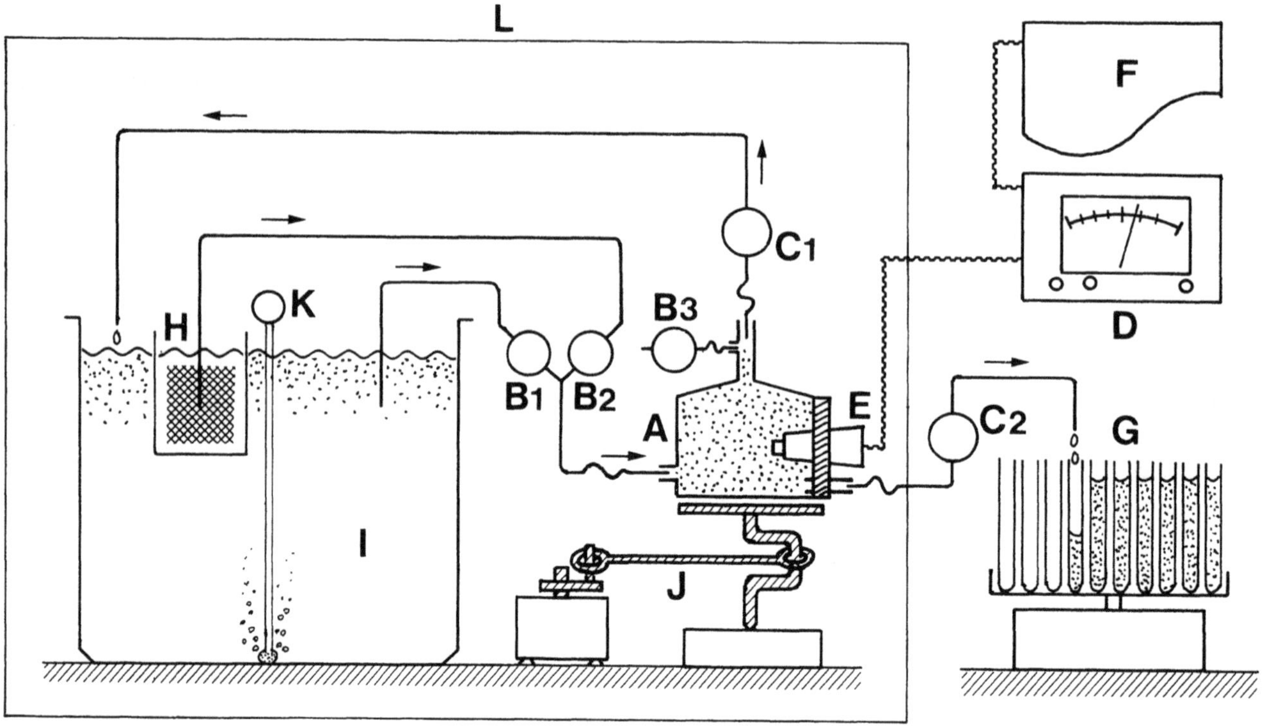

Fig. 1. Schematic diagram of the apparatus for the continuous minitoring of rotifer oxygen consumption.

A respiration chamber
B valve
C peristaltic pump
D D.O. meter
E D.O. probe
F chart recorder
G fraction collector
H plankton net filter
I culture tank
J semi rotary shaker
K air
L incubater

This is a modification of the semi-rotary shaking and intermittent flow-through method of Yamasaki & Hirata [1982]. Rotifers and medium were introduced into the respiration chamber (A) through valve (B1) by a peristaltic pump (C1). The water exchange rate of the chamber (capacity 30 ml) was twice per minute. Dissolved oxygen (D.O.) level within the chamber was measured with a YSI 57 D.O. meter (D), and a chart recorder (F).

Upon termination of the D.O. recording, a water sample was taken directly from the respiration chamber with a fraction collector (G) and peristaltic pump (C2). Rotifer and algal food densities were determined from this sample. The medium was then filtered through a 60 μm mesh plankton net (H) and introduced into the chamber through valve (B2) by pump (C1). Changes in D.O. of this medium alone was measured as a control.

Each series of measurements (input of water sample, measurement of D.O., and water sampling) was repeated automatically by incorporation of three timer switches. The time interval between introduction of water into the chamber and water sampling was adjusted to 2 minutes in both feeding designs. Duration of each measurement of D.O. were 56 minutes in the duplicate feeding design, and 26 minutes in the constant feeding design.

The experiment was conducted at 20‰ salinity in an incubator at 20 °C. Illumination was provided by fluorescent lamps. The light intensity within the respiratory chamber was controlled to about 30 lux, with a photoperiod of L:D = 15:9.

The oxygen consumption rate of *B. plicatilis* was calculated as follows:

oxygen consumption rate $= (a-b)/c$ nl ind^{-1}h^{-1}

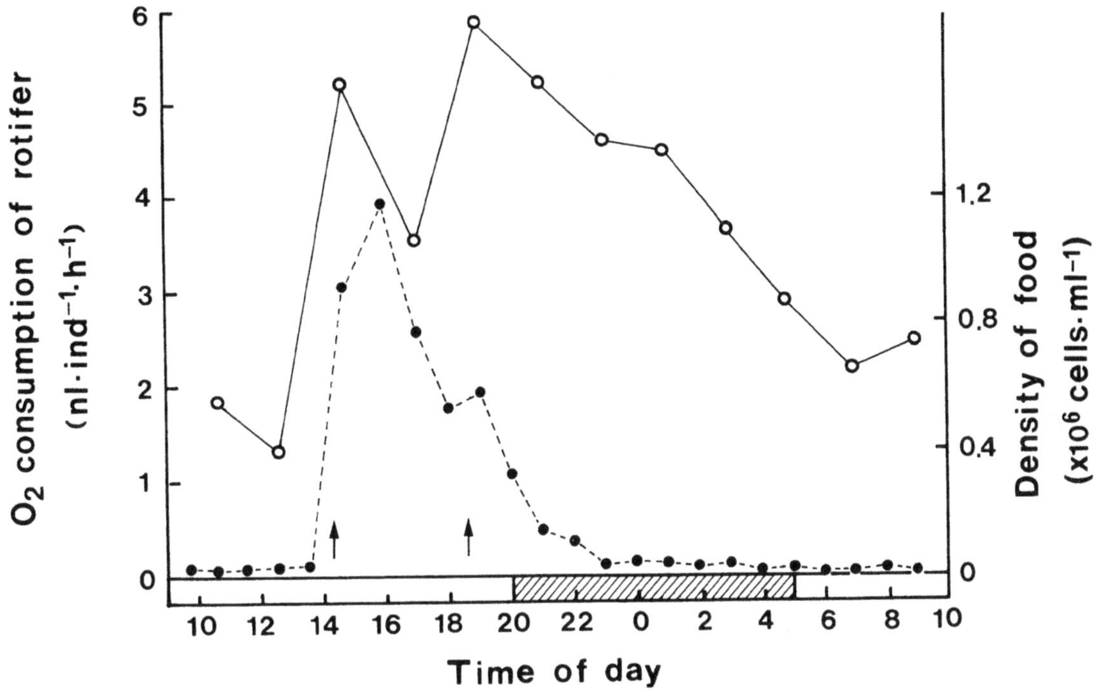

Fig. 2. A typical 24 h profile of rotifer oxygen consumption (o – o) and food density (• -- •) from the duplicate feeding experiments. Arrows indicate the addition of food. Shaded area shows the period of darkness.

where: a = decrease in D.O. per hour in the rotifer culture medium, b = decrease in D.O. per hour in culture medium without rotifers, c = density of rotifers.

Feeding ratio was calculated by dividing the food density 2 hours after feeding with that 2 hours before feeding, and converting to logarithms. Oxygen consumption ratio was calculated similarly, but without logarithmic transformation.

Results and Discussion

The oxygen consumption rate of rotifers in the duplicate feeding trials ranged from 1 to 7 nl $ind^{-1}h^{-1}$, and average 3.1 + 1.3 nl $ind^{-1}h^{-1}$ (n = 137). Figure 2 shows a typical 24 hour oxygen consumption profile for these experiments. During the first feeding, oxygen consumption increased from 1.3 to 5.9 nl $ind^{-1}h^{-1}$, about 4 times the rate before feeding. However, during the second feeding the oxygen consumption rate increased only 1.6 times, although it rose to 5.9 nl $ind^{-1}h^{-1}$. This was due to the already high food densities and oxygen consumption rates before feeding.

Figure 3 shows the relationship between oxygen consumption ratio and food density ratios in these

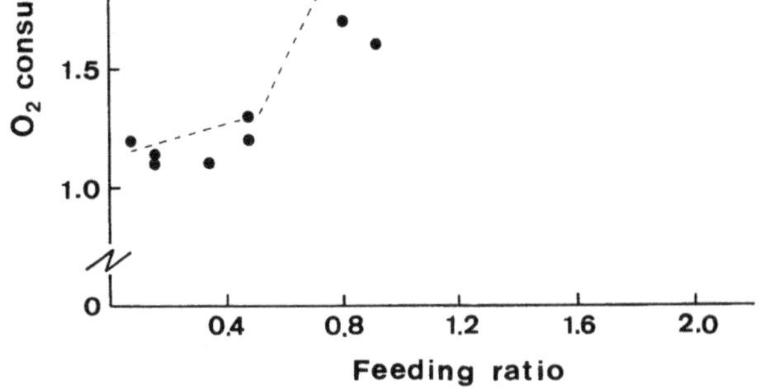

Fig. 3. Relationship between feeding ratio and oxygen consumption ratio in *Brachionus plicatilis.*

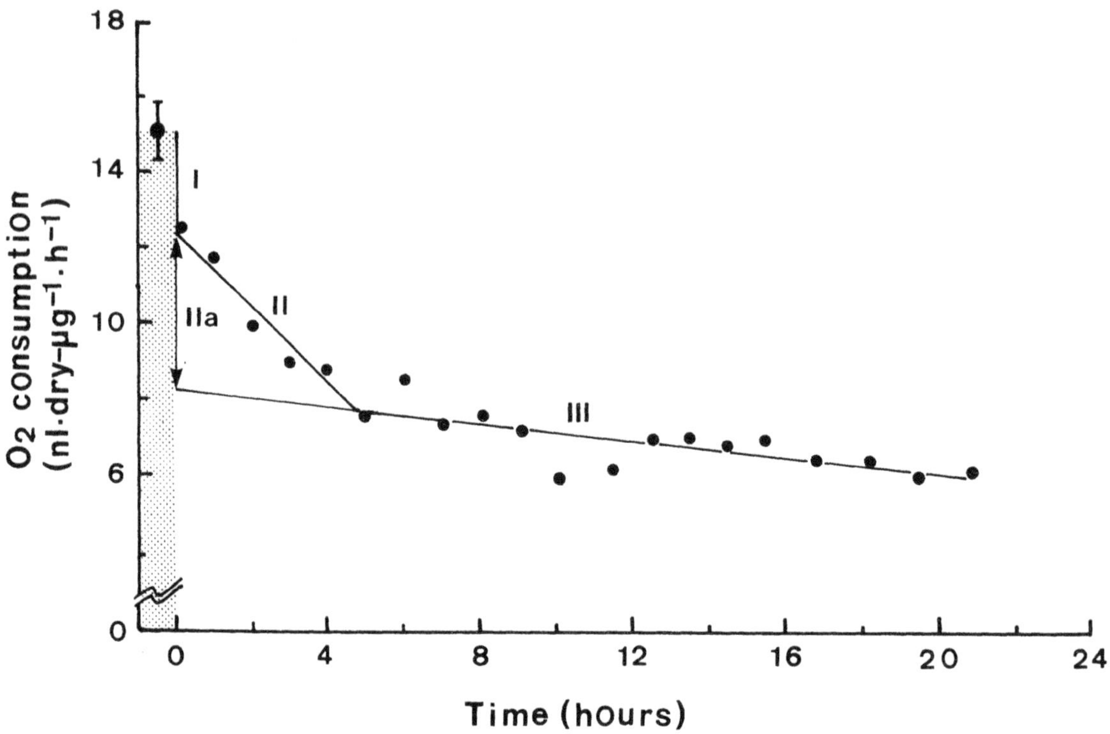

Fig. 4. Oxygen profile for *Brachionus plicatilis* from the constant feeding experiments. The shaded area shows the total oxygen consumption rate at a food density of 3×10^6 cells ml^{-1}. The unshaded area shows the three phases of decline in oxygen demand recorded after food removal.

trials. High oxygen consumption ratios of about 2.5 were observed at feeding ratios above 1.0. In contrast, much lower oxygen consumption ratios of about 1.2 were recorded at feeding ratios below 0.4.

In the constant feeding experiments, the oxygen consumption of rotifer from three trials at a food density of 3×10^6 cells ml^{-1} were 15.3 + 1.3, 14.1 + 1.0, and 15.1 + 0.8 nl μg dry $wt^{-1}h^{-1}$, respectively. Figure 4 shows a typical oxygen consumption profile for one of these experiments. Immediately following food removal, the oxygen demand decreased by about 20% (see Fig. 4, phase I). It then decreased more gradually (0.87 nl μg dry $wt^{-1}h^{-2}$) over the next 4 hours (see Fig. 4, phase II) and finally fell even more slowly at 0.17 nl μg dry $wt^{-1}h^{-2}$, in phase III (see Fig. 4).

Feeding movements in *B. plicatilis* have been observed to increase in the presence of food, and decrease when that food is taken away [Hirayama, 1983]. The initial decrease in oxygen consumption (phase I) following food removal, in the constant feeding experiments, probably corresponds to this decline in feeding activity. As such, it indicates the rate of oxygen uptake associated with the processes of filtration, mastication and locomotion during feeding.

The remaining oxygen consumption during food removal consists of that used in SDA, routine metabolism, and egg formation. SDA includes the processes of digestion and absorption of food, and assimilation and storage of materials in the body. This could last for about 4 hours, and probably corresponds to phase II of the oxygen consumption profile shown in Figure 4, moreover, in phase II and III, a constant decrease in the number of eggs carried by the rotifers was observed during the food removal period. It is assumed that total activity including egg production decreases during the period by consumption of preserved materials in the body for egg production and energy supply [Scott & Baynes, 1978]. Thus, the oxygen demand of SDA can be estimated as the decline of IIa. This is calcu-

Table 1. Oxygen consumption rate (nl μg dry wt^{-1} h^{-1}) of *Brachionus plicatilis* during feeding on *Chlorella saccharophila* at a density of 3×10^6 cells ml^{-1}.

	Trial 1	Trial 2	Trial 3	Mean
Filtration, mastication and locomotion during feeding	3.54 (23%)	3.04 (22%)	2.75 (18%)	3.11 (21%)
Specific dynamic action	2.29 (15%)	2.70 (19%)	4.11 (27%)	3.03 (20%)
Egg formation and routine metabolism	9.45 (62%)	8.36 (59%)	8.21 (55%)	8.67 (59%)
Total	15.28 (100%)	14.10 (100%)	15.07 (100%)	14.82 (100%)

lated at 3.03 nl μg dry $wt^{-1}h^{-1}$, or 20% of the total oxygen demand at a food density of 3×10^6 cells ml^{-1}.

Accordingly, oxygen consumption rate at a food density of 3×10^6 cells ml^{-1} could be categorized into three parts: 1) that used in the processes of filtration, mastication, and locomotion during feeding; 2) that used in SDA; 3) that used in egg formation and routine metabolism. The oxygen consumption rate for each of these categories is shown in Table 1 as 3.11, 3.03, and 8.67 nl μg dry $wt^{-1}h^{-1}$, respectively. The corresponding percentages of the total oxygen consumed were 21%, 20%, and 59%, respectively.

In the constant feeding experiments the elevation of oxygen consumption during feeding was calculated to be about 1.7 times. However, in practical mass cultures, which are fed only once or twice per day, the oxygen consumed by routine metabolism and egg formation may decrease to a very low value immediately before feeding. In these circumstances, it is possible for the oxygen consumption rate to more than double when the cultures are fed, as observed in the duplicate feeding trials.

In conclusion, the oxygen consumption rate of this rotifer increased during the feeding period. This increase was categorised into three parts. The first part is involved in the processes of filtration, mastication, and locomotion during feeding; the second part is used in SDA; the third part is used in egg formation and routine metabolism.

Acknowledgements

The authors gratefully thank Dr. Linda May and Dr. W. D. Nagata for critically reading the manuscript.

References

Brett, J. R. & T. D. D. Groves, 1979. Physiological energetics. In: W. S., Hoar, D. J. Randall, J. R. Brett (eds), Fish Physiology Vol. VIII, Academic Press, New York; San Francisco; London: 279–352.

Doohan, M., 1973. An energy budget for adult Brachionus plicatilis Müller (Rotatoria), Oecologia, 13: 351–362.

Epp, P. W. & W. M. Lewis, Jr., 1980. Metabolic uniformity over the environmental temperature range in Brachionus plicatilis (Rotifera). Hydrobiologia. 73: 145–147.

Epp, P. W. & P. W. Winston, 1978. The effect of salinity and pH on the activity and oxygen consumption of Brachionus plicatilis (Rotatoria). Comp. Biochem. Physiol. 59: 9–12.

Hirata, H., S. Yamasaki, T. Kawaguchi & M. Ogawa, 1983. Continuous culture of the rotifer Brachionus plicatilis fed recycled algal diets. Hydrobiologia. 104: 71–75.

Hirayama, K., 1982. Physiology of population growth. In: The rotifer Brachionus plicatilis, Biology and Mass Culture. Japan. Soc. Sci. Fish. Series 44: 52–68. (in Japanese)

Imada, O., 1982. Feeding and environmental condition. In: The rotifer Brachionus plicatilis, Biology and Mass Culture. Japan. Soc. Sci. Fish. Series 44: 129–155. (in Japanese)

Ito, T., 1971. Studies on the artificial seed production of Ayu fish – LXXX. Oxygen consumption rate of rotifer and its lethal level. Artificial Propagation of Ayu Fish, Mie Univ., (1): 279–295. (in Japanese)

Jobling, M. & P. S. Davies, 1980. Effects of feeding on metabolic rate and the specific dynamic action in Plaice, Pleuronectes platessa L., J. Fish. Biol. 16: 629–638.

Machida, Y., 1981. Studies of specific dynamic action in some freshwater fishes. Rep. Usa Mar. Biol. Inst., Kouchi Univ. 3: 1–50.

Murtaugh, R. A., 1985. The influence of food concentration and feeding rate on the gut residence time of Daphnia. J. Plankton Res. 7: 415–420.

Scott, A. P. & S. M. Baynes, 1978. Effect of algal diet and tem-

perature on the biochemical composition of the rotifer, Brachionus plicatilis. Aquaculture, 14: 247–260.
Seto Inland Sea Farming Fisheries Association (SISFFA), 1964. Cultivation of live food organisms at the Yashima Station. Newsletter of Saibai Gyogyo, 2–4: 4- (in Japanese)
Starkweather, P. L. & J. J. Gilbert, 1977. Radiotracer determination of feeding in Brachionus calyciflorus: The importance of gut passage times. Arch. Hydrobiol. Beih. 8: 261–263.
Yamasaki, S. & H. Hirata, 1982. An electrode respirometer for planktonic organisms. Mem. Fac. Fish., Kagoshima Univ. 31: 141–144.
Yamasaki, S. & H. Hirata, 1983. Test on the use of electrode respirometer for the rotifer, Brachionus plicatilis. Mem. Fac. Fish., Kogoshima Univ. 32: 91–96. (in Japanese)

Hydrobiologia 147: 289–296 (1987)
May, L., Wallace, R. & Herzig, A. (eds), Rotifer Symposium IV

A laboratory study of feeding and assimilation in *Euchlanis dilatata lucksiana*

R. D. Gulati*, J. Rooth* & J. Ejsmont-Karabin**

* *Limnological Institute, 'Vijverhof' Laboratory, Rijksstraatweg 6, 3631 AC Nieuwersluis, The Netherlands*

** *Polish Academy of Sciences, Institute of Ecology, Department of Hydrobiology, 05-150 Komianki, Dziekanów Lésny K. Warszawy, Poland*

Keywords: *Euchlanis,* Rotifera, feeding, ^{14}C-technique, clearance rates, food concentration, assimilation efficiency

Abstract

Feeding of *Euchlanis dilatata lucksiana,* a brachionid rotifer isolated from the Loosdrecht Lakes (The Netherlands), was examined in the laboratory using ^{14}C-labelled food. The gut-filling time at a food concentration of 9.6 μg C ml^{-1} was about 15 minutes. Animals which were fed on 3 size fractions of the lake seston (< 7 μm, 7–15 μm, and 15–33 μm) showed a clear preference (Jacobs' selectivity index) for the largest size fraction. This fraction was composed predominantly of filamentous cyanobacteria. For animals weighing 0.37–0.49 μg C ind.$^{-1}$ the daily ration (daily food consumption per unit body weight $\times$ 100) ranged from 50 to 100% at food levels of 2 mg C l^{-1} and below, but increased to 150–250% at food concentrations of 5 mg C l^{-1} and above. The assimilation efficiency was 100% up to 5 mg C l^{-1} of food, but decreased to about 80% at higher food levels.

Introduction

Recent studies on the zooplankton of the lakes in the Loosdrecht area of The Netherlands have shown a tenfold increase in rotifer densities since 1955 [Gulati, 1984]. This increase seems to have occurred in response to the eutrophication of these lakes [Gulati, 1983]. During their maximal abundance, rotifer biomass has been as much as five times higher than that of the microcrustaceans. However, in spite of their apparent importance in lake trophic dynamics [Stemberger, 1981], rotifers have not received adequate attention. This is due mainly to their small size, and the related difficulties of isolating, culturing and maintaining them under laboratory conditions, and determining their importance in lake food chains.

This paper reports a series of short-term experiments to determine the feeding and assimilation rates of *Euchlanis dilatata lucksiana,* using ^{14}C-labelled lake seston as food [Gulati *et al.*, 1982; Duncan & Gulati, 1983]. Several studies on rotifer grazing [Bogdan & Gilbert, 1984], particularly on *Brachionus* sp. [Starkweather, 1980], have been published recently. However, there has been no work on feeding of *Euchlanis* sp. since King's [1967] investigation. Therefore, our's the first study on feeding in *E. dilatata lucksiana.*

Materials and methods

Euchlanis *species*

The species studied was isolated from the Loosdrecht Lakes and identified as *E. dilatata lucksiana* (Hauer) (hereafter referred to as *Euchlanis*). This form is morphologically similar to *E. lucksiana* Hauer, reported by Carlin [1943] from the Motalaström in Sweden. Being planktonic, *E. dilatata lucksiana* differs in its habitat from *E. dilatata* which is commonly a member of the littoral fauna of lakes [King, 1967]. Because of the shallowness and almost continuous, wind-induced mixing of the Loosdrecht Lakes, the presence of a littoral rotifer in the pelagic region of these lakes would not be surprising. However, the morphological differences between the two forms [Carlin, 1943], and the virtually uninterrupted presence of *Euchlanis* in the pelagic region from May to late October indicates that our form is probably the limnetic form.

Laboratory cultures of Euchlanis

Euchlanis was cultured in the laboratory at 18–19 °C in lake water filtered with a 33 μm mesh filter. Three species of cyanobacteria dominated the lake water: *Aphanizomenon flos-aquae*, $3-6 \times 10^3$ filaments ml^{-1}; *Lyngbya*, $19-47 \times 10^3$ filaments ml^{-1}; and *Oscillatoria limnetica*, $17-117 \times 10^3$ filaments ml^{-1}. Green algae and diatoms were sparse, but nanoflagellates reached maximal densities of $3-12 \times 10^3$ organisms ml^{-1}. Biomass estimates computed from biovolumes indicated that cyanobacteria formed about 90% of the total sestonic mass (< 150 μm). Food concentrations in cultures and experimental vessels generally varied between 5 and 8 μg C ml^{-1}, these concentrations were similar to those in natural lake water.

Animals used in our feeding studies were removed from cultures by using a 60 μm or 125 μm sieve. Wet weights were calculated from body measurements using the geometrical formula of Ruttner-Kolisko [1977]. Dry weights were estimated from length and dry weight regression derived by Dumont *et al.* [1975]. Dry weight to wet weight ratio was 20–22%. This is similar to that reported for many crustaceans [Peters & Downing, 1984]. Dry weights were converted to carbon biomass by multiplying dry weight by a factor of 0.427, this being the mean carbon/dry weight ratio determined for *Euchlanis* by COD method [Gulati *et al.*, 1982].

Feeding experiments

Radioactively (^{14}C) labelled assemblage of lake food was manufactured in essentially the same way as reported in Gulati *et al.* [1982]. Using this food as tracer and laboratory cultured *Euchlanis*, three types of experiments were carried out: gut filling time, selective feeding and effect of food concentration on feeding and assimilation rates.

Gut filling time was investigated in two separate experiments conducted at a food concentration of 9.6 μg C ml^{-1}. The durations of feeding on ^{14}C-food in the first experiment were 7.5 and 15 minutes and that in the second 7.5, 15, 22.5 and 30 minutes (Fig. 1), with 4 to 6 observations for each of the feeding durations.

Food selection by *Euchlanis* was studied in 3 simultaneous experiments each with one of the 3 food sizes, namely < 7 μm, 7–15 μm or 15–33 μm, as tracer food. The total food concentration in all three experiments was 8.8 μg C ml^{-1}, i.e. the same natural assemblage of sestonic particles (< 33 μm) as in the lake water (Table 1). Jacobs' [1974] selectivity index: $D = (r - p)/(r + p - 2rp)$, in which p and r are the proportions of the three fractions in food offered and food ingested, respectively, was calculated and used as a measure of food size selection (Table 1).

Feeding and assimilation experiments were carried out for 15 and 60 minutes, respectively, in 1-litre vessels containing 500 ml lake water and about 700 individuals of *Euchlanis*. These animals had been acclimated to the experimental conditions for about 30 minutes, prior to the experiment. One hour after adding the tracer food, the animals for the assimilation rate experiment were transferred to non-active food for a further 45 minutes to allow them to evacuate their guts of undigested labelled food. The experiments were terminated by filtering the animals through a 60 μm mesh sieve, rinsing them thoroughly in non-active, filtered lake water and killing them in hot water (95 °C). The animals were then fixed in 4% formalin and stored until sorting and counting of radioactivity.

The amount of radioactivity retained by the animals was determined according to the procedures of Gulati *et al.* [1982]. Loss of radioactivity from the animals is related to the methods of killing, preservation, storage and handling [Holtby & Knoechel, 1981]. In our study, this loss of radioactivity in the preservative, i.e. on storage before sorting of the animals into scintillation vials, was measured and the results corrected accordingly.

In the first two of the three experiments on feeding and assimilation of *Euchlanis* (Fig. 2), gross clearance rate (CR), i.e. the volume of water swept clear of food particles by one animal, and net clearance rate (namely volume of water cleared of food particles which are assimilated) were measured. In the third experiment (Fig. 3) both gross CR and net CR were measured simultaneously using same series of food concentrations. Food consumption rates (f) and food assimilation rates (A) were calculated by multiplying the gross CR and net CR by the food concentrations used.

The daily ration and net daily ration are ratios (%) of f per day and animal's body weight and of A per day and animal's body weight, respectively. The assimilation efficiency is a ratio of f:A expressed as %.

The data on the rates of clearance, feeding and assimilation were regressed on the food concentrations employed. Several regression models were tested with the help of a Hewlett Packard calculator 9815A programmed to compute 'family regressions'. The power model, with regression of the form $Y = aX^b$ (transformed to $\ln Y = \ln a + b \ln X$ for computational purposes), gave the highest correlation coefficient r of all the models tested. This model is used here (Table 2) to analyse the effects of food concentration (X) on the different rates (Y) measured.

Results and discussion

Gut-filling time

In the feeding studies on rotifers using a radioisotope as tracer, a prior knowledge of the gut-filling time is important. Starkweather and Gilbert [1977] found this to be about 20 minutes in *Brachionus calyciflorus.* The literature cited by these authors shows that the gut passage time can be highly variable depending on species and external conditions, from 2 to 30 minutes. The clearance rates (CR) recorded in experiments on gut-filling time of *Euchlanis* measured by us are difficult to explain (Fig. 1), especially at the shorter (7.5 min) duration. Clearance rates at 7.5 min differed widely in the two experiments. Contamination of animals by labelled food, or elevated feeding behaviour, could have contributed to the extremely high CR found in the first experiment, but this is unlikely, because none of the replicates were very variable. Further, there is no evidence of hyperactivity in the feeding durations of 15 minutes or longer.

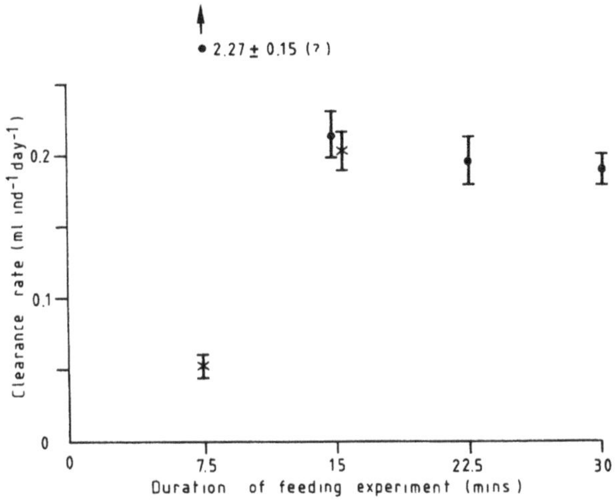

Fig. 1. Clearance rates of *Euchlanis dilitata lucksiana* (mean ± S.D.) measured over different time intervals in two separate experiments indicated by •: and x, respectively.

Interestingly, CRs are nearly constant in the experiments of 15 minutes or greater. This means either a lack of egestion of undigested tracer food – at least in the first 30 minutes – or a 100% assimilation efficiency. The latter appears broadly to be the case in *Euchlanis* (see further). Thus, errors, because of long feeding time in the experiments normally will be negligible. A feeding period of 15 minutes was adopted for subsequent experiments on food selection and feeding.

Selective feeding

Sestonic particles in the Loosdrecht Lakes are com-

Table 1. Food selection by *Euchlanis dilatata lucksiana* on different size fractions of a natural assemblage of seston from Loosdrecht Lakes; n = 3 in each case. For calculation procedure of Jacobs' selectivity index D see in 'materials and methods' (N.B. abbreviation 'r' has a different meaning from 'r' in Table 2).

Food		Proportion in food (p)	Clearance rate (ml ind^{-1} d^{-1})	Feeding rate (μg C ind^{-1})	Proportion ingested food (r)	Jacobs' selectivity index (D)
Size fraction (μm)	Concentration (μg C ml^{-1})					
15 – 33	0.77	0.087	0.121	0.093	0.233	+0.52
7 – 15	1.93	0.219	0.062	0.119	0.297	+0.20
<7	6.11	0.694	0.031	0.188	0.469	−0.44
>33	8.8	1.000	–	0.400	1.000	

prised essentially of filamentous cyanobacteria 175 – 250 μm in length, and 2 – 5 μm in diameter (Mrs. Boesewinkel-de Bruyn, pers. comm.). So, fractionation of sestonic food using sieves will be only partial. Only some species of centric diatoms, nanoflagellates and chlorococcales pass through screens of 7 μm, while the separation of filamentous forms into size classes, as we attempted here, may not be reproductible because it will be influenced by the extent of coiling and lumping of the filaments, and of their adherence to detritus.

In each of the three different CR experiments to determine food selection a clear preference for the 15 – 33 μm fraction was found (Table 1). Interestingly, the smallest size fraction (< 7 μm) which was the most abundant was the least preferred. Calculations of Jacobs' selectivity index D [Jacobs, 1974] confirm that *Euchlanis* selected the relatively coarser food particles. Thus, the selection of food by *Euchlanis* as well as by other rotifers [Starkweather, 1980] appears to be inversely proportional to the relative abundance of these food particles in their natural environment. Besides the preference of *Euchlanis* for the 15 – 33 μm food fraction, these results show that this species can exploit situations in which cyyanobacteria dominate. Dumont [1977] suggested that rotifers show a clear preference for nanoflagellates, especially species of *Cryptomonas, Rhodomonas,* and chrysomonads, and for small chlorococcales, detrital particles and bacteria. Myers [1930] found *Euchlanis* sp. to feed on diatoms, desmids, bacteria and organic detritus. However, from another earlier study [Carlin, 1943] there is some anecdotal evidence of a strong correlation between the occurrence of *Oscillatoria* sp. and of other cyanobacteria and *E. dilatata.*

Feeding and assimilation rates

The CRs of *Euchlanis* were measured using a natural assemblage of lake food, in a range of concentrations varying by an order of magnitude (Fig. 2). Two experiments, the first on feeding and the second on assimilation, were carried out under similar conditions, although the animals in the feeding experiment were larger.

The CR in the first experiment initially decreased sharply as food concentration increased from 1.1 μg C ml^{-1} tot 2.2 μg ml^{-1}. This was followed by a more steady decline in clearance rate with increasing food concentration at higher food densities (Fig. 2a). A similar trend was recorded for the net clearance rates (Fig. 2c). Feeding rates in experiment 1 (Fig. 2b) and assimilation rates in experiment 2 (Fig. 2d) were very similar even though the animals in the latter experiment weighed about 20% less than in the former case. This can be explained by the very high assimilation efficiencies irrespective of the increase in food concentrations. Also, a relatively high daily ration of the smaller animals in experiment 2 can contribute to what may apparently look like high assimilation rates, but this appears to have been of minor importance.

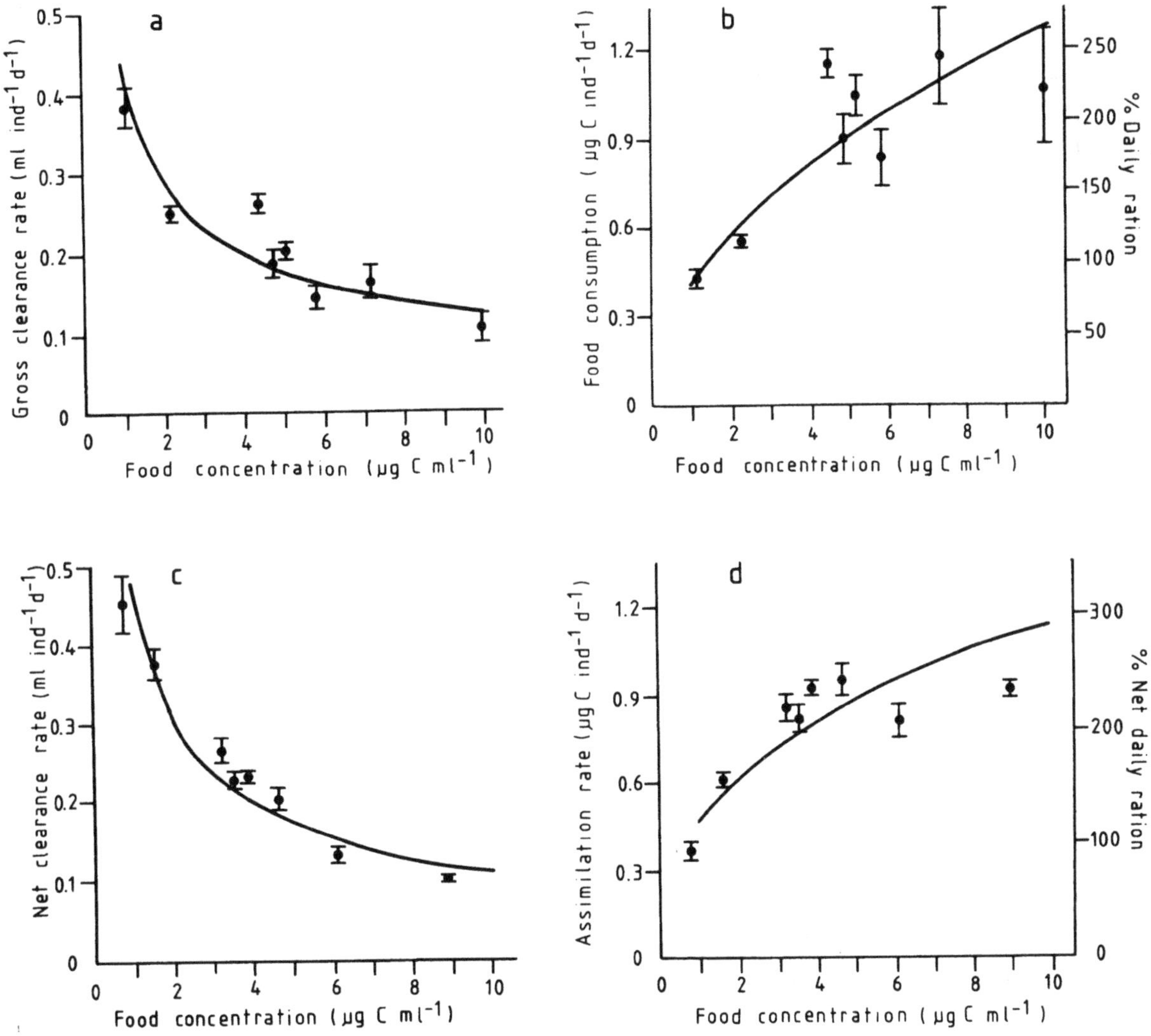

Fig. 2. The effect of food concentration on gross clearance rates and consumption rates (Expt. 1: panels a & b) and on net clearance rates and assimilation rates (Expt. 2: panels c & d) in *Euchlanis dilatata lucksiana*. Mean rates ± S.D. are given. The curves are calculated from a regression model of the form Y = aX^b. Size of experimental animals: 0.49 µg C ind^{-1} and 0.39 µg C ind^{-1}, respectively.

Interestingly, the high assimilation rates found, do indicate that all the food particles below 33 µm ingested were incorporated equally well.

This observation was confirmed in a subsequent experiment with animals of the same size. Both consumption and assimilation rates were measured simultaneously using five food concentrations, doubling stepwise from 1.17 to 18.75 µg C ml^{-1} (Fig. 3). The assimilation rates were again high. Assimilation efficiencies were 100%, but decreased to about 79% at food concentrations of 9.4 µg C ml^{-1} and greater (Fig. 4).

The daily ration and the net daily ration, i.e. the weight-specific assimilation rate, reached 100% at a food concentration of 6–7 µg C ml^{-1} and 8–9 µg C ml^{-1}, respectively (Fig. 3). Thus, even at these high food levels a further increase in food concentration appears to be exploited efficiently by *Euchlanis*. This may help this species maintain a favourable energy balance between energy put to growth and reproduction and energy respired, i.e. by diverting major part of the assimilated energy to growth and reproduction. Rotifers appear to be, in general, superior to cladocerans because of their

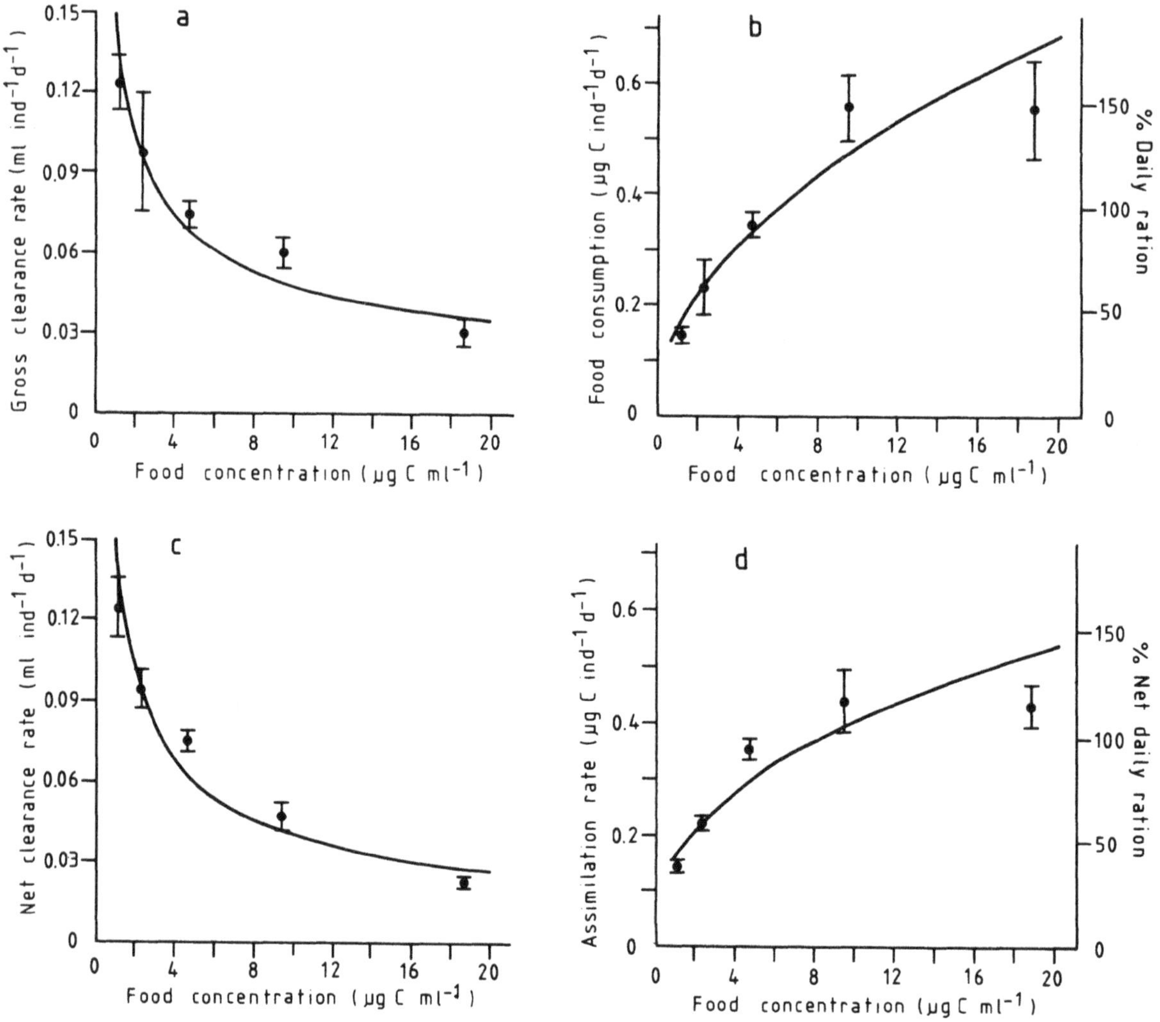

Fig. 3. Legend as for Fig. 2. Results of Expt. 3 in which both the clearance rates and net clearance rates were measured simultaneously. Size of experimental animals: 0.37 µg C ind^{-1}.

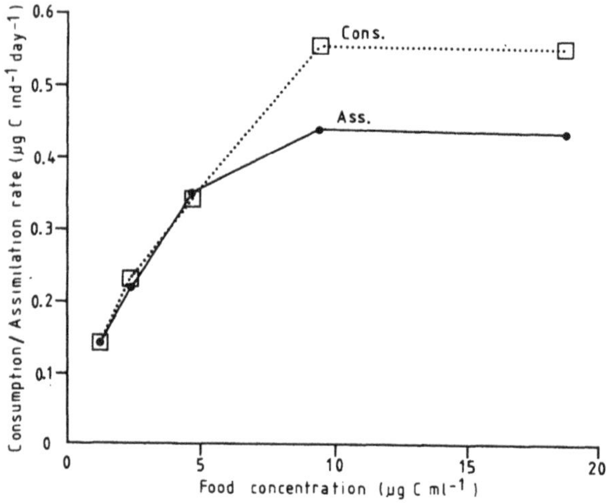

Fig. 4. The consumption rate (□ ... □) and assimilation rate (● – ●) of *Euchlanis dilatata lucksiana* at different food concentrations (Expt. 3). (N.B. the two rates are similar up to a food concentration of ca. 5 µg C ml^{-1}, indicating 100% assimilation efficiency at these levels).

faster multiplication rates [Erman, 1962]. This may be due to the former's high assimilation efficiencies [Dumont, 1977]. The high incorporation rates of *Euchlanis* on a weight-specific basis can be explained by high respiration rates, but these did not exceed 25% of the body weight (J. Rooth, unpub-

Table 2. Coefficients of the regression equation of the form ($Y = aX^b$) relating the dependent variables (Y), i.e. gross clearance rate (F), food consumption (f), net clearance rate (F_a) and assimilation (A) rate and the independent variable (X), i.e. food concentration; r is the correlation coefficient, n is the number of measurements and P the confidence limit.

Exp. No.	Parameter	a	b	r	n	P (<)
1	F	0.419	−0.51	−0.92	8	0.001
	f	0.419	0.49	0.91	8	0.001
2	F_a	0.474	−0.62	−0.95	8	0.001
	A	0.474	0.38	0.89	8	0.002
3	F	0.146	−0.48	−0.97	5	0.002
	f	0.146	0.52	0.97	5	0.002
	F_a	0.155	−0.59	−0.97	5	0.001
	A	0.155	0.41	0.95	5	0.001

lished data). A direct transmittance to the embryo of food particles, as suggested by Dumont [1977], may also play a role in *Euchlanis.*

Regression coefficients (b) for F and f, as a function of food concentrations, were around − 0.5 and + 0.5, respectively (Table 2). Thus, in *Euchlanis* CRs declined and feeding rates rose in proportion to the square root of food concentration. Such a generalization was found also by Peters & Downing [1984] for the feeding rates of cladocera but not for the filtering or clearance rates, which declined proportionally to the inverse of the cube root.

Regression coefficients for net clearance (F_a) and assimilation (A) rates as a function of food concentration were − 0.62 and − 0.59 and + 0.38 and + 0.41, respectively (Table 2). These coefficients differ considerably from those for F and f. These differences give the impression that F_a falls more sharply and A increases less than the respective changes in F and f, in response to the increases in food concentration. However, the intercept values (a) for the F_a and A are relatively higher (Table 2). This may considerably neutralize the effect of the lower regression coefficient b in decreasing the rates, especially at the lower food concentrations, 1−3 μg C ml^{-1}. At low food concentrations the assimilation activities are very high leading to assimilation efficiencies of 100% (Fig. 4).

Acknowledgements

We thank Drs. B. Z. Salomé and Mr. K. Siewertsen for reading and improving the manuscript, and the latter also for his help in preparing the illustrations. We also thank Mrs. H. Spijker-Roede for typing the manuscript and Mr. E. Mariën for doing the photographic work.

References

Bogdan, K. G. & J. J. Gilbert, 1984. Rotifer grazing: in situ studies lectivity and rates. In: D. G. Meyers & J. R. Strickler (eds.). Trophic Interactions within aquatic ecosystems 97−133. AAAS Selected Symposium 85. Westview Press, Colorado.

Carlin, B., 1943. Die Planktonrotatorien des Motalaström. Medd. Lunds Univ. Limnol. Inst. 5: 1−260.

Dumont, H. J., 1977. Biotic factors in the population dynamics of rotifers. Arch. Hydrobiol. Beih. 8: 98−122.

Dumont, H. J., I. van de Velde & S. Dumont, 1975. The dry weight estimate of biomass in a selection of Cladocera, Copepoda and Rotifera from the Plankton, Perifyton and Benthos of Continental Waters. Oecologia 19: 75−97.

Duncan, A. & R. D. Gulati, 1983. Feeding studies with natural food particles on tropical species of planktonic rotifers. In: F. Schiemer, (ed.) Limnology of Parakrama Samudra − Sri Lanka: a case study of an ancient man-made lake in the tropics. Developments in Hydrobiology. 12: 117−125. Junk, The Hague.

Erman, L. A., 1962. On the quantitative aspects of feeding and selectivity of food in the plankton rotifer Brachionus calyciflorus Pall. Zool. Zh. 41: 34−38.

Gulati, R. D., 1983. Zooplankton and its grazing as indicators of trophic status in Dutch lakes. Environm. Monitor. Assessm. 3: 343−354.

Gulati, R. D., 1984. The zooplankton and its grazing as measures of trophy in the Loosdrecht Lakes. Verh. int. Ver. Limnol. 22: 863−867.

Gulati, R. D., K. Siewertsen & G. Postema, 1982. The zooplankton: its community structure, food and feeding and its role in the ecosystem of Lake Vechten. Hydrobiologia 95: 127−163.

Holtby, L. B. & R. Knoechel, 1981. Zooplankton filtering rates: error due to loss of radioisotope label in chemically preserved samples. Limnol. Oceanogr. 26: 774−780.

Jacobs, J., 1974. Quantitative measurements of food selection. A modification of forage ratio and Ivlev's Electivity Index. Oecologia 14: 413−417.

King, C. E., 1967. Food, age and the dynamics of a laboratory population of rotifers. Ecology 48: 111−128.

Myers, F. J., 1930. The rotifer fauna of Wisconsin. V. The genera Euchlanis and Monommata. Trans. Wisconsin Acad. 25: 353–413.

Peters, R. H. & J. A. Downing, 1984. Empirical analysis of zooplankton filtering and feeding rates. Limnol. Oceanogr. 29: 763–784.

Ruttner-Kolisko, A., 1977. Suggestions for biomass calculation of plankton rotifers. Arch. Hydrobiol. Beih. 8: 71–76.

Starkweather, P. L., 1980. Aspects of feeding behaviour and trophic ecology of suspension-feeding rotifers. Hydrobiologia 73: 63–72.

Starkweather, P. L. & J. J. Gilbert, 1977, Radiotracer determination of feeding in Brachionus calyciflorus: the importance of gut passage times. Arch. Hydrobiol. Beih. 8: 261–263.

Stemberger, R. S., 1981. A general approach to the culture of planktonic rotifers. Can. J. Fish. aquat. Sci. 38: 721–724.

Hydrobiologia 147: 297–301 (1987)
May, L., Wallace, R. & Herzig, A. (eds), Rotifer Symposium IV

The potential for population growth of *Ascomorpha ecaudis*

Richard S. Stemberger
Department of Biological Sciences, Dartmouth College, Hanover, New Hampshire 03755, USA

Keywords: zooplankton, rotifers, *Ascomorpha*, symbiosis, threshold food level, mucus feeding

Abstract

The intrinsic rate of natural increase (r_m) of the rotifer *Ascomorpha ecaudis* was found to be a hyperbolic function of food concentration. The threshold food concentration and $r_{max}/2$-food concentration determined for this species were significantly lower than values predicted from allometric models. These growth characteristics may be related to the mucus house in which *Ascomorpha* lives and/or the symbiotic algae living in its body tissues. The maximum rate of population growth recorded (0.71 d^{-1}) was similar to that of other soft-bodied rotifers of similar body mass. These population growth characteristics and the resistance of this species to invertebrate predation should allow it to become a dominant member of freshwater zooplankton communities. However, field observations suggest that it is not. Reasons for this are suggested.

Introduction

A species' threshold food concentration is defined as the food concentration at which the intrinsic rate of natural increase (r_m) is zero. It is a predictable, positive function of rotifer body mass and a potentially important parameter affecting the distribution and species composition of rotifers in nature [Stemberger & Gilbert, 1985a]. Species departing from the predicted threshold food concentration may do so because of special adaptations to their environment. For example, *Keratella crassa* has a very high threshold food requirement for its body mass [Stemberger & Gilbert, 1985a]. This may be because the species is protected against predation by a strongly thickened lorica which is probably energetically costly to produce or maintain. This lorica may also increase the animal's specific gravity; if so, swimming costs would also be high to offset faster sinking rates. In contrast, other species may have very low threshold food requirements and show adaptations for living in low resource environments. Elongated spines (e.g. *Kellicottia longispina*) and mucus sheaths, for example, may increase buoyancy, reduce sinking rates and consequently reduce swimming costs.

This study looks at the population growth response curve of *Ascomorpha ecaudis* as a function of food density and determines the threshold food concentration and the food concentration at half the maximum population growth rate ($r_{max}/2$-food concentration) for this species. These results are compared to values predicted from allometric equations [Stemberger & Gilbert, 1985a].

Methods

Ascomorpha ecaudis was isolated from Star Lake, Norwich Vermont, U.S.A., in the fall of 1984. A clone was cultured in organic medium (MBL) diluted 1:1 (v:v) with double distilled water and fed *Cryptomonas erosa* var. *reflexa* [Stemberger, 1981]. All cultures were maintained at 19°C under low light intensity (350 lux) with a light cycle of LD = 16:8. Steady-state population growth rates were determined for eight food concentrations ranging from 0.05 $\mu g\ ml^{-1}$ to 15.0 $\mu g\ ml^{-1}$ dry mass of *C. erosa*. The concentration of *Cryptomonas* batch cultures was determined with an electronic particle counter (Particle Data, Inc.) and diluted with medium to the desired cell density for each experiment.

For food densities of 0.05 μg ml^{-1}, population growth rates were determined by least squares linear regression [Stemberger & Gilbert, 1985a, b]. Population growth rates for the remaining concentrations (0.3, 0.6, 1.0, 3.0, 8.0, and 15 μg ml^{-1}) were calculated as follows. Five adults were placed into each of three disposable polystyrene petri dishes containing 20 ml of medium and the appropriate food concentration. Each day the rotifers were transferred to fresh medium and food. The initial population density of each growth experiment was chosen after 4–5 days of acclimation to the experimental food concentration. The final population number was determined after 24 h or 48 h. The population growth rate (r_m) was calculated using the exponential growth model [Edmondson, 1960].

Twenty individuals from each of three replicate containers at each food concentration were then transferred to fresh medium and food and the population growth rate was determined again. This procedure was sometimes repeated a third time, and the mean and standard deviation of all values of r_m for each food concentration was calculated. This method was used to prevent the population from overgrazing their algal food supply.

The threshold food concentration was determined graphically [Stemberger & Gilbert, 1985b] by randomly pairing each of the three replicate estimates of r_m occurring just above and below the zero isopleth of population growth (0.05 and 0.1 μg ml^{-1}) and connecting a line between each pair. This procedure gave three separate estimates of $r_m = 0$. Variance was estimated using a bootstrap technique [Efron & Gong, 1983]. The variance of the concentration was similarly determined using six independent estimates of r_{max} measured at 0.6 μg ml^{-1} of food.

The threshold food concentration and the $r_{max}/2$-food concentration of *Ascomorpha* was compared to the values predicted by the following equations [Stemberger & Gilbert, 1985a]:

$$T = 1.824\ M^{0.832}$$
$$H = 14.149\ M^{1.223}$$

where T = threshold food concentration (μg ml^{-1}), H = $r_{max}/2$-food concentration, and M = rotifer dry mass (μg adult^{-1})

The maximum rate of population growth (r_{max} d^{-1}) also was compared using the equation [Stemberger & Gilbert, 1985a]:

$$\dot{r}_{max} = 1.019\ M^{0.381}$$

Ascomorpha and *C. erosa* dry masses were determined with a Cahn electrobalance as described by Stemberger & Gilbert [1985a].

Results

Ascomorpha ecaudis was found to live within a spherical mucus sheath (Fig. 1) which can be more than 1 mm in diameter. Its corona and trophi are not directly exposed to food resources in the environment. The animal passively collects algae and other particles on the mucus and feeds upon this material from within the house. The cells are pulled through the outer mucus layer, probably by ciliary currents, and sucked into the mouth.

The population growth curve of *Ascomorpha* showed a hyperbolic relationship to food concentration. The intrinsic rate of population growth reached its maximum at the relatively low food density of 0.6 μg ml^{-1} (Fig. 2). It then remained more or less constant with increasing food levels up to 8 μg ml^{-1} and finally decreased slightly at 15 μg ml^{-1}.

The threshold concentration for *Ascomorpha* was 0.064 ± 0.028 μg ml^{-1} (95% C.I.). This was six times lower than the predicted value of 0.42 μg ml^{-1} for its body mass of 0.17 μg [Stemberger & Gilbert, 1985a]. The 95% C.I. about the predicted threshold food level (0.21–0.95 μg ml^{-1}) was well above the upper bound of the 95% C.I. for the measured value of 0.092 μg ml^{-1}. Similarly, the $r_{max}/2$-food concentration 0.22 ± 0.02 μg ml^{-1} (95% C.I.), was more than 7 times lower than the predicted value of 1.62 μg ml^{-1}. The lower bound of the predicted value (0.78 μg ml^{-1}) was also well above the upper bound of the measured value. Finally, the maximum rate of population increase

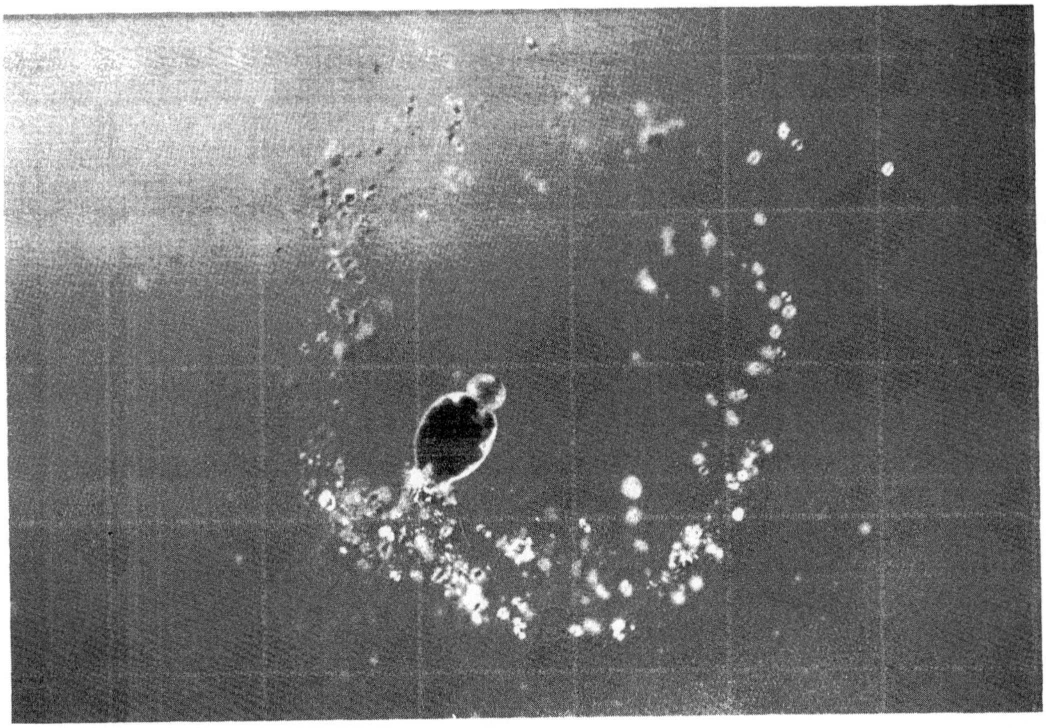

Fig. 1. Ascomorpha ecaudis feeding on *Cryptomonas* trapped on the surface of a mucus house 900 μm in diameter.

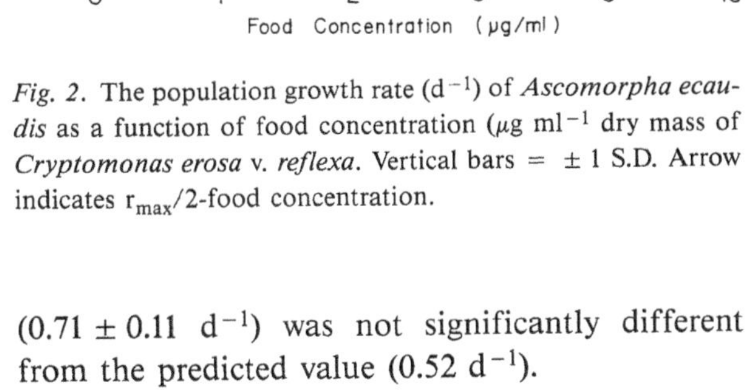

Fig. 2. The population growth rate (d^{-1}) of *Ascomorpha ecaudis* as a function of food concentration (μg ml^{-1} dry mass of *Cryptomonas erosa* v. *reflexa*. Vertical bars = ± 1 S.D. Arrow indicates $r_{max}/2$-food concentration.

(0.71 ± 0.11 d^{-1}) was not significantly different from the predicted value (0.52 d^{-1}).

Discussion

This is the first report of the presence of a mucus sheath in *Ascomorpha ecaudis. Ascomorpha* may swim out of the sheath when disturbed during collection and the mucus may dissolve in samples preserved in formalin. These factors may explain why the sheath has gone unnoticed. However, one other investigator [Dr. L. May, pers. comm.] has also observed mucus sheaths in individuals from live samples and in laboratory cultures.

The threshold food concentration and $r_{max}/2$-food concentration determined from these experiments for *A. ecaudis* were significantly lower than those predicted by the allometric models. This is probably due to the mucus sheath and/or the symbiotic algae living in the animal's body. The mucus sheath has three important functions which probably result in considerable savings in energy for *A. ecaudis.* (1) It passively protects the rotifer from predation by copepods and *Asplanchna* [Stemberger, 1985; Gilbert, pers. commun.]. The texture or taste of the mucus appears to repel potential predators. (2) The mucus sheath traps algae on its surface which are then available to the rotifer as food. Although mucus feeding has been reported in some marine zooplankton [Jorgensen, 1966; Harbinson & Gilmer, 1976; Alldredge, 1977], it was previously unknown in freshwater zooplankton and has not been described in *Ascomorpha.* (3) The mucus sheath probably reduces the rotifer's specific gravity and consequently its sinking rate. This would significantly reduce the cost of swimming which is especially high in rotifers because of the inefficient transformation of chemical energy into kinetic energy in cilia [Sleigh & Blake, 1977; Epp & Lewis, 1984].

The symbiotic algae may benefit *Ascomorpha* in several ways. The animal may utilize the photosynthate for metabolism, growth or mucus production. Individuals cultured in the dark for about a week lose their mucus sheaths and become free-swimming. They secrete mucus again when exposed to light for a short period ($\simeq$20 min.) [Stemberger, unpublished data]. The algae may also utilize and hence remove waste products which accumulate in the cavity of the mucus house.

The threshold food concentration and $r_{max}/2$-food concentration of *Ascomorpha* are similar to those of much smaller rotifer species such as *Keratella cochlearis, Synchaeta oblonga* and *Polyarthra remata* [Stemberger & Gilbert, 1985a]. In contrast, *Ascomorpha*'s r_{max} (0.71 d^{-1}) is much higher than that of these smaller species (0.28 – 0.34 d^{-1}). The low $r_{max}/2$-food concentration suggests an adaptation to low resource environments [Stemberger & Gilbert, 1985a].

The combination of the population growth characteristics described above and the protection conferred by the mucus sheath against predation should allow *Ascomorpha* to become a highly competitive and potentially dominant member of plankton communities. Field studies suggest that this is not the case, although it often occurs when predator abundance is high and food resources are low. *Ascomorpha* occurred in 30 of the 43 lakes in the northern portion of Michigan's lower peninsula and was never dominant in any of the 500 samples collected. The species never occurred in winter months [Stemberger, unpubl. data; Fuller *et al.*, 1977] and was absent from the deep and cold offshore environment of the Laurentian Great Lakes [Stemberger *et al.*, 1979].

Five factors may limit the distribution of *Ascomorpha ecaudis.* (1) The species may not be able to control its depth and thus exploit food-rich strata or actively locate new patches of food. Mucus sheaths in algae, for example, do not provide neutral buoyancy and sinking inevitably occurs, albeit at a slow rate [Walsby & Reynolds, 1980]. (2) The species may sink to the bottom of the lake where population densities may be high but are not adequately sampled with a plankton net. Hudson and Gosse [1886] and Pejler [1962] found the species living on the bottom of ponds and on the submerged vegetation. (3) If symbiotic algae are critical to *Ascomorpha*'s success then low light environments would be detrimental to its survival. Therefore, one might expect the species to occur only in shallow lakes or in deeper lakes which mix to the bottom thereby keeping *Ascomorpha* suspended and in adequate light. In a survey of Michigan lakes *Ascomorpha* was not detected in lakes having a mean depth greater than 9 metres [Stemberger, unpubl. data]. (4) Colder water temperatures may have a significant negative effect on population growth rates. (5) The mucus sheath may be inefficient for collecting non-motile algae which could exclude *Ascomorpha* from environments where microflagellates are sparse.

The distribution and abundance of *Ascomorpha* and its unique and potentially complex association with symbiotic algae and mucus warrant further study.

Acknowledgements

I thank J. J. Gilbert for comments which improved the manuscript. This research was supported by NSF Grant DEB-8121639 to J. G. Gilbert.

References

Alldredge, A. L., 1977. House morphology and mechanisms of feeding in the Oikopleuridae (Tunicata, Appendicularia). J. Zool. Lond. 181: 175 – 188.

Edmondson, W. T., 1960. Reproductive rates of rotifers in natural populations. Mem. Ist. Ital. Idrobiol. 12: 21 – 77.

Efron, B. & G. Gong, 1983. A leisurely look at the bootstrap, the jacknife, and cross-validation. Amer. Stat. 37: 36–48.

Epp, R. W. & W. M. Lewis, Jr., 1984. Cost and speed of locomotion for rotifers. Oecologia 61: 289–292.

Fuller, D. R., R. S. Stemberger & J. E. Gannon, 1977. Limnetic rotifers as indicators of trophic change. J. E. Mitchell Sci. Soc. 93: 104–113.

Harbinson, G. R. & R. W. Gilmer, 1976. The feeding rates of the pelagic tunicate, Pegea confederata and two other salps. Limnol. Oceanogr. 27: 1395–1403.

Hudson, C. T. & P. H. Gosse, 1886. The Rotifera or Wheel-Animalicules. 2 vols. Longmans, Green, London, 144 pp., 30 plates.

Jorgensen, C. B., 1966. The biology of suspension feeding. Pergamon Press, Oxford, 357 pp.

Pejler, B., 1962. On the taxonomy and ecology of benthic and periphytic Rotatoria. Investigations in northern Swedish Lapland. Zool. Bidr. Upps. 33: 327–422.

Sleigh, M. A. & J. R. Blake, 1977. Methods of ciliary propulsion and their size limitations. In J. R. Pedly (ed.), Scale effects in animal locomotion. Academic Press, N.Y.: 243–256.

Stemberger, R. S., 1981. A general approach to the culture of planktonic rotifers. Can. J. Fish. Aquat. Sci. 38: 721–724.

Stemberger, R. S., 1985. Prey selection by the copepod Diacyclops thomasi. Oecologia 65: 492–497.

Stemberger, R. S., J. E. Gannon & F. J. Bricker, 1979. Spatial and seasonal structure of rotifer communities in Lake Huron. EPA-600/3-79-085. 160 p.

Stemberger, R. S. & J. J. Gilbert, 1985a. Body size, food concentration and population growth in planktonic rotifers. Ecology 66: 1151–1159.

Stemberger, R. S. & J. J. Gilbert, 1985b. Assessment of threshold food levels and population growth in planktonic rotifers. Arch. Hydrobiol. 21: 269–275.

Walsby, A. E. & C. S. Reynolds, 1980. Sinking and floating. In: I. Morris (ed.), The Physiological Ecology of Phytoplankton. Blackwell, Oxford: 371–412.

Hydrobiologia 147: 303–306 (1987)
May, L., Wallace, R. & Herzig, A. (eds), Rotifer Symposium IV

Further nutritional studies on the marine rotifer *Encentrum linnhei*

J. M. Scott
Scottish Marine Biological Association, Dunstaffnage Marine Research Laboratory, Oban, Argyll, Scotland

Keywords: rotifer, nutrition, tryptophan, flagellate, *Encentrum*

Abstract

Evidence is given that the colourless flagellate *Oxyrrhis marina* is a natural source of tryptophan in the diet of the rotifer *Encentrum linnhei,* which inhabits coastal rockpools in western Scotland. The rotifer feeds on this flagellate in addition to the photosynthetic alga *Brachiomonas submarina.* This study also shows that tryptophan excreted by one alga can be taken up by another and made available to *Encentrum,* a process which allows the rotifer to grow on combinations of algae which, individually, are nutritionally deficient.

Introduction

Encentrum linnhei is a marine rotifer which inhabits a sheltered group of rock pools on the west coast of Scotland. Previous experiments have shown that, when cultured on the alga *Brachiomonas submarina* [Scott, 1983], this species requires a dietary supplement of the amino acid tryptophan. In natural waters, *Encentrum* also feeds on small ciliates and flagellates. This study investigates the possibility that, in the natural situation, *Encentrum* derives its necessary supply of tryptophan from the flagellate *Oxyrrhis marina* Dujardin, which is abundant in the rockpools during the summer months.

Small amounts of tryptophan have been detected in several marine algae, including *Brachiomonas* [Cowey & Corner, 1966; Chau *et al.*, 1967], which suggests that this amino acid is synthesized by these algae. However, there is normally insufficient tryptophan in *Brachiomonas* to satisfy the nutritional requirements of *Encentrum.* Either tryptophan production is so tightly controlled that little of the free amino acid accumulates in the cells, or excess tryptophan is rapidly excreted. *Encentrum* can survive equally well on a diet of *Brachiomonas* in a tryptophan enhanced medium, or on *Brachiomonas* supplemented with an additional alga. This suggests that, in both cases, *Brachiomonas* takes up sufficient tryptophan from its environment to satisfy the needs of the rotifer; it has already been shown that this alga is able to take up exogenous tryptophan from solution and make it available to *Encentrum* [Scott, 1983]. Under the latter feeding conditions, the only possible source of this amino acid was that excreted by the other alga. This study tests the hypothesis that *Brachiomonas* can take up tryptophan excreted by another alga and make it available to *Encentrum.*

Materials and methods

Bacteria-free cultures of the rotifer *Encentrum linnhei* were maintained in S88 culture medium [Droop, 1968] and fed on *Brachiomonas* supplemented with 100 $\mu g\ ml^{-1}$ of tryptophan. These were subcultured every 2–3 weeks and tested for sterility using enriched culture medium. To reduce the risk of transferring tryptophan into the experimental cultures, intermediate subcultures of *Encentrum* were grown in tryptophan-free medium for about two weeks before each experiment.

Bacteria-free cultures of *Brachiomonas submarina* S.M.B.A. strain 44, *Dunaliella primolecta* S.M.B.A. strain 57, *Isochrysis galbana* S.M.B.A. strain 58, and *Oxyrrhis marina* S.M.B.A. strain 18 were maintained similarly in S88 culture medium with added vitamin B_{12} and thiamine. In the case of *Oxyrrhis, Dunaliella* was used as the food organism.

To determine whether the flagellate *Oxyrrhis marina* might provide a source of dietary tryptophan for the rotifer *Encentrum*, experiments were carried out in which *Brachiomonas* was supplied as food, with and without additional *Oxyrrhis.* Test tubes containing 5 ml of sterile S88 culture medium were used for each experiment, with three to five replicates for each condition. Two tryptophan-free rotifers, from the intermediate subcultures, were aseptically inoculated into each tube. *Brachiomonas* and *Oxyrrhis* were added to the tubes, as required, such that half of the tubes contained *Brachiomonas* alone, while the other half also contained *Oxyrrhis.* The experimental tubes were kept at 20 °C, in subdued light, for two to three weeks. The rotifers in each tube were then counted. The experiment was repeated three times. A similar set of experiments were performed using *Isochrysis* with and without *Oxyrrhis* as a food source. In addition, *Oxyrrhis* alone was also offered as food.

Further experiments were carried out to determine whether tryptophan excreted by one alga could be taken up by another and made available to *Encentrum. Dunaliella* was grown aseptically in flasks containing S88 culture medium. After 9 days the algal cells were removed by filtration (whatman glass-fibre (GF/C) filters) and the filtrate was resterilized by filtering through a 0.2 μm membrane filter (Millipore). Four drops of the sterile filtrate were then added to a number of replicate test tubes containing 5 ml of sterile medium which had been aseptically inoculated with *Brachiomonas* and two rotifers. Control cultures, without filtrate, were also established. The cultures were incubated at 20 °C under subdued light for 13–14 days before the rotifers in each tube were counted. The experiment was repeated 4 times. A similar series of experiments were carried out with the alga *Isochrysis.*

Results

In each of the three experiments carried out to compare the growth rates of *Encentrum* fed on *Brachiomonas,* with and without additional *Oxyrrhis,* the cultures with *Oxyrrhis* showed a greatly increased rate of growth compared to those without *Oxyrrhis* (Table 1). Similarly, the cultures fed on *Isochrysis* also showed considerably enhanced growth when the flagellate was present (Table 2).

Table 1. Response of *Encentrum* to *Oxyrrhis* as a supplement to *Brachiomonas.*

Diet	Rotifer counts per tube		
	Expt 1	Expt 2	Expt 3
Brachiomonas	9	14	16
	18	45	44
	17	12	14
	92	23	8
	7	85	38
Brachiomonas plus *Oxyrrhis*	780	323	284
	–	363	190
	918	–	231
	810	457	257
	870	548	160
Duration of experiment	20 days	15 days	13 days

Table 2. Response of *Encentrum* to *Oxyrrhis* as a supplement to *Isochrysis galbana.*

Diet	Rotifer counts per tube		
	Expt 1	Expt 2	Expt 3
Isochrysis	8	35	8
	27	41	4
	14	22	10
Oxyrrhis		2	7
		2	–
		3	5
Isochrysis plus *Oxyrrhis*	106	113	270
	190	81	177
	219	151	150
Duration of experiment	16 days	16 days	15 days

Table 3. Response of *Encentrum* to filtrates of *Dunaliella* and *Isochrysis* as supplements to *Brachiomonas* cells.

Diet	Rotifer counts per tube			
	Expt 1	Expt 2	Expt 3	Expt 4
Brachiomonas	53 64 Median 53 43 –	2 16 17 Median 16 16 27	42 34 28 Median 31 19	32 69 Median 56 56
Brachiomonas plus *Dunaliella* filtrate	128 154 Median 134 166 89	12 27 57 Median 39 52	41 32 Median 46 51 58	115 118 37 Median 116 121
Brachiomonas plus *Isochrysis* filtrate	223 32 Median 136 136 210		83 38 Median 80 92 77	95 132 Median 113 89 131
Duration of experiment	14 days	14 days	14 days	13 days

However, when fed on *Oxyrrhis* alone, the results showed conclusively that *Encentrum* could not reproduce successfully (Table 2).

The results of the experiments to investigate the effect of adding algal excreted tryptophan to cultures of *Encentrum* fed entirely on *Brachiomonas* are given in Table 3. The addition of filtrate from either the *Dunaliella* or *Isochrysis* cultures produced, in most cases, a rotifer count at least double that of the controls, indicating a considerable increase in the growth rate of these cultures.

Discussion

Scott [1983] demonstrated experimentally that *Encentrum linnhei* has a dietary requirement for the amino acid tryptophan. This posed the question of where the rotifer obtained a natural supply of this nutrient. The results from these experiments clearly show that *Encentrum* can readily obtain an adequate supply of this amino acid from the flagellate *Oxyrrhis marina*, which is abundant in the rock-pools it inhabits. The relatively large counts which occurred in some of the controls indicates that there was still sufficient residual tryptophan in some of the rotifers to allow the growth of a few generations, but this did not affect the overall difference between the two nutritional conditions.

Encentrum is particularly able to utilize this supplementary food source because, in addition to its ciliary feeding currents which passively collect algal cells, it also has forcipate trophi which can seize flagellates and small ciliates.

Since these experiments were conducted the nutritional benefit of *Oxyrrhis*, over a period of more than six months, has been established.

This study has also shown that the filtrate from mature algal cultures can also contain sufficient excreted tryptophan to support the growth of rotifer cultures when another alga is supplied as food. In previous experiments it had been noted that *Encentrum* would not grow on *Brachiomonas, Dunaliella* or *Isochrysis* alone, but would grow successfully on any combination of these species. Undoubtedly, the explanation of these phenomenon is that the tryptophan excreted by one alga was subsequently taken up by the other and made availabe to the rotifer; there is plenty of evidence that amino acids are excreted by algae, though not specifically tryptophan

[Hammer & Brockmann, 1983; Hellebust, 1974]. It also seems likely that a given alga is somehow inhibited from taking up its own excreted tryptophan.

There is a likelihood that more than one amino acid behaves in this manner, and possibly other nutrients too. For example, *Encentrum* will not grow on *Dunaliella* or *Isochrysis* alone, even when supplemented with tryptophan, although these two algae together support good rotifer growth. This suggests that both of these algae are deficient in more than one nutrient necessary to *Encentrum*. Further experiments on filtrates from these algae are necessary to confirm this.

It is likely that there will be free amino acids in the rock pool which *Encentrum* inhabits, but the rotifer is unable to make use of these directly. An intermediary organism which can take up these nutrients, and no which *Encentrum* can feed, is necessary to make these nutrients available to the rotifer. *Encentrum* has become adapted to this situation and probably feeds on combinations of an alga and a flagellate, or two or more algae to overcome these problems in its natural environment.

References

Chau, Y. K., L. Chuecas & J. P. Riley, 1967. The component combined amino acids of some marine phytoplankton species. J. Mar. biol. Ass. U. K. 47: 543–554.

Cowie, C. B. & E. D. S. Corner, 1966. The amino acid composition of certain unicellular algae, and of the faecal pellets produced by Calanus finmarchicus when feeding on them. In: H. Barnes (ed.) Some contemporary studies in marine science. George Allen & Unwin Ltd., London: 225–231.

Droop, M. R., 1968. Vitamin B_{12} and marine ecology. IV. The Kinetics of uptake, growth and inhibition in Monochrysis lutheri. J. mar. biol. Ass. U.K. 48: 689–733.

Hammer, K. D. & U. H. Brockmann, 1983. Rythmic release of dissolved free amino acids from partly synchronized Thalassiosira rotula under nearly natural conditions. Mar. Biol. 74: 305–312.

Hellebust, J. A., 1974. Extracellular products. In: W. D. P. Stewart (ed.) Algal physiology and biochemistry. Blackwell, Oxford: 838–863.

Scott, J. M., 1983. Rotifer nutrition using supplemented monoxenic cultures. Hydrobiologia 104: 155–166.

Hydrobiologia 147: 307–317 (1987)
May, L., Wallace, R. & Herzig, A. (eds), Rotifer Symposium IV

Planktonic rotifers and temperature

G. A. Galkovskaja
Institute of Zoology, Byelorussian Academy of Sciences, Academytcheskaya 27, Minsk 220072, USSR

Keywords: Rotifera, planktonic, temperature, food, embryonic development, respiration rate, ingestion, population dynamics

Abstract

The influence of temperature (t) upon rotifer embryonic development rate (D_e) has been analysed using data from the literature, and the author's own results from experimental and natural populations. For *Keratella cochlearis* (Gosse), within the temperature range of 1–28 °C, this relationship is best expressed by the equation:

$$1/D_e = 0.002 + 0.00025t + 0.000065t^2.$$

For *Brachionus calyciflorus* Pallas, between 8 °C and 35 °C, the best relationship is given by the equation:

$$1/D_e = 0.005 + 0.00013t + 0.00013t^2.$$

Increasing the incubation temperature to 37–40 °C resulted in a decrease in development rate and a sharp reduction in life length.

Analysis of the relationship between respiration rate and temperature in experimental and natural populations of *Brachionus calyciflorus* and *Hexarthra mira* (Hudson) showed that the maximum rate of oxygen consumption occurred at 32–33 °C.

The effects of temperature upon the ingestion rates of rotifers is greatly influenced by food concentration. Consequently, this factor also influences the secondary production of experimental populations at different temperatures.

Introduction

Most species of planktonic rotifers have a global distribution [Pejler, 1977; Dumont, 1983; de Ridder, 1984] and are characterised by wide temperature tolerances. They are able to reproduce over a wide temperature range, providing that other factors are not limiting. It is difficult to determine the effect of temperature on an individual or a population in the natural environment because (1) temperature influences other lake processes which, in turn, affect the rotifers, (2) temperature varies in time and space in most fresh-water bodies and is difficult to estimate and (3) the rate of biological processes is seldom influenced by temperature alone, but also by a number of other factors. The effect of changing temperature on rotifers can only be treated in a time-space dimension for both the environmental temperature and the rotifer population. It is almost impossible to separate the direct and indirect effects of the environmental factor of temperature. At the same time, it is essential to know something of the temperature characteristics of animal responses and the limits of the temperature acclimation possibilities, particularly when the heat loading is greatly increased.

This paper attempts to analyse the relationships between temperature and several kinds of rate processes in rotifers, using data on experimental and natural populations from the literature, and from our own data.

Materials and methods

Clonal cultures of *Brachionus calyciflorus* Pallas were grown at 10, 15, 20, 25, 30, 35, 37 and 39 °C in 0.8 litre culture vessels and fed on *Chlorella* sp. cultivated on Tamiya medium. The algae were centrifuged and washed in distilled water before feeding to the rotifers. The rotifers were grown in settled artesian water. The method adopted for the individual culture of rotifers is described by Galkovskaya & Mityanina [1983].

Embryonic development times were determined by the methods of Edmondson [1960] and Amrén [1964]. Oxygen consumption rates were measured by the closed-bottle technique, using a micro-Winkler titration. The experimental bottles were 35–38 ml in volume and rotifers were added to give a final density of about 50 individuals ml^{-1}. The exposure period was 2–8 h for cultured animals and 4 h for animals collected from natural populations.

Ingestion rate was estimated from the change in algal concentration in 10–20 ml aliquots of food suspension over varying periods of time. The density of grazing rotifers and the exposure time were chosen according to the nature of the experiment. Population growth rates were calculated during the exponential growth phase and all experiments were carried out in the dark. Data was processed using the statistical methods of Rokitsky [1964].

Embryonic development rate

The duration of embryonic development depends upon the temperature at which oogenesis took place [Pourriot & Hillbricht-Ilkowska, 1969]. In addition, Pourriot & Deluzarches [1971] showed that a thermophilic clone of rotifers collected from an astatic pond, where they developed only in summer, and subsequently cultured in the laboratory at 20–22 °C, differed in their response to changing temperature from a eurythermic clone of the same species, collected in spring from another water body, and grown in the laboratory at 10–12 °C. Animals reared at low temperatures had longer embryonic development times at higher temperatures than animals adapted to high temperatures. The embryos of animals grown at high temperatures developed more slowly at low temperatures than those of animals adapted to low temperatures. Similar effects have been observed by many investigators, particularly when working with planktonic forms colllected from a range of temperature conditions [Herzig, 1983]. Clearly, acclimation is an important factor in the experimental determination of the effect of temperature on the rate of biological processes in planktonic animals. The necessity for separating data from acclimated and non-acclimated animals is now recognised [Galkovskaja & Suschchenya, 1978; Ivleva, 1981; Mityanina, 1983]. Data used in this paper came from either acclimated experimental animals, or from field populations at their natural temperature.

Within each temperature condition, the duration of embryogenesis is determined by egg size. Pourriot [1973] gives a linear relationship between egg size and embryonic development rate for *Brachionus calyciflorus.* Egg size is a function of female body size in many zooplankton [copepods: McLaren, 1965; cladocerans: Green, 1956; rotifers: Sudzuki, 1955; Pourriot, 1973]. This relationship appears to remain constant even though females of the same species may attain maturity at considerably different body sizes under various natural or experimental conditions. The size and mass of both egg and female animal reflect the cumulative effect of a whole complex of exogenous and endogenous factors, the most important of which are food and temperature.

The relationship between the rate of embryonic development ($1/D_e$, d^{-1}) and temperature (°C) published by Morozov [1983] for *Brachionus calyciflorus* is given as a straight line for the temperature range 15–35 °C. A linear equation was also published by Galkovskaya & Mityanina

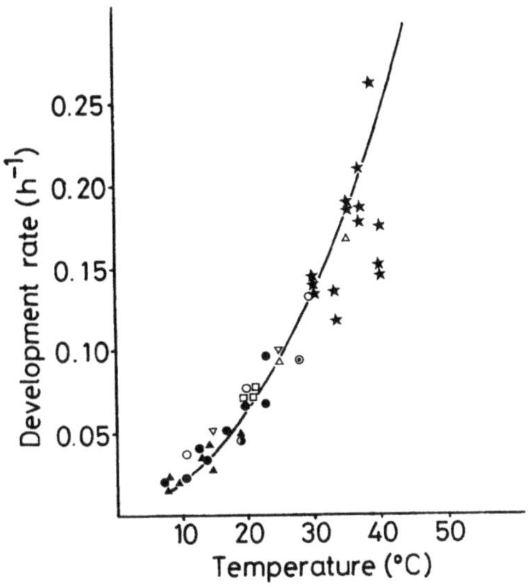

Fig. 1. Embryonic development rate in relation to temperature for *Brachionus calyciflorus* Pallas. Sources of data: Duncan, 1983 (⊙); Pourriot & Deluzarches, 1971 (○); Halbach, 1970b (▼); Pourriot, 1973 (▲); Morozov, 1983 (△); Kryutchkova & Rybak, 1974 (□); Le Douarin, cit. Duncan, 1983 (◑); present author (field data: ●, laboratory cultures: ★).

[1985] for field populations of *Keratella cochlearis* Gosse from two Belorussian lakes. In the latter case, values for the embryonic development rates were obtained by applying Edmondson's method to eggs simultaneously sampled from different depth strata whose temperatures were 18 °C, 12 °C and 6 °C, respectively. Embryonic development rate and female body size were different in the two lakes, but the temperature relationship was linear in both cases.

Figure 1 shows embryonic development rate plotted against temperature for *Brachionus calyciflorus*. The data are derived from the literature, from our own observations on eggs collected from the field, and from cultures reared in the laboratory. An overall increase in the rate of embryonic development with increasing temperature is evident over the range 8–35 °C. However, most values for temperatures between 37 °C and 40 °C are less than would be expected by extrapolation of the data collected at the lower temperatures. Three types of mathematically generated curves were fitted to the available data. Table 1 shows that equation (3) described this relationship most effectively.

Table 1. Statistical parameters of equations describing the dependence of the embryonic development rate ($1/D_e$, h^{-1}) on temperature (t, °C) in *Brachionus calyciflorus* and *Keratella cochlearis*. n = number of observations; a, b, c = constants of the curve fitting equations; $\delta 1/D_e$ is a measure of the deviation of the empirical values from the generated curve.

Equation	n	a	b	c	$\delta 1/D_e$
B. calyciflorus					
(1) $1/D_e = a + bt$	50	−0.0536	0.0064	-	0.0180
(2) $1/D_e = a\ e^{bt}$	50	0.0121	0.0787	-	0.0162
(3) $1/D_e = 1 + bt + ct^2$	50	0.0050	0.00013	0.000131	0.0142
K. cochlearis					
(1) $1/D_e = a + bt$	44	−0.0037	0.0018	-	0.0092
(2) $1/D_e = a\ e^{bt}$	44	0.0036	0.1110	-	0.0085
(3) $1/D_e = a + bt + ct^2$	44	0.0020	0.00025	0.000065	0.0060

The relationship between temperature and embryo development rate for *Keratella cochlearis* was treated similarly (Fig. 2 & Table 1). Again the empirical values were best described by the parabolic curve of equation (3).

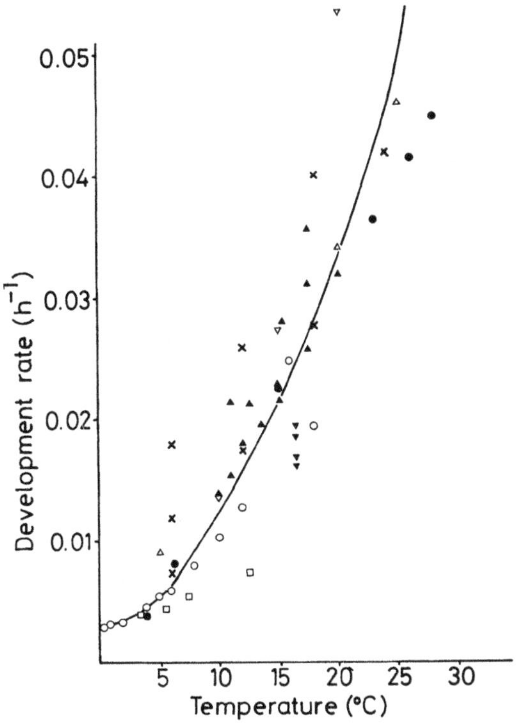

Fig. 2. Embryonic development rate in relation to temperature for *Keratella cochlearis* (Gosse). Sources of data: Walz, 1983a (△); Edmondson, 1965 (▲); Rigler *et al.*, 1974 (○); Makarewicz & Likens, 1979 (▼); Magnien & Gilbert, 1983 (●); Vasilyeva & Nikulin, 1974 (□); Baker, 1979 (▽); present author (×).

Table 2. Some parameters characterizing the reproduction of *Brachionus calyciflorus, Brachionus urceus* and *Epiphanes brachionus* under individual cultivation of cohorts (third generation). D = duration of embryonic development (embryogenesis); D = duration of post embryonic development (juvenile phase); T = mean generation time; D_l = duration of life span; r_c = potential reproductive rate; R_o = net reproductive rate.

Temperature °C	Concentration of *Chlorella* (× 10^7 cells ml^{-1})	$D_e \pm \delta$ (h)	$D_j \pm \delta$ (h)	$D_e + D_j$ (h)	T_c (h)	$D_l \pm \delta$ (h)	r_c (h^{-1})	R_o
B. calyciflorus								
30	0.5	6.7 ± 2.63	9.3 ± 2.07	16.0	34.2	62.7 ± 12.6	0.065	9.2
30	5.0	8.3 ± 1.51	7.3 ± 1.63	15.7	32.8	65.3 ± 17.6	0.091	20.0
37	0.5	4.0 ± 1.26	4.7 ± 1.63	8.7	20.8	52.7 ± 16.3	0.118	11.7
37	5.0	4.3 ± 1.07	4.3 ± 0.82	8.6	24.2	42.0 ± 20.0	0.123	16.7
40	0.5	6.7 ± 3.00	5.5 ± 2.52	12.2	18.3	36.0 ± 9.7	0.091	5.2
40	5.0	8.2 ± 2.39	5.2 ± 1.67	14.4	18.3	27.8 ± 13.2	0.081	4.4
B. urceus								
37	0.5	7.6 ± 4.44	10.7 ± 1.32	18.3	30.7	55.9 ± 8.2	0.094	16.9
37	5.0	4.6 ± 2.05	6.6 ± 1.59	11.2	27.8	42.2 ± 18.5		
40	0.5		7.9 ± 1.74					
40	5.0	6.2 ± 3.30	9.0 ± 2.71	15.2				
E. brachionus								
37	0.5	3.0 ± 1.41	16.0 ± 2.83	19.0		31.5 ± 3.9		
37	5.0	7.3 ± 3.15	25.3 ± 4.25	32.6		44.0 ± 5.4		

Biological features of rotifers at high temperatures

Individuals of the same age and belonging to three species of rotifers, *Brachionus calyciflorus, Brachionus urceus* (Linnaeus) and *Epiphanes brachionus* (Ehrenberg), were cultured at two concentrations of *Chlorella* sp. (5 × 10^6 and 5 × 10^7 cells ml^{-1}), and at various high temperatures, in order to study biological variability in three consecutive generations. The parental female was replaced by a daughter of the second brood and the third generation was investigated throughout its whole life span (D_l, h). The total number of offspring, duration of embryogenesis (D_e, h) and duration of the juvenile phase (D_j) were recorded. The results (Table 2) show that the durations of all of these ontogenetic phases were reduced as the temperature rose from 39 °C to 37 °C, although in no regular manner. The unpredictable relationship between the durations of these phases suggests that similar effects take place throughout ontogenesis under the stress of high temperatures. As the temperature was raised from 37 °C to 40 °C, the durations of both the embryonic and juvenile phases were prolonged, whereas the total life span was shortened; this resulted in a sharp decline in the ratio of life span to the sum of the durations of the embryonic and juvenile phases. Comparison of the effects of the two food concentrations on developmental duration did not indicate any food effect on D_e and D_j at either 30 °C or 37 °C. At 40 °C and 5 × 10^7 cells ml^{-1} there was some prolongation of both D_e and D_j compared with the values obtained at 40 °C and 5 × 10^6 cells ml^{-1}; it seemed that development was prolonged by both high temperature and high food concentration.

At each age interval (x) of two hours, the survivorship (l_x), age specific birth rate (m_x and age specific fecundity rate ($l_x m_x$) were determined and v-curves [Fisher, 1930] were drawn (Fig. 3). From this, it was possible to calculate a theoretically expected generation time as the population parameter (T_c). It was assumed that death and birth rates of

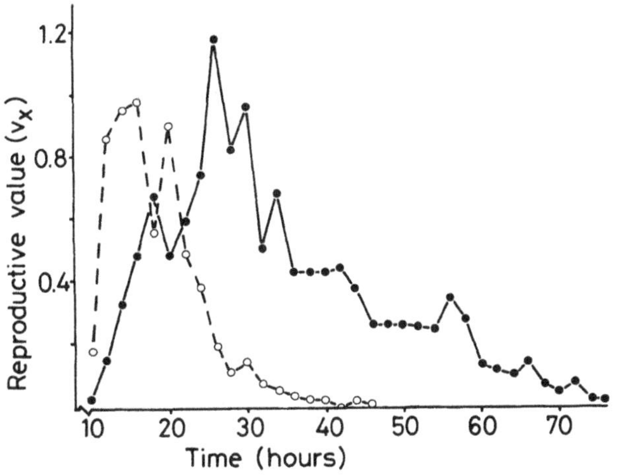

Fig. 3. Reproductive value (v_x) of *Brachionus calyciflorus* Pallas at 30 °C (●) and 40 °C (⊆).

the next generation were the same as those of the previous one. The mean generation time, or the mean age of females producing offspring, [Laughlin, 1965] was calculated as follows:

$$T_c = \frac{\Sigma l_x m_x \cdot x}{\Sigma l_x m_x}$$

As shown by King [1967], for *Euchlanis dilatata*, the mean generation time was shortened when the egg was laid earlier in the life cycle and prolonged when it was laid later; so, T_c was dependent upon the duration of the post-embryonic or juvenile phase. Table 2 shows that the mean generation time was more then twice as long as the sum of D_e and D_j. It can also be seen from Table 2 that the mean generation time became shorter as the temperature increased from 30 °C to 40 °C, and that its ratio to $(D_e + D_j)$ became smaller. The net reproduction rate (R_o) was also calculated for each temperature-food condition (Table 2). At 39 °C R_o increased by a factor of two as the food concentration increased from 5×10^6 to 5×10^7 cells ml^{-1}. The food effect was less at 37 °C; R_o at the lower food level (5×10^6 cells ml^{-1}) was slightly higher at 37 °C than at 30 °C. There was a considerable reduction in the value of R_o at 40 °C at both food levels due to increased mortality and the shift of the reproduction maximum to the earliest stages of maturity. Reproductive effort, calculated as the ratio of R_o to life span, tended towards constant values: 0.182 at 40 °C and 0.191 at 30 °C. Figure 3 illustrates the way in which reproductive effort is maintained at 30 °C and 40 °C. The values obtained are the highest known so far; they may reflect an extremely good food supply and are partly related to the high incubation temperatures.

Respiration and temperature

Bearing in mind the possibility of acclimation effects during respiration experiments on planktonic animals [Galkovskaja & Sushchenya, 1978; Eryomova, 1982], our experiments were conducted only with acclimated animals. The rotifers were taken from continuous cultures kept at 10 °C, 15 °C, 20 °C, 25 °C, 30 °C and 35 °C, respectively. The relationship between individual (R) and weight-specific (R/W) rates of oxygen consumption and temperature in *Brachionus calyciflorus* showed a positive correlation and could be defined as follows:

$\ln (R) = -(3.0597 \pm 0.394) + (0.0895 \pm 0.0156)t$
($r = 0.868$; $n = 41$; $P < 0.001$)
$\ln (R/W) = (2.835 \pm 0.5379) \pm (0.0856 \pm 0.0214)t$
($r = 0.899$; $n = 41$; $P < 0.001$)

where R is in ml $\times 10^{-5}$ O_2 $inD.^{-1}$ h^{-1}; R/W is in μg O_2 mg^{-1} d.w. h^{-1}; t is in °C.

However, the difference between the mean values at 30 °C and 35 °C (Fig. 4) was not significant.

Figure 4 shows rates of oxygen consumption for laboratory populations of *Brachionus calyciflorus*, and natural populations of *B. calyciflorus* and *Hexarthra mira* (Hudson) collected from a pond. The temperature variation in the ponds when the animals were collected ranged from 22.3 °C to 32.0 °C and the respiratory experiments were carried out at 22–23 °C, 27 °C, 32 °C and 37 °C. Maximal respiratory rates for both species were observed at 32 °C and the rate decreased at 37 °C.

Exponential equations of the form $R = ae^{bt}$ (where R is respiratory rate, t is temperature and a,

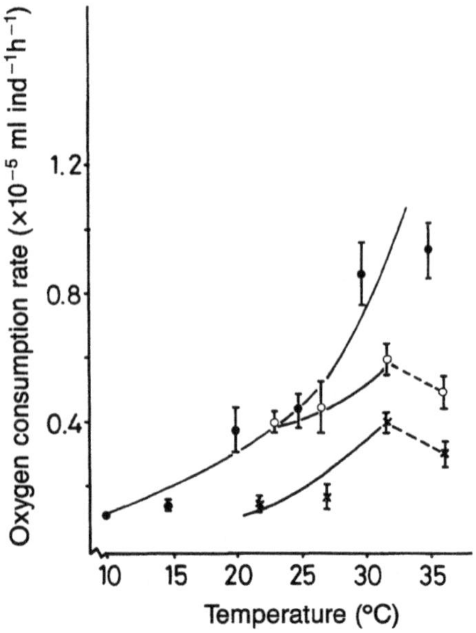

Fig. 4. Oxygen consumption rate ($R \times 10^{-5}$ ml $inD.^{-1}$ h^{-1}) in relation to temperature for experimental (●) and natural (○) populations of *Brachionus calyciflorus* Pallas and *Hexarthra mira* (Hudson) (×).

b and constants) were calculated for the exponential parts of the respiratory curves (Table 3, Fig. 4). In the laboratory populations of *B. calyciflorus* the empirical values of respiratory rate measured at 35 °C differed significantly from the values predicted by the corresponding equation given in Table 3 ($t = 3.06$; $P = 0.001$).

Similarly, the empirical mean respiration rates for the natural populations of *B. calyciflorus* and *H. mira* differed significantly from the predicted values at 37 °C ($t = 4.4$, $P = 0.001$, $t = 6.8$, $P = 0.001$, respectively) (see Table 3 & Fig. 4 for details).

By analysing the complete data set for each of the three groups studied, using second and third order parabolic regression equations, it was possible to determine more precisely the temperatures at which respiratory rates were maximal in each case. These were 33.4 °C for the experimental population of *Brachionus calyciflorus*, 32.6 °C for the field populations of this species and 32.9 °C for the field populations of *Hexarthra mira*. Hence, the overall maximal temperatures for oxygen consumption fell within the range 30 °C to 35 °C.

Table 3. Parameters of equations of the form $R = a^{bt}$ describing the relationship between oxygen consumption rate (R, $\times 10^{-5}$ ml ind^{-1} h^{-1}) and temperature (t, °C) in *Brachionus calyciflorus* and *Hexarthra mira* from experimental (exp. pop.) and natural (nat. pop.) populations. n = number of observations; δ_R is a measure of the deviation of the empirical values from the generated curve.

Rotifer	Equation	n	δ_R	Temperature range for regression calculation (°C)
B. calyciflorus (exp. pop.)	$R = 0.040e^{0.098t}$	35	0.173	10 – 30
B. calyciflorus (nat. pop.)	$R = 0.117e^{0.049t}$	26	0.155	23 – 32
H. mira (nat. pop.)	$R = 0.0095e^{0.114t}$	0.070	0.070	22 – 32

Food consumption and temperature

The ingestion of *Chlorella* sp. by *Brachionus calyciflorus* was studied at food concentrations of 0.3×10^6 to 245×10^6 cells ml^{-1} and at 20 °C, 25 °C, 30 °C, 35 °C, 37 °C and 39 °C. Table 4 shows that the maximal ingestion rate increased form 1.2×10^4 cells $inD.^{-1}$ h^{-1} at 20 °C to 3.6×10^4 cells $inD.^{-1}$ h^{-1} at 35 °C. The food concentration at which maximal ingestion rates were achieved increased with temperature throughout the temperature range tested. Figure 5 shows that maximal feeding, i.e. the saturation point, was not reached at the food concentrations studied and that the relationship between the logarithms of ingestion rate and food level is linear. The mean ingestion rates at the highest food concentrations studied was

Table 4. Maximal ingestion rates recorded at the temperatures indicated for *Brachionus calyciflorus* over a range of food densities ($0.3 \times 10^6 - 2.45 \times 10^8$ *Chlorella* cells ml^{-1}).

Temperature (°C)	Number of observations	C_{max} ($\times 10^4$ cells ind^{-1} h^{-1})
20	19	1.2
25	18	1.7
30	46	2.6
35	38	3.6

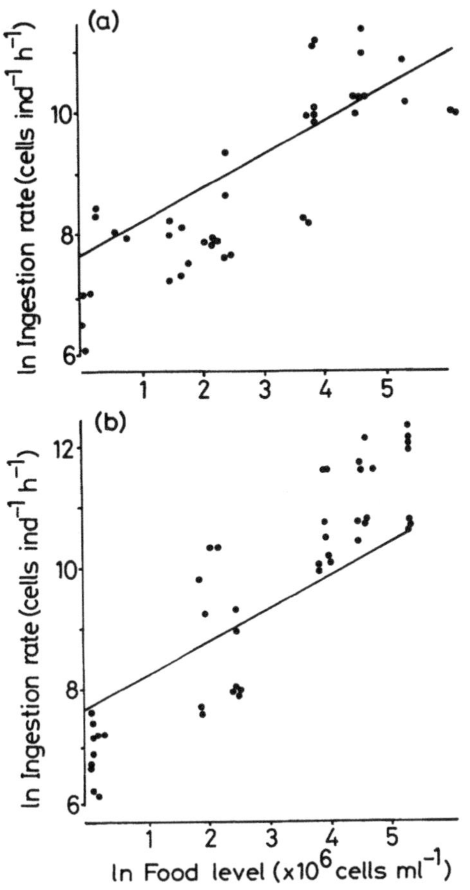

Fig. 5. Ingestion rate of *Brachionus calyciflorus* Pallas in relation to food concentration at (a) 37 °C and (b) 39 °C. The relationships between ingestion rate (c, cells inD.$^{-1}$ h^{-1}) and food concentration (C × 10^6 cells ml^{-1}) are described by the following equations (a) ln (c) = 6.9441 + 0.6375 ln (C) and (b) ln (c) = 6.9288 + 0.7943 ln (C).

3.4×10^4 cells inD.$^{-1}$ h^{-1} at 37 °C and 8.0×10^4 cells inD.$^{-1}$ h^{-1} at 39 °C. Assuming that the dry weight of an individual rotifer is 0.2 µg and that of 1×10^6 cells of *Chlorella* sp. is 6.23 µg, the daily ration at 37 °C will be 25 times that of the body weight. Duncan [1984] has published a value of 33.5 times the body weight for tropical *B. calyciflorus* living at temperatures of 26–28 °C. It is difficult to understand the biological meaning, importance or relevance of such ingestion values without comparing them to the rotifer's mass growth. Galkovskaya [1983] has shown that the potential reproductive rate (r_c) can exceed 0.091 h^{-1} which is equivalent to a doubling every 7.6 hours. Under such conditions, the reproduction rate would be 9.38 offspring per day. In this case, the possibility that the daily food intake is more than 25 times the animals body mass seems realistic, assuming that 40% of assimilated food is used for growth.

The ingestion rate at 37 °C was comparable to that at 35 °C, but the rate at 39 °C was three times larger. However, there was a sharp decrease in growth rate at 39 °C compared to that at 35 °C and 37 °C. This phenomenon probably reflects an imbalance between food ingested and food digested. In general, data for rotifers reared at such high temperatures is difficult to obtain because mortality is greatly increased in succeeding generations of animals cultured at 39 °C.

Experimental population dynamics at different temperatures

Temperature is one of the main factors affecting population parameters such as birth rate and growth rate. As early as 1967, King used Edmondson's series of papers, published between 1945 and 1965, to conclude that the birth rate of rotifer populations was most influenced by temperature, food quantity and food quality, in that order. This conclusion has been confirmed subsequently by a number of authors [Halbach, 1970a, 1970b; Hillbricht-Ilkowska & Pourriot, 1970; King, 1972; Pourriot, 1973; Galkovskaja, 1979; Walz, 1983b; Pascual & Yufera, 1983]. It is far more difficult to identify the direct effect of temperature on population growth parameters than to estimate its effect on the growth parameters of individuals. Studies on experimental populations of *Brachionus caluciflorus* have recorded maximal values for instantaneous growth (r) at temperatures of 30–35 °C [Galkovskaja, 1979], when cultured on *Chlorella* sp. at a concentration of 10^6 cells ml^{-1}. The experimental temperatures ranged from 10 °C to 37 °C. The maximal value recorded was r = 0.75 d^{-1}. Later, it was shown that above 20 °C the instantaneous growth rate increased with food concentration up to concentration of 5×10^7 cells ml^{-1}; during the growth period, r could reach 2.0 d^{-1}, but averaged 0.9 d^{-1} at 25 °C and 1.5 d^{-1}

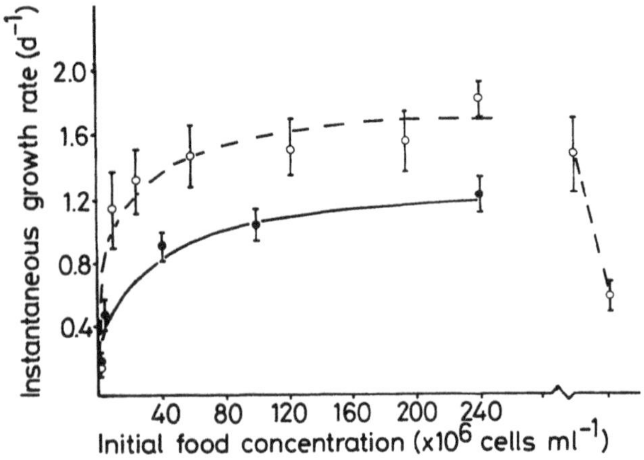

Fig. 6. Instantaneous growth rate (r, d^{-1}) of *Brachionus calyciflorus* Pallas at 25 °C (●), and 33 °C (o), grown on *Chlorella* sp. at a range of food densities.

at 33 °C (Fig. 6). Since r-values are known to depend upon population density, r-values can be compared only if determined under the same initial animal densities.

Figure 7 shows that the daily production of *B. calyciflorus* at 33 °C was 2.5–3 times greater than

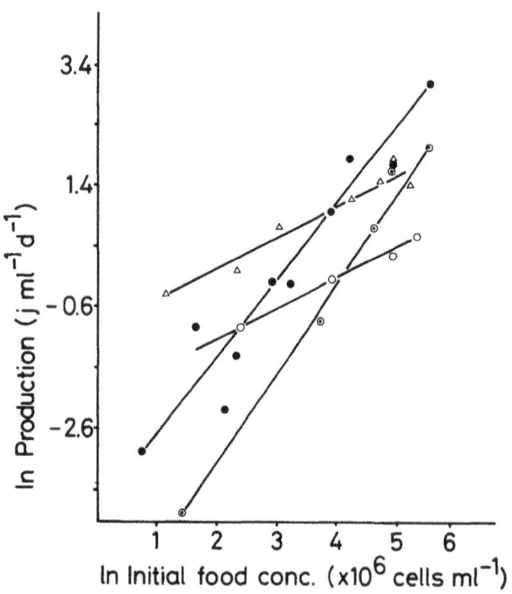

Fig. 7. Daily production of *Brachionus calyciflorus* Pallas populations in relation to initial food concentration at 25 °C (⊙); 33 °C (●); 35 °C (△); 37 °C (o). Initial rotifer density was 5–7 inD. ml^{-1}.

at 25 °C, over the whole range of food concentrations studied. However, at temperatures higher than 33 °C, the relationship between production and algal concentration changed. This resulted in production at 35 °C being greater than that at 33 °C at comparatively low food concentrations (i.e. $\leq 5 \times 10^7$ cells ml^{-1}). At higher food concentrations (i.e. $= 5 \times 10^7$ cells ml^{-1}), production at 35 °C was less than at 33 °C. Thus, production increased sharply with food concentration at 25 °C and 33 °C, and increased much more slowly with food concentration at 35 °C and 37 °C, exceeding the 25 °C and 33 °C rates only when food concentrations were less than 1×10^7 cells ml^{-1}. At lower temperatures, and with a good food supply (0.2×10^6 cells inD.$^{-1}$), growth and assimilation of food were enhanced. With a greater supply of food (2.0×10^6 cells inD.$^{-1}$), not all of the food was assimilated and an appreciable amount died, rather than being eaten, and decomposed having a detrimental effect on the growth of the experimental populations, especially at high temperatures.

Discussion

It is well known that the development of rotifer populations in water bodies is mainly determined by trophic conditions, pH, temperature and the presence of predators [Edmondson, 1960, 1965; Gallagher, 1963; Baker, 1979; Moore, 1981]. Often, temperature is considered to be the most important factor determining the population dynamics of rotifers [Ruttner-Kolisko, 1975; Hofmann, 1977]. Rotifer communities of 'cold' waters (i.e. arctic and subarctic), ranging from oligotrophic through to eutrophic, are generally far more similar to each other than warm-water and cold-water communities from the same waterbody [Nauwerck, 1978]. Many so-called cold-water/oligotrophic species also inhabit eutrophic waters in winter, although the optimum temperature for these animals, as estimated from the periods of maximum population density, is apparently a little higher in eutrophic waters.

Horkan *et al.* [1977] studied the rotifer community of a Canadian lake which was subject to an overall increase in temperature due to thermal effluent. The maximum temperature recorded before 'heating' was 20.4 °C; after 'heating' a maximum temperature of 29.4 °C was recorded. The authors observed that *Notholca acuminata* (Ehrenberg) and *Filinia longiseta* (Ehrenberg) disappeared at 22.4 °C. *Polyarthra vulgaris* Carlin, *Keratella earlinae* Ahlstrom and *Conochilus unicornis* Rouss. were eliminated at 22.6 °C, 26.5 °C and 25.6 °C, respectively. In contrast, *Keratella hiemalis* Carlin, *Keratella cochlearis* (Gosse), *Kellicottia longispina* (Kellicott), and *Asplanchna priodonta* Gosse remained in the plankton, even at 29 °C. This shows that temperature affects the survival of different rotifer species in different ways.

Analysis of the latitudinal distribution of the Brachionidae [Pejler, 1977] revealed no evidence to suggest that temperature alone limits the geographical distribution of this family; the differences in distribution were more closely associated with trophic condition. However, there is a tendency for the distributions of at least some species to be related to temperature. *Brachionus* species, for example, seem to prefer subtropical zones, while *Notholca, Argonotholca* and *Kellicottia* are more commonly found in temperate zones. Recently, it has been suggested that species of *Notholca, Argonotholca* and *Kellicottia* can exhibit an extended range of temperature tolerance in running and saline waters e.g. *Notholca acuminata* was recorded in the Rakhad river of Sudan, Africa, by de Ridder [1984]. Nevertheless, only two of the 23 recorded *Notholca* species (including the 13 species endemic to lake Baikal) have been found in subtropical and tropical zones.

The effect of temperature on rate processes of individual rotifers, or of rotifer populations, is not well studied. Most of the published data concerns the effect of temperature on embryonic development time and has largely been summarised by Vinberg & Galkovskaja [1979]. Morozov [1983] found a linear relationship between embryonic development rate and temperature, within the range 15 °C to 35 °C, for temperature acclimated clonal cultures of *Brachionus calyciflorus.* Similarly, Edmondson [1965] found a linear relationship between these parameters in field populations. Other authors have shown that the relationship between embryonic development rate and temperature in laboratory and field populations of *Keratella cochlearis* is also best described by a straight line [Walz, 1983a; Galkovskaja & Mityanina, 1985]. When our own data on these two species was added to that found in the literature, the resulting data set covered a wide range of temperatures and the relationship between embryonic development rate and temperature was then better described by a second order parabolic equation.

Using a stepwise method of acclimation we were able to rear *B. calyciflorus* cultures at temperatures of up to 40 °C. However, at 39–40 °C these cultures were short-lived, although the brackish-water rotifer *Brachionus plicatilis* Müller can be successfully cultivated at these high temperatures [Pascual & Yufera, 1983]. It is, perhaps, worthwhile to note that between 37 °C and 40 °C increasing temperature no longer accelerated the embryonic and juvenile development phases in *B. calyciflorus*, as occurred at lower temperatures; indeed, at 40 °C a pronounced decrease in the rate of embryonic development occurred. The rate of embryonic development over the whole range of temperature tolerance [Pattee *et al.*, 1973] of *B. calyciflorus* is probably best described by a parabolic curve with a maximum at 30–35 °C. A sharp reduction in life expectancy was also recorded at these higher temperatures.

Within the same high temperature range, the maximum oxygen consumption rate and production of rotifer populations were also determined. However, these rate processes, for an individual and for a population, are affected by trophic conditions. For example, feeding rotifers can consume 6 times as much oxygen as animals which are not feeding [Galkovskaya, 1963]. In addition, the direct effect of temperature on instantaneous growth rate in rotifers is also difficult to determine, since it is mainly affected by population density and food supply. Even with similar food supplies at the same concentrations, different relationships between r and temperature are often recorded. The main problem in studying the effect of temperature and feeding conditions on the rate of processes of rotifers lies in the fact that feeding conditions are themselves also a function of temperature. Continuous cultivation in chemostats is probably the only experimental method which will improve our understanding and quantitative measurement of these interactions.

Acknowledgements

Grateful thanks to Nan Duncan, Alois Herzig and Linda May who reviewed and revised the manuscript.

References

Amrén, H., 1964. Ecological studies of zooplankton in some ponds on Spitsbergen. Zool. Bidr. Upps. 36: 161–191.

Baker, R. L., 1979. Birth rate of planktonic rotifers in relation to food concentration in a shallow, eutrophic lake in Western Canada. Can. J. Zool. 57: 1206–1214.

De Ridder, M., 1984. A review of the rotifer fauna of the Sudan. Hydrobiologia 110: 1113–130.

Dumont, H. J., 1983. Biogeography of rotifers. Hydrobiologia 104: 19–30.

Duncan, A., 1983. The influence of temperature upon the duration of embryonic development of tropical Brachionus species (Rotifera). In F. Schiemer (ed.)., Limnology of Parakrama Sumadra-Sri Lanka: A case study of an ancient man-made lake in the trophics. Dr. W. Junk Publishers, The Hague: 107–115.

Duncan, A., 1984. Assessment of factors influencing the composition, body size and turnover rate of zooplankton in Parakrame Sumadra an irrigation reservoir in Sri Lanka. Hydrobiologia 113: 201–215.

Edmondson, W. T., 1960. Reproductive rates of rotifers in natural populations. Mem. Ist. ital. Idrobiol. 12: 21–77.

Edmondson, W. T., 1965. Reproductive rates of planktonic rotifers as related to food and temperature in nature. Ecol. Monogr. 35: 61–112.

Eryomova, N. G., 1982. Metabolic adaptation of planktonic crustacea to temperature with the example of Daphnia magna Straus. Autoeferat diss. kand. biol. nauk Minsk, 22 pp. (In Russian).

Fisher, R. A., 1930. The genetical theory of natural selection. Claredon Press, Oxford, 272 pp.

Galkovskaja, G. A., 1963. The relationship between food utilization and the growth and maximum production of the rotifer Brachionus calyciflorus Pallas. Zool. Zh. 43: 506–512 (In Russian).

Galkovskaja, G. A., 1979. The acclimation temperature and instantaneous rate of increase of experimental populations of the rotifer Brachionus calycifllorus Pallas. Zh. Obshch. biol. 40: 734–739. (In Russian).

Galkovskaya, G. A., 1983. On temperature acclimation in an experimental population of Brachionus calyciflorus. Hydrobiologia 104: 225–227.

Galkovskaya, G. A. & L. M. Suschchenya, 1978. Growth of aquatic animals under variabe temperature conditions. Minsk, Nauka i Tekhnika, 142 pp. (In Russian).

Galkovskaja, G. A. & I. F. Mityanina, 1983. Capacity for parthenogenetic reproduction in the rotifer Brachionus calyciflorus Pallas at high temperatures. DAN BSSR 27: 184–186 (In Russian).

Galkovskaja, G. A. & I. F. Mityanina, 1985. Birth rate characteristics of natural populations of the rotifer Keratella cochlearis (Gosse). DAN BSSR: v pechati. (In Russian).

Gallagher, J. J., 1963. Field verification of two general factors influencing population variation in euplanktonic rotifers. Proc. Louisiana Acad. Sci. 26: 58–65.

Green, J., 1956. Growth, size and reproduction in Daphnia (Cladocera Crustacea). Proc. zool. Soc. Lond. 126: 173–204.

Halbach, U., 1970a. Einfluss der Temperatur auf die Populationsdynamik des planktischen Rädertieres Brachionus calyciflorus Pallas. Oecologia 4: 176–207.

Halbach, U., 1970b. Die Ursachen der Temporalvariation von Brachionus calyciflorus Pallas (Rotatoria). Oecologia 4: 262–318.

Herzig, A., 1983. The ecological significance of the relationship between temperature and duration of embryonic development in planktonic freshwater copepods. Hydrobiologia 100: 65–91.

Hillbricht-Ilkowska, A. & R. Pourriot, 1970. Production of experimental population Brachionus calyciflorus Pallas (Rotatoria) exposed to artificial predation of different rates. Pol. Arch. Hydrobiol. 17: 241–248.

Hofmann, W., 1977. The influence of abiotic environmental factors on population dynamics in planktonic rotifers. Arch. Hydrobiol. Beih. 8: 77–83.

Horkan, H., G. Gallup & P. Nursall, 1977. Effects of thermal effluent upon the planktonic Rotifera – survival and egg production. Arch. Hydrobiol. Beih. 8: 84–87.

Ivleva, I. V., 1981. Environmental temperature and the rate of energetic exchange of aquatic animals. Kiev, Naukova dumka; 231 pp. (In Russian).

King, C. E., 1967. Food, age and dynamics of a laboratory population of rotifers. Ecology 48: 111–128.

King, C. E., 1972. Adaptation of rotifers to seasonal variation. Ecology 53: 408–418.

Kryutchkova, N. M. & V. Kh. Rybak, 1974. Development time, life length and fecundity of 4 species of rotifers. Kiev, Naukova dumka, 231 pp. (In Russian).

Laughlin, R., 1965. Capacity for increase: a useful population statistic. J. anim. Ecol. 34: 77–91.

Magnien, R. E. & J. J. Gilbert, 1983. Diel cycles of reproduction and vertical migration in the rotifer Keratella crassa and their influence on the estimation of population dynamics. Limnol. Oceanogr. 28: 957–969.

McLaren, I. A., 1965. Some relationships between temperature and egg size, body size, development rate, and fecundity of the Copepod Pseudocalanus. Limnol. Oceanogr. 10: 525–538.

Mityanina, I. F., 1983. The growth and development of daphnids and rotifers, ontogenesis and consecutive parthenogenetic generations, in relation to temperature. Avtoref. diss. kand. biol. nauk. Minsk, 24 pp. (In Russian).

Moore, J. W., 1981. Influence of temperature, photoperiod and trophic conditions on the seasonal cycles of phytoplankton and zooplankton in two deep Subarctic lake of Northern Canada. Int. Revue ges. Hydrobiol. 66: 745–770.

Morozov, A. M., 1983. Thermo-resistance and development rate of the rotifer Brachionus calyciflorus Pallas in relation to temperature. Vesti An BSSR, ser. biyal. navuk: 93–94. (In Russian).

Nauwerck, A., 1978. Notes on the planktonic rotifers of Lake Ontario. Arch. Hydrobiol. 84: 269–301.

Pascual, E. & M. Yufera, 1983. Crecimento en cultivo de una cepa de Brachionus plicatilis O. F. Müller en funcion de la temperatura y la salinidad. Invest. pesq.: 151–159.

Pattee, E., C. Lascombe & R. Delolme, 1973. Effect of temperature on the distribution of Turbellarian Triclads. In W. Wieser (ed.), Effects of temperature on ectothermic organisms, Berlin: 201–207.

Pejler, B., 1977. On the global distribution of the family Brachionidae (Rotatoria). Arch. Hydrobiol. Supplement 53: 255–306.

Pourriot, R., 1973. Rapports entre la temperature, la taille des adultes, la longeur des oeufs et le taux de development embryonnaire chez Brachionus calyciflorus Pallas (Rotifera). Ann. Limnol. 4: 103–115.

Pourriot, R. & A. Hillbricht-Ilkowska, 1969. Recherches sur la biologie de quelques rotifères planctoniques. I. Resultats preliminaire. Bull. Soc. zool. Fr. 94: 111–118.

Pourriot, R. & M. Deluzarches, 1971. Recherches sur la biologie des rotifères. Influence de la temperature sur la durée du developpment embryonnaire et postembryonnaire. Ann. Limnol. 7: 25–52.

Rigler, F. H., MacCallum & J. C. Roff, 1974. Production of zoo-plankton in Char Lake. J. Fish. Res. Bd Can. 31: 637–646.

Rokitsky, P. F., 1964. Biological statistics. Minks, Vysheishaya shkola: 328 pp. (In Russian).

Ruttner-Kolisko, A., 1975. The influence of fluctuating temperature on plankton rotifers. A graphical model based on life data of Hexarthra fennica from Neusiedlersee, Austria. Symp. Biol. Hung. 15: 197–204.

Sudzuki, M., 1955. Life history of some Japanese rotifers. I. Polyarthra trigla Ehr. Sci. rep. Tokyo Kyoiku Daig., 8B 118: 41–63.

Vasilyeva,a G. L. & V. L. Nikulin, 1974. Production of Baikal rotifers. Zool. zh. 53: 1125–1132. (In Russian).

Vinberg, G. G. & G. A. Galkovskaya, 1979. Relationship between the development rate of rotifers and temperature. V kn. Obshchyie osnovy izuchieniya vodnykh ekosistiem. L, Nauka: 149–155.

Walz, N., 1983a. Individual culture and experimental population dynamics of Keratella cochlearis (Rotatoria). Hydrobiologia 107: 35–45.

Walz, N., 1983b. Continuous culture of the pelagic rotifers Keratella cochlearis and Brachionus angularis. Arch. Hydrobiol. 98: 70–92.

Hydrobiologia 147: 319–322 (1987)
May, L., Wallace, R. & Herzig, A. (eds), Rotifer Symposium IV

Effect of algal diet and temperature on the embryonic development time of the rotifer *Brachionus plicatilis* in culture

M. Yúfera
Instituto de Ciencias Marinas de Andalucía (C.S.I.C.) Polígono Río San Pedro s/n, Apartado oficial Puerto Real, Cádiz, Spain

Keywords: embryonic development, algal food, temperature, *Brachionus plicatilis*, culture

Abstract

The embryonic development times of two strains of *Brachionus plicatilis* (Bs and S-1) cultured on three different algal diets (*Nannochloris oculata, N. maculata* and *Nannochloropsis gaditana*), have been determined at 20 °C, 25 °C and 30 °C. As expected, the embryonic development times decreased with increasing temperature in all cases. However, embryos from adults fed on *N. gaditana* tended to develop more slowly than those of individuals fed on the other algal species. Mean egg volume was also affected by diet, larger eggs being produced by females fed on *N. gaditana.* No obvious relationship between egg size and temperature was detected.

Two principal factors seemed to affect the embryonic development time. The first was temperature which acts through its well known effect on metabolic rates. The second was maternal diet which probably affects development time through its effect on yolk content, as reflected in the size of the egg.

Introduction

It is well known that embryonic development time in rotifers, as in other zooplankton, is highly correlated with temperature [for recent review see Duncan, 1983; Herzig, 1983]. However, there was no evidence that other factors could affect this process, until recent studies detected a possible influence of food quantity [Walz, 1983] and quality [Yúfera & Pascual, 1984] in *Brachionus* spp.

The literature reports a relationship between several parameters involved in the developmental process, particularly between embryonic development time, egg volume and maternal size [Pourriot, 1973; Duncan, 1983]. Some authors have also reported that food type may influence the size of females in cultures [Yúfera, 1982; Snell & Carrillo, 1984]. This suggests a possible association between food and embryonic development time.

In the present work, the influence of diet on the embryonic development time of rotifers in mass culture has been studied and compared to the effect of temperature. Variations in egg volume under different conditions have also been examined, with particular reference to the relationship between egg size and development time.

Materials and methods

Two strains of *Brachionus plicatilis* (strains Bs and S-1) were used for the experiments. The characteristics of both strains, and the culture methods for algae and rotifers, are described by Yúfera [1982] and Yúfera *et al.* [1983]. The algal species used as food were: *Nannochloris oculata* (Cambridge, 251/6), *Nannochloris maculata* (Cambridge, 251/13) and *Nannochloropsis gaditana* (Cádiz, B-3).

The embryonic development time of the rotifers was determined at 20 °C, 25 °C and 30 °C for each algal species. Females were removed from 1 litre culture flasks (after acclimatization to the experimental food and temperature) during the exponential growth phase. The animals were placed in small petri dishes with ≃5 ml of algal suspension at a concentration of 50–100 μg d.w. ml^{-1}. The algae were also taken from cultures in the exponential phase of growth. Cultures were examined at half hourly intervals during the critical periods before laying and hatching. Thus, each determination has an error of ±0.5 hours. Embryo development time was determined 20 times for each set of culture conditions.

Size of females [following the method of Yúfera, 1982] and egg volume (0.524 × length × $width^2$) were measured with an ocular micrometer with 2 μm precision.

Results

Figure 1 shows the relationship between embryonic development time and temperature for each alga tested. The data has been fitted to an exponential regression (see Table 1). Both *Brachionus* strains showed the expected decrease in embryonic development time with increasing temperature, but the values were different for each strain. Embryos from strain S-1 consistently developed at a slower rate than those from strain Bs, for any given temperature. In addition, populations fed on *Nannochloropsis gaditana* tended to have longer development times than those fed on *Nannochloris oculata*

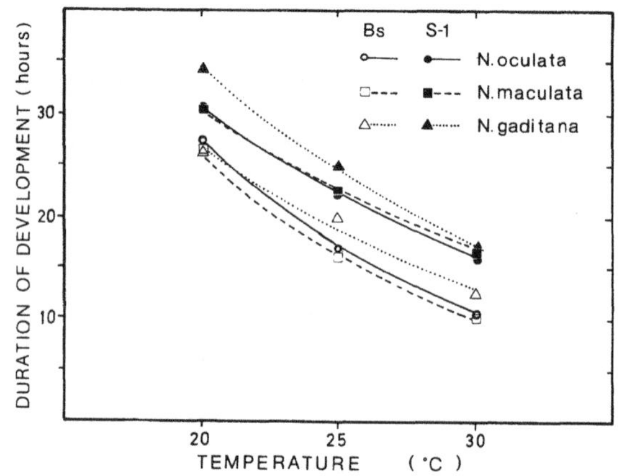

Fig. 1.. Relationship between embryonic development time and temperature for two strains of *Brachionus plicatilis* (Bs and S-1) fed on three different algal species.

and *N. maculata*. This was true of both rotifer strains over all temperatures tested, except for strain Bs at 20 °C and strain S-1 at 30 °C.

An analysis of variance showed a highly significant influence of temperature, algal diet and rotifer strain ($P < 0.001$) on embryonic development time (Table 2). The two-way interactions (algae × strain) and (temperature × algae) as well as the three-way interaction were significant ($P < 0.001$). These data showed that embryonic development time was influenced primarily by temperature and genotype. The effect of diet was less and depended upon the genetic characteristics of the rotifers, and on their response to temperature when fed on each algae.

Temperature did not sifnificantly affect egg volume (Table 3). Only strain Bs showed any response – a reduction in egg size at 30 °C. However, for a given

Table 1. Coefficients of an exponential equation of the form $y = ae^{bx}$ relating embryonic development time to incubation temperature for two strains of *Brachionus plicatilis* (Bs and S-1) fed on three algal diets.

Alga	*Brachionus plicatilis* strain					
	Bs			S-1		
	a	b	r	a	b	r
N. oculata	181.79	−0.095	−0.994	110.74	−0.064	−0.995
N. maculata	173.13	−0.095	−0.996	98.70	−0.059	−0.991
N. gaditana	115.67	−0.073	−0.979	141.66	−0.075	−0.991

Table 2. Factorial ANOVA of the effects of temperature, algal food species, and rotifer strain on embryonic development time. Fixed model with 20 replicates.

Source	df	SS	SM	F	P
Temperature	2	13986.05	6993.03	13194.4	0.001
Algae	2	295.80	147.90	282.83	0.001
Strain	1	2667.78	2667.78	5033.55	0.001
Temperature × algae	4	55.36	13.84	26.11	0.001
Temperature × strain	2	1.70	0.85	1.60	ns
Algae × strain	2	31.80	15.90	29.94	0.001
Temperature × algae × strain	4	152.42	38.11	71.76	0.001
Error	342	181.50	0.53		

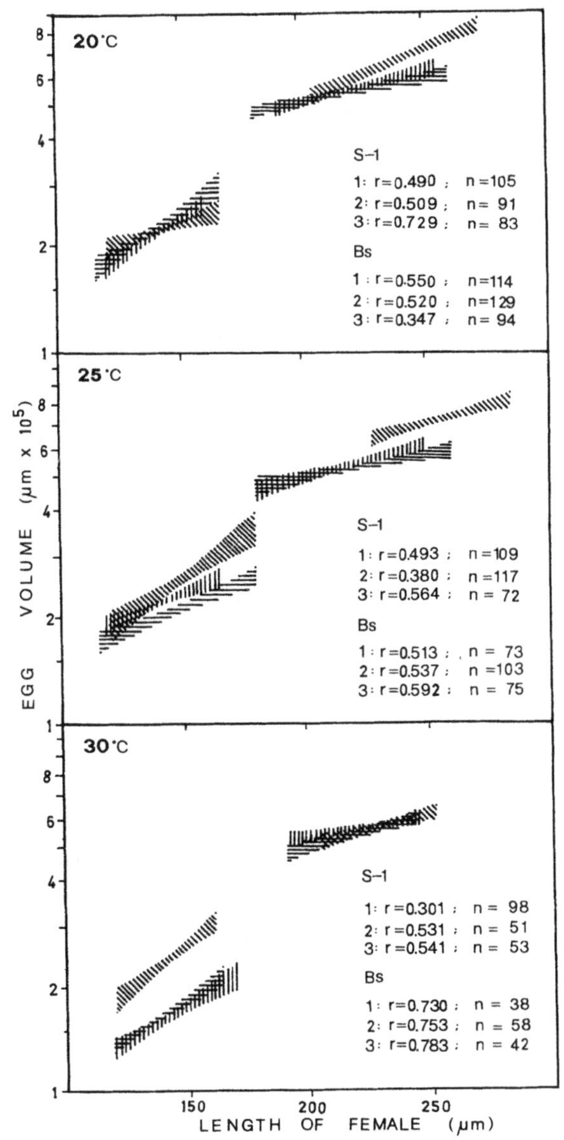

Fig. 2.. Relationship between egg volume and length of the mother individual. Shaded areas indicate 95% confidence intervals for animals fed *Nannochloris oculata* (vertical shading), *Nannochloris maculata* (horizontal shading) and *Nannochloropsis gaditana* (oblique shading). For each temperature the regressions shown on the right correspond to strain S-1, on the left to strain Bs.

temperature, egg volume did vary with diet. Animals fed *N. gaditana* tended to produce larger eggs. Figure 2 shows the relation between egg volume and maternal size. The correlation coefficients found (with the logarithmic transformed egg volumes) were significant in all cases ($P < 0.001$; except in strain S-1 fed on *N. oculata* at 30 °C: $P < 0.005$). However, the correlation coefficients were low (0.3 to 0.8) and only partially accounted for variability in egg volume.

Discussion

The results confirm that diet can affect the egg development time of *B. plicatilis* in mass culture. In this study, a change of diet to *N. gaditana* had a similar effect on embryonic development rate as a decrease in temperature of about 2 °C. This may be due to a nutritional deficiency since rotifers show limited ingestion rate with this alga [Yúfera & Pascual, 1985].

Table 3. Egg volume ± S.D. ($\mu m^3 \times 10^3$) and egg development time ± S.D. (hours) under three different food conditions and three incubation temperatures.

Strain:	Bs			S-1		
Temperature:	20	25	30	20	25	30
Egg volume						
N. oculata	219 ± 33	220 ± 24	166 ± 21	556 ± 65	545 ± 69	565 ± 50
N. maculata	227 ± 39	210 ± 40	164 ± 21	539 ± 65	530 ± 62	535 ± 49
N. gaditana	223 ± 28	258 ± 52	244 ± 37	691 ± 110	708 ± 78	572 ± 62
Development time						
N. oculata	27.4 ± 1.3	16.9 ± 0.6	10.6 ± 0.5	30.6 ± 0.6	22.2 ± 0.7	16.1 ± 0.5
N. maculata	26.0 ± 0.7	16.2 ± 0.6	10.1 ± 0.4	30.2 ± 0.7	22.5 ± 0.3	16.7 ± 0.9
N. gaditana	25.8 ± 1.0	19.8 ± 0.8	12.4 ± 0.5	34.2 ± 0.8	25.0 ± 0.9	16.9 ± 0.8

In contrast to results obtained by Pourriot [1973] for *Brachionus calyciflorus*, the present study did not show a clear relationship between egg volume and temperature. Further experiments over a wider range of temperature would be necessary to confirm this. However, the results did show that larger eggs and longer embryonic development times were associated with algal diets inducing lower egg per female ratios [Yúfera & Pascual, 1984]. This is similar to the findings of Wals [1983] who reported a good correlation between clutch size and egg volume in *Brachionus angularis*. Although the relationship between egg volume and development time is clear, the nature of this relationship is not entirely understood.

In summary, two main factors seem to affect embryonic development time in rotifers. The first, and probably the most important factor, is temperature which influences egg development rate through its effect on metabolic rates. The second is egg volume (i.e. yolk content) which is determined by the feeding conditions of the mother.

Acknowledgements

The author gratefully acknowledges the linguistic revision of the editors.

References

Duncan, A., 1983. The influence of temperature upon duration of embryonic development of tropical Brachionus species (Rotifera). In F. Schiemer (ed.), Limnology of Parakrama Samudra – Sri Lanka: a case study of an ancient man-made lake in the tropics. Junk, The Hague, Dev. Hydrobiol. 12: 107–115.

Herzig, A., 1983. Comparative studies on the relationships between temperature and duration of embryonic development of rotifers. Hydrobiologia 104: 237–246.

Pourriot, R., 1973. Rapports entre la température, la taille des adultes, la longeur des oeufs et la taux de développement embryonnaire chez Brachionus calyciflorus Pallas (Rotifère). Ann. Hydrobiol. 4: 103–115.

Snell, T. W. & K. Carrillo, 1984. Body size variation among strains of the rotifer Brachionus plicatilis. Aquaculture 37: 359–367.

Walz, N., 1983. Continuous culture of the pelagic rotifers Keratella cochlearis and Brachionus angularis. Arch. Hydrobiol. 98: 70–92.

Yúfera, M., 1982. Morphometric characterization of a small sized strain of Brachionus plicatilis in culture. Aquaculture 27: 66–61.

Yúfera, M. & E. Pascual, 1984. Influencia de la dieta sobre la puesta del rotífero Brachionus plicatilis en cultivo. Inv. Pesq. 48: 549–556.

Yúfera, M. & E. Pascual, 1985. Effects of algal food concentration on feeding and ingestion rates of Brachionus plicatilis in mass culture. Hydrobiologia 122: 181–187.

Yúfera, M., L. M. Lubián & E. Pascual, 1983. Efecto de cuatro algas marinas sobre el crecimiento poblacional de dos cepas de Brachionus plicatilis (Rotifera: Brachionidae) en cultivo. Inv. Pesq. 47: 325–337.

Hydrobiologia 147: 323–327 (1987)
May, L., Wallace, R. & Herzig, A. (eds), Rotifer Symposium IV

Factors influencing the occurrence of males in natural populations of *Synchaeta* spp.

Sif Johansson
Askö Laboratory, Institute of Marine Ecology, and Department of Zoology, University of Stockholm, S-106 91 Stockholm, Sweden

Keywords: *Synchaeta* spp., Baltic Sea, sexual reproduction

Abstract

Changes in the temporal abundance of females and males of *Synchaeta* spp. were followed for one year (1983) at two stations in the brackish northern Baltic proper. Males were found only when the abundance of females was increasing and the phytoplankton biomass was dominated by large diatoms.

Introduction

Most monogonont rotifers follow a pattern of parthenogenetic reproduction alternating with periods of sexual reproduction. The factors, which induce sexual reproduction have been the subject of several studies, exhaustively reviewed by Birky & Gilbert [1971], Gilbert [1974, 1977], and Pourriot & Snell [1983].

This study looks at the temporal occurrence of males of *Synchaeta* spp. in two areas of the northern Baltic proper with different primary productivity. The results are discussed in relation to changes in environmental conditions.

Materials and methods

Investigated area

The two stations investigated were situated in the southern Stockholm archipelago, Sweden. Station A, situated in a sheltered bay, Himmerfjärden, was directly influenced by the discharges from a modern sewage treatment plant. Station B was located in an area open to the Baltic Sea. In 1983, the annual primary production was estimated to be 256 and 148 g C m^{-2} $year^{-1}$ at stations A and B respectively (Larsson pers. comm.). Although phytoplankton and zooplankton biomass was substantially higher at station A than at station B, no major differences in species composition or the pattern of seasonal succession were recorded (unpubl. data).

In 1983, the salinity ranged from 6.3 to 7.6$^0/_{00}$ S at station A and from 6.5 to 8.3$^0/_{00}$ S at station B. The difference between the stations was usually less than 0.5$^0/_{00}$ S, but increased during autumn to a maximum of 1.1$^0/_{00}$ S in November.

From May to November the mean temperature in the trophogenic layer (stn. A:0–15 m, stn B:0–20 m) was 0.5 to 2 °C higher at station A than at station B. Maximum surface temperatures were reached at both stations in August (stn. A:17.6 °C, stn. B:17.1 °C).

Sampling

In 1983 the two stations were sampled weekly during the spring bloom and biweekly from May until late autumn. A total of 20 zooplankton samples were taken at each station in conjunction with an extensive pelagic sampling programme.

A tube sampler (length 1.5 m, volume 23 litres) was used to sample at 5 meter intervals from the surface to the sediments (stn. A:0–25 m, stn. B: 0–35 m). Plankton from the entire water column was pooled and concentrated by filtration through 90 and 35 μm mesh nets. The animals were narcotized with carbonated water and preserved in 4% buffered formaldehyde [Dybern *et al.,* 1976].

The samples were split using a Kott subsampler [Kott, 1953], and at least 500 specimens were counted [Hernroth & Viljamaa, 1979] with an inverted microscope. The genus *Synchaeta* (Ehrenberg) was dominated by *S. baltica* (Ehrenberg), but at least two additional species, *S. monopus* (Plate) and *S. curvata/fennica* (Lie-Pettersen/Rousselet) occurred. The species are difficult to distinguish in preserved samples, so all specimens of the genus *Synchaeta* have been grouped together as *Synchaeta* spp.

The phytoplankton biomass was divided into two size-classes (< 30 μm and > 30 μm) based on morphometric characters (mainly cell and/or chain size) (Hobro pers. comm.).

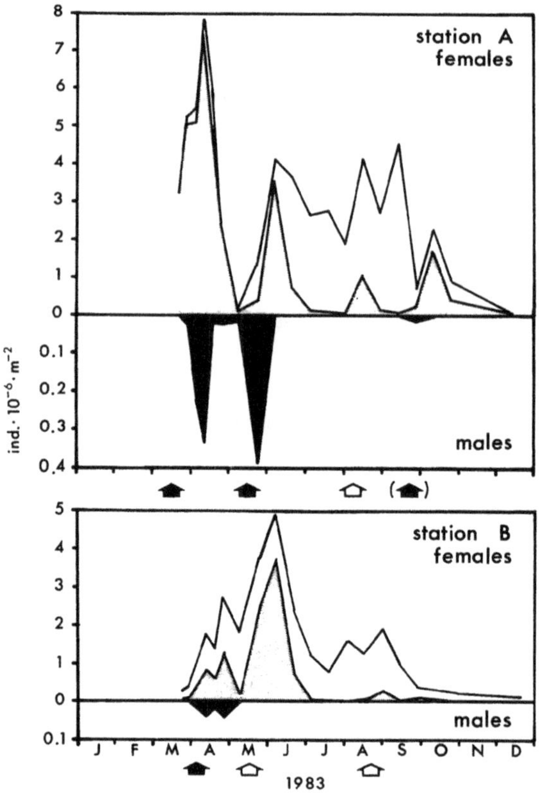

Fig. 1. Abundance of total net zooplankton, *Synchaeta* spp. females (lightly shaded) and males (heavily shaded) in 1983. Arrows indicate periods when *Synchaeta* spp. females started to increase in abundance. Filled arrows show periods when males also occurred.

Results

The seasonal abundance of female *Synchaeta* spp. in relation to the total abundance of net zooplankton at stations A and B is shown in Fig. 1. Peaks of abundance of *Synchaeta* spp. females were recorded in spring/early summer at both stations. Distinct peaks followed in August and October at station A, but at station B only one minor peak occurred in August/September. Males were present in low numbers during the first peak at both stations, and at station A during the second spring/early summer peak and at the autumn peak. Males occurred towards the beginning of a population increase, and only during periods when the phytoplankton biomass was dominated by large (> 30 μm) diatoms (*Skeletonema costatum* ((Grev.) Cleve), *Thalassiosira baltica* (Grunow), *Chaetocerus wighami* (Brightwell), *C. holosaticum* (Schutt) and *Coscinodiscus granii* (Gough), cf. Hobro [1979], Hobro pers. comm.) (Fig. 2).

The abiotic parameters changed very little during the periods (one to two weeks) preceeding the appearance of males, that is the periods when mictic females were induced. At station A the salinity was constantly 7.1‰ S and 6.5‰ S during the pre-male periods in April and September respectively. At station B a slight decrease was measured during the pre-male period in April, from 7.2 to 7.0‰ S. The temperature increased slowly in April, from 0.6 to 0.8 °C (+ 0.03 °C day^{-1}) at station A and from 1.1 to 1.9 °C (+ 0.06 °C day^{-1}) at station B. No vertical stratification of the water mass was detected at either station. In May, during the pre-male period at station A, the temperature increased from 4.2 to 5.9 °C (+ 0.13 °C day^{-1}) and in autumn it decreased from 14.8 to 12.7 °C (−0.15 °C day^{-1}). Daylength increased from 11 hours in mid-March to 18.5 hours at the end of June.

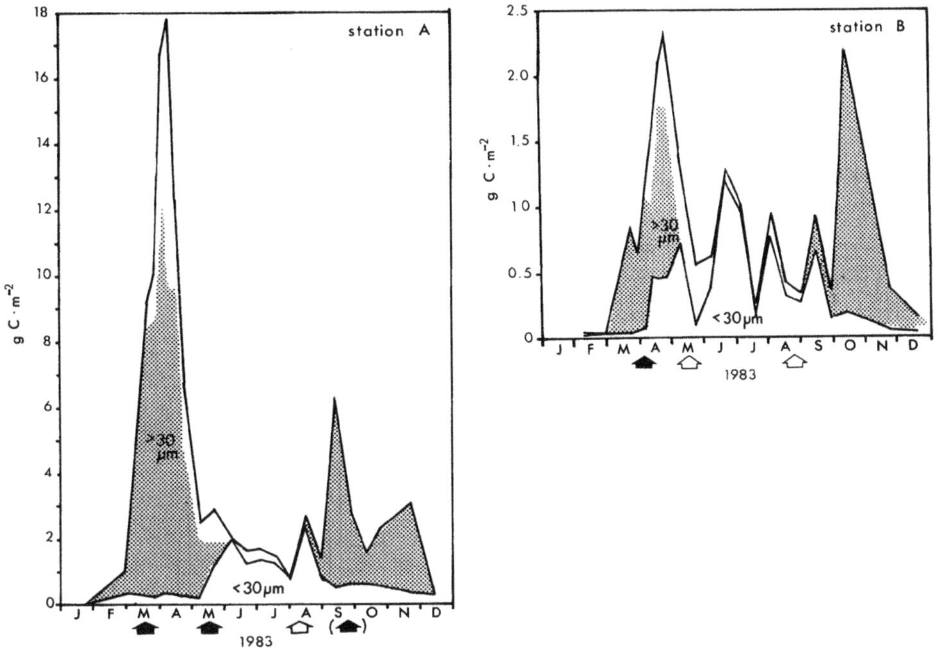

Fig. 2. Total biomass of phytoplankton divided into two size-classes (smaller and larger than 30 μm). Shaded area shows biomass of diatoms larger than 30 μm. Arrows as in Fig. 1.

Discussion

In experimental studies, changes in abiotic factors have induced sexual reproduction in rotifers. Reductions in salinity (from 100% to 50% and 25% seawater in the culture media) have led to mixis in *Brachionus plicatilis* [Lubzens *et al.,* 1980]. Temperature changes have been shown to influence the outcome of sexual reproduction [Halbach & Halbach-Keup, 1972; Hino & Hirano, 1984], but are not known to induce sexual reproduction except in the case of sudden cold shock [Ruttner-Kolisko, 1964]. Long daylength (>14 hours) was found to induce sexual reproduction in *Notommata copeus* [Pourriot, 1963]. In *Synchaeta* spp. in the Baltic proper, however, it seems unlikely that daylength alone induced sexual reproduction since males were found only at station A during late May/early June, while females were abundant at both stations. It is also improbable that changes in salinity and temperature induce mixis in these populations, since only very small changes were recorded during periods of mictic female production, i.e. the periods just preceding the occurrence of males. Thus, abiotic factors seem less likely to have induced the observed sexual reproduction in these natural populations.

Food quality and quantity have been shown to influence the induction of sexual reproduction (discussed by Pourriot & Snell [1983]). In *Asplanchna sieboldi* and *A. brightwelli* sexual reproduction was induced by a specific dietary substance, tocopherol [Gilbert, 1975]. Gilbert [1977] suggested that tocopherol acts as a cue indicating high phytoplankton biomass and, thus, favorable conditions for an increase of small herbivores, suitable as food for *Asplanchna.* In the present study, *Synchaeta* spp. males occurred when diatoms larger than 30 μm dominated the phytoplankton, potentially providing a rich food resource for this species. *Synchaeta* spp. are known to feed upon nannoplankton, different kinds of monads and centric diatoms from a few μm in size to larger than 50 μm [Naumann, 1923; Nauwerck, 1963; Pourriot, 1977; Stemberger, 1981]. In the laboratory *Synchaeta baltica* was observed to ingest ^{14}C-labelled *Skeletone-*

ma costatum (unpubl. data), usually the dominant diatom in this area. When large diatoms were abundant few other zooplankton were present to compete with *Synchaeta* spp. for food (Fig. 1).

Although high densities of *Brachionus* have been shown to induce sexual reproduction in laboratory cultures [Gilbert, 1963], males in natural populations have often been found before and during the abundance peaks [Wesenberg-Lund, 1930; Carlin, 1943; Nauwerck, 1963; Amrén, 1964; King & Snell, 1980] as was also the case in this study. This suggests that mictic female production is induced during the initial phase of the population increase. This is consistent with experimental observations on *Brachionus calyciflorus, B. plicatilis* and *Notommata copeus*. In these species, age of the parental female was shown to be inversely correlated to the production of mictic daughters [see Pourriot & Snell, 1983]. During a population increase there will be a relatively high proportion of young individuals, since fecundity decreases as the population reaches its maximum abundance [Hofmann, 1977]. King [1980] emphasized that the timing of sexual reproduction will also affect the genetic variation among the resting eggs produced. Since genetic variation decreases in a population reproducing parthenogenetically, sexual reproduction occurring early during the population increase may be advantageous, since it will produce resting eggs with a larger genetic variation.

In this study males were not found during some abundance peaks, indicating that sexual reproduction was not induced, despite favorable conditions for a populations increase. Experimentally, Snell [1986] showed that asexual and sexual females of *Brachionus plicatilis* gave different reproductive responses when exposed to a range of temperatures, salinities, and food concentrations. The general conclusion was that asexual females, in comparison with sexual, reproduced over a broader range of environmental conditions.

The results from the present study indicate that sexual reproduction in natural populations of *Synchaeta* spp. is induced at an early stage of a population increase if phytoplankton biomass is dominated by large diatoms. These are suggested to be favorable food conditions.

Acknowledgements

My sincere thanks to U. Larsson, B. Ganning and L. May for critical reading the manuscript, C. Hill for linguistic corrections, and to all my collegues at the Askö Laboratory for their support. This study was financed by the Swedish Environmental Protection Board.

References

Amrén, H., 1964. Ecological and taxonomical studies on zooplankton from Spitsbergen. Zool. Bidr. Upps. 36: 209–276.

Birky, C. W. Jr. & J. J. Gilbert, 1971. Parthenogenesis in rotifers: The control of sexual and asexual reproduction. Am. Zool. 11: 245–266.

Carlin, B., 1943. Die Planktonrotatorien des Motalaström. Medd. Lunds Univ. Limnol. Inst. 17: 1–163.

Dybern, B. I., H. Ackefors & R. Elmgren, 1976. Recommendations on methods for marine biological studies in the Baltic Sea. Baltic mar. Biol. 1, 98 pp.

Gilbert, J. J., 1963. Mictic female production in the rotifer Brachionus calyciflorus. J. Exp. Zool. 153: 113–124.

Gilbert, J. J., 1974. Dormancy in rotifers. Trans. am. micros. Soc. 93: 490–513.

Gilbert, J. J., 1975. Polymorphism and sexuality in the rotifer Asplanchna, with special reference to the effects of prey-type and clonal variation. Arch. Hydrobiol. 75: 442–483.

Gilbert, J. J., 1977. Mictic female production in monogonont rotifers. Arch. Hydrobiol. Beih. 8: 142–155.

Halbach, U. & G. Halbach-Keup, 1972. Einfluss von Aussenfaktoren auf den Fortpflanzungsmodus hetergoner Rotatorien. Oecologia 9: 203–214.

Hernroth, L. & H. Viljamaa, 1979. Recommendations on methods for marine biological studies in the Baltic Sea. Mesozooplankton biomass assessment. Baltic mar. Biol. 6, 15 pp.

Hino, A. & R. Hirano, 1984. Relationship between water temperature and bisexual reproduction rate in the rotifer Brachionus plicatilis. Bull. Jap. Soc. Sci. Fish. 50: 1481–1485.

Hobro, R., 1979. Annual phytoplankton succession in a coastal area in the northern Baltic. In E. Naylor & R. G. Hartnoll (eds.), Cyclic phenomena in marine plants and animals. Pergamon Press, Oxford, N.Y. pp. 3–10.

Hofmann, W., 1977. The influence of abiotic environmental factors on population dynamics in planktonic rotifers. Arch. Hydrobiol. 24: 329–341.

King, C. E., 1980. The genetic structure of zooplankton populations. In W. C. Kerfoot (ed.), Evolution and Ecology of Zooplankton Communities. The University Press of New England, Hanover (N.H.); Lond.: 315–328.

King, C. E. & T. W. Snell, 1980. Density-dependent sexual reproduction in natural populations of the rotifer Asplanchna girodi. Hydrobiologia 73: 149–152.

Kott, P., 1953. Modified whirling apparatus for the subsampling of plankton. Austr. J. Mar. Freshwat. Res. 4: 387–393.

Lubzens, E., R. Fishler & V. Berdugo-White, 1980. Induction of sexual reproduction and resting egg production in Brachionus plicatilis reared in sea water. Hydrobiologia 73: 55–58.

Naumann, E., 1923. Spezielle Untersuchungen über die Ernährungsbiologie und die naturliche Nahrung der Copepoden und der Rotiferen des Limnoplanktons. Lunds Univ. Årsskrift N.F. 2, 19: 1–17.

Nauwerck, A., 1963. Die Beziehungen zwischen Zooplankton und Phytoplankton in See Erken. Symb. Bot. Upsal. 17: 1–163.

Pourriot, R., 1963. Influence du rythme nycthemeral sur le cycle sexual de quelques rotiferes. Comptes Rendu Acad. Sci., Paris 256: 5216–5219.

Pourriot, R., 1977. Food and feeding habits of Rotifera. Arch. Hydrobiol. Beih. 8: 243–260.

Pourriot, R. & T. W. Snell, 1983. Resting eggs in rotifers. Hydrobiologia 104: 213–224.

Ruttner-Kolisko, A., 1964. Uber die labile Periode im Fortpflanzungszyklus der Rädertiere. Int. Rev. ges. Hydrobiol. 49: 473–482.

Snell, T. W., 1986. Effect of temperature, salinity and food level on sexual and asexual reproduction in Brachionus plicatilis (Rotifera). Mar. Biol. 92: 157–162.

Stemberger, R. S., 1981. A general approach to the culture of planktonic rotifers. Can. J. Fish. Aquat. Sci. 38: 721–724.

Wesenberg-Lund, C., 1930. Contributions to the biology of the Rotifera. II. The periodicity and sexual periods. Mem. Acad. Roy. Sci. et Lett. de Danmark. Sec. Sci. Serie 9. 2: 1–230.

Hydrobiologia 147: 329–334 (1987)
May, L., Wallace, R. & Herzig, A. (eds), Rotifer Symposium IV

Fertilization and male fertility in the rotifer *Brachionus plicatilis*

Terry W. Snell[1] & Frank H. Hoff[2]
[1]*Division of Science, University of Tampa, Tampa, FL 33606, USA*
[2]*Florida Aqua Farms, 5532 Old Dade City Road, Dade City, FL 33525, USA*

Keywords: fertilization, sperm number, sperm motility, ageing, diet, Rotifera

Abstract

The role of males in fertilization in the rotifer *Brachionus plicatilis* was examined. Neonate mictic females sometimes escaped fertilization even when sperm were present in their pseudocoeloms. Males made a major contribution to this fertilization failure through temporary infertility. As males aged, sperm numbers and motility decreased. Maternal diet was found to play a primary role in determining the fertilizing capacity of the F_1 males. Females from log phase populations produced males with higher sperm counts and motility than females from stationary phase populations. A conditioning factor in the medium may increase sperm number and fertilization rate.

Introduction

A common observation in cultures of the rotifer *Brachionus plicatilis* is the appearance of high densities of males, but only a few fertilized females bearing resting eggs. This population structure suggests that mictic females are present in large numbers, but that most have somehow escaped fertilization and produce only males; mictic females produce males if they are unfertilized and resting eggs if fertilized. All mictic females are assumed to be fertilizable, but there have been a few observations which suggest that this may not be so. Age has been shown to strongly influence a mictic female's susceptibility to fertilization. Buchner *et al.* [1967] found that mictic *Asplanchna sieboldi* females were susceptible to fertilization for only about four hours after birth. After that time, even if insemination occurred, their haploid eggs would not be fertilized and instead would develop into males. Ruttner-Kolisko [1985] has reported that in some instances, *B. plicatilis* mictic females are not fertilized and produce males, even though sperm are observed in their pseudocoeloms.

If it is true that not all mictic females are fertilizable, this raises questions about how and why fertilization might fail. There are at least four possibilities. (1) At low population densities, the probability of male-female encounter is so low that males and females are simply unlikely to meet. (2) Female susceptibility to fertilization may fluctuate with environmental conditions such as population density and the quality or quantity of food. (3) Females may become fertilized, but lack the resources to form resting eggs. (4) There may be a loss of male fertility due to factors such as maternal diet or temperature extremes. In this paper, we focus upon male fertility and its role in controlling fertilization.

Gilbert [1983] has reviewed spermatogenesis and sperm function in rotifers and points out that little is known about the factors influencing male fertility. Which male traits are likely to be good measures of male fertility also has not been established. In this paper we report observations on two important aspects of male fertility, sperm number and sperm motility, and discuss how these traits are influenced by male age and maternal diet.

Methods

A series of experiments was conducted to examine copulation, insemination, and fertilization in *Brachionus plicatilis* (Müller). Experiments utilizing the Russian strain (RUS) were initiated by placing 3–5 young males and one neonate female in 50 μl of F medium [Gulliard, 1983]. Rotifers were observed for several minutes and the number of male-female encounters, time until copulation, and duration of copulation was recorded. After 24 h, female rotifers were transferred individually to a well of a multiwell plate and cultured until they produced at least one offspring. This allowed the females to be classified as amictic, unfertilized mictic, or fertilized mictic. Insemination was verified by examining (at 1000× magnification) the pseudocoelomic contents of several females for the presence of sperm.

Sperm counts were made on neonate males within four hours of their births. This was accomplished by placing individual males on a microscope slide and applying gentle pressure to the coverslip so that the contents of the seminal vesicle were evacuated into the surrounding medium. Motile and non-motile spermatozoa and rods [see Gilbert, 1983] were then counted.

Population growth experiments were conducted at 25 °C, 20‰ salinity and 4000 lux (constant illumination). Populations were initiated by hatching approximately 100 resting eggs and transferring the females to 2 litres of F medium containing 10^5 cells ml^{-1} each of *Chlorella vulgaris* and *Tetraselmis tetrathele*. Rotifer and algal population growth followed a repeatable pattern which consisted of an extended log phase, short stationary phase and rapid population decline when the algal food supply was exhausted.

Results

The evidence that mictic females can escape fertilization is presented in Table 1. In four separate experiments, a total of 444 neonate females were exposed to males using the protocol described above. Of these, 127 were mictic and potentially fertilizable. However, none of these females were fertilized; all produced male offspring. Fertility of RUS strain males and females has been demonstrated repeatedly over the past two years by prolific resting egg production at various times before and after these experiments. Consequently, the failure of fertilization is not a genetic defect, but a temporary loss of fertility due to some unidentified environmental factors.

Table 1. A summary of four fertilization experiments. Each neonate female was exposed to three males in 50 μl and copulation was observed in every case.

Neonates isolated	444
Neonates surviving to reproduce	404
Number amictic	277 (69%)
Number mictic	127 (31%)
Number fertilized	0

After screening several environmental factors, it appeared that diet played an important role in determining the rate of fertilization of mictic females. Many types of algal diets were investigated, two of which were especially interesting and the results are presented in Table 2. Mictic females fed a mixed algal diet of *C. vulgaris* and *T. tetrathele* (Chl + Tet) were fertilized at a rate of 17.6%. This compares to a 16.0% fertilization rate for mictic females fed a unialgal diet of *C. vulgaris* (Chl). While the mictic female fertilization rate was similar on either diet, the standard errors for both of these means are high. This is because, for unexplained reasons, no

Table 2. Summary of fertilization experiments from 3/6/83 to 6/19/84. Female diet is for RUS strain mictic females, male maternal diet refers to the diet fed the mictic females that gave birth to the experimental males. The percent fertilization is the mean ± standard error proportion of mictic females fertilized for all experiments. Chl = *Chlorella vulgaris,* Tet = *Tetraselmis tetrathele.*

Female diet	Male maternal diet	Percent fertilization of mictic females	Number of experiments
Chl + Tet	Chl + Tet	17.6 ± 7.5	7
Chl + Tet	Chl	0	7
Chl	Chl	0	6
Chl	Chl + Tet	16.0 ± 13.6	6

fertilization occurred on either diet in some of the experiments. In contrast to females, males were only capable of fertilization when their mothers were fed a Chl + Tet diet. Fertilization was never observed in any of the 6 or 7 experiments, representing at least 540 males screened, when the male maternal diet was only Chl. These data therefore strongly suggest that diet-related male infertility is a major contributor to fertilization failure in *B. plicatilis.*

The obvious question is what happens to males to render them temporarily infertile? We have explored a number of male traits in an attempt to identify the cause of their infertility. Males were normal in size and produced normally sized sperm and rods, which were 37.3 × 1.1 μm and 14.5 × 1.1 μm, respectively (Fig. 1). Copulation behavior appeared consistent with that reported in the literature. With our experimental protocol, copulation was observed within 5.8 minutes of female introduction into the 50 μl volume containing 3 males. The probability that a male-female encounter resulted in copulation was 0.74 for this strain, a high value when compared to other strains [Snell & Hawkinson, 1983]. The site of copulation was remarkably consistent. Eighty-five percent of all copulations occurred in the coronal region, 11% in the foot opening and only 4% at other sites on the body. Clearly, males avoid the lorica and choose soft parts for copulation. The duration of copulation was 1.23 ± 0.07 minutes, probably long enough for sperm transfer to occur. Insemination was confirmed in 90% of all the neonate females with this mating protocol by observing sperm in their pseudocoeloms.

Since copulation and insemination appeared to occur normally, we explored other features of males that could account for their infertility. Male traits possibly influencing fertility include sperm number and sperm motility. We describe below the results of experiments on the effects of male age and both qualitative and quantitative aspects of maternal diet on sperm number and sperm motility.

As males age, both sperm number and sperm motility fall (Fig. 2). Neonate males maintain their original number of sperm until they are at least 8 h old. During this interval, sperm count averages 29.5 ± 0.98 per male. By the time males are 24 h old, mean sperm count has fallen to 19.9 ± 1.92 which is significantly lower than the neonate sperm count ($t = 4.45$, $P < 0.001$). At 25 °C, a 24 h old male has lived approximately one third of his lifespan. In 48 h old males, mean sperm number is even lower at 10.9 ± 1.44 which is also significantly less than that of 24 h old males ($t = 3.75$, $P < 0.01$). The data presented are for virgin males, so this reduction in sperm number cannot be attributed to sperm loss due to insemination.

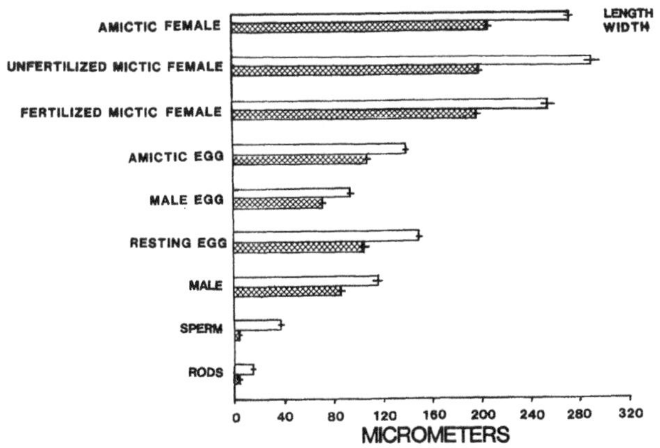

Fig. 1. Comparative sizes of females, eggs, males, sperm and rods. Horizontal lines at the center of each bar indicate standard errors.

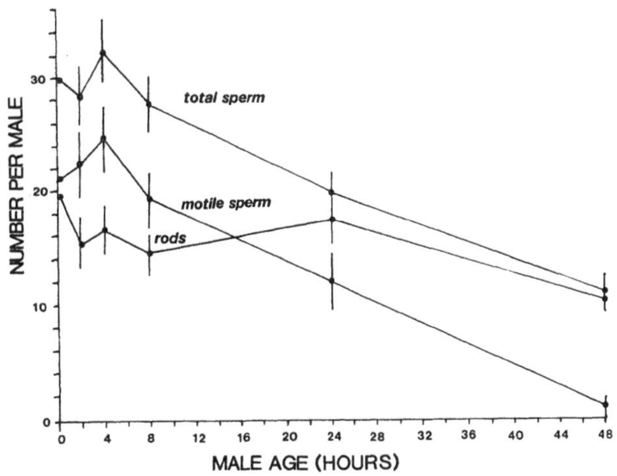

Fig. 2. Effect of male age on the number of total sperm, motile sperm and rods. Vertical lines indicate standard errors.

Table 3. Effect of maternal diet on male fertility. The percent neonate males with motile sperm and the average neonate sperm count on two consecutive days. Means ± standard errors are listed. Sample size: n = 30 for percent motility and n = 10 for sperm count. (Chl = *Chlorella vulgaris*, Tet = *Tetraselmis tetrahele*).

Diet	Percent motile on 7/11	Percent motile on 7/12	Average sperm count on 7/11	Average sperm count on 7/12
Chl + Tet	83	57	20.2 ± 2.22	24.7 ± 2.65
Chl	70	90	18.6 ± 2.52	18.4 ± 2.96
Chl with B-12	87	80	25.0 ± 3.62	21.8 ± 1.48

Sperm motility follows a similar pattern with age. There are no significant differences in number of motile sperm among 0, 2, 4, or 8 h old males. Their mean number of motile sperm is 21.9 ± 1.05, which represents about 75% motility. The number of motile sperm is reduced to 12.1 per male at 24 h of age, representing 60% motility, and finally to 0.9 motile sperm per male, 10% motility, at 48 h. Rods also decline with male age, but not as rapidly as sperm.

Effects of maternal diet on sperm number and motility in populations over two consecutive days can be seen in Table 3. A comparison of three algal diets reveals no significant differences in the percentage of males with motile sperm on July 11 (this is a different measure of male fertility than the number of motile sperm per male that was used above). There is a significant difference for the Chl-Tet treatment between July 11 and July 12 when a significantly lower proportion of males had motile sperm ($G = 5.22$, $P < 0.03$). This reduction from 83% to 57% in the Chl-Tet treatment was associated with the rotifer population entering stationary phase, an interesting observation that is expanded upon below. There are no significant differences in sperm counts on any of the three diets for either day.

Changes in sperm number and motility through a population growth cycle were also investigated. A typical cycle in a 2 litre batch culture is shown in Fig. 3. As this population grew, males and females were removed at regular intervals for analysis of sperm number and fertilization rates. The results of

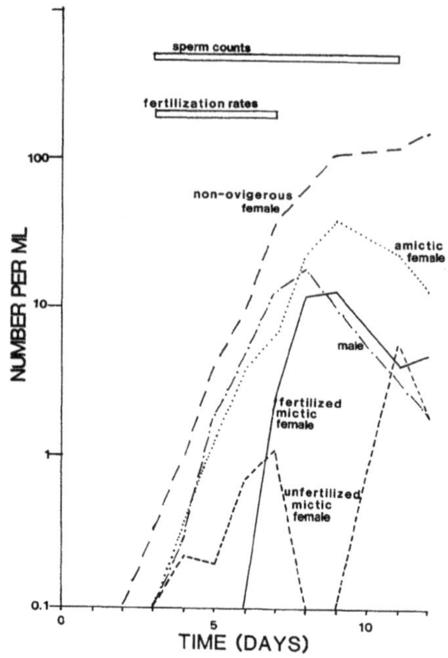

Fig. 3. Amictic and mictic reproduction through a typical population growth cycle. The two bars at the top of the graph indicate the period during which sperm counts and fertilization rates were recorded.

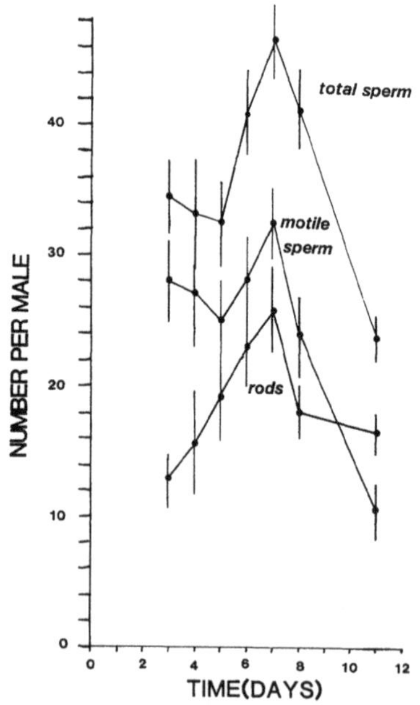

Fig. 4. Changes in total sperm, motile sperm and rods through a population growth cycle. All males used for these observations were neonates and were obtained from the population shown in Fig. 3. Vertical lines indicate standard errors.

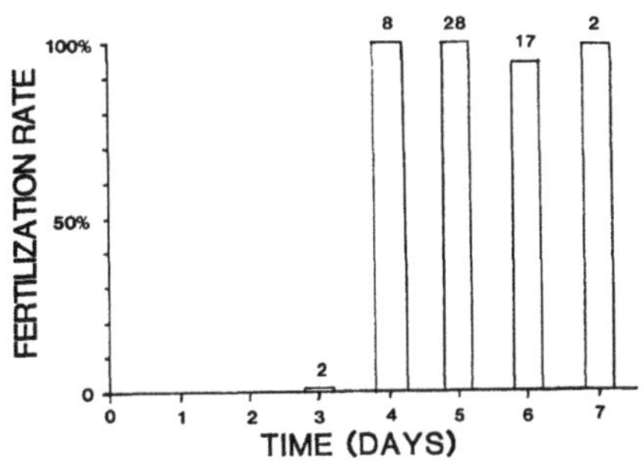

Fig. 5. Fertilization rate of mictic females from the population shown in Fig. 3. The numbers over the bars indicate the number of mictic females exposed to males using the mating protocol.

sperm counts during different phases of population growth are shown in Fig. 4. In early log phase (days 3, 4, 5), mean neonate sperm count was 33.5 ± 0.58, with 80% of these sperm motile. In late log phase (days 6, 7, 8), mean neonate sperm count was significantly higher at 43.0 ± 1.72 ($t = 5.23$, $P < 0.01$), with 66% motility. Even though the motility percentage was lower in late log phase, the higher sperm count yielded two more motile sperm per neonate male. In stationary phase (day 11), mean sperm count was 23.6 ± 1.72 which was significantly lower than either early or late log phase ($t = 3.00$, $P < 0.02$). The percentage of motile sperm per neonate male was also lower at 42%, so that each male was carrying an average of 13.3 motile sperm at birth.

Fertilization rates were monitored in the same population, but all observations were made during log phase growth (Fig. 5). These fertilization rates represent estimates of the fertilizability of neonate mictic females by neonate males. On day 3, population densities were very low and mictic females were just beginning to appear in the population. Consequently, only two mictic females were isolated and neither of these were fertilized when exposed to males. This escape from fertilization was dramatically reversed the next day, however, when 8 mictic females were isolated and all were fertilized. Fertilization rate continued at 100% on day 5 while sample size grew to 28 mictic females. Fertilization rate dropped to 94% on day 6, but only 1 of 17 mictic females escaped fertilization. Sample size fell to 2 on day 7 and both of these mictic females were fertilized. Although sample size is small, the data suggest that marked shifts in fertilization rate can occur through a population growth cycle.

Discussion

Our observations on copulation were consistent with other reports for brachionids [Gilbert, 1963; Ruttner-Kolisko, 1983; Snell & Hawkinson, 1983] with one exception. Ruttner-Kolisko [1983] stated that 'copulation can occur in any place in the body wall'. However, our observations on the site of copulation in *B. plicatilis* do not support this view. The data presented in this paper clearly show that males avoid the lorica by copulating in the coronal region or through the foot opening.

Males have a rather low number of sperm. The sperm counts that we recorded are close to the 30–40 sperm reported for *B. plicatilis* by Whitney [1917]. A consequence of such low sperm counts is that the number of inseminations that a male is capable of accomplishing in his lifetime is severely limited.

One of the most interesting conclusions from these data is that not all spermatozoa are motile. The maximum percentage of motile sperm that we observed in males was 80%. So, even at birth, 20% of a male's sperm are non-motile. Furthermore, percent motility decreases as males age. Whether these non-motile sperm can participate in fertilization is an intriguing question for future research.

Our measurements of sperm and rod size are in reasonably close agreement with values reported in the literature. Koehler [1965] found spermatozoa of *Asplanchna brightwelli* to be 15–20 μm long, which is somewhat shorter than the 37.3 μm dimensions for *B. plicatilis*. Rods in *A. brightwelli* [Koehler & Birky, 1966] were more similar to *B. plicatilis*: 10–15 μm vs. 14.5 μm, respectively.

The effect of age on the fertilizing capacity of *B. plicatilis* males has been examined by Ruttner-Kolisko [1983]. She reports that males less than 24 h old fertilize mictic females more frequently than males older than 24 h. In the E strain, these figures were 45% and 8% in young and old males, respectively. Our data suggest that this reduction in fertilizing capacity in older males may be due to their lower sperm number and motility.

We have demonstrated that mictic females can escape fertilization in some circumstances, even with sperm in their pseudocoeloms. This failure of fertilization is, at least in part, due to males and appears to be related to maternal diet. Mictic females from stationary phase populations produce males with about half the number of motile sperm as males from log phase populations. Stationary phase females are presumably food limited, so it is reasonable to hypothesize that undernourished females produce males with fewer motile sperm and limited fertilizing potential.

Conditioning of the medium may have an effect on both sperm number in males and susceptibility of females to fertilization. We observed significantly higher sperm counts in males from late log phase populations where there were moderately high population densities. Mictic females in this environment may have been exposed to high concentrations of a conditioning factor, but still had adequate food for reproduction. Consequently, they were able to produce especially virile males. How a conditioning factor might enhance sperm number is unclear since spermiogenesis is probably before birth [Gilbert, 1983]. Rate of mictic female fertilization may also be enhanced by a conditioning factor. On day 3 of the population growth experiment, two mictic females were isolated from a very low density population. Neither of these females were fertilized after exposure to males. On the next day, fertilization rate was 100% and remained high, even on day 7 when again only two mictic females were isolated and both were fertilized. More data are obviously needed, but these preliminary results suggest that both sperm number and fertilization rate may be increased by a conditioning factor.

Acknowledgements

We acknowledge with appreciation the expert technical assistance of Emily Boyer and Michael Childress. This material is based upon work supported by the National Science Foundation under Grant OCE-8314717.

References

Buchner, H., C. Mutschler & H. Kiechle, 1967. Die Determination der Mannchen- und Dauereiproduktion bei Asplanchna sieboldi. Biol. Zbl. 86: 599–621.

Gilbert, J. J., 1963. Contact chemorecption, mating behavior, and sexual isolation in the rotifer genus Brachionus. J. exp. Biol. 40: 625–641.

Gilbert, J. J., 1983. Rotifera. In: Reproductive Biology of Invertebrates. Vol. II: Spermatogenesis and Sperm Function. K. G. Adiyodi & R. G. Adiyodi (eds.), John Wiley & Sons, New York, pp. 181–193.

Gulliard, R. R. L., 1983. Culture of phytoplankton for feeding marine invertebrates. In: C. J. Berg, Jr. (ed.), Culture of Marine Invertebrates. Hutchinson-Ross, Stroudsberg, PA, pp. 108–132.

Koehler, J. K., 1965. An electron microscope study of the dimorphic spermatozoa of Asplanchna (Rotifera). I. The adult testis. Z. Zellforsch. 67: 57–76.

Koehler, J. K. & C. W. Birky Jr., 1966. An electron microscope study of the dimorphic spermatozoa of Asplanchna (Rotifera). II. The development of 'atypical spermatozoa'. Z. Zellforsch. 70: 303–321.

Ruttner-Kolisko, A., 1983. The significance of mating processes for the genetics and for the formation of resting eggs in monogonont rotifers. Hydrobiologia 104: 181–190.

Ruttner-Kolisko, A., 1985. Results of individual cross-mating experiments in three distinct strains of Brachionus plicatilis (Rotifera). Verh. int. Ver. Limnol. 22: 2979–2982.

Snell, T. W. & C. A. Hawkinson, 1983. Behavioral reproductive isolation among populations of the rotifer Brachionus plicatilis. Evolution 37: 1294–1305.

Whitney, D. D., 1917. The production of functional and rudimentary spermatozoa in rotifers. Biol. Bull. 33: 305–315.

Hydrobiologia 147: 335–338 (1987)
May, L., Wallace, R. & Herzig, A. (eds), Rotifer Symposium IV

Effect of incubation temperature on the hatching of rotifer resting eggs collected from sediments

Linda May
Institute of Terrestrial Ecology, Bush Estate, Penicuik, Midlothian EH26 0QB, Scotland

Keywords: Rotifera, resting eggs, temperature, sediment

Abstract

Sediment samples from Loch Leven, Scotland, were incubated at 5 °C, 10 °C and 15 °C to induce hatching of rotifer resting eggs. The emergent animals were identified and counted. The temperature which induced hatching varied among the nine species studied. These results are discussed in relation to seasonality and temperature preferences previously recorded for the most abundant species. Resting egg densities of 2.2–13.9 eggs cm^{-3} were recorded in the upper 5 cm of sediment.

Introduction

Many rotifers are seasonal in occurrence and survive periods of unfavourable conditions as resting eggs in the sediments of lakes and ponds. When favourable conditions return, these resting eggs hatch and the rotifer community is quickly re-established. The stimulus for hatching varies among species, but probably includes changes in light temperature and salinity [Pourriot & Snell, 1983].

It has already been shown that a complete species list for a given freshwater site can be compiled from sediment samples collected in early spring and incubated at various temperatures [May, 1986]. This paper examines more closely the effect of temperature on the emergence of rotifers from sediments. The results are discussed in relation to the seasonality and temperature preferences of these rotifers, which have been previously described by May [1983].

Methods

The sampling sites, collection of the sediment cores and sub-sampling techniques have been described in detail by May [1986]. A sediment core was collected from each of six open water sites on 16 February 1984, when the water temperature was about 2 °C and very few planktonic rotifers were present. The top 5 cm of mud from each core was thoroughly mixed and a 20 ml subsample was placed into a 250 ml Erlenmeyer flask. The mud suspension was diluted with filtered loch water to give a final volume of 200 ml. The vessels were covered with plastic film and incubated at 5 °C, 10 °C or 15 °C under artificial light, LD = 12:12.

At 3–7 day intervals the clear surface water in each conical flask was decanted and replaced by fresh, filtered lake water. The rotifers which had emerged from each sample were identified and counted with a Nikon inverted microscope. When no further emergence had been recorded on three consecutive occasions the incubation temperature of the sediment samples was changed. The '5 °C' samples were changed to 10 °C, the '10 °C' samples to 15 °C and the '15 °C' samples to 5 °C. The experiment was continued until no further emergence had been recorded on three consecutive counts. The incubation temperatures were changed again and the process was repeated.

Results

Densities of resting eggs in each sample varied from species to species. Table 1 shows the maximum, mean and minimum densities recorded for the most abundant species, assuming that all resting eggs were induced to hatch. If some eggs remained unhatched, then the actual resting egg densities could be higher. The mean densities of resting eggs in the sediments ranged from 0.01 to 2.23 eggs cm^{-3} for a single species, and 2.2 to 13.9 eggs cm^{-3} in total. This is equivalent to 5×10^{-4} to 1×10^{-2} eggs mm^{-2} for a single species and 0.11 to 0.7 eggs mm^{-2} for all species.

The effect of incubation temperature on the hatching of resting eggs varied from species to species. *Keratella cochlearis* (Gosse), *Synchaeta kitina* Rousselet and *Synchaeta grandis* Zacharias emerged at all three incubation temperatures (Figs. 1 a–c). However, the number of individuals emerging at each temperature varied widely; *S. kitina* and *K. cochlearis* seemed to emerge most successfully at 10 °C, while *S. grandis* was most successful at 15 °C. In all three cases, faster emergence rates were associated with higher incubation temperatures.

In *Asplanchna priodonta* Gosse, *Trichocerca* sp. Lamarck and *Pompholyx sulcata* (Hudson) high rates of emergence were recorded at 10 °C and 15 °C, after a short lag phase (Figs. 1 d–f). However, at 5 °C these species showed a much longer lag phase, 20–25 days, followed by a slow rate of emergence.

In contrast, *Polyarthra dolichoptera* Idelson, *Notholca squamula* (Müller) and *Conochilus unicornis* (Rousselet) seemed to hatch more successfully at 5 °C and 10 °C (Figs. 1 g–i). At 15 °C emergence was inhibited until the incubation temperature was lowered. *P. dolichoptera* and *N. squamula* hatched when the temperature was reduced to 5 °C, while *C. unicornis* remained in dormancy until the incubation temperature reached 10 °C.

Table 1. Estimates of resting egg densities from the sediments of Loch Leven.

Species	Resting egg densities per cm^3		
	Max.	Mean	Min.
Keratella cochlearis	1.2	0.16	0
K. quadrata	0.1	0.01	0
Notholca squamula	1.3	0.54	0.05
Trichocerca pusilla	1.3	0.24	0
Trichocerca sp.	6.0	1.60	0
Asplanchna priodonta	0.35	0.07	0
Polyarthra dolichoptera	2.0	0.65	0
P. euryptera/vulgaris	0.25	0.05	0
Synchaeta kitina	5.5	2.23	0.6
S. grandis	1.5	0.48	0.05
Pompholyx sulcata	4.6	1.66	0.2
Filinia longiseta	0.1	0.01	0
Conochilus unicornis	0.65	0.19	0

Discussion

The estimated densities of resting eggs in the sediments of Loch Leven were generally lower than those reported by other workers. Snell *et al.* [1983] report a resting egg density of 2 mm^{-2} for *Brachionus plicatilis* in a brackish water pond. Similarly, Ito [1985] records egg densities of up to 9.66 eggs mm^{-2} and Nipkow [1961] gives densities of 1–6 eggs mm^{-2} for 7 species and 30–44 eggs mm^{-2} for *Synchaeta oblonga.*

There are two possible reasons for these differences in density. The methods used in these three studies are very different. Ito [1985] simply picked eggs from the sediments, while Snell *et al.* [1983] used a sugar flotation method. The present study used a hatching technique and estimated egg densities from the number of emergent animals. These methods need to be directly compared before the results can be assessed realistically.

The second reason for the observed differences in density estimates may be that the number of resting eggs in the sediments of Loch Leven are, indeed, lower than at the other localities. This is possible since the number of resting eggs laid depends upon the maximum population densities achieved by each species in previous years [Herzig, 1987]. If the population densities were generally lower in Loch Leven than at the sites studied by Ito [1985], Nipkow [1961] and Snell *et al.* [1983] then the resting

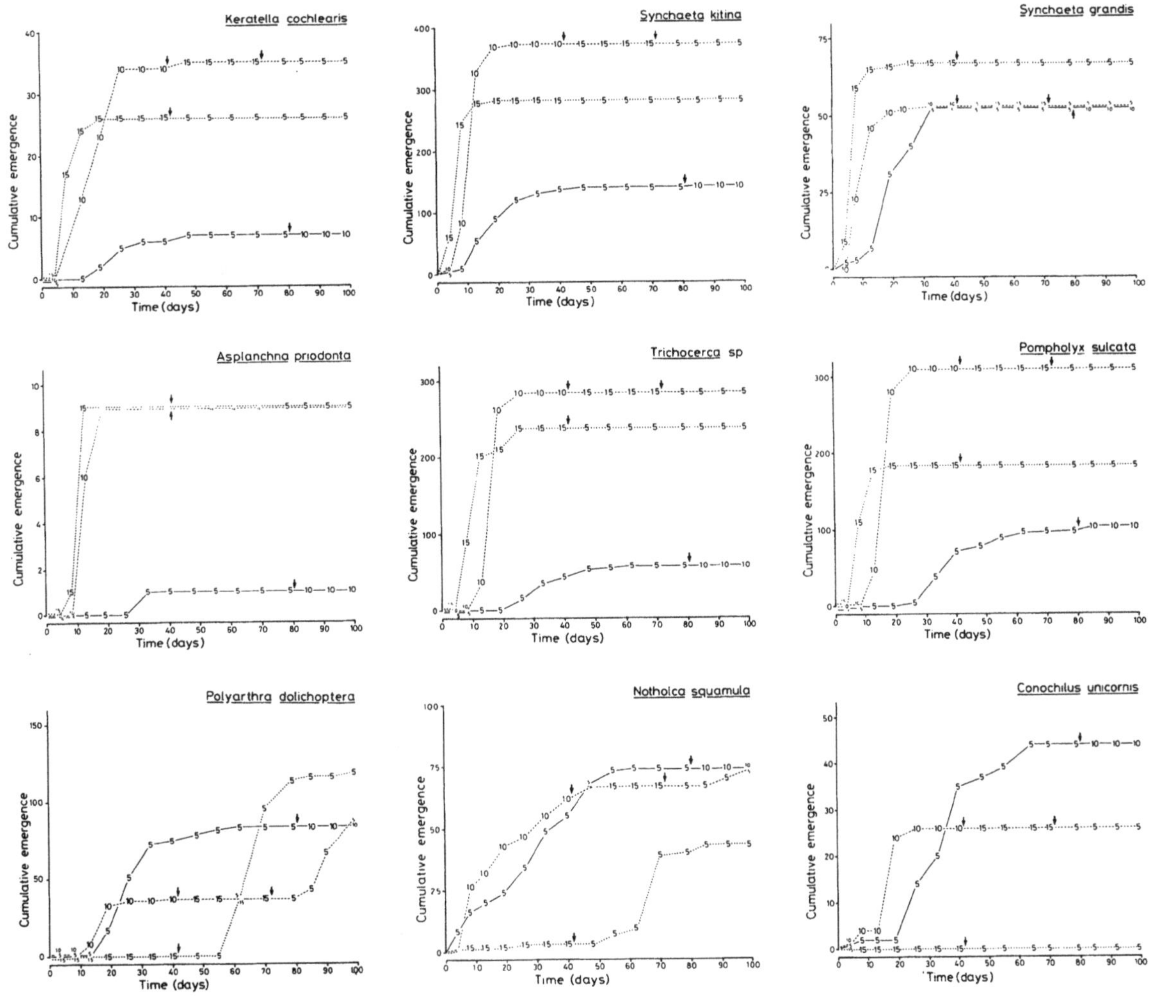

Fig. 1a–i. Cumulative emergence of rotifers from the incubated sediment samples. Numbers indicate incubation temperatures. Arrows indicate the points at which the incubation temperature was changed.

egg densities in the sediments would also be lower.

The species specific and temperature dependent emergence rates recorded in this study strongly reflected the seasonality and temperature preferences previously recorded for these nine species [May, 1983]. Therefore, it seems likely that changes in water temperature stimulate mass emergence from the sediments in stenothermal species. For example, *Keratella cochlearis, Synchaeta grandis* and *S. kitina*, which are all more or less perennial species in Loch Leven [May, 1983 & unpubl. data], hatched from the sediments over the whole range of incubation temperatures (5–15 °C). In contrast, the late spring and summer species *Asplanchna priodonta, Trichocerca* sp., and *Pompholyx sulcata* [May, 1983 & unpubl. data], hatched more readily at 10 °C and 15 °C. At 5 °C hatching was very slow and there were less emergent individuals than at higher temperatures. The hatching of winter species, generally known to be cold stenotherms, was inhibited by temperatures of 15 °C. These species hatched readily at lower temperatures. The results of this study are consistent with the suggestion of Pourriot & Snell [1983] that the optimal temperature for resting eggs to hatch is within the thermal preference of that population.

References

Herzig, A., 1987. The analysis of rotifer populations: A plea for long term investigations. Hydrobiologia 147: 163–180.
Ito, T., 1958. Studies on the 'Mizukawari' in eel culture ponds. X. The density of dormant eggs of rotifer on bottom deposits in eel culture ponds. Rep. Fac. Fish. Prefect. Univ. Mie. 3: 170–177.
May, L., 1983. Rotifer occurrence in relation to water temperature in Loch Leven, Scotland. Hydrobiologia 104: 311–315.
May, L., 1986. Rotifer sampling – a complete species list from one visit? Hydrobiologia 134: 117–120.
Nipkow, F., 1961. Die Rädertiere im Plankton des Zürichsee und ihre Entwicklungsphasen. Schweiz Z. Hydrol. 23: 398–461.
Pourriot, R. & T. W. Snell, 1983. Resting eggs in rotifers. Hydrobiologia 104: 213–214.
Snell, T. W., B. E. Burke & S. D. Messur, 1983. Size and distribution of resting eggs in natural population of the rotifer Brachionus plicatilis. Gulf Research Reports 7: 285–287.

Hydrobiologia 147: 339–359 (1987)
May, L., Wallace, R. & Herzig, A. (eds), Rotifer Symposium IV

Movements in rotifers: correlations of ultrastructure and behavior

Pierre Clément
Equipe Neuro-Ethologie, bâtiment 401-C Université Lyon 1, 69622 Villeurbanne Cédex, France

Keywords: rotifers, neuro-ethology, etho-ecology, locomotion, feeding behavior, ultrastructure, muscles, cilia, trajectometry

Abstract

In this review, original observations on rotifer cilia and muscles are used to explain the mechanism of their movements. The following topics are discussed: (1) cilia and infraciliature of the cingulum, (2) muscle insertions into the infraciliature, (3) innervation of the muscles and non-innervation of the ciliated cells, (4) recent data on the location of gap junctions which link some muscles, (5) cilia and infraciliature of the cirri of the pseudotrochus of *Brachionus calyciflorus*, and the insertion of muscles onto their infraciliature, (6) the relation between muscles and integument in the foot of *Philodina roseola*, (7) the relation between muscles and integument in the joint present at the base of the tail of *Trichocerca rattus*, and (8) hypotheses on the first articulated (muscle-integument) movements which exist alongside ciliary motion in certain rotifers.

Neuro-ethological summaries, based on data of the effectors and on observations on behavior and the sensitivity of rotifer sensory receptors (reviewed in Clément *et al.*, 1983), are attempted. A more complete analysis is provided in the following instances: (1) swimming and the mechanism of turning, (2) retraction of the rotatory apparatus, (3) feeding behavior in *Trichocerca rattus* involving the succession swimming, crawling and nutrient intake.

Other behavioral sequences, for which the ethological and ultrastructural data are incomplete, are presented and briefly discussed.

Lastly, the argument is put forward that rotifers constitute an exceptional model for comparative, neuroethological studies which may enable us to better understand the mechanisms of evolution. Carrying out detailed observations, aimed at understanding how certain movements occur, should enable us to gain insights into the question of why they exist.

1. Introduction

To the naked eye a rotifer, if it can be seen at all, appears to be a fixed point in suspension that is moved about only by the water currents. One is led to ask if it has any movements of its own and what their utility might be, besides keeping the animal afloat as it is being carried along by the currents. However, as soon as one begins to observe the behavior of rotifers more precisely, the efficiency of their movements becomes evident, e.g. the positive phototaxis of certain planktonic species.

How and why do rotifers move? The question of 'how' is the easier of the two to take up, and the bulk of this review paper attempts to answer it. I will do that by establishing correlations between the motor sequences of behaviors (the movements) and their morphological basis, analyzed at the ultrastructural level. The question of 'why' is more complex and more philosophical, and it depends upon the point of view of the observer. Further, it is difficult to avoid anthropomorphisms when considering this question.

The question of 'why' deals with adaptation and,

consequently, with the various geneses of a rotifer: its phylogenesis, its ontogenesis, its individual life history and that of its parents and grandparents. All of this occurs within one or several precise ecosystems which change in the course of these geneses.

The movements of a rotifer depend upon its perception of the environment (cf. Clément *et al.*, [1983]: a review of rotifer sensory perception). However, sensory perception varies with the movements of the animal and, what is more, certain perceptions are only possible as a result of movement [Campan & Lambin, 1984; Clément, 1984]. Therefore, it is somewhat artificial to separate movement (motor sequence) from sensory perception in the analysis of behavior. After I have described the ultrastructure of the motor basis of a behavior, I shall place it in a larger context which includes the sensory receptors, the nervous system, etc.

Neuro-ethological analysis is one of the two essential points of this work [see also Clément & Amsellem, in press]. Rotifers provide a particularly suitable model in this respect since they are characterized by a nervous system consisting of less than 200 neurons, very specialized sensory receptors and muscular effectors made up of only a few cells, eutely, and diploid parthenogenesis.

The second point is etho-ecological and it is also directed at the mechanisms of evolution. Behavior seems to act as a motive force in evolution [Waddington, 1974; Piaget, 1976]. A comparative study of the behavior of various rotifer species and clones (and of the ultrastructure of the morphological basis of the same) might interest ecologists as well as those who are interested in the evolution of rotifers.

2. Swimming (Plates I, II, IVa; Figs. 1 and 2)

Swimming characteristics of rotifers vary with different environmental conditions [Clément *et al.*, 1983]. The present review describes more precisely the cilia of the cingulum and the muscles inserted onto their infraciliature. I then discuss data regarding the effectors in light of observations on sensory receptors already published [Clément, 1977a, 1977b, 1984a; Clément *et al.*, 1983] in order to provide some hypotheses concerning the mechanisms of swimming.

a) Coordination of ciliary beat in the cingulum

Coordination of ciliary beat makes the cingulum looks like wheels at the anterior end of rotifers. Several workers have taken advantage of this feature and have measured the frequency of ciliary beat using a stroboscope [Lansing & Lamy, 1961a; Luciani, 1982; Kutikova, 1983].

The mechanism by which ciliary coordination is achieved is not known, but we may infer that the infrastructure of the cilia of the cingulum plays a major role in it, for four reasons. 1) There are always horizontal connections between the bases of adjacent cilia in the cingulum (Plates I and II). 2) The ciliary rootlets are always associated with mitochondria and the endoplasmic reticulum [Plates I and II; Clément, 1977a; Luciani, 1982; Cornillac, 1982]. 3) Horizontal components of the infra-ciliature, present either at the apex of the cell (Plate IIb) or below the apex (Plate IIa), are inserted onto epithelio-epithelial desmosomes which assure the mechanical coherence of the infraciliature of several adjacent cells in the cingulum (Plate IIa). 4) Elsewhere, the epithelial cells of the cingulum are coupled by gap junctions (P. Clément, J. Amsellem and E. Wurdak, unpublished observations on *Trichocerca, Philodina*, and *Asplanchna*; Cornillac, [1982], and Luciani, [1982] on *Brachionus*).

The cilia of the cingulum show all the properties associated with motile cilia [Lansing & Lamy, 1961a; Clément, 1977a, b; Plates I & II]. Their movements are coordinated, without a doubt, by the infraciliature and by the plasma membrane which is also the membrane investing the cilia and which is in close contact with the axoneme (the tubulin-dynein complex responsible for ciliary movement) at the base of each cilium.

b) Mechanism of turning in Brachionus *and* Asplanchna

Plates I and II show that some muscles (cingulum

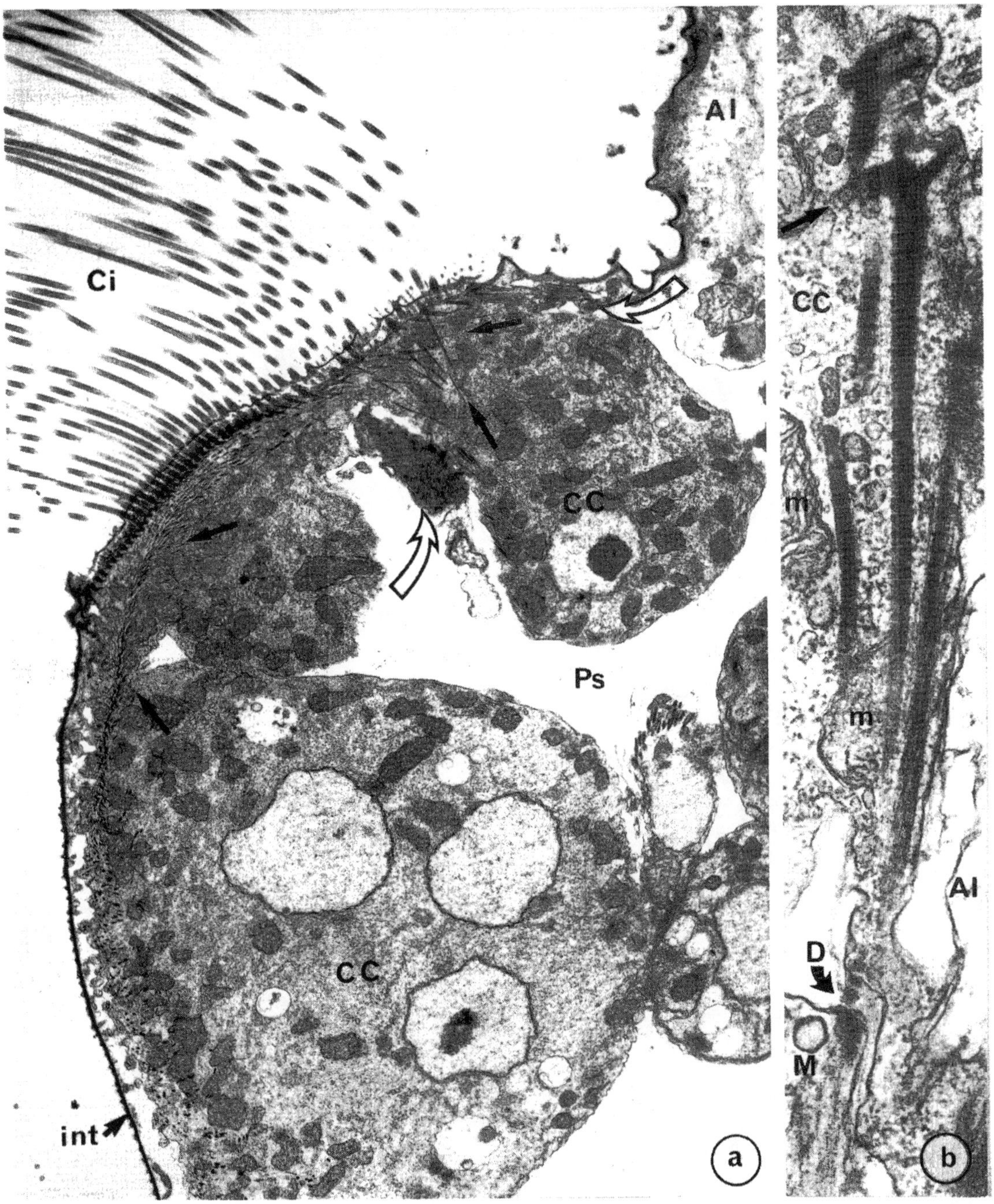

Plate I. The cingulum of *Asplanchna brightwelli*

a. ×4300. Section through the ciliated cells of the cingulum (CC) containing several nuclei, glycogen particles, and mitochondria which are especially numerous near the ciliary rootlets. The rootlets bifurcate at the base of each cilium (solid arrows). The tip of each of these branches inserts onto a muscle (large open arrows). (AI – apical integument; Ci – cilia of cingulum; int – syncytial integument of trunk; Ps – pseudocoel).

b. ×42000. Details of the insertion of a ciliary rootlet onto a musculo-epithelial desmosome (D). The cingulum cell (CC) is very close to the apical integument (AI). (M = muscle ending). Note the classical striation of the ciliary rootlets and their association with mitochondria (m) and endoplasmic reticulum vesicles. Electron micrograph courtesy of Wurdak and Clément (unpublished).

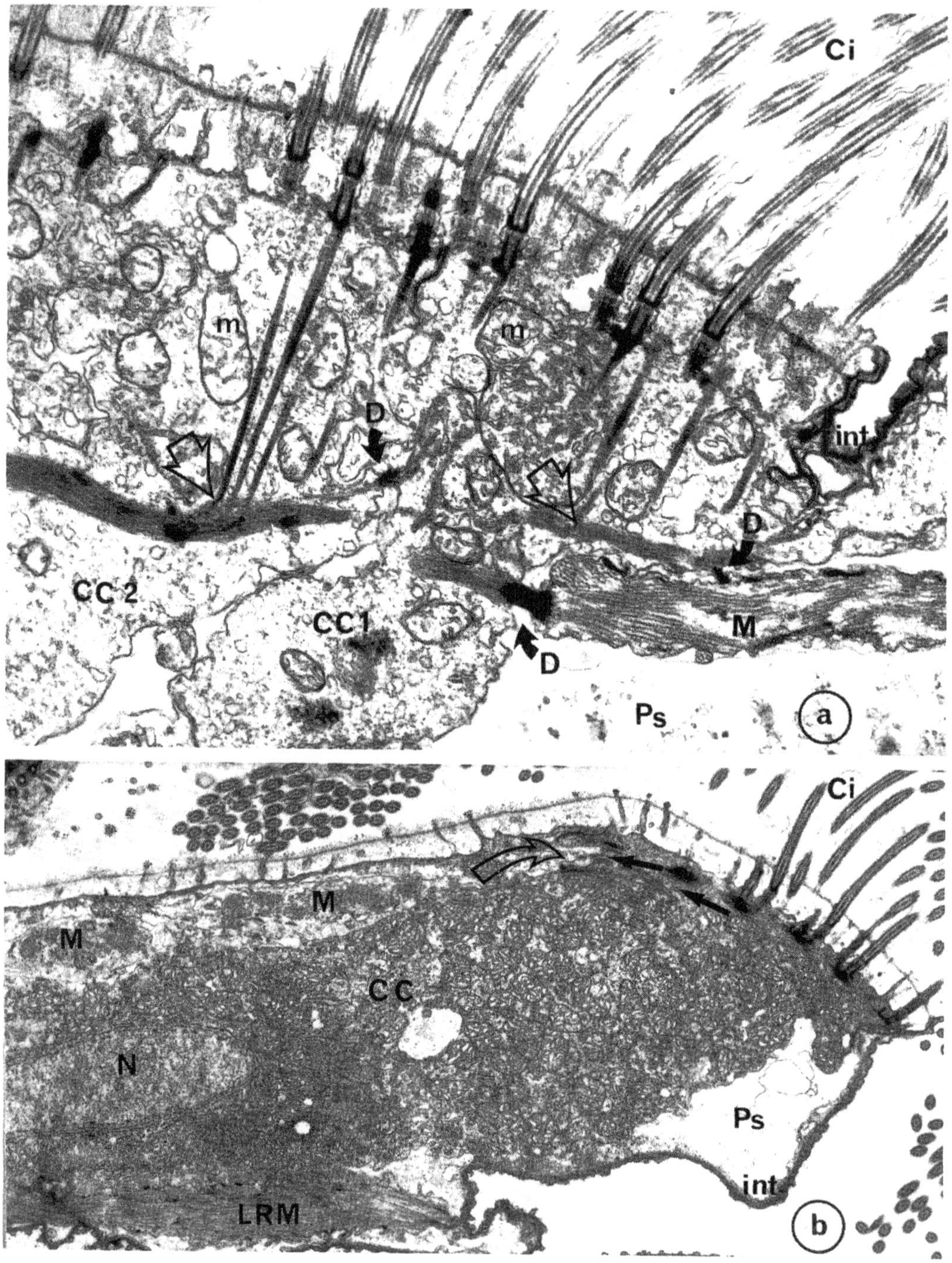

Plate II. The cingulum of *Brachionus*

a. ×18 000. *Brachionus plicatilis.* The striated, vertical ciliary rootlets are inserted (empty, large arrows) onto the horizontal bundles of filaments which are, in turn, inserted onto desmosomes (D). Cilia borne by different cingulum cells (CC_1 CC_2) are connected to the cingulum muscle (M) in this manner. (m = mitochondrion; Ps = pseudocoel).

b. ×7 800. *Brachionus calyciflorus.* In this species the striated ciliary rootlets lie close to the surface of the cell and extend horizontally, i.e. parallel to the surface (solid arrows). They attach to muscle ending (M) by means of a musculo-epithelial desmosome (large empty arrow). One of the lateral retractor muscles (LRM), cut axially, inserts onto the ventral integument (int). N = nucleus of the cingulum cell (CC). Electron micrograph courtesy of Amsellem, Clément, Cornillac and Luciani (Figure IIa from Cornillac, 1982 & Luciani, 1982).

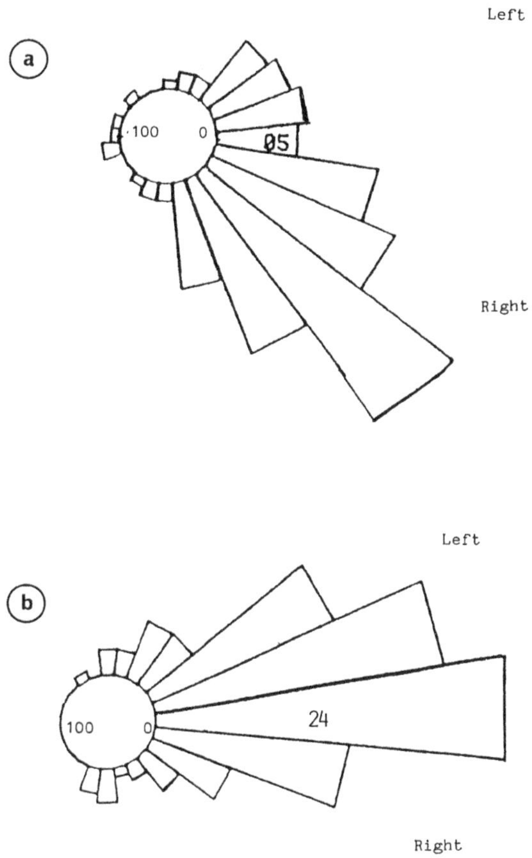

Fig. 1.. Histogram of the angles of deviation in a trajectory of *Brachionus plicatilis* (a) and in a trajectory of *Asplanchna brightwelli* (b). The X and Y coordinates of the animal's position are registered 12 times per second with an automatic tracking system. Measurements of angles are carried out by taking every sixth point [Coulon *et al.*, 1985; Clément *et al.*, 1984].

muscles) have their insertions on the rootlets of the cingulum. The insertion of muscles onto ciliary rootlets have already been described, but it has not been mentioned whether the rootlets belong to the cingulum or the pseudotrochus [Clément & Fouillet, 1970; Clément, 1977a, b]. Other authors [Lansing & Lamy, 1961a, b; Scholtyseck & Danneel, 1962] have described the structure of ciliary rootlets, but they did not observe these musculo-epithelial desmosomes. In *Trichocerca rattus* there is a pair of cingulum muscles (Fig. 2): each of them passes through a nerve ganglion and it is richly innervated. Several synaptic buttons coming from different nerve endings have been observed [Clément, 1977a, b]. Two of these synaptic buttons are shown on Plate IVa; they enclose synaptic vesicles

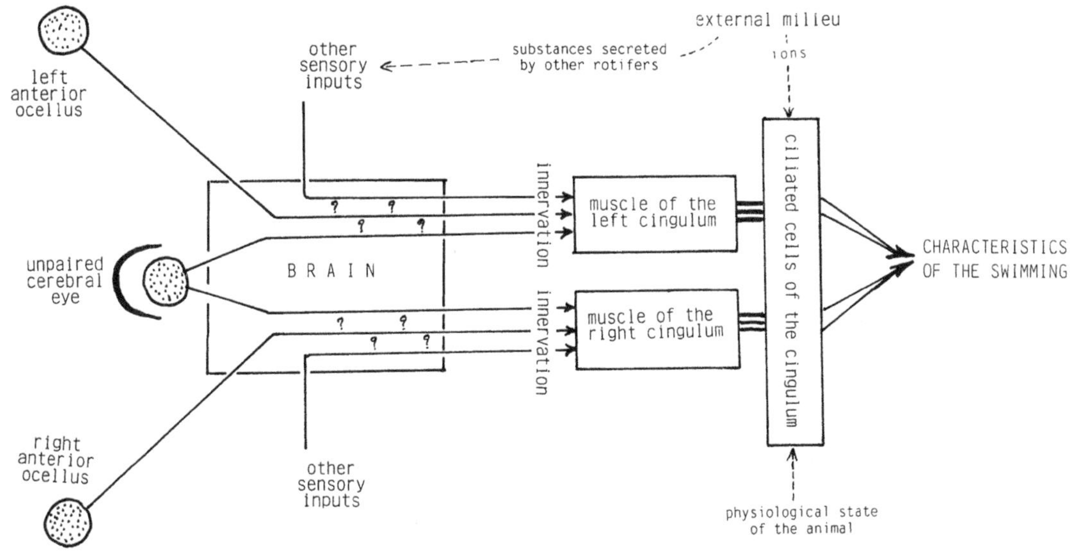

Fig. 2. Neuro-ethological schematic diagram of the regulation of swimming in response to illumination in a rotifer. Hypothetical interactions among sensory inputs in each half of the brain (left and right) are indicated by question marks. The three lines which connect the muscle to the cells of the cingulum represent musculo-epithelial desmosomes joined to the infraciliature of the cilia of the cingulum that are responsible for swimming [From Clément & Amsellem, in press].

which are similar in appearance to vesicles containing acetylcholine. However, other nerve fibers contain different vesicles [Villeneuve & Clément, 1971; Clément, 1977a, b].

I have never observed a direct innervation of cells making up the cingulum in any species (including *T. rattus* whose structure is quite well known). Therefore, one must conclude that the cingulum muscles and the infraciliature serve as intermediaries in the neural control of swimming.

Observations of swimming behavior in *Brachionus* and *Asplanchna*, carried out with a recently refined automatic tracking device [Coulon *et al.*, 1983, 1984; Clément *et al.*, 1984, 1985], have demonstrated a consistent, positive correlation between amplitude of turning and swimming speed. Swimming speeds decrease as the animal turns: the wider the angle of turning, the more marked the decrease in speed. Therefore, turning seems to be the result of the slowing down of either the right (right-hand turn) or the left (left-hand turn) half of the cingulum. This interpretation is supported by observations made with the aid of a stroboscope which show differences in ciliary beat between the right and the left half of the cingulum. It is also supported by observations on *Notomata copeus*. This animal is able to retract its two auricles independently of each other thereby illustrating the autonomous action of the different cingulum muscles.

c) Helicoidal swimming

The swimming pattern of most rotifers is helicoidal. In *Brachionus* and *Trichocerca*, for example, the cingulum muscle is inserted onto the infraciliature which is arranged in a horizontal plane (i.e., perpendicular to the base of the cilia). This plane is apical in *Brachionus calyciflorus* (Plate IIb). It is made up of fairly voluminous, horizontal, striated ciliary rootlets that insert onto a musculo-epithelial or epithelio-epithelial (integument or anterior ciliated cells) desmosome [Clément, 1977a, b; Plate IIb]. This plane is situated at greater depth in the epithelial cells of *Brachionus plicatilis*. In this species it is composed of bundles of actin-like microfilaments, that also insert onto desmosomes (Plate IIa) located either between two epithelial cells or between an epithelial cell and the cingulum muscle. In this arrangement, the cingulum muscle controls the ciliary rootlets of the entire cingulum by means of a single zone of insertion.

Muscular control is, at the same time, unilateral. This accounts for the fact that in *Brachionus*, the mean turning angle lies around 45° and it is to the right (Fig. 1). Almost all turns are toward the right; very rarely do we observe a left-hand turn. Therefore, we are probably dealing with turns where the animal is in a ventral position in its helicoidal swimming path.

*d) Non-helicoidal swimming (*Asplanchna*)*

Ciliary rootlets in *Asplanchna* differ from those of *Brachionus* (Plate I). There is no horizontal plane perpendicular to the base of the cilia. Instead the base of each cilium bifurcates and extends into the cytoplasm as two ciliary rootlets which form an inverted V. A muscle inserts onto each extremity of the V. Therefore, each cilium is under control of two opposing muscles (Plate Ia, b). The resulting ciliary beat has two characteristics: 1) The pattern of swimming is not helicoidal. 2) A frequency histogram of turning angles in *Asplanchna* (Fig. 1) shows a distribution around 0°. There is no systematic tendency to turn toward the riht and the largest class of turning angles is in the −7.5° to +7.5° range. Outside of this range, frequency of turning angles diminishes as their amplitude increases.

e) Factors influencing swimming path

One of the objectives of my research and that of colleagues has been to identify external stimuli that bring about a change in swimming path [cf. Clément *et al.*, 1983]. Luciani *et al.* [1983] have shown that swimming characteristics in *B. plicatilis* vary with age. Luciani [1982] has noted other age-related variations in the same species (e.g., decrease in ciliary beat, loss of nutrient reserves in cells of the cingulum, decrease in ATP content of whole animals, an increase followed by a decrease in size and number of mitochondria present in cells of the cingulum, lysosome accumulation, and progressive vesiculation of cells of the cingulum).

Luciani *et al.* [1983] also have shown that the presence of calcium in the medium can influence some of these age-related changes in swimming behavior. These results are more difficult to interpret; for example, is this a direct influence of calcium on the cilia, and/or the influence of calcium present within the animal, notably within the cells of the cingulum? Experimental results on the influence of inherited calcium on longevity support the second hypothesis [C. E. King, pers. comm.].

Wallace [1980] likewise has demonstrated that swimming speed varies with age in the larvae of the sessile rotifer *Ptygura beauchampi*. In this case, however, the animal is still young and no degradative mechanisms are likely to be involved. Perhaps this decrease in swimming speed is the result of depleted energy reserves or a developmental effect on swimming behavior.

f) Photokinesis: Acceleration and deceleration of ciliary beat

Viaud [1943c] studied ciliary beat in a number of phototaxic rotifers with the aid of a stroboscope. He showed that variations in illumination have no effect on the rhythm of ciliary beat, but they do have an effect on its amplitude. This effect probably explains changes in swimming speed first noted by Viaud [1943a, b] and later by Clément [1977a], Cornillac [1982] and Clément & Wurdak [1984] that follow changes in the intensity or quality of light. However, recent results differ from those of Viaud [1943a, b] in certain respects, particularly in the influence of wavelength, but they also confirm the existence of a photokinesis. It would be desirable to repeat Viaud's stroboscopic investigations using isoquantic, monochromatic light.

One of the aims of this research would be to identify the photoreceptor (eye or ocellus: Clément *et al.* [1983]; Clément & Wurdak, [1984]) responsible for photokinesis, another would be to determine whether the same sensory-motor pathways are involved in photokinesis as in phototaxis (Fig. 2). Generally speaking, acceleration or deceleration of cilia of the cingulum following stimulation (light, currents etc.) might be the result of a slight contraction of the two cingulum muscles simultaneously (Figs. 2 and 3).

g) Synthesis (Fig. 2)

Figure 2 summarizes the observations discussed above. The right cingulum muscle is innervated independently of the left cingulum muscle. The sinuosity [Bovet, 1984] of swimming depends on the autonomy of right and left halves of the cingulum.

Figure 2 also summarizes information concerning the brain and the sensory receptors. The brain shows a clear bilateral symmetry, which had been observed with both the light [Nachtwey, 1925; Remane, 1929–32] and the electron [Ware, 1971; Clément, 1977a, b] microscope.

The sensory receptors themselves are also symmetrical. There are paired organs, unpaired organs consisting of paired neurons (e.g., dorsal antenna), and unpaired organs consisting of unpaired neurons that send two axons, one to each half of the brain (cerebral eye of *Asplanchna brightwelli* and *Brachionus calyciflorus*) [Clément *et al.*, 1983; Clément & Wurdak, 1984; Clément, 1984a]. There seem to be, therefore, nerve tracts that originate at sensory receptors and course independently through the right or left half of the animal all the way to the right or left cingulum muscle.

The central point of neuro-ethological analysis currently lies in the nervous network. In this regard several important questions need to be addressed. (1) Is there interaction among different sensory modalities at the level of the brain? (2) Are some modalities pre-eminent over others? (3) Why is there multiple innervation of each muscle of the cingulum? The study of behavior (different sensory influences on swimming) combined with studies of neuro-anatomy may yield answers to these questions.

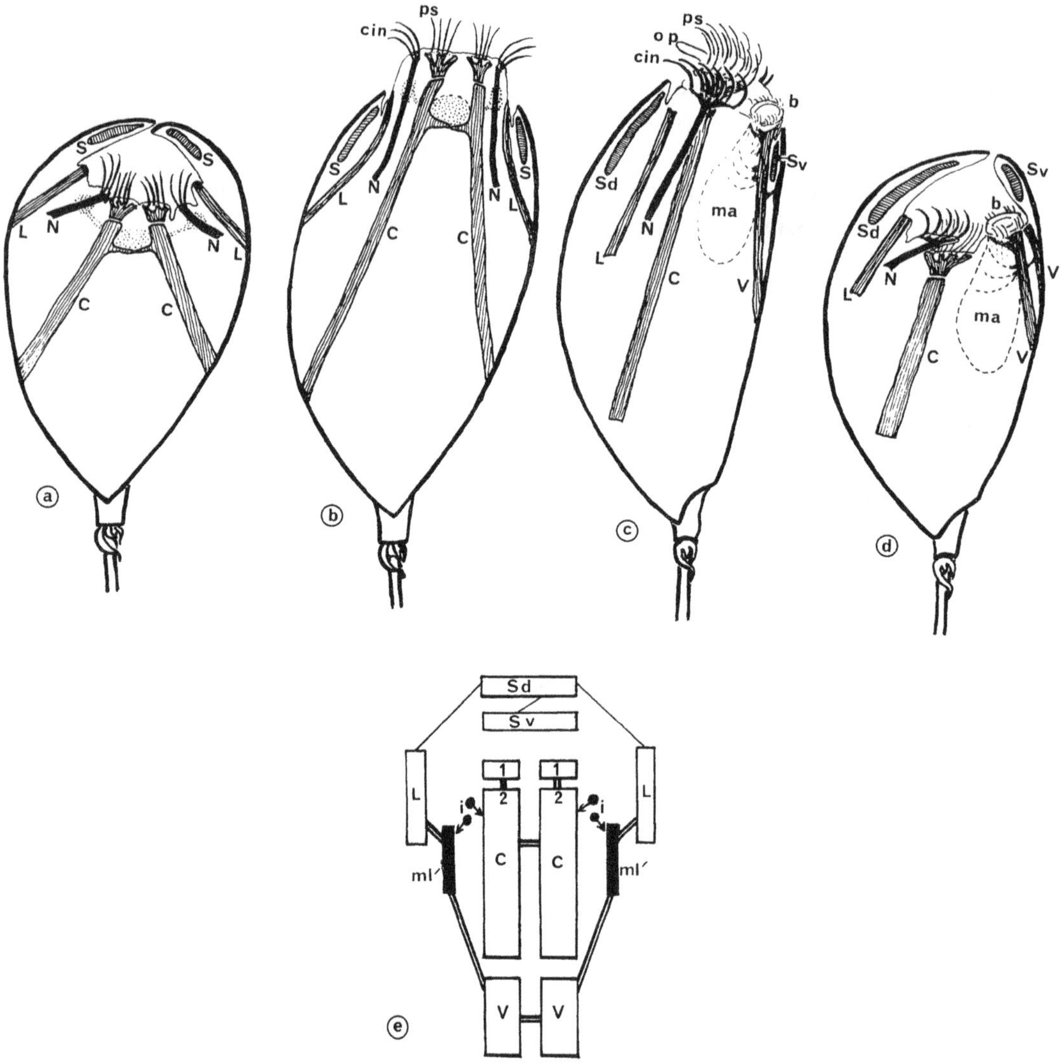

Fig. 3. A diagram of the muscles involved in withdrawal of the corona in *Trichocerca rattus*. a and b: dorsal views, the brain is stippled; c and d: lateral views; b and c: relaxed animals; a and d: contracted animals; e: diagram of the gap junctions present between the muscles. A double line between 2 cells indicates that the junctions have been observed; a single line indicates that such junctions are very likely to be present (the plasma membranes are in close contact) but have not been actually observed (the angle of sectioning was unfavorable). Abbreviations: b = mouth and ciliated buccal field; C = central retractor muscles (cells 1 and 2); cin = cilia of the cingulum; L = lateral retractor muscles; ma = mastax; N = muscles of the cingulum; op = palpar organ; ps = cilia of the pseudotrochus; S = anterior sphincter made up of 2 circular muscles, one dorsal (Sd) and one ventral (Sv). [From Clément and Amsellem, in press].

3. Retraction of the rotatory apparatus

Retraction of the rotatory apparatus is a widespread and frequent behavior among rotifers, which is especially spectacular in bdelloids and sessile rotifers. The only exceptions to this are those species that possess body wall expansions (spines or paddles) at the anterior boundary of the integument of the trunk (cf. part 4 of this review).

The majority of rotifers withdraw the rotatory apparatus into the interior of the trunk at the slightest provocation (e.g., touch, chemicals, vibration, etc.). Certain stimuli such as coming into contact with an obstacle, trigger only a slight retraction, while others cause a more complete retraction with a final closure of the opening which separates the rotatory apparatus from the external medium. The trunk integument of the anterior end of the animal consequently serves as a protective covering over the rotatory apparatus and its numerous sensory receptors.

a) Muscular insertions

Jacqueline Amsellem and I have studied the ultrastructure of the muscles involved in this behavior using serial sections of *Trichocerca rattus* [Amsellem & Clément, 1977 and unpublished data; Clément & Amsellem, in press; Amsellem, 1985].

The only muscles concerned with this movement are 3 pairs of longitudinal muscles (central, ventral and lateral retractors) and one pair of anterior circular muscles (one member of the pair being dorsal and the other ventral) (Fig. 3). All the retractors are inserted onto the trunk integument at their base. For an initial explanation of their function their apical insertions must be examined in detail. (1) The central retractors have their insertions under the central region of the apical field and the pseudotrochus. (2) The ventral retractors insert into the peripheral muscles of the mastax, the mouth, and the ciliated cells surrounding the mouth. (3) The lateral retractors have their insertions on the soft integument anterior to the trunk.

The anterior circular muscles are inserted strictly onto the integument of the trunk at various points corresponding to the *Z* lines. Plate Vc shows the ultrastructure of one of these insertions in detail. A desmosomal plate, consisting of dense bodies which are the sites of attachment of the thin (actin) myofilaments exists on the muscle side of the insertion. On the epithelial side there are only filaments which join the intracytoplasmic skeletal lamina to the desmosome. Such muscle-integument junctions have already been described in the Rotifera [Koehler, 1965; Clément, 1969, 1977a, b, 1985; Storch & Welsh, 1969]. A precise knowledge of all the muscular insertions permits us to have an understanding of the respective functions of all 3 pairs of retractors and that of the 2 anterior circular muscles which form an incomplete sphincter.

b) Correlations between muscle cytology and function

All the muscles in question, except the central retractors, are unicellular. The central retractors are bicellular. Therefore, they represent an interesting system for a detailed ultrastructural study (quantitative cytology included) of the entire muscle. These cytological results and other observations, such as, the extent of contraction of these muscles following fixation [Amsellem & Clément, 1977], the position of the retracted organs, and the dynamics of the process of retraction, are good indicators of their physiological characteristics. Data that have been obtained are too voluminous to be presented or even summarized here. They lead to the following conclusions [Amsellem, 1985]. 1) The central retractors contract very rapidly over a long distance. They are the longest and they are often found in a supercontracted state. These muscles are likely to be the least enduring (i.e., the contraction is of short duration). 2) The shorter ventral retractors seem to contract somewhat more slowly and to be a little more enduring. 3) The lateral retractors and the anterior circular muscles have a definitely slower and more protracted contraction.

From this information it is possible to reconstruct the sequence of events which lead to the contraction of the corona. First the apical part of the rotatory apparatus is pulled into the trunk by the central retractors. The buccal zone and the mastax are pulled in at about the same time, but somewhat slower and not as far, by the contraction of the ventral retractors. Finally, the lateral retractors and the anterior circular muscles bring about a prolonged retraction.

Termination of retraction is not due to contraction of antagonistic muscles, but to the progressive relaxation of the same 4 pairs of muscles which initiated it (especially the most enduring ones: lateral retractors and anterior circular muscles). The internal pressure is then sufficient for the reappearance of the corona.

Plate III. The pseudotrochus of *Brachionus calyciflorus*

a. ×80000. Transverse section through the cilia of one of the cirri of the pseudotrochus. The cilia are arranged in rows and they beat in unison perpendicularly to the plane of the row, which is determined by the position of the central tubules.

b. ×4000. Transverse section through one of the 3 bundles of cirri. Each cirrus consists of 4–8 rows of cilia in close registry.

c. ×800. Dorso-apical view of the rotatory apparatus under the scanning electron microscope. The pseudotrochus (P) consists of 3 bundles of cirri in the inner corona.

d. ×24000. The striated rootlets of the cilia of the pseudotrochus are inserted onto musculo-epithelial desmosomes (arrow). Mitochondria (m) lie in close proximity to the rootlets. M = muscle. Electron micrograph courtesy of Amsellem, Cornillac & Clément (unpublished).

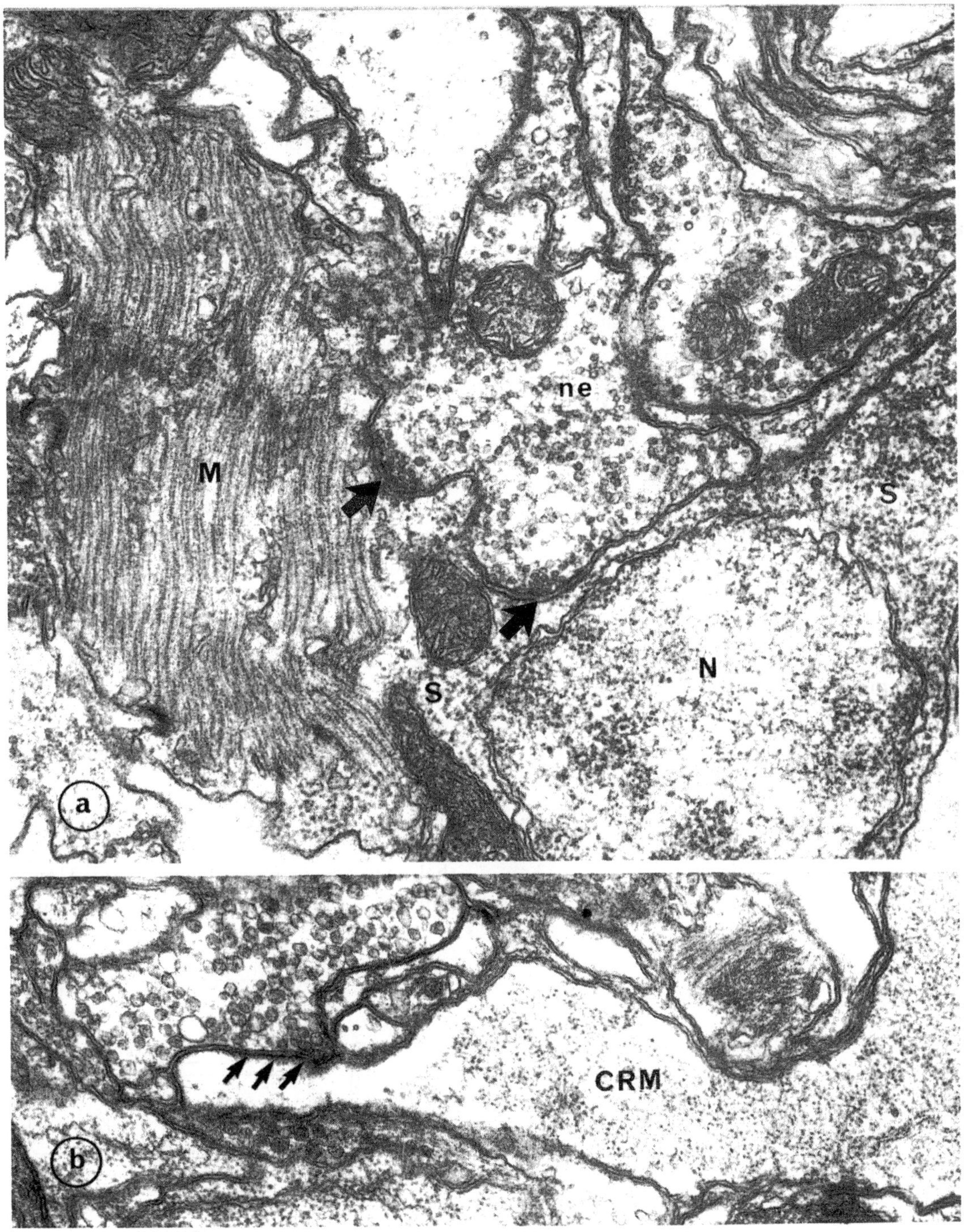

Plate IV.

a. Innervation of the cingulum muscle in *Trichocerca rattus*. ×27 000. The two synaptic boutons (arrows) in this section come from one nerve ending (ne) and innervate the muscle near its nucleus (N). (M = myoplasm; S = sarcoplasm).

b. Innervation of the central retractor muscle (CRM). ×32 000. The arrows point to the post-synaptic membrane at the upper part of the sarcoplasmic extension (CRM) which penetrates into the brain. Note the size of the vesicles present in different nerve fibers. The nerve fiber which makes synaptic contact with the muscle contains larger vesicles than those seen in the previous micrograph. Electron micrographs by Amsellem and Clément (from Clément, 1977a).

c) Innervation and gap junctions (Plate IVb; Plate Va; Fig. 2e; Fig. 4)

An ultrastructural study, carried out with the aid of serial sections, allowed Clément & Amsellem [1985] to localize the innervation of these muscles and the gap junctions which unite them. Some important observations of that study are shown on Figs. 2e and 4. They may be summarized as follows. 1) The two central retractors are clearly coupled (Plate Va), as are the ventral retractors and the 2 cells making up each central retractor. 2) Only the central retractors are innervated, each by nerve fibers which extends into the brain and receive several synaptic endings (Plate IVb). 3) None of the other 3 pairs of muscles involved in the retraction of the corona seems to be innervated. However, they are connected via gap junctions to each other and to a small lateral muscle (ml′), which is innervated. Hence there is an electrical coupling among them.

Two important conclusions regarding the process of retraction may be drawn from these observations. 1) The left and right sides of an animal are always coupled in retraction, whereas during swimming they are not. 2) Two sets of muscular effectors act as a unit: the central retractors, on one hand, and the 3 other pairs of muscles, on the other.

d) Synthesis (Fig. 4)

The diagrammatic summary of retraction (Fig. 4) differs from that of swimming (Fig. 2). There is always a total response (left and right sides acting in unison) even if the stimulus reaches only one of the two paired sensory receptors involved.

The two effector units are innervated laterally (left and right). This means that there is probably no communication between the two halves of the brain, at least as far as this behavior is concerned. Coupling is at the level of the muscular effectors. It is possible that initiation of this behavior (in response to a brief contact between the animal and an obstacle) corresponds to the beginning of the contraction of the central retractors. Only prolonged stimulation would bring about supercontraction of the central muscles and the contraction of all the other muscles involved in retraction. Synapses in the brain would be the basis for this sensory-motor integration. Unfortunately they have not been mapped yet.

4. Musculo-integumentary movements (Plates Vb, VI & VII)

Ciliary swimming is possible in rotifers because of their small size [Epp & Lewis, 1984]. However, certain species of the genera *Polyarthra* and *Filinia*

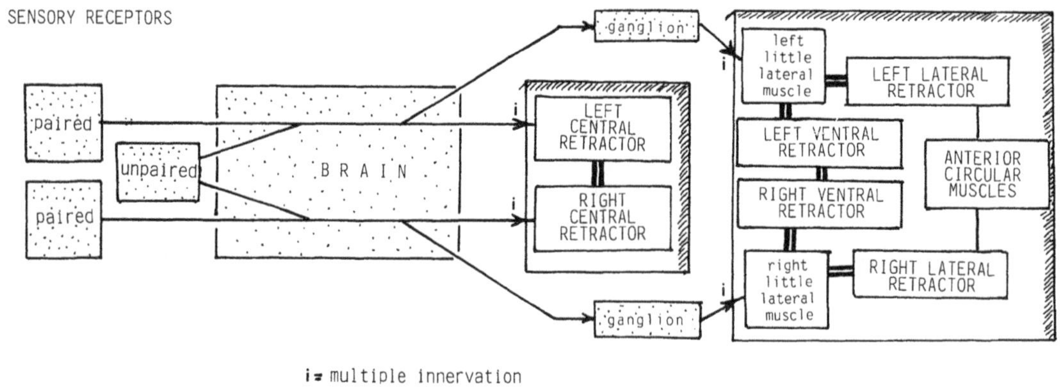

Fig. 4. Neuro-ethological schematic diagram of the withdrawal behavior in *Trichocerca rattus*.

The stimulation may come from one side only; however, the response is always global. If only the central retractors respond, a slight retraction results; if all the muscles shown respond, total retraction follows. The connections between muscles are indicated by double lines (observed gap junctions) or single lines (probably gap junctions). [From Clément & Amsellem, in press].

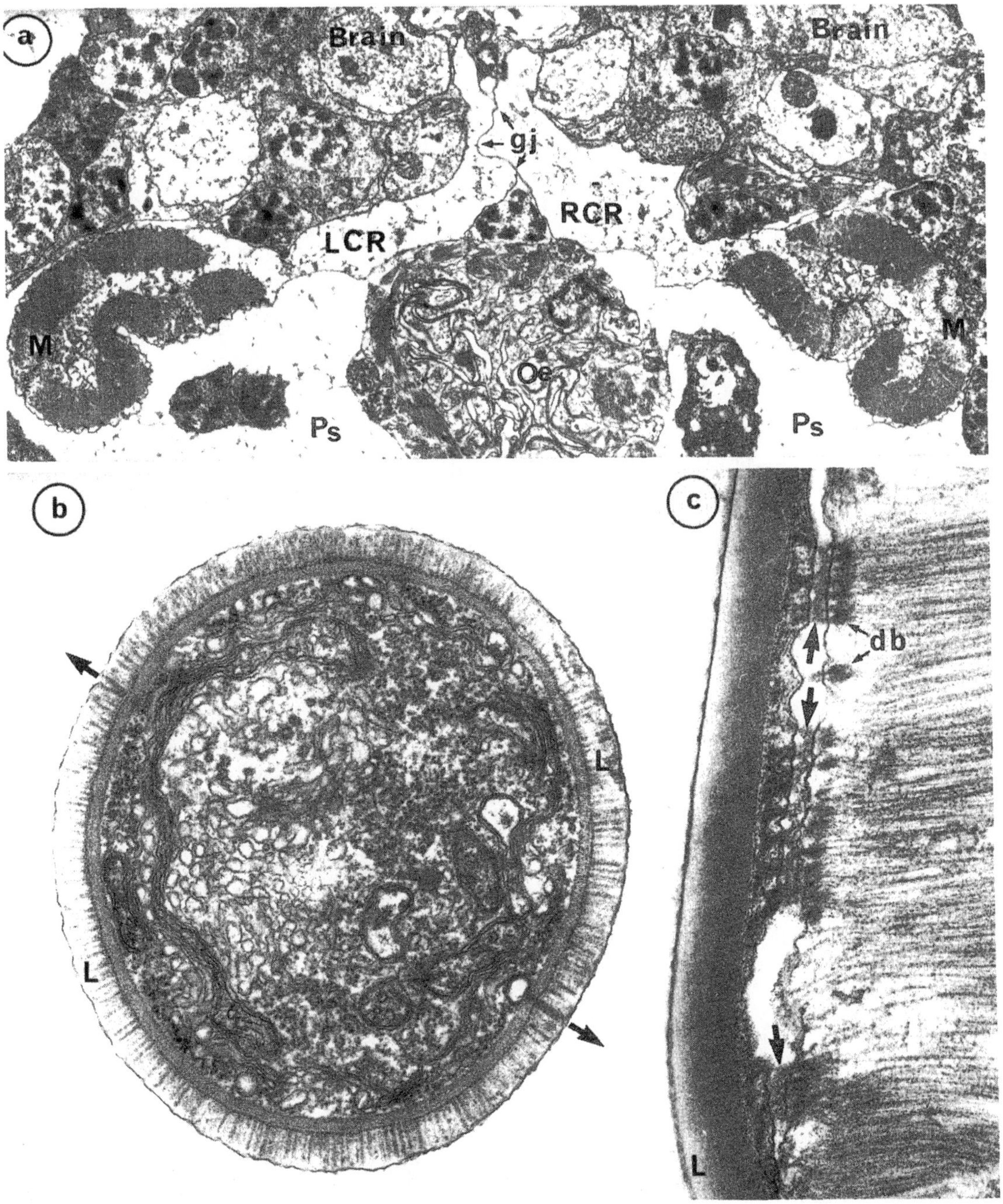

Plate V.

a. Gap junction (gj) between the sarcoplasm of the left central retractor (LCR) and the right central retractor (RCR) in a transverse section of *Trichocerca rattus.* ×7200 (Oe = esophagus, M = myoplasm, Ps = pseudocoel).

b. Spines in *Brachionus calyciflorus.* ×26500 – Transverse section of the posterolateral spine of a *Brachionus calyciflorus* female whose mother had been exposed to water in which *Asplanchna brightwelli* has been cultured. Note that the intracytoplasmic skeletal lamina (L) is quite thick and has the typical structure described in *Brachionus* [Clément, 1969]. There is no pseudocoel in the spine at this level. The cytoplasm of the integument contains a well developed Golgi apparatus, which participates in the processing of integumentary secretions (arrows).

c. Insertion of the anterior circular muscle on the integument of *Trichocerca rattus.* ×50000. Arrows point to the central stratum present in the intercellular space at the desmosomes. Note the dense bodies (db) in the muscle cell cytoplasm which are aligned next to the plasma membrane at the desmosome. Myofilaments of actin are inserted on these dense bodies. In the integumentary cytoplasm dense bundles of filaments join the plasma membrane to the intracytoplasmic skeletal lamina (L). Electron micrographs by Amsellem, Clément and Cornillac [*V*a from Amsellem, 1985].

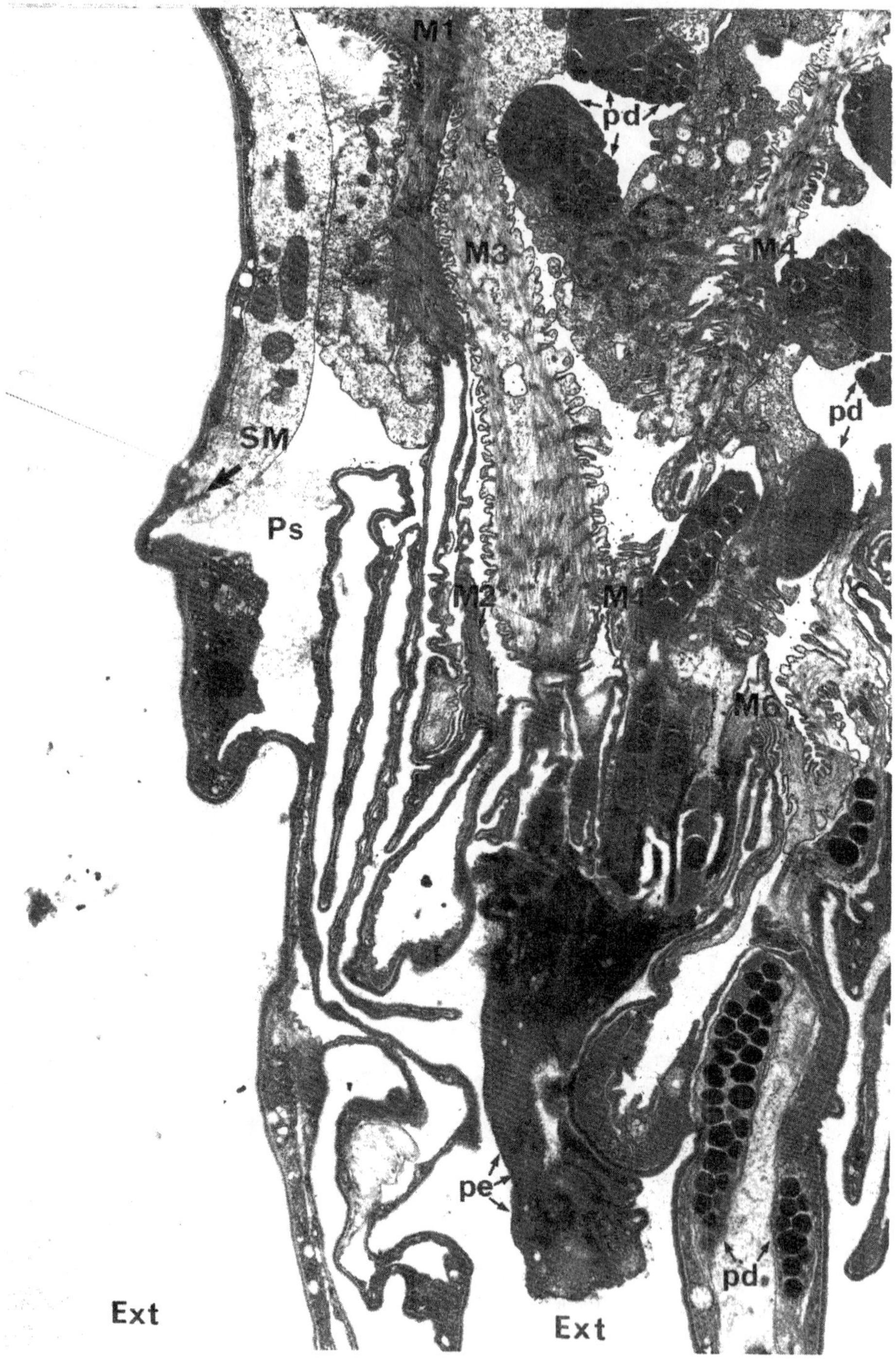

Plate VI. Axial section through the foot of *Philodina roseola*. ×6600. Oblique, transverse, and axial sections of the pedal ducts show at least 2 kinds of secretions which are discharged into the external medium (Ext) [pe = external pedal secretions]. The pseudocoel (Ps) contains longitudinal sections through several paired muscles (M_1, M_2, M_3, M_4, M_6) which are inserted on the soft integument of the foot and its invaginations. A longitudinal skeletal muscle (SM), sectioned axially, is inserted on the more rigid integument of the trunk (arrow) at the upper left. Electron micrograph by Amsellem and Clément [unpublished].

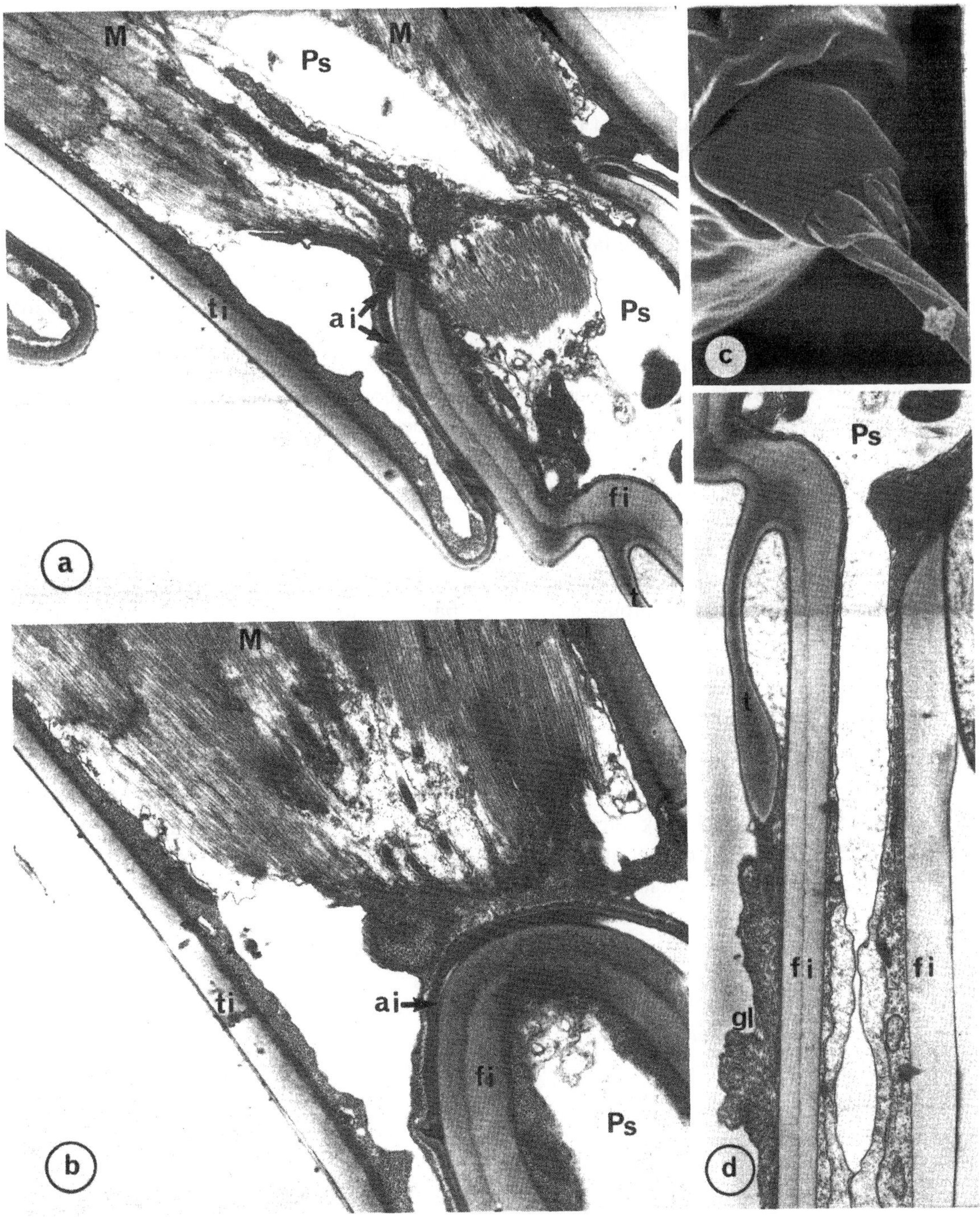

Plate VII. Axial sections through the foot of *Trichocerca rattus.*

a. ×18 000 and b ×22 000. The principal muscle fibers (M) are inserted at the junctions of the rigid foot integument (ti) and the soft articular integument (ai). The integument at the base of the tail (fi) is twice as thick as that of the trunk and of the base of the foot (ti). Ps = pseudocoel.

c. ×1 750. Scanning electron micrograph of the posterior extremity of the trunk and the base of the foot which bears the tail and several small toes.

d. ×22 000. Axial section through the base of the tail and one tow. The tip of the toe is embedded in the glue (gl) secreted by the pedal gland of *T. rattus*. The tail integument is (fi) twice as thick as that of the trunk and of the base of the foot. The pseudocoel reaches into the base of the tail. The tip of the tail is filled by the cytoplasm of the integument (ti = integument of the trunk and of the foot, fi = integument of the tail). Electron micrographs by Amsellem and Clément [VIIc from Clément, 1977a].

utilize, on occasion, another mode of locomotion to quickly move away from predators. This movement is a sudden jump and is accomplished by powerful muscles acting upon expansions of the integument which articulate with the trunk integument at their base. Gilbert [1985, 1987] has analyzed the jumps of *Polyarthra* through high speed cinematography. However, at present, we lack precise ultrastructural observations that would assist us in understanding the mechanism of these movements. Nonetheless, they have considerable adaptive significance, since they permit displacements which are very rapid in *Polyarthra*: 92 mm per second versus 0.7 mm per second in normal swimming [Gilbert, 1985].

However, we currently have some ultrastructural data that may help us in understanding the phylogenesis of these crude beginnings of articulated movements in rotifers. I will consider them briefly before I return to the mechanism of jumping movements in *Polyarthra* and *Filinia*.

a) Retraction of the corona and foot

The beginning of an articulation is present in two areas, corona and foot. The rigid trunk integument lies next to the soft integument which connects the trunk to the rotatory apparatus. The lateral retractors are inserted directly onto the soft integument which folds inward (Plate IIb; Fig. 3).

Retraction of the foot, particularly in bdelloids (Plate VI), illustrates how articulated movements and telescoping structures are made possible by alternation of rigid and soft integument. Here again, various paired muscles of the foot are inserted on the soft articular integument at each telescoping zone (Plate VI).

b) Movements of the tail of Trichocerca rattus *(Plate VII)*

These movements, though built on the same principle described above, show a more elaborate articulation. The tail is simply a very long cylinder made up of syncytial integumentary epithelium which has an intracytoplasmic, dense, skeletal lamina that is double the usual thickness. At its base the tail bears a ball and socket joint. Powerful foot muscles are inserted onto the soft part of the integument at several sites in the joint. This arrangement explains the possible movements of the tail of this rotifer.

c) Spines of Brachionus calyciflorus *(Plate Vb)*

These spines appear only in the descendants of females that have been exposed to water in which their predator, *Asplanchna* has been cultured [de Beauchamp, 1952a, 1952b; Pourriot, 1964, 1974; Gilbert, 1967]. They are simple extensions of the syncytial integument (Plate Vb).

Analysis of filmed sequences [Clément & Pourriot, 1982] show that their spines articulate at their base. Upon coming into contact with an *Asplanchna* female, *B. calyciflorus* retracts its rotatory apparatus and, at the same time, extends its spines in a direction perpendicular to its body. It remains immobile in this manner until swimming is resumed at the end of coronal retraction. With the resumption of swimming, the spines are returned to their initial position, parallel to the trunk.
It is possible that the retraction of the rotatory apparatus causes spine movement by an increase of the internal pressure.

d) Paddles of Polyarthra *and* Filinia

In contrast to the previous example, movement of these expansions of the integument does not accompany retraction of the corona, but replaces it. In this case, most or all of the longitudinal retractor muscles might have a modified anterior insertion that enables them to participate in the articulated motion. An ultrastructural study on *Filinia longiseta* which is currently in progress by Clément and Cornillac might bring us the answer to this question.

5. Crawling and feeding in rotifers which graze on filamentous algae (Figs. 5 & 6)

Feeding behavior of *Notommata copeus* has al-

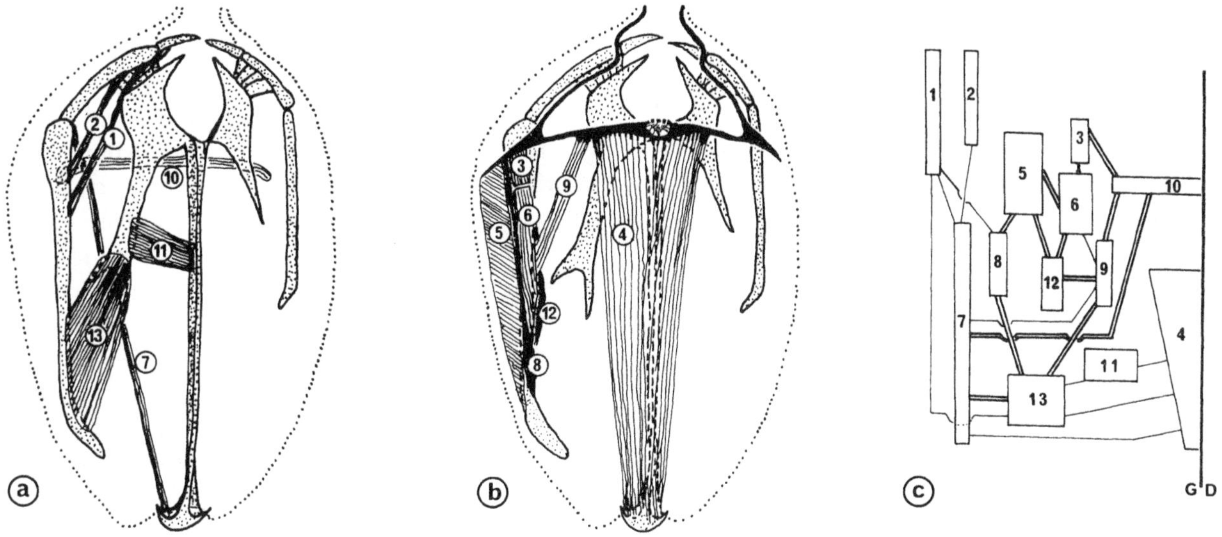

Fig. 5. The mastax muscles of *Trichocerca rattus*. Each muscle cell is numbered; 6 and 3 are the only bicellular muscles present; all the others are unicellular. (a and b) Location of the muscles: they insert on the trophi (stippled) or on the walls of the mastax cavity: the piston muscle (4) on the floor, cells 3, 5, 8 and 12 on the lateral walls. (c) Diagram of the gap junctions present between these muscles. The same symbols are used as in figure 3e. [From Clément & Amsellem, in press].

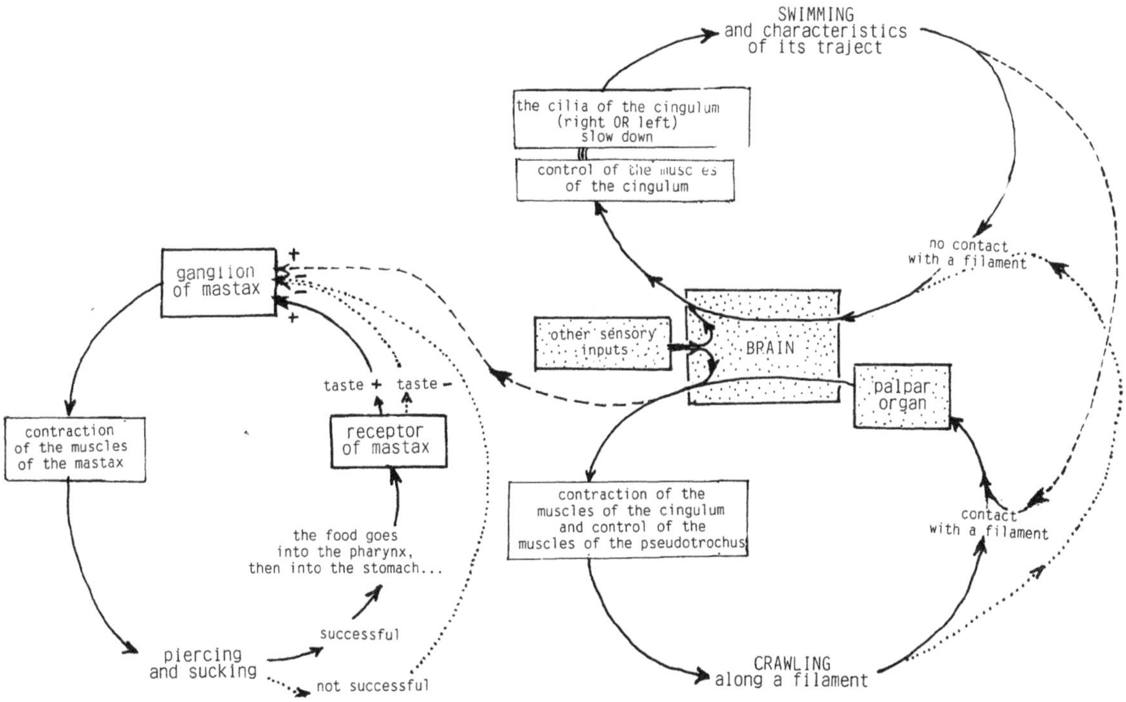

Fig. 6. Neuro-ethological schematic diagram of the feeding behavior in *Trichocerca rattus*, with the succession of three sequences: swimming, crawling, ingestion of food (see text). [From Clément & Amsellem, in press].

ready been analyzed [Clément *et al.*, 1983]. Since then, the study has been extended and an in-depth ultrastructural analysis of the basis of movements in *Trichocerca rattus* has been completed. Here, I shall present only the essential conclusions of a recent summary of this investigation [Clément & Amsellem, 1985].

a) Coordination of movements in the mastax of T. rattus

All the mastax movements in *T. rattus* seem to be linked (Fig. 5). Clément *et al.* [1983] distinguished 2 separate behavioral sequences: 'attempt at piercing the algal filament' and 'sucking by mastax movements'. With the aid of ultra-thin serial sections, it is possible to count the muscle cells in the mastax (11 paired and 2 unpaired), to pinpoint their insertions, and to detect definite and probable gap junctions between these cells (Fig. 5).

From these observations, J. Amsellem and I have concluded that all the muscle cells seem to be coupled. Therefore, the mastax is only capable of a few specific types of coordinated movements, which translate into piercing or non-piercing of the alga. If piercing is successful it is immediately followed by aspiration of the cell's contents. Usually, several sucking movements are required before the migration of chloroplasts into the pharynx can be seen.

b) Succession of 3 behavioral sequences: swimming, crawling, and feeding (Fig. 6)

Retraction commences whenever the rotatory apparatus of *T. rattus* comes into contact with an algal filament. If the palpar organ also comes into contact with the algal filament, the animal begins to crawl along it. The palpar organ is a highly specialized tactile receptor [Clément, 1977a, b; Clément *et al.*, 1983]. Crawling involves at least a partial retraction of the cingulum. In *N. copeus* the retraction is made evident by invagination of the auricles which bear the cingulum. In *T. rattus*, there is another muscle (an extension of the central retractor or perhaps a small independent muscle?) inserted onto cells of the cingulum beside the cingulum muscle which is inserted on the infraciliature (see above and Fig. 2).

Crawling is always accompanied by an attempt at piercing the filament, after a period of time which varies according to several parameters (e.g., age of the animal and its nutritional state). This implies that there is a passage of information from the brain to the mastax ganglion. A nerve which may work in this regard has been described at the level of the light microscope by Remane (1929–32). The mastax then pierces the filament and sucks out its contents. The sucking movements are repeated only if the animal was successful in the first place and the mastax receptors [described by Clément, 1977a; Clément *et al.*, 1983] were stimulated by algal material. If the mastax receptor is not stimulated (because the wall was not pierced, or the algal cell was empty, or the contents of the cell were not appropriate) the sucking movements cease and the animal resumes crawling or, in case of loss of contact with the filament, swimming.

6. Other rotifer movements

The purpose of this section is to discuss briefly the motor sequences of other behaviors which have not been analyzed to the same extent as the ones described already.

a) The feeding behavior of Brachionus calyciflorus *and other rotifers having a malleate or ramate mastax*

Gilbert and Starkweather [1977] have described three mechanisms by which the ingestion of suspended particles is regulated in *B. calyciflorus*. The cirri of the pseudotrochus can assume one of 2 positions: extension, allowing particles to enter, or screen formation, preventing particles (even as small as yeast) from entering. Plate III shows that each cirrus is composed of dozens of closely packed cilia which have a highly developed infraciliature. The latter consists of striated ciliary rootlets which extend deep into the epithelial cell and serve for the insertion of muscles (Plate IIId).

The assemblage of muscles, infraciliature, and cilia can partly explain the control of the quality of ingested food in the case of the pseudotrochus. Nevertheless, the scheme suggested by Clément and Pourriot [1982] and Clément [1983] remains to be verified through the identification of the sensory receptors involved and of the muscles that insert onto the pseudotrochus.

The two other mechanisms observed by Gilbert and Starkweather still await confirmation and analysis. They involve a rejection of particles from the oral canal 'presumably by a change in ciliary beat' and a rejection by the jaws which push the particles back into the buccal field.

The motor mechanisms involved in feeding, doubtless show similarities in all rotifers having a grinding (malleate or ramate) type of mastax. The presence of a buccal velum which we have observed in *B. calyciflorus, B. plicatilis, Philodina roseola* and, most recently, *Filinia longiseta* supports this hypothesis.

b) Substract attachment by means of the foot

To my knowledge, motor sequences leading to the attachment of a rotifer to the substrate have not been described with precision. However, the sensory determination of attachment has been studied by Viaud [1947] in *Brachionus rubens* and by Wallace [1975, 1978, 1980] in sessile rotifers [reviewed in Clément *et al.*, 1983].

Electron microscopic examination of the foot has shown that the ducts of the pedal glands are surrounded by a muscular envelope [Clément, 1977a, b] and that these muscles are innervated [Amsellem & Clément, unpubl. obs.]. In *T. rattus*, the pedal glue is secreted by the toes at the base of the tail. At first, it accumulates in a reservoir whose walls are lined by muscles characterized by a slow, protracted, contraction [Clément, 1977a, b; Clément & Amsellem, unpubl. obs.].

c) Copulatory behavior

Ethological or neuro-ethological studies are lacking for this behavior. There seem to be some sensory receptors at the tip of the penis [Clément *et al.*, 1983]. The copulatory behavior of male *B. calyciflorus* has been described by Gilbert [1963] in the presence and absence of the female and filmed by Clément and Pourriot [1982] along with those of *N. copeus* and *A. brightwelli.* Males and females come into contact at random as they move about. In the case of *B. calyciflorus* the chances of contact are increased through a chemoklinotaxis [Gilbert, 1963]. However, it is not known what motor sequences intervene between the initial contact and the attachment of the penis to the integument of the female.

d) Oviposition

Oviposition has been described and filmed [Pourriot & Clément, 1982]; we don't know which muscles are involved in this activity. A muscular sphincter around the cloacal opening has not been detected using electron microscopic techniques [Clément *et al.*, 1983]. In *T. rattus* there is a longitudinal muscle whose contraction seems to aid opening the cloacal aperture. It is not known which glands secretes the filament that anchors eggs to the foot of the female in those species that carry their eggs around.

In *N. copeus*, egg laying is followed by a distinctive behavior sequence [Clément, 1977a; Pourriot & Clément, 1982; Clément *et al.*, 1983]. A neuroethological analysis of this behavior is lacking. Such an analysis would be all the more interesting since, of all the rotifer activities described to date, this is one of the longest (aroudn 20 minutes) and most elaborate.

7. Conclusions

Analysis of rotifer behavior is still in its infancy. Therefore, it is very risky to ask such questions as: Why do these behaviors exist? and Why do these movements take place? Of course, the very fact that these behaviors exist indicates that they have some adaptive significance as they have withstood natural selection. As I have mentioned in the introduction, these behaviors are sufficiently well adapted to the environment so that the species or clones inhabiting that environment are able to survive.

Rather than succumbing to the illusion of providing an answer by evoking the term 'optimalization', a modern substitute for finalism, I wish to turn the question around and ask 'why not?' as a challenge to rotiferologists and biologists in general. Investigators should make use of the extraordinary simplicity of rotifer structure and behavior, and examine the problems of evolution by means of a comparative study of rotifer behavior. These animals possess highly advantageous characteristics, particularly isogeny resulting from diploid parthenogenesis [King, 1977]. They merit consideration by a joint team of ecologists, population geneticists, ethologists and neurobiologists. Such a collaborative venture ought to result in a description of behavior in terms of the life history of the individuals, of the populations and of the species.

Acknowledgements

This work was supported by CNRS grants RCP-657 and GIS-14 and by Lyon 1 University (Neuro-ethological team and, previously, by the Laboratory of Histology). The electron microscopical observations were made by a joint effort among several team members including, Jacqueline Amsellem (*Trichocerca rattus, Brachionus plicatilis*), Elizabeth Wurdak (*Asplanchna brightwelli*) Annie Cornillac (*Brachionus calyciflorus*), and Anne Luciani (*B. plicatilis*). The observations were carried out at the Centre de Microscopie Electronique Appliquée à la Biologie et à la Géologie (CMEABG, Lyon 1 University). My ideas concerning the behavioral and neuro-ethological observations reported in this review were improved by the valuable discussion I had with members of the SFECA (Société Francaise d'Etude du Comportement Animal), particularly those with Dr. Raymond Campan. The most striking results presented in this review owe their existence to the enthusiastic and capable efforts of Jacqueline Amsellem. I also wish to thank Elizabeth Wurdak for graciously undertaking the unexpected and arduous task of translating this text into English.

References

Amsellem, J., 1985. Analyse cytologique multifactorielle de muscles d'Invertébrés (Cténaires, Rotifères, Mollusques). Thèse Doctorat Etat, Univ. Nice, France.

Amsellem, J. & P. Clément, 1977. Correlation between ultrastructural features and contraction rates in rotiferan muscles. I. Preliminary observations on longitudinal retractor muscles in Trichocerca rattus. Cell Tissue Res. 181: 81–90.

Beauchamp, P. de, 1952a. Un facteur de la variabilité chez les Rotifères du genre Brachionus. C.R. Acad. Sci., Paris 234: 573–575.

Beauchamp, P. de, 1952b. Variation chez les Rotifères du genre Brachionus. C.R. Acad. Sci., Paris 235: 1355–1356.

Bovet, P., 1984. Modeles clinocinétiques. In: La vision chez les Invertébrés, C.N.R.S. ed., Fr. 232–236.

Clément, P., 1977a. Introduction à la photobiologie des Rotifères dont le cycle reproducteur est controlé par la photopériode. Approches ultrastructurale et expérimentale. Thèse Doctorat Etat, Univ. Lyon 1, Fr. 7776, 262 pp.

Clément, P., 1977b. Ultrastructural research on rotifers. Arch. Hydrobiol. Beih., 8: 260–297.

Clément, P., 1977c. Phototaxis in rotifers (action spectra). Arch. Hydrobiol. Beih. 8: 67–70.

Clément, P., 1984a. Yeux et ocelles des Rotifères. In: La Vision chez les Invertébrés, C.N.R.S. ed., Fr.: 82–88.

Clément, P., 1984b. Accords et désaccords sur la définition de la vision: les Invertébrés voient-ils? In: La Vision chez les Invertébrés, C.N.R.S. ed., Fr.: 317–331.

Clément, P. & J. Amsellem, in press. Ultrastructures et comportements: neuro-éthologie des Rotifères. Comportements, in press.

Clément, P., J. Amsellem, A. M. Cornillac, A. Luciani & C. Ricci, 1980a. An ultrastructural approach to feeding behavior in Philodina roseola and Brachionus calyciflorus (Rotifers). I. The buccal velum. Hydrobiologia 73: 127–137.

Clément, P., J. Amsellem, A. M. Cornillac, A. Luciani & C. Ricci, 1980b. An ultrastructural approach to feeding behavior in Philodina roseola and Brachionus calyciflorus (Rotifers). III. Cilia and muscles, conclusions. Hydrobiologia 73: 137–141.

Clément, P. & X. Fouillet, 1970. Les jonctions cellulaires d'um organisme (rotifère). 7ème Congr. internat. Micro. Electr., Grenoble, Fr. 1: 7–8.

Clément, P., A. Luciani, A. Cornillac, J. L. Chassé, P. Y. Coulon, J. P. Charras, M. Nougaret & A. Fournier, 1985. Un système de trajectographie automatique pour étudier la nage d'animaux planctoniques (Rotifères, miracidiums de Plathelminthes etc....). Verh. Internat. Verein. Limnol. 22: 3002–3006.

Clément, P. & R. Pourriot, 1982. Biology of fresh-water rotifers. 1. Swimming and feeding. Movies CNRS Audio-visuel, Fr., 16 mm., 25 minutes. R.C.P. C.N.R.S. 657.

Clément, P., 1983. Un dispositif de suivi automatique des déplacements d'animaux. Premiers résultats et perspectives d'utilisation. Bull. int. S.F.E.C.A., Fr. 2: 119–130.

Clément, P. & E. Wurdak, 1984. Photoreceptors and photoreceptions in Rotifers. In: Photoreception and vision in Invertebrates, NATO-ASi, Plenum Press, N.Y. and London: 241–288.

Clément, P., E. Wurdak & J. Amsellem, 1983. Behavior and ultrastructure of sensory organs in Rotifers. Hydrobiologia 104: 83–129.

Cornillac, A. M., 1982. Yeux cérébraux et réponses motrices à la lumière chez Brachionus calyciflorus et Asplanchna brightwelli (Rotifères). Thèse doctorat 3ème cycle, Univ. Lyon 1, Fr., 1212, 78 pp.

Coulon, P. Y., J. L. Charras, P. Clément, A. M. Cornillac, A. Luciani & E. Wurdak, 1983. An experimental system for the automatic tracking and analysis of Rotifer swimming behavior. Hydrobiologia 104: 197–196.

Epp, R. W. & W. M. Lewis, Jr., 1984. Cost and speed of locomotion for rotifers. Oecologia 61: 289–292.

Gilbert, J. J., 1963. Contact chemoreception, mating behavior, and sexual isolation in the rotifer genus Brachionus. J. Exp. Biol. 40: 625–641.

Gilbert, J. J., 1967. Asplanchna and postero-lateral spine production in Brachionus calyciflorus. Arch. Hydrobiol. 64: 1–62.

Gilbert, J. J., 1985. Escape response of the rotifer Polyarthra: a high-speed cinematographic analysis. Oecologia 66: 322–331.

Gilbert, J. J., 1987. The Polyarthra escape response: reduction of interference from Daphnia. Hydrobiologia.

King, C. E., 1977. Genetics of reproduction, variation and adaptation in Rotifers. Arch. Hydrobiol. Beih. 8: 187–201.

Koehler, J. K., 1965. A fine structure study of the rotifer integument. J. Ultrastruc. Res. 12: 113–134.

Kutikova, L. A., 1983. Parallelism in the evolution of rotifers. Hydrobiologia 104: 3–8.

Lansing, A. I. & F. Lamy, 1961a. Fine structure of the cilia of rotifers. J. Biophysic. and Biochem. Cytol. 9: 799–812.

Lansing, A. I. & F. Lamy, 1961b. Localization of ATPase in rotifer cilia. J. Biophysic. and Biochem. Cytol. 9: 498.

Luciani, A., 1982. Contribution à l'étude du vieillissement chez le Rotifère Brachionus plicatilis: nage, cils et battements ciliares, métabolisme énergétique. Thèse 3ème cycle Univ. Lyon 1, 1211, 79 pp.

Luciani, A., J. L. Chassé & P. Clément, 1983. Aging in Brachionus plicatilis: the evolution of swimming as a function of age at two different calcium concentrations. Hydrobiologia 104: 141–146.

Nachtwey, R., 1925. Untersuchungen uber die Keimbahn Organogenese und Anatomie von Asplanchna priodonta Gosse-Zeitschr. Wiss. Zool. 126: 239–492.

Piaget, J., 1976. Le Comportement moteur de l'Evolution. Gallimard (ed.) Coll. Idées, 190 pp.

Pourriot, R., 1964. Etude expérimentale de variations morphologiques chez certaines espèces de Rotifères. Bull. Soc. Zool., Fr. 89: 555–561.

Pourriot, R., 1974. Relations prédateur-proie chez les Rotifères: influence du prédateur (Asplanchna brightwelli) sur la morphologie de la proie (Brachionus bidentata). Ann. Hydrobiol. 5: 43–55.

Pourriot, R. & P. Clément, 1982. Biology of freshwater rotifers. II. Reproduction and reproductory cycles. Movies CNRS, Audio-visuel, Fr., 16 mm, 25 minutes.

Remane, A., 1929–1933. Rotatoria. In: M. G. Bronn (ed.) Klassen und Ordnungen des Tier-Reichs IV (Vermes), 2 (Aschelminthes), 1 (Rotatorien, Gastrotrichen und Kinorhynchen) 3: 1–448.

Scholtysek, E. & R. Danneel, 1962. Die Feinstruktur des Wimperapparates von Brachionus urceolaris (Rotatoria). Protoplasma 56: 99–108.

Storch, V. & U. Welsch, 1969. Uber den Aufbau des Rotatorienintegumentes. Z. Zellforsch. 95: 405–414.

Viaud, G., 1940. Recherches expérimentales sur le phototropisme des Rotifères: I. Bull. Biol. Fr. Belg. 74: 249–308.

Viaud, G., 1943a. Recherches expérimentales sur le phototropisme des Rotifères: II. Bull. Biol. Fr. Belg. 77: 68–93.

Viaud, G., 1943b. Recherches expérimentales sur le phototropisme des Rotifères: III. Stroboscopie des mouvements ciliaires et phototropisme. Bull. Biol. Fr. Belg. 77: 224–242.

Viaud, G., 1947. Recherches expérimentales sur les tropismes des Rotifères. L'oscillorhéotropisme des Brachionus rubens Ehr., cause de la fixation de ce Rotifère phorétique sur les Daphnies et autres crustacés d'eau douce. Ann. Sci. nat. Zool. 9: 39–62.

Villeneuve, J. & P. Clément, 1971. Le neuropile du cerveau de rotifère: observations ultrastructurale préliminaires. J. Microscopie, Fr. 11: 100.

Waddington, C. J., 1975. The evolution of an evolutionist. Edinburgh University Press, 328 pp.

Wallace, R., 1975. Larval behavior of the sessil rotifer Ptygura beauchampi (Edmondson). Verh. int. Ver. Limnol. 19: 2811–2815.

Wallace, R., 1978. Substrate selection by larvae of the sessile rotifer Ptygura beauchampi. Ecology 59: 221–227.

Wallace, R., 1980. Ecology of sessile Rotifers. Hydrobiologia 73: 181–193.

Ware, R. W., 1971. Computer aided nerve tracing in the brain of the Rotifer Asplanchna brightwelli. Doctorat Ph.D. M.I.T., USA, 213 pp.

Hydrobiologia 147: 361–371 (1987)
May, L., Wallace, R. & Herzig, A. (eds), Rotifer Symposium IV

Ultrastructure and histochemistry, of the stomach of *Asplanchna sieboldi*

Elizabeth S. Wurdak
Department of Biology, Saint John's University Collegeville, MN 56321

Keywords: rotifer, stomach cell, food vacuole, ultrastructure

Abstract

Stomach cells of female *Asplanchna sieboldi* are specialized for absorption and intracellular digestion of nutrients. Evidence is presented to show that electron-opaque colloidal substances, present in the medium and within digestive vacuoles of the prey (*Paramecium*), are taken up by the stomach cells at the apical cell membrane and sequestered within food vacuoles which contain hydrolases working in both the acid and alkaline pH range. The stomach cells are also implicated in the absorption of molecules below the resolving power of the electron microscope. In rotifers possessing a complete digestive tract, this task is presumed to be handled by the intestine.

Introduction

Female *Asplanchna sieboldi* subsist on a diet composed of protozoans and small metazoans, including planktonic crustaceans and other rotifers [Gilbert, 1973; Salt *et al.*, 1978].

Their alimentary tract consists of mouth, pharynx, salivary glands, esophagus, gastric glands, and stomach [Beauchamp, 1909; Waniczek, 1930]. An outstanding feature of the genus *Asplanchna* is the total absence of an intestine and anus. The stomach is a blind pouch; undigested food remnants are brought back to the pharynx and rejected through the mouth with the assistance of the jaws [Dalrymple, 1849; Hudson & Gosse, 1886; Beauchamp, 1909].

The principal role of the mouth, pharynx and esophagus is the ingestion of prey. Digestion and absorption of nutrients are accomplished through the combined action of the salivary glands, the gastric glands and the stomach. The exact contribution of each of these three organs to the processing of food cannot be assessed on the basis of published observations. Due to the small size of the organism, physiological and biochemical data are difficult to obtain. Electron microscopy, histochemistry and electron opaque tracers have been applied to the study of aging in *Philodina* [Lansing, 1964; Harold & Meadow, 1970].

In the present report, the same techniques are used to examine the cellular stomach of *A. sieboldi*, in order to determine more specifically its role in the breakdown and assimilation of food. Particular attention is paid to the formation, fate and enzymatic composition of inclusions in the stomach cells, which were dubbed 'food vacuoles' by Lansing [1964] on account of their similarity to protozoan organelles of the same name.

Materials and methods

Culture

The rotifers used in this study were drawn from clones 10C6 and 12C1 of *Asplanchna sieboldi* maintained at Dartmouth College by Dr. John Gilbert. Both clones were derived from the stocks of Dr. C. W. Birky Jr. The culture conditions were the same as described for *Asplanchna brightwelli* by Gilbert [1968].

Histochemistry

I. Acid phosphatase. The presence of the enzyme was demonstrated at the level of the light microscope with the azo-coupling dye technique [Barka & Anderson, 1962]. Amictic female rotifers were fixed in 10% formalin, buffered with 0.5 M phosphate buffer, pH 7.5 for 1 h at 4 °C. They were rinsed in 3 changes of buffer, dehydrated in glycol methacrylate in the cold and embedded in a mixture of glycol and butyl methacrylate [Leduc & Bernhard, 1967]. The specimens were placed in gelatin capsules (size 00) with liquid monomer mixture and polymerized under an ultraviolet light source in a cold room at 4 °C. Sections were cut at 4 μm with glass knives and transferred to depression slides. The sections were reacted for acid phosphatase using naphtol-AS-TR-phosphoric acid as substrate and hexazotized pararosanilin [Davis & Ornstein, 1959] as the capture reagent [Lojda *et al.*, 1964]. The incubation was carried out at 37 °C for 30 minutes. Controls were run with the omission of substrate or with the addition of 1 mg ml^{-1} sodium fluoride to the incubation mixture.

II. Alkaline phosphatase. The reaction for alkaline phosphatase was also carried out on formalin-fixed material embedded in methacrylate. Sections (4 μm) were incubated at room temperature for 15–30 minutes in incubation mixtures containing sodium alpha-naphthyl phospate and Fast Red TR (the stable diazotate of 5-chloro-o-toluidine) buffered to pH 9.9 with 0.1 M Tris buffer [Pearse, 1968]. Controls were run with omission of substrate. The solutions were filtered prior to use.

III. Nonspecific (carboxyl) esterase. The demonstration of nonspecific esterase was also carried out on 4 μm sections of material which had been fixed in formalin and embedded in methacrylate. The incubation mixture was made up by dissolving 10 mg alpha naphthyl acetate in 0.25 ml acetone and bringing the solution to 20 ml with 0.1 M phosphate buffer at pH 7.4. After shaking, 50–100 mg Fast Blue B salt was added. The solution was filtered prior to use. The incubation was carried out at room temperature for 15–30 minutes. Controls were treated in an identical fashion with the omission of substrate.

Electron microscopy

For routine scanning and for the observation of the normal morphology of the female digestive system, rotifers were fixed in 2.5% – 3.5% glutaraldehyde in 0.1 M phosphate buffer (pH 7.3 – 7.5) for 1–4 h at 4 °C. The fixed animals were washed with three 20-minute changes of buffer and then postfixed in 1% osmium tetroxide in 0.1 M phosphate buffer, pH 7.3 – 7.5, for 1 h over ice. Dehydration was carried out in a graded ethanol series at room temperature. While the animals were in 70% ethanol, the hypodermis was punctured with a tungsten needle to enhance infiltration of the embedding medium.

Embedding was done either in Epon 812 according to the procedure of Luft [1961], with or without the propylene oxide rinse, or in the Spurr low-viscosity epoxy mixture [Spurr & Harris, 1968; Spurr, 1969]. In the latter case the animals were not punctured.

Amictic females used for thorium dioxide labelling were obtained free of preformed food vacuoles by isolating the gravid mothers into fresh rotifer medium free of paramecium. Newborn female offspring were picked out within 2–3 minutes after birth and exposed to rotifer medium containing paramecium and Thorotrast, in a concentration of 1 part Thorotrast to 50 parts medium, for lengths of time ranging from 3 minutes to 2 hours and then fixed and embedded as above.

Results

Light microscopy

In a well-nourished adult animal the stomach is a prominent saccular organ (Fig. 1a). Its walls are composed of a single layer of cells, which are ciliated along their inner margins. In living animals the entire gut contents are steadily rotated in the same direction. The cells are capable of extensive deformation. Their general outline may be columnar or

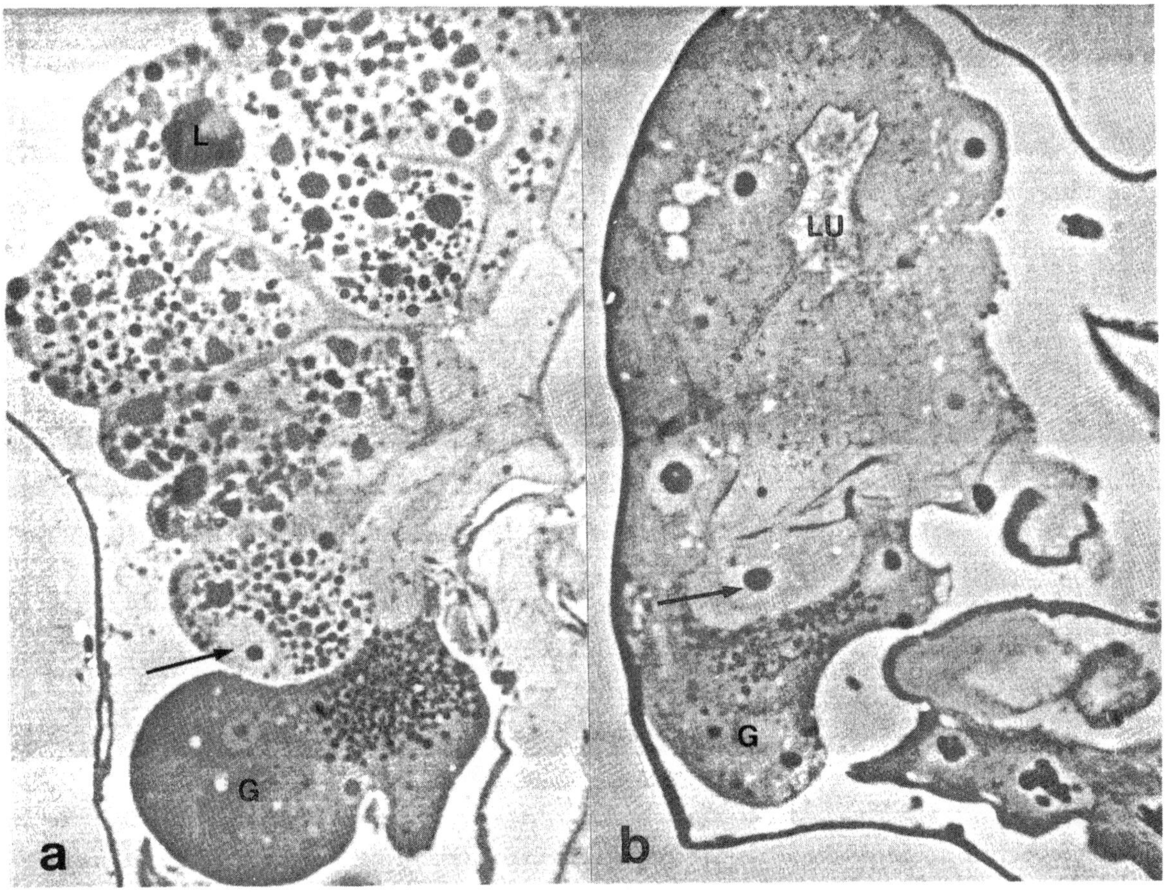

Fig. 1. Light micrographs of the stomach and gastric gland of female *Asplanchna sieboldi*.
(a) ×1800 Cluster of stomach cells and gastric gland (G) of a well-nourished adult. The cells are packed with food vacuoles. Arrow points to the nucleus of one of the stomach cells. A large lipid droplet (L) is present in the cell at the upper right. Epon. Phase contrast optics.
(b) ×1800 Section through the entire stomach and one gastric gland of a newborn female. The lumen (LU) of the stomach is small and food vacuoles are lacking. Arrow indicates stomach cell nucleus. The gastric gland (G) of the newborn resembles that of the adult. Epon. Phase contrast optics.

horizontally flattened, depending on how far the lumen is engorged with food. The position and shape of the single nucleus varies accordingly. The gastrodermal cells are packed with spherical inclusions. They show considerable variation in size; some even attain the dimensions of the nucleus. Superficially, they resemble the granules present in the gastric gland. The stomach granules are collectively known as food vacuoles.

In a well-fed individual each stomach cell contains a single, very large fat droplet, 10 μm or more in diameter (Fig. 1a). Commonly, however, the fat is distributed as a number of smaller globules.

The stomach of a newly-born female is devoid of food vacuoles; the nuclei stand out sharply against a light cytoplasm. The lumen of the stomach is narrow. The entire organ appears considerably smaller than that of the adult. The gastric gland contains abundant secretory granules at birth which are indistinguishable from those of the adult (Fig. 1b).

I. Acid phosphatase (Figs. 2a, b). Reaction product for acid phosphatase was localized within particulate components of the gut which correspond to food vacuoles in the control sections (Fig. 2a). Some food vacuoles, particularly those of smaller

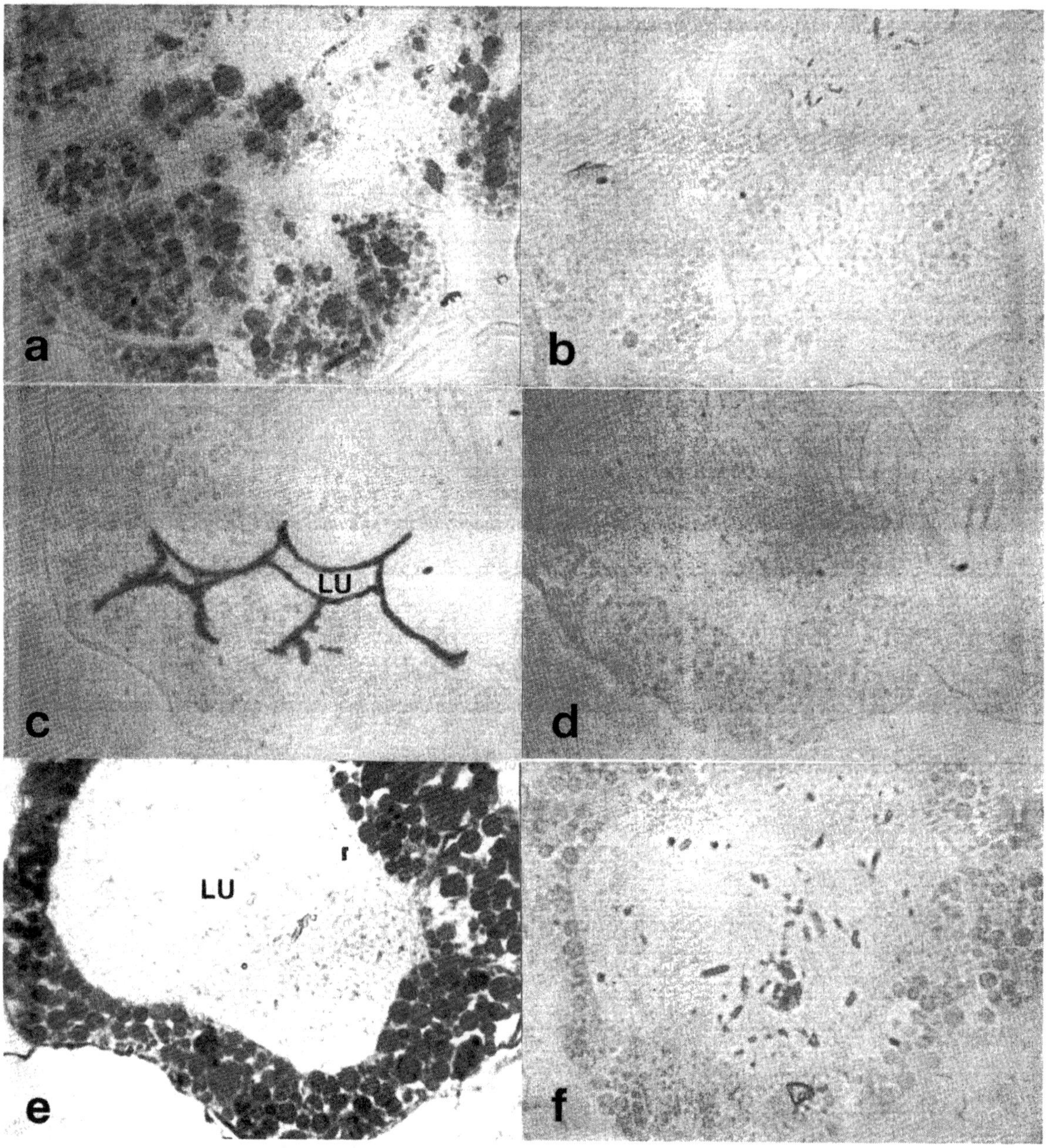

Fig. 2. Histochemistry of the stomach of female *A. sieboldi* (a) ×1100 Acid phosphatase test on a section through the stomach of an actively feeding adult female shows that many food vacuoles, especially large vacuoles with irregular contours, contain reaction product. Azo coupling method. Bright-field optics.

(b) ×1100 Acid phosphatase control. Bright-field optics.

(c) ×1100 Alkaline phosphatase assay on the stomach of the female. Cell membrane next to the lumen (LU) stains heavily. Fast Red TR stain. Bright-field optics.

(d) ×1100 Alkaline phosphatase control. Bright-field optics.

(e) ×1100 Localization of non-specific (alpha-naphthyl) esterase in the stomach. All vacuoles show reaction product. Stomach lumen (LU) is non-reactive. Fast Blue B stain. Bright-field optics.

(f) ×1100 Esterase control. Bright-field optics.

dimensions, remained unstained. Staining over the gastric gland was light or absent altogether. Control sections in which either the substrate was lacking or NaF was added to the incubation mixture were uniformly nonreactive (Fig. 2b). The results were reproducible over three separate assays.

II. Alkaline phosphatase (Figs. 2c, d). A strong reaction for alkaline phosphatase was found at the edge of the cells bordering the lumen of the stomach and in the grooves formed where two cells meet at the apex (Fig. 2c). The opposite, pole of the cell remained unstained. It is unlikely that the staining is due to a bacterial enzyme since there was no reaction in the lumen itself.

The gastric gland contains no detectable levels of alkaline phosphatase. Control sections lacking substrate were entirely negative (Fig. 2d).

II. Nonspecific esterase (Figs. 2e, f). All of the food vacuoles in the stomach contained reaction product for esterase (Fig. 2e). The addition of NaF to the incubation mixture did not eliminate the staining. This finding is in agreement with the morphological evidence that the enzyme is in a bound form, i.e., lysosomal.

The gastric gland contained no detectable level of enzyme. Staining was absent in control sections incubated with the omission of substrate (Fig. 2f).

Electron microscopy

I. Ultrastructure of the cellular stomach. The stomach is organized into discrete cells which form a single layer around the lumen (Fig. 3). The lumen contains partially digested food, including remnants of the mitochondria and trichocysts of *Paramecia* (Fig. 5a). The stomach cells show a definite polarity. On the side next to the pseudocoelom, each cell is bordered by a basement membrane, consisting of the plasma membrane and a thin (40 nm) layer of granular material. The surface of the cell facing the lumen bears cilia and microvilli (Figs. 3, 4a). A distinctive junctional specialization is found where two cells meet at the apex. It consists of two separate regions: a zonula adherens and a septate desmosome (Fig. 3, arrow).

The height of the microvillous border is 0.15 – 0.25 μm. The microvilli are unbranched and they have dense tips (Fig. 4a). In some sections they are seen to contain a fibrous core which continues into the terminal web. The terminal web is a loose fibrillar zone lying immediately beneath the cell membrane (Fig. 4a).

The internal organization of the cilia conforms to the general pattern of 9 double outer fibers plus 2 central fibers. Ciliary basal bodies or kinetosomes are present at the cell margins. There are no ciliary rootlets. (Fig. 5b).

Between the microvilli, the plasma membrane invaginates to form characteristic, goblet-shaped inpocketings (Fig. 4a) or 'caveolae intracellularis' [Yamada, 1955]. The caveolae are surrounded by densely-staining collars in the region of the terminal web (Fig. 4a). They are occasionally connected with bristle-coated vesicles through a slightly constricted neck (Figs. 4a, 5b). Free coated vesicles are present in the surrounding cytoplasm.

The most conspicuous organelle present in the stomach cell is the food vacuole (Fig. 4b, 5c). Food vacuoles are highly pleomorphic. Their outline may be neatly circular or show numerous extensions into the surrounding cytoplasm. The diameter of one of the larger vacuoles was measured to be 5 μm. The vacuole matrix may be finely granular or contain vesicles, short membrane fragments and particles, similar in size to free ribosomes or beta glycogen, against a background which varies in density from one vacuole to another. These variations probably reflect different stages in the breakdown of nutrients.

All food vacuoles are surrounded by a single membrane which is like the plasma membrane, the membrane surrounding the coated vesicles and the tubular bodies in their vicinity. The food vacuole membrane shows two dense outer layers separated by a light inner layer following osmium fixation and lead staining (Fig. 4b).

II. The uptake of thorium dioxide (thorotrast) into food vacuoles. Uptake of dietary thorium dioxide by newborn females is extremely rapid. A 3-minute

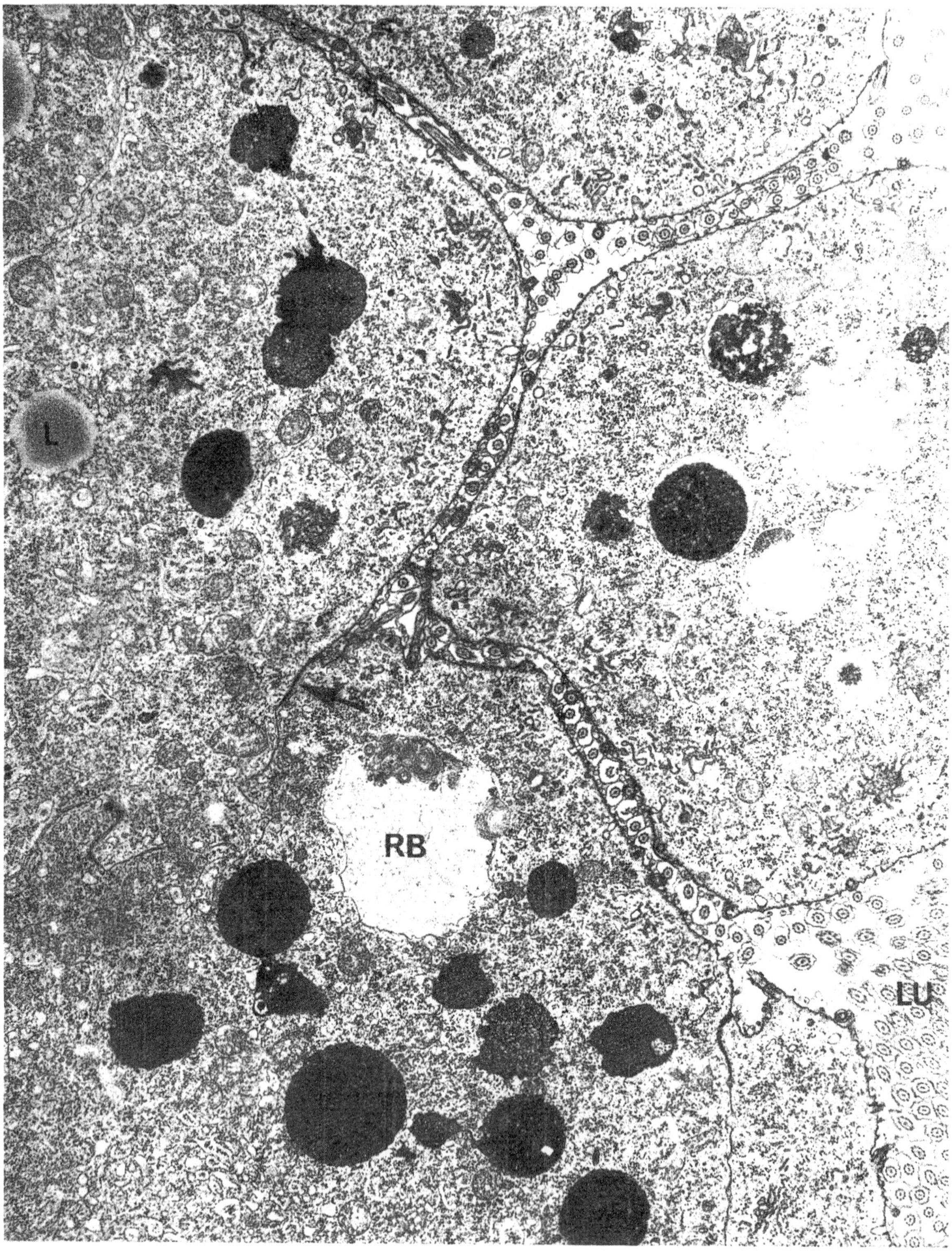

Fig. 3. ×9000 Portion of the stomach of an actively feeding female showing the lumen (LU) with numerous cilia and microvilli. The junctional complex present where adjacent cells meet at the apex is indicated by an arrow. The most prominent inclusions in the stomach cells are lipid droplets (L), food vacuoles with a dense matrix, and residual bodies (RB) containing whorls of membranes. Epon.

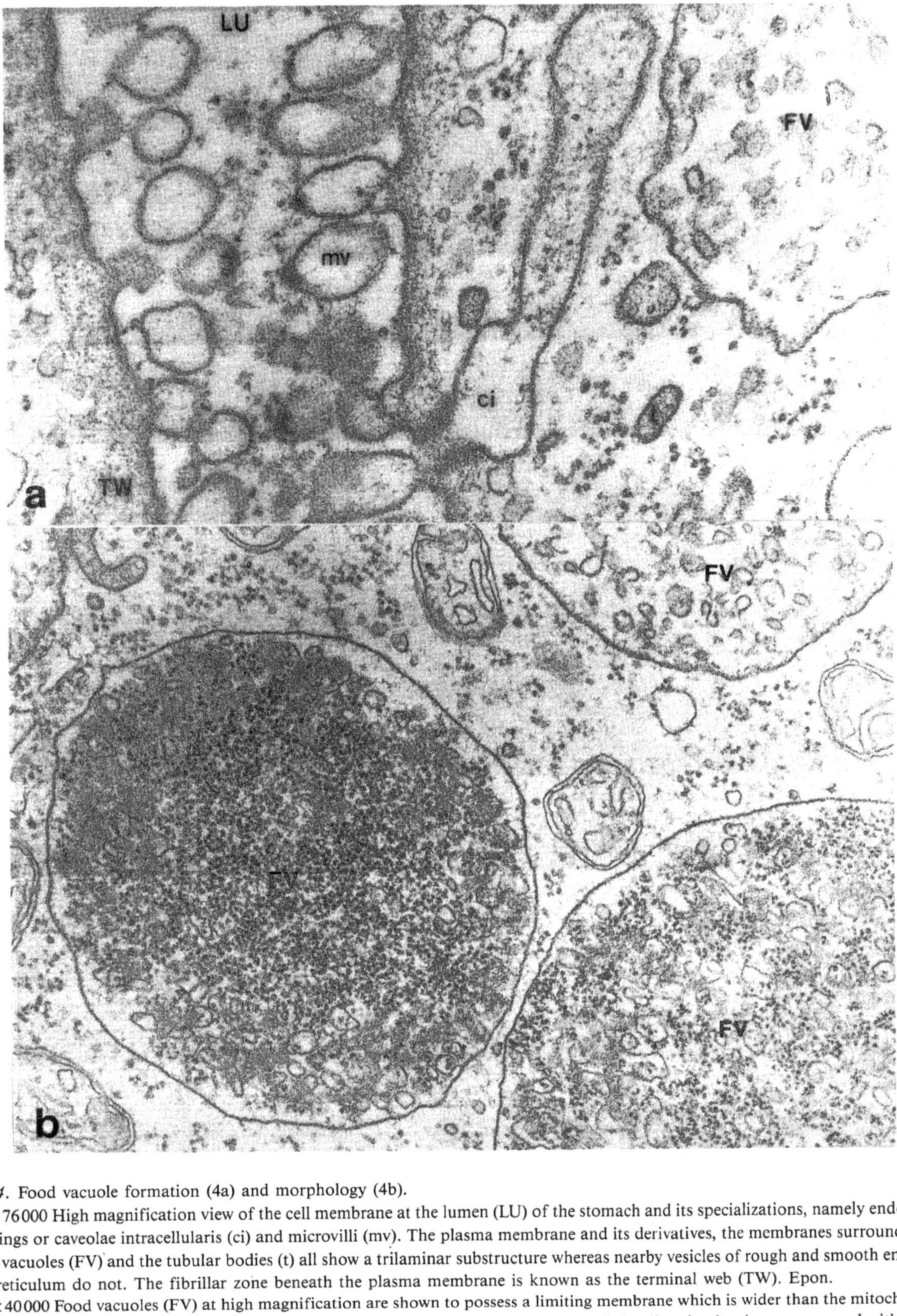

Fig. 4. Food vacuole formation (4a) and morphology (4b).
(a) ×76000 High magnification view of the cell membrane at the lumen (LU) of the stomach and its specializations, namely endocytotic openings or caveolae intracellularis (ci) and microvilli (mv). The plasma membrane and its derivatives, the membranes surrounding the food vacuoles (FV) and the tubular bodies (t) all show a trilaminar substructure whereas nearby vesicles of rough and smooth endoplasmic reticulum do not. The fibrillar zone beneath the plasma membrane is known as the terminal web (TW). Epon.
(b) ×40000 Food vacuoles (FV) at high magnification are shown to possess a limiting membrane which is wider than the mitochondrial membranes. Judging by the relatively intact vesicles and granules in the interior, intracellular digestion has just commenced within these vacuoles. Epon.

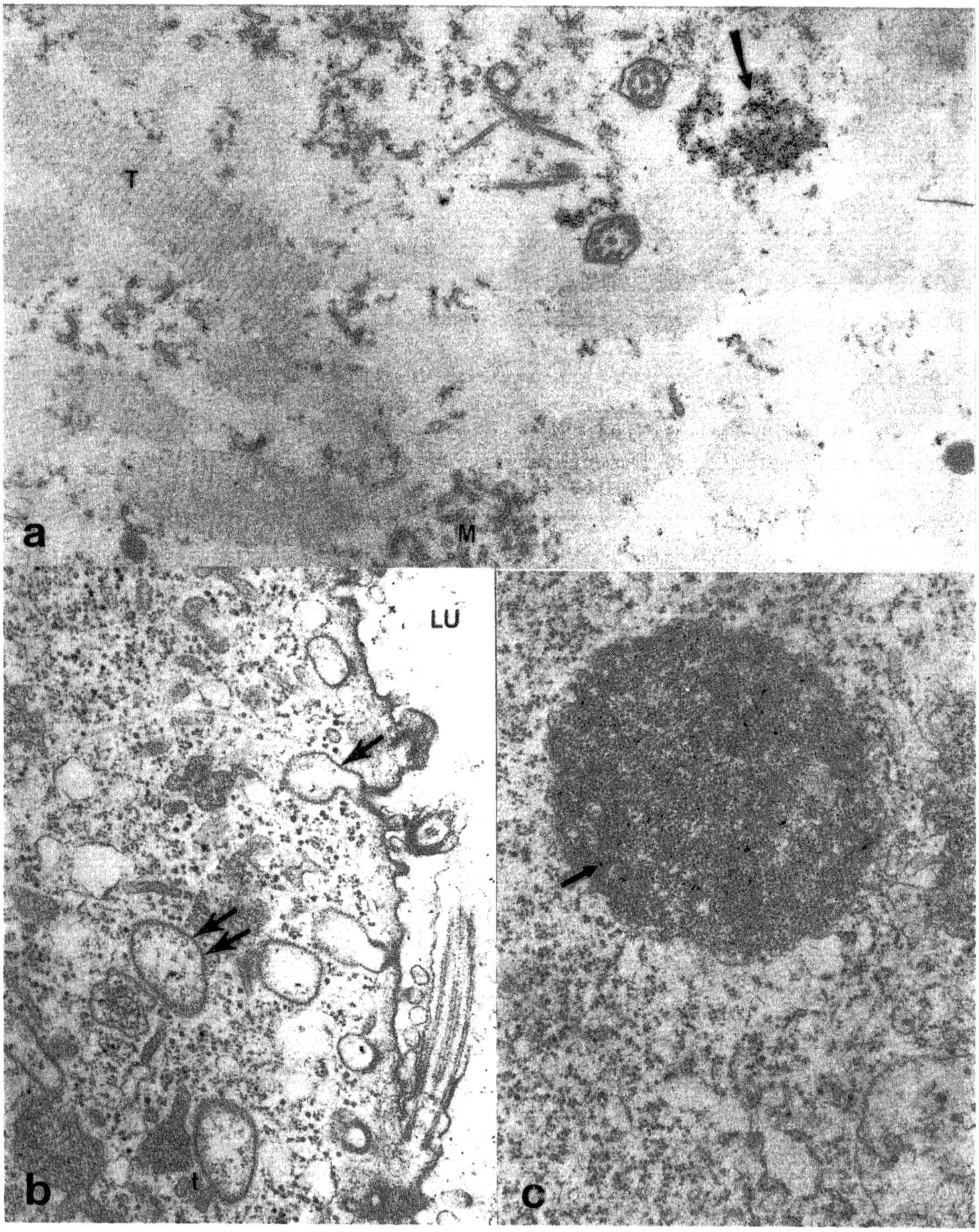

Fig. 5. Uptake of thorium dioxide (Thorotrast) by the stomach cells.

(a) ×21 000 After a 3-minute exposure to Thorotrast, disperse particles and aggregates (arrow) of this material are found in the stomach lumen. Remnants of trichocysts (T) and mitochondrial cristae (M) are recognizable among the stomach contents. Spurr.

(b) ×28 000 The various steps of the uptake process can be seen in this micrograph of an animal exposed to paramecium – ThO_2 mixture for 15 minutes. Thorotrast particles are taken into the cell by endocytosis (single arrow) and transported into the interior within coated vesicles (double arrow) which pinch off the caveolae intracellularis. Coated vesicles fuse with other vesicles and tubular bodies (t) to form food vacuoles. Spurr.

(c) ×28 000 After a 10-minute exposure to paramecium–Thorotrast suspension, numerous Thorotrast particles (arrow) are present within sizeable food vacuoles. Spurr.

exposure is sufficient to permit free label to enter the stomach cells. Thorotrast particles first appear in goblet-shaped endocytosis pits, in coated vesicles and in nascent food vacuoles in the cytoplasm bordering the lumen (Fig. 5b). Nascent food vacuoles have a coarsely granular matrix of intermediate density and an irregular outline due to the attachment of several tubular bodies at the periphery. Sizeable, rounded food vacuoles containing Thorotrast particles are present in animals fixed after 10 minutes of exposure (Fig. 5c). Last to be labelled are irregularly-shaped vacuoles with a light matrix.

The Thorotrast is confined entirely to food vacuoles in all animals; it does not penetrate into the cytoplasm between the vacuoles and, unlike Lansing's gold particles [1964], it does not appear in the pseudocoelom surrounding the stomach. The secretory bodies and other inclusions present in the gastric gland do not contain Thorotrast at any of the stages examined.

Discussion

In *Asplanchna*, the initial phase of digestion is extra-cellular. This phase may be extremely rapid. Only one out of many specimens, picked at random for examination under the electron microscope, contained an entire paramecium. However, some intact prey may have been regurgitated since the fixative causes an immediate contraction of the body, thereby bringing the stomach up to the mouth.

Extracellular digestion in the stomach of *Asplanchna* appears to be extensive. Digestive enzymes and nutrients, agitated by the movement of cilia, mix freely in the lumen of the stomach. The prey is reduced to small vesicles and membrane fragments. Particles of macromolecular dimensions are taken up by the gastrodermis through endocytotic openings and further degraded intracellularly.

In *Asplanchna*, endocytotic openings are fairly uniform in size and shape and they probably do not admit the entrance of particles whose diameter exceeds 0.1 μm, since the constricted neck of the openings measures 0.1 μm across. Ingested molecules are transported into the interior of the stomach cell by coated vesicles which bleb off the inner aspect of the endocytosis bulbs. Coated vesicles have been implicated in the uptake of protein [Roth & Porter, 1964]. Coated vesicles fuse with other vesicles and tubules in their vicinity and acquire digestive hydrolases, notably acid phosphatase and esterase. Ruffling and blebbing at the membrane diminishes, and the vacuoles, properly designated as food vacuoles, migrate toward the interior of the cell.

The fate of ingested particles and of the food vacuole itself, from this point on, is largely conjectural. Assuming that the food vacuole possesses a full complement of lysosomal enzymes, ingested proteins would be broken down to free amino acids and dipeptides, which can cross the vacuole membrane into the cytoplasm [Gordon, 1973] to take part in synthetic processes.

Acid phosphatase is thought to play a role in the initial stages of digestion of exogenous and endogenous materials [Rosenbaum & Rolon, 1960; Tomlinson & Warren, 1960; Jennings, 1962; Osborne & Miller, 1963; Ryder & Bowen, 1975]. The physiological function of alkaline phosphatase is unknown [Reid & Wilson, 1971; Saleem & Alikhan, 1974]. The currently accepted hypothesis is that the enzyme is involved in active transport of nutrients across the cell membrane [Fernley, 1971; Vyas & Dev, 1972; Ryder & Bowen, 1974].

Nonspecific esterase is thought to be synthesized and transported via the same route as acid phosphatase (Kukletova *et al.*, 1974] and it is frequently found at the same intracellular site as the latter, namely, in the lysosomes [Holt 1963; Deimling & Madreiter, 1972; Vladutin & Rose, 1974]. It has been suggested repeatedly that the enzyme plays a role in digestion [Clements, 1967; Ryberg, 1973] specifically in fat mobilization [Bongers & Nogge, 1970; Sudderruddin & Tan, 1973; Geering & Freyvogel, 1974].

In *Asplanchna* the three enzymes are found at the same subcellular loci as in other organisms. This correlation lends credence to the hypothesis that the enzymes are involved in the absorption of nutrients and their intracellular digestion.

An examination of the ultrastructure of the

stomach cells for signs that they also participate in the secretion of the extracellular enzymes, which are active in the stomach lumen, yields negative results. Morphological evidence for the synthesis of proteins destined for export includes a well developed rough endoplasmic reticulum, intense Golgi activity and secretory droplets. All of these are absent in the stomach cells. The proteinases and lipases responsible for the extracellular phase of digestion, which precedes intracellular digestion, probably originate in the salivary glands and the gastric glands.

References

Barka, T. & P. J. Anderson, 1962. Histochemical methods for acid phosphatase using hexazonium pararosanilin as coupler. J. Histochem. Cytochem. 10: 741–753.

Beauchamp, P. M. de, 1909. Recherches sur les rotifères. Les formations tegumentaires et l'appareil digestif. Arch. Zool. exp. gén., ser 4, 10: 1–410.

Bongers, J. & G. Nogge, 1970. Unspezifische Esterase und Proteinase der Speicheldrusen und des Darmes von Hypoderma bovis. J. Insect Physiol. 16: 2127–2136.

Clements, A. N., 1967. A study of soluble esterases in Pieris brassicae. J. Insect Physiol. 13: 1021–1030.

Dalrymple, J., 1849. Description of an infusory animalcule allied to the genus Notommata of Ehrenberg, hitherto undescribed. Phil. Trans. r. Soc., Lond. 139: 331-346.

Davis, B. J. & L. Ornstein, 1959. High resolution enzyme localization with a new diazo reagent 'hexazonium pararosaniline'. J. Histochem. Cytochem. 7: 297–298.

Deimling, O. & H. Madreiter, 1972. Esterase. II. A new method for the electron microscopical demonstration of nonspecific esterase in animal tissues. Histochemie 29: 83–96.

Fernley, H. N., 1971. Mammalian alkaline phosphatases. In: P. D. Boyer (ed.) The Enzymes. Vol. IV. Academic Press, N.Y.: 417–447.

Geering, K. & T. A. Freyvogel, 1974. The distribution of acetylcholine esterase in the midgut of Aedes aegypti L. Comp. Biochem. Physiol. 49B: 775–784.

Gilbert, J. J., 1968. Dietary control of sexuality in the rotifer Asplanchna brightwelli Gosse. Physiol. Zool. 41: 14–43.

Gilbert, J. J., 1973. The adaptive significance of polymorphism in the rotifer Asplanchna. Humps in males and females. Oecologia 13: 135–146.

Gordon, A. H., 1973. The role in lysosomes in protein catabolism. In: J. T. Dingle (ed.) Lysosomes in Biology and Pathology. Vol. III, North Holland Publishing Co., Amsterdam: 89–137.

Herold, R. C. & N. D. Meadow, 1970. Age related changes in the ultrastructure and histochemistry of rotiferan organs. J. Ultrastruct. Res. 33: 203–318.

Holt, S. J., 1963. Esterases in lysosomes. In: A. V. S. de Reuck & M. P. Cameron (eds) The CIBA Foundation Symposium on Lysosomes. Little, Brown and Co., Boston: 114–125.

Hudson, C. T. & P. H. Gosse, 1886. The Rotifera or Wheel-Animalcules, both British and Foreign. Longmans, Green & Co., London: Vol. I, 1886: I–IV + 1–128; Vol. II, 1889: 1–144.

Jennings, J. B., 1962. Histochemical study of digestion and digestive enzymes in the rhyncoelan Lineaus ruber (O. F. Muller). Biol. Bull. Mar. Biol. Lab., Woods Hole 122: 63–72.

Kukletova, M., Z. Lukas & K. Mazanec, 1974. Submicroscopic localization of nonspecific esterase in the oocyte and 2-cell stage of the rat ovum. Z. Mikrosk. Anat. Forsch. Leipzig, 88: 455–464.

Lansing, A. I., 1964. Age variations in cortical membranes of rotifers. J. Cell. Biol 23: 403–421.

Leduc, E. H. & W. Bernhard, 1967. Recent modifications of the glycol methacrylate embedding procedure. J. Ultrastruct. res. 19: 196–199.

Lojda, Z., B. Vecerek & H. Pelichova, 1964. Some remarks concerning the histochemical detection of acid phosphatase by azo coupling reaction. Histochemie 3: 428–454.

Luft, J. H., 1961. Improvements in epoxy embedding methods. J. Biophys. Biochem. Cytol. 9: 409–427.

Osborne, P. J. & A. T. Miller, 1963. Acid and alkaline phosphatase changes associated with feeding, starvation and regeneration in planarians. Biol. Bull. Mar. Biol. Lab., Woods Hole 124: 285–292.

Pearse, A. G. E., 1968. Histochemistry. Theoretical and Applied. Churchill, J. and A., London: 713-714.

Reid, T. W. & L. B. Wilson, 1971. E. coli alkaline phosphatase. In: P. D. Boyer (ed.) The Enzymes. Vol. IV. Academic Press, N.Y.; London: 373–415.

Rosenbaum, R. M. & C. I. Rolon, 1960. Intracellular digestion and hydrolytic enzymes in the phagocytes of planarians. Biol. Bull. Mar. Biol. Lab., Woods Hole 118: 315–323.

Roth, T. F. & K. R. Porter, 1964. Yolk protein uptake in the oocyte of the mosquito Aedes aegypti L. J. Cell Biol. 20: 313.

Ryberg, E., 1973. The localization of cholinesterases and nonspecific esterase in the echinopluteus. Zool. Scr. 2: 163–170.

Ryder, T. A. & I. D. Bowen, 1974. The fine structural localization of alkaline phosphatase in Polycelis tenuis (Iijima). Protoplasma 79: 19–29.

Ryder, T. A. & I. D. Bowen, 1975. The fine structural localization of acid phosphatase activity in Polycelis tenuis (Iijima). Protoplasma 83: 79–90.

Saleem, M. & M. A. Alikhan, 1974. The distribution of phosphomonoesterases in the digestive gut of Porcellio laevis Latreille (Porcellionidae: Isopoda). Comp. Biochem. Physiol. 48: 375–385.

Salt, George W., G. F. Sabbadini & M. L. Commins, 1978. Trophi morphology relative to food habits in six species of rotifers (Asplanchnidae). Trans. am. microsc. Soc. 97: 469–485.

Spurr, A. R., 1969. A low viscosity epoxy resin embedding medium for electron microscopy. J. Ultrastruct. Res. 26: 31–43.

Spurr, A. R. & W. M. Harris, 1968. Ultrastructure of chloroplasts and chromoplasts in Capsicuum annuum. I. Thylakoid membrane changes during fruit ripening. Am. J. Bot. 55: 1210–1224.

Sudderruddin, K. T. & K. H. Tan, 1973. Some hydrolases and their involvement in insecticide resistance. Pans 19: 24–35.

Tomlinson, N. & R. A. J. Warren, 1960. Lingcod muscle phosphomonoesterases. I. Acid Phosphatases that hydrolyse p-nitrophenyl phosphate. Can. J. Biochem. 38: 605–612.

Vladutin, G. D. & N. R. Rose, 1974. Intracellular distribution of a primate-specific esterase in cultured cells and tissues. J. Cell. Biol. 62: 560–566.

Vyas, I. & B. Dev, 1972. Localization of alkaline phosphatase in different components of the alimentary canal of the common earthworm Pheretima posthuma and its bearing on their function. Fol. Histochem. Cytochem. 10: 151–154.

Waniczek, H., 1930. Untersuchungen über einige Arten der Gattung Asplanchna Gosse (A. girodi de Guerne, A. brightwellii Gosse, A. priodonta Gosse). Ann. Mus. Zool. Pol. 8: 109–322.

Yamada, E., 1955. The fine structure of the gall bladder epithelium of the mouse. J. biophys. biochem. Cytol. 1: 445–458.

Hydrobiologia 147: 373 (1987)
May, L., Wallace, R. & Herzig, A. (eds), Rotifer Symposium IV

Neuropharmacology of rotifer feeding, oviposition, and anesthesia

Thomas Nogrady
Department of Biology, Queen's University, Kingston, Ont. K7L 3N6, Canada

Keywords: Rotifer, neuropharmacology, cholinergic, acetylcholine, feeding, *Brachionus, Philodina,* oviposition, anesthesia, insecticide

Abstract of Communication

Effect of acetylcholine and anticholinergic drugs on feeding, oviposition, and anesthesia in rotifers was investigated. Neurotransmitter as well as antagonist drugs inhibited feeding in *Brachionus calyciflorus* in a dose-dependent manner. Most antagonist drugs caused an oscillating tachyphylaxis (drug habituation): the drug effect wore off and returned several times within an hour. Acetylcholine inhibited oviposition in *Philodina acuticornis,* and this effect was antagonized by all groups of anticholinergic drugs. The strongest antagonism was caused by neuromuscular blockers, and thus the cause of oviposition inhibition may be a cloacal sphincter spasm. Acetylcholinesterase inhibitory insecticides also antagonize the acetylcholine effect. Acetylcholine potentiates the anesthetic activity of ionizing local anesthetics (procaine, lidocaine) as well as that of atropine and the beta-adrenergic blocker propranolol. Muscarinic antagonists (atropine, benactyzine) and propranolol caused foot paralysis in *B. calyciflorus,* which is also potentiated by acetycholine. Further details of these results are given by Nogrady and Keshmirian (1986a, b).

References

Nogrady, T. & J. Keshmirian, 1986a. Rotifer neuropharmacology I. Cholinergic drug effects on oviposition of *Philodina acuticornis* (Rotifera, Aschelminthes). Comp. Biochem. Physiol. 83C: 335–338.

Nogrady, T. & J. Keshmirian, 1986b. Rotifer neuropharmacology II. Synergistic effect of acetylcholine on local anesthetic activity in *Brachionus calyciflorus* (Rotifera, Aschelminthes). Comp. Biochem. Physiol. 83C: 339–344.

Hydrobiologia 147: 375–378 (1987)
May, L., Wallace, R. & Herzig, A. (eds), Rotifer Symposium IV

Combined influences of particulate and dissolved factors in the toxicity of *Microcystis aeruginosa* (NRC-SS-17) to the rotifer *Brachionus calyciflorus*

Peter L. Starkweather & Penelope E. Kellar
Department of Biological Sciences, University of Nevada, Las Vegas, Nevada 89154, U.S.A.

Keywords: rotifers, cyanobacteria, blue-greens, feeding, population dynamics, *Brachionus*, *Microcystis*

Abstract

The rotifer *Brachionus calyciflorus* is subject to the combined influence of particulate and dissolved factors derived from the toxic cyanobacterium *Microcystis aeruginosa*. Toxicity manifests itself through reduction of cohort survivorship and probably though reduction of cohort reproduction. Toxic effects are seen only when there is a particulate food material available in suspension, either *M. aeruginosa* itself or another food type which may act as a carrier for a dissolved toxic factor. Regardless of exact mechanism, the presence of a toxic strain of this blue-green may lead to diminution of population growth of co-occurring rotifer populations.

Introduction

Rotifers, as do all members of the freshwater zooplankton community, come into contact with a large number of different potential food sources. These include eukaryotic algae, heterotrophic bacteria, detrital particles and, in some instances, various cyanobacteria (commonly called the blue-green algae). Members of this last group appear to pose special difficulties for some zooplankton species on the basis of their large size, resistance to digestion and often potent toxicity [Porter, 1977]. One of the cyanobacterial taxa prominently included in any list of such unsatisfactory foods is *Microcystis aeruginosa. M. aeruginosa* is a common, often colonial, bloom-forming species responsible for development of toxic population levels in many lakes, ponds and slow-moving rivers [see reviews in Carmichael, 1981].

The toxins of *Microcystis* were originally isolated over 25 years ago [Hughes *et al.*, 1958; Bishop *et al.*, 1959] and have now been identified as a family of cyclic peptides with primarily hepatotoxic effects in mammalian assay [Gorham & Carmichael, 1979; Carmichael, 1981; Bates *et al.*, 1982; Adams *et al.*, 1985]. Various combinations of these peptides probably constitute what was previously known as a 'fast death factor' (FDF) [Hughes *et al.*, 1959] derived from this species. It has been suggested [Collins, 1978] that cell disruption normally may be necessary for manifestation of *M. aeruginosa* toxic effects. This is not a universal property of blue-greens, however. Snell [1980] reported that the rotifer *Asplanchna girodi* was affected by cyanobacteria in suspension, even though direct consumption of the cyanobacteria by these rotifers was unlikely. Clearly, cyanobacteria might conceivably affect rotifer populations either directly, through ingestion of cyanobacteria, or indirectly, due to sensitivity to material released from the same cells.

Brachionus calyciflorus consumes *M. aeruginosa* in large quantities, at least when the latter is present in non-colonial form [Starkweather & Kellar, unpubl.]. Effects of this cyanobacterium on the population dynamics of the rotifer depend upon the specific *M. aeruginosa* strain used in laboratory culture. One strain, NRC-SS-17, appairs to have potent toxic effects on *B. calyciflorus*, whether or not the cyanobacterium is present in pure suspension or mixed with an alternative food [Starkweather & Kellar, unpubl.]. Early experiments did not show, however, if the toxic effect was due to the action of the cyanobacterial cells after they had been ingested by the rotifers, or if the presence of dissolved materials released by the *M. aeruginosa* might have been sufficient to produce a deleterious effect. This paper describes an experiment which tests the relative effects of *M. aeruginosa* cells and materials leached from those cells in modifying the survival and reproduction of cohorts of *B. calyciflorus*.

Materials and methods

We reared *B. calyciflorus* [strain S-4, Starkweather, 1981] for over 100 generations using *Euglena gracilis* [strain 753, Starr, 1978] as the sole food. Unlike our previous work on cyanobacteria/rotifer interactions [Starkweather & Kellar, 1983], we grew the rotifers for an initial two generations using a vitamin-enriched medium (ASM-1-TR) originally designed for culturing cyanobacteria [Carmichael & Gorham, 1974]. We cultured *E. gracilis* and *M. aeruginosa* as described before [Starkweather & Kellar, 1983], with the euglenoid grown on a commercial agar medium and the cyanobacterium grown in ASM-1-TR). We used two medium preparations during this experiment. 'Fresh' medium was autoclaved and filtered (Whatman GF/C) ASM-1-TR which was stored for daily use at 4° C. 'Conditioned' medium was ASM-1-TR in which *M. aeruginosa* had been grown for 6–8 days. We removed the cells by gentle centrifugation and filtered the supernatant through GF/C filters, preparing new conditioned medium on each day of the experiment. We used fresh and conditioned media to make up the food cell suspensions for experimental culture of the matched rotifer cohorts. All groups were held at 20–22° C under dim, continuous illumination.

As in our previous experiments [Starkweather & Kellar, 1983], we reared triplicate cohorts of *B. calyciflorus* neonates, with individuals isolated in 0.8 ml culture volumes of food suspension and ASM-1-TR. We fed the rotifers equivalent dry weight biomass quantities (20 μg ml^{-1}) of either *E. gracilis* or *M. aeruginosa*; a third experimental set received only filtered medium. We based our life table calculations on the methods of Birch [1948], determining cohort median survivorship ($0.5 l_x$), net reproductive rate (R_0), generation time (T) and rate of population increase (r). We compared treatment effects using the non-parametric Kruskal-Wallis test corrected for ties [Sokal & Rohlf, 1981].

Results and discussion

Table 1 shows the combined effects of food type and medium preparation on the median survivorship of *B. calyciflorus* cohorts. It is clear that, for a given medium, the effect of food type is important in determining cohort survival. Groups of rotifers fed *E. gracilis* lived more than twice as long as groups fed either *M. aeruginosa* or no food at all, irrespective of medium conditioning. This result, naturally, could be the product of the positive benefits of *Euglena* as a food, the negative effects of *Mycrocystis* or a combination of both. Within a food treatment and comparing media, the results are more variable. Animals exposed only to filtered medium show no significant difference in median survivorship between fresh and conditioned medi-

Table 1. Comparison of median survivorship ($0.5\ l_x$) values for *B. calyciflorus* fed 20 μg ml^{-1} of *E. gracilis* (E-753), 20 μg ml^{-1} of *M. aeruginosa* (SS-17) or no food in two media; Fresh: newly prepared and filtered ASM, and Conditioned: ASM in which *M. aeruginosa* strain SS-17 had grown for 6–8 d, cells removed and medium filtered. Values shown are means for three replicate cohorts ± 1 s.d.

Medium	Food type			
	E-753	SS-17	No food	p(K.-W.)
Fresh	8.14 ± 0.29	3.08 ± 0.15	2.93 ± 0.05	<0.05
Conditioned	6.94 ± 1.78	2.57 ± 0.27	3.08 ± 0.21	<0.05
p(K.-W.)	0.05 < p < 0.10	<0.05	N.S.	

um. On the other hand, if fed *M. aeruginosa* NCR-SS-17, the animals have significantly reduced survivorship on conditioned medium. We find an intermediate result with *B. calyciflorus* fed *Euglena*; here we see a substantial, but only marginally significant diminition of cohort survival when the animals are exposed to the cyanobacteria-treated (conditioned) medium. Taken together, these results lead to the conclusion that food type plays the principal role in determining the life table parameters of *B. calyciflorus* with this design. However, medium condition also has an effect, at least when there is particulate food available in suspension. The curious aspect of the *Euglena*-fed rotifer survivorship results is that the negative factor appears to be associated with particulate feeding. As noted, animals in filtered medium alone show no survival differences; it is only when a food is available that a substantial medium effect can be demonstrated. Additionally, as shown by the *Microcystis*-fed groups, dissolved and particulate factors may be complementary in their influence.

One should note here that the exact cause of the between-media differences in survivorship is unclear. At least two possibilities are obvious: either the conditioned medium carries some toxic factor released by the *M. aeruginosa* which affects the rotifers, or the *Microcystis* modifies the original inorganic or organic character of the medium in such a way as to make it less suitable for rotifer culture. Since ASM-1-TR is a fairly rich medium (no constituent should be significantly depleted in 6–8 days of dilute culture) and since *M. aeruginosa* NRC-SS-17 has a history of animal toxicity, we suspect that the former possibility is more likely.

The negative effects of conditioned medium may be limited to reducing cohort survival in *B. calyciflorus.* Table 2 shows life table parameters for the rotifer fed *E. gracilis* and compares the influence of the two media. Values for R_0, T and r (all based on reproduction as well as survivorship) show no significant differences between medium treatment groups. As noted also in Table 1, survivorship differences between medium groups are only marginally significant. The reproduction-based results are consistent, but they suffer from unusually high variability among replicate cohorts

Table 2. Comparison of life table parameters for *B. calyciflorus* fed 20 μg ml^{-1} *E. gracilis* in two media; Fresh: newly prepared and filtered ASM, and Conditioned: ASM in which *M. aeruginosa* strain SS-17 had grown for 6–8 d, cells removed and medium filtered. Values shown are means for three replicate cohorts ±1 s.d.

Medium	Life table parameter			
	$0.5\ l_x$	R_0	T	r
Fresh	8.14±0.29	10.95±1.2	3.80±0.11	0.63±0.02
Conditioned	6.94±1.78	8.24±3.3	4.14±0.36	0.49±0.14
p(K.-W.)	0.05<p<0.10	N.S.	N.S.	N.S.

exposed to conditioned medium. Mean values intuitively suggest a medium conditioning influence on reproduction, a suggestion made more compelling by examination of Fig. 1. Early reproduction (age specific fecundity – m_x) appears to be higher in the fresh medium group than in the conditioned

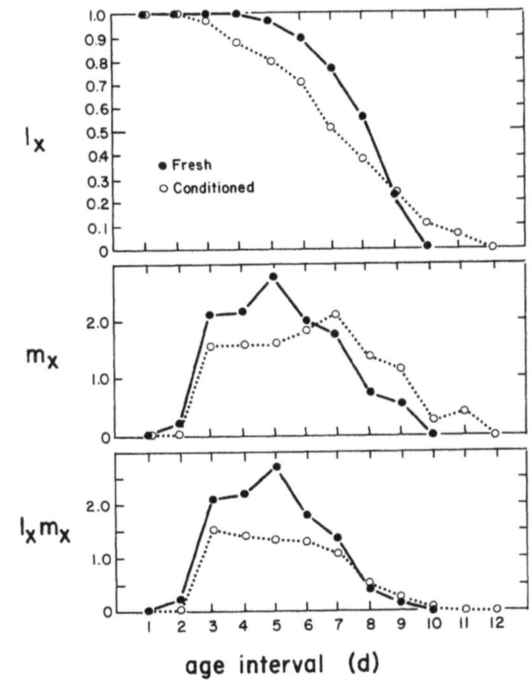

Fig. 1. Age-specific survivorship (l_x), fecundity (m_x) and the product of those parameters for cohorts of *Brachionus calyciflorus* fed 20 μg ml^{-1}) *Euglena gracilis* and grown in two medium treatments, *fresh* – closed circles and *conditioned* – open circles; see text for explanation.

medium treatment group; note, however, that the pattern of late reproduction may be reversed. A reproductive differential is seen also in the survivorship-weighted $l_x \cdot m_x$ patterns, in which the early fecundity advantage of fresh medium groups looks substantial. We speculate that the failure to show significant differences in the life table analysis is due to high variability of one of the experimental groups.

Summary

When exposed to a toxic strain of the cyanobacterium *Microcystis aeruginosa*, B. *calyciflorus* appears to suffer from the double influence of particulate and dissolved negative factors. Both these are likely related to toxic materials synthesized by the cyanobacterium, and are delivered to the rotifers through direct ingestion of *Microcystis* cells, through indirect uptake of related materials absorbed to other (non-toxic) food materials in the diet or through passive uptake of dissolved toxins during ingestion of non-toxic foods. In all three instances, *M. aeruginosa* is anticipated to reduce the population growth of a major cyanobacterium predator, thereby enhancing the changes of its own long-term survival.

Acknowledgements

We thank Elizabeth Walsh for assistance in the laboratory and Dr. Wayne Carmichael for supplying the toxic cyanobacterium. Supported by National Science Foundation Grant DEB-8105724 to P.L.S.

References

Adams, W. H., R. D. Stoner, D. G. Adams, D. N. Slatlein & H. W. Siegelman, 1985. Pathophysiologic effects of a toxic peptide from Microcystis aeruginosa. Toxicon 23: 441–448.

Birch, L. C., 1948. The intrinsic rate of natural increase of an insect population. J. Anim. Ecol. 17: 15–26.

Bishop, C. T., E. Anet & P. R. Gorham, 1959. Isolation and identifcation of the fast-death factor in Microcystis aeruginosa NRC-1. Can. J. Biochem. Physiol. 37: 453–471.

Botes, D. P., H. Kruger & C. C. Viljoen, 1982. Isolation and characterization of four toxins from the blue-green alga Microcystis aeruginosa. Toxicon 20: 945–954.

Carmichael, W. W. (ed.), 1981. The water environment: algal toxins and health. Plenum Press, New York.

Carmichael, W. W. & P. H. Gorham, 1974. An improved method of obtaining axenic clones of planktonic blue-green algae. J. Phycol. 10: 283–240.

Collins, M., 1978. Algal toxins. Microbiol. Rev. 42: 725–746.

Gorham, P. R. & W. W. Carmichael, 1979. Phycotoxins from blue-green algae. Pure and Appl. Chem: 165–174.

Hughes, E. O., P. R. Gorham & A. Zehnder, 1958. Toxicity of a unialgal culture of Microcystis aeruginosa. Can. J. Microbiol. 4: 225–236.

Porter, K. G., 1977. The plant-animal interface in freshwater ecosystems. Am. Sci. 65: 159–170.

Snell, T. W., 1980. Blue-green algae and selection in rotifer populations. Oecologia. 46: 343–346.

Sokal, R. R. & F. J. Rohlf, 1981. Biometry. 2nd ed. W. H. Freeman. San Francisco, 859 pp.

Starkweather, P. L., 1981. Trophic relationships between the rotifer Brachionus calyciflorus and the blue-green alga Anabaena flos-aquae. Verh. int. Ver. Limnol. 21: 1507–1514.

Starkweather, P. L. & P. E. Kellar, 1983. Utilization of cyanobacteria by Brachionus calyciflorus: Anabaena flos-aquae (NRC-44-1) as a sole or complementary food source. Hydrobiologia 104: 373–377.

Starr, R. C., 1978. The culture collection of algae at the University of Texas at Austin. J. Phycol. 14/Suppl.: 47–100.

Hydrobiologia 147: 379–381 (1987)
May, L., Wallace, R. & Herzig, A. (eds), Rotifer Symposium IV

A centrifugation method for measuring the relative density (specific gravity) of planktonic rotifers (Rotifera), with values for the relative density of *Polyarthra major* Burckhardt and *Keratella cochlearis* (Gosse)

A. P. Saunders-Davies & Rosalind M. Pontin
Department of Zoology, Royal Holloway College, Egham, England

Keywords: Relative density, centrifugation, Ficoll, *Polyarthra, Keratella*

Abstract

The relative density of *Polyarthra major* and *Keratella cochlearis* was measured by centrifugation in solutions of different densities. Ficoll was used as the solute. The mean values for the relative density of living specimens of *P. major* was found to be 1.0277 ± 0.0062. The value for preserved animals was lower at 1.0160 ± 0.0029. The relative density of fixed specimens of *K. cochlearis* was measured and found to be approximately 1.0450 ± 0.0175.

Introduction

Rotifers form a large part by number of freshwater zooplankton and, in spite of their small size, must account for a considerable proportion of the zooplankton biomass. Their small size (ca. 100 to 500 μm in length) makes them difficult to weigh directly, although mean dry weights have been obtained for a number of species by Dumont *et al.* [1975], Doohan and Rainbow [1971] and Bottrell *et al.* [1976]. Bottrell *et al.* [1976] also published a table, compiled from various authors, of wet weights derived using calculated volumes. The authors comment that the formulae employed are not always made clear in the original papers, and point out that another possible cause of the inaccuracies reported in weight calculations lay in the use of a relative density value assumed to be 1.0 (i.e. water). They suggested that measurements of relative density could be made using a non-osmolite, such as Ficoll, in a density gradient.

The density of dead or preserved rotifers is obviously greater than that of water since they tend to sink in still water. It is probable that living rotifers also have a density greater than 1.0 and maintain their position in the water column by swimming. However, the relative density of living rotifers may differ from that of dead animals whose osmotic control has been destroyed. Variations in relative density may also occur under different physiological or environmental circumstances.

The current study was undertaken to develop a convenient and simple method of measuring the relative density of planktonic rotifers. Centrifugation was employed, using solutions of different relative densities. Animals centrifuged in a medium of relative density less than their own will settle to the bottom of the tube, whilst animals in a medium of relative density equal to or greater than their own will remain in suspension. Thus, by using a graded series of solutions of increasing relative density it should be possible to determine the solution in which the animals just fail to settle out. A similar method was employed by Alekseev [1978] to determine the relative densities of some Copepoda, using saline solution of known density. However, saline was considered unsuitable for the present study because of its osmotic effects.

Materials and methods

The experimental rotifers *Polyarthra major* Burckhardt and *Keratella cochlearis* (Gosse) were collected from a pond in the grounds of the Zoology Department, Royal Holloway College, Egham. Only females were present in the sample.

Ficoll was used to provide solutions of different relative density. It is a sugar polymer of high molecular weight (ca. 400000) and negligible osmotic effects which dissolves slowly but completely in water. Stock solutions were prepared up to a relative density of 1.07, buffered with Hepes solution and calcium chloride, and adjusted to a ph of 7.4 with sodium hydroxide.

The stock solution was diluted to provide a series of experimental solutions of increasing relative density. The concentrations were chosen to provide the approximate relative density required after adding the water containing the sample animals.

Experimental methods were first tried using freshly collected animals fixed in 5% formalin. 2 ml of this suspension were added to 10 ml of the appropriate Ficoll solution in glass centrifuge tubes and centrifuged for 15 min. at 3000 r.p.m. The upper (supernatant) 10 ml were carefully separated from the bottom 2 ml using a siphon with a stop valve to prevent mixing. The relative density of the supernatant was determined as described below and the total numbers of the two main species, *P. major* and *K. cochlearis*, in each fraction were counted under the microscope.

The same technique was then applied to living animals. These proved viable in the Ficoll solution. Indeed animals were observed actively swimming after centrifugation was completed which presented an experimental difficulty as they began to swim up the tube again as soon as the centrifuge slowed sufficiently. This problem was solved by the use of perspex funnels fitted into the centrifuge tube, operating on the same principle as the common fish trap. The funnel allowed rotifers to pass through to the bottom of the tube during centrifugation but tended to prevent them swimming back up the tube when the centrifuge had stopped. A small perspex collar was used to space the funnel from the bottom of the tube such that the bottom fraction remained at 2 ml. A control experiment showed that there was no significant effect of the funnels on the determinations. Metal centrifuge tubes were then used to avoid disintegration of the glass tubes when used with the funnels. These were slightly larger than the glass tubes and the supernatant and pellet fraction were 12 ml and 2 ml respectively.

The relative density of the experimental and supernatant Ficoll solutions were determined by weighing. Conventional relative density bottles or open tubes allowed too much evaporation to give consistent results. These effects were avoided by drawing the solution up into a modified hypodermic syringe with a ground glass barrel; a lapped metal piston and a stop behind the piston ensured that precisely the same amount of solution filled the syringe for each weighing. The solutions and the distilled water used for the reference were allowed to equilibrate to room temperature before weighing. The precision of the method was checked by multiple determinations of the two solutions representing the range of densities.

Results and discussion

The numbers of animals found in the supernatant was expressed as a percentage of the total for each species. These values were plotted as a function of the relative density of the solution (Figs. 1 & 2).

Data expressed in this manner are difficult to analyse statistically in the conventional way, since each data point represents the sum of a number of individuals of a certain relative density. However, assuming a normal distribution for the parameter relative density, in common with similar population parameter, the sigmoid curve that fits the plotted data is equivalent to the cumulative percentile curve of a normal distribution. Thus the median and the mean can be read off at the 50% point. Similarly one standard deviation above and below the mean can be read off at 84.13% and 15.87% respectively. These initial estimates were used in an iterative computer program using the Simplex method to find the mean and standard deviation which gave the best fit to the data for a normal dis-

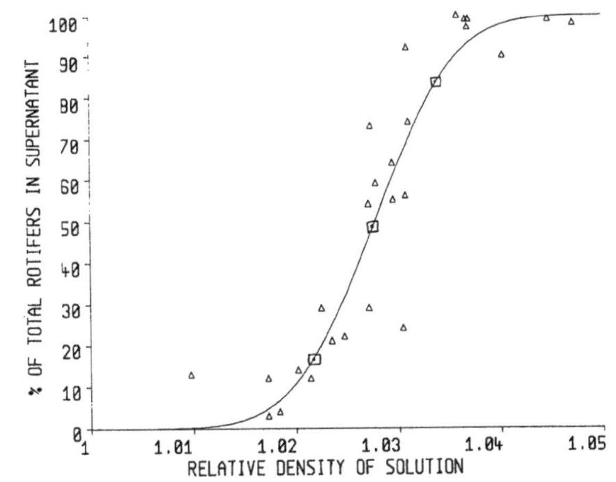

Fig. 1. Plot of the percentage of fixed *Polyarthra major* in supernatant fraction as a function of the relative density of the solution. Boxed points mark the mean and one standard deviation either side of the mean.

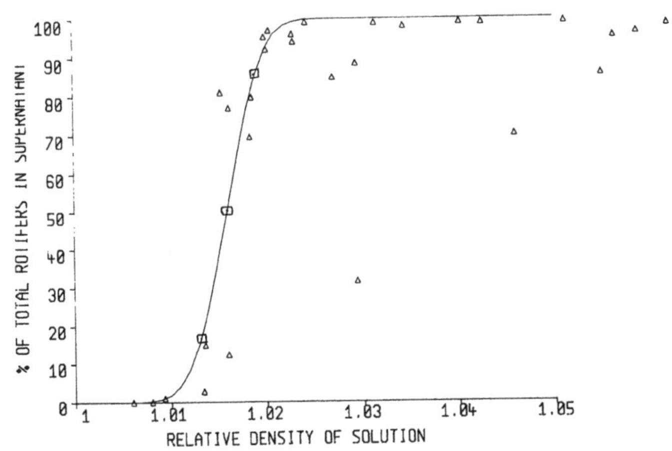

Fig. 2. Plot of the percentage of live *Polyarthra major* in supernatant fraction as a function of the relative density of the solution. Boxed points mark the mean and one standard deviation either side of the mean.

tribution. These are shown in the Table 1.

The initial results for *K. cochlearis* were much more variable than those for *P. major* and it became clear that very high concentrations of Ficoll would be necessary to obtain the same near complete separation of this rotifer that was achieved for *P. major*. Experiments also indicated that the viability of *K. cochlearis* in these high concentrations was doubtful.

The results for *P. major* are very consistent which suggest that, in conjunction with volumes deter-

Table 1. Relative densities of rotifers derived from centrifugation experiments with solutions of different relative density.

Date	Species	Regime	Mean	S.D.
Nov/Dec 1982	Fixed *P. major*	no funnels	1.0160	0.0029
Feb 1983	Live *P. major*	+ funnels	1.0277	0.0062
Feb 1983	Fixed *P. major*	no funnels	1.0200	0.0043
Feb 1983	Fixed *P. major*	+ funnels	1.0170	0.0043
Nov/Dec 1982	Fixed *K. cochlearis*	no funnels	1.0450	0.0175

mined by the methods of Ruttner-Kolisko [1977], more accurate estimates of relative rotifer biomass can be established for non-loricate forms. The relatively greater variability of the results for *K. cochlearis* may reflect greater variability in the population. This may have been due to variations in the proportion of juveniles in the experimental populations. Immature individuals of this species could have less dense loricas than older animals.

Acknowledgements

We wish to thank Professor C. Lewis, Zoology Dept., Royal Holloway College, for kindly providing facilities for us. We also thank Mr Ray Jalland for his help and advice, and Dr A. Wroote of Royal Holloway College Computing Department for his assistance in analysing the data.

A. Saunders-Davies was in receipt of a grant from the Crowther Fund of the Open University.

References

Alekseev, V. R., 1978. Determination of copepoda body density with the aid of standard solutions. Informatsionnyi Byull. Biol. Vnutr. Vod. 40: 74–76 (in Russian).

Bottrell, H. H., A. Duncan, Z. M. Gliwicz, E. Grygierek, A. Herzig, A. Hillbricht-Ilkowska, H. Kurasawa, P. Larsson & T. Weglenska, 1976. A review of some problems in zooplankton production studies. Norw. J. Zool. 24: 419–456.

Doohan, M. & V. Rainbow, (1971). Determinations of dry weights of small aschelminthes (0.1 μg). Oecologia 6: 380–383.

Dumont, H. J., I. Van de Velde & S. Dumont, 1975. The dry weight estimate of biomass in a selection of Cladocera, Copepoda and Rotifera from the plankton, periphyton and benthos of continental waters. Oecologia 19: 75–97.

Ruttner-Kolisko, A. 1977. Suggestions for biomass calculations of plankton rotifers. Arch. Hydrobiol. Beih. 8: 71–76.

www.ingramcontent.com/pod-product-compliance
Lightning Source LLC
Chambersburg PA
CBHW080934240926
51849CB00003BA/456
* 9 7 8 9 0 6 1 9 3 6 4 5 9 *